Advances in Polymer Science

Volume 288

Aims and Scope

The series Advances in Polymer Science presents critical reviews of the present and future trends in polymer and biopolymer science. It covers all areas of research in polymer and biopolymer science including chemistry, physical chemistry, physics, and material science.

The thematic volumes are addressed to scientists, whether at universities or in industry, who wish to keep abreast of the important advances in the covered topics.

Advances in Polymer Science enjoys a longstanding tradition and good reputation in its community. Each volume is dedicated to a current topic, and each review critically surveys one aspect of that topic, to place it within the context of the volume. The volumes typically summarize the significant developments of the last 5 to 10 years and discuss them critically, presenting selected examples, explaining and illustrating the important principles, and bringing together many important references of primary literature. On that basis, future research directions in the area can be discussed. Advances in Polymer Science volumes thus are important references for every polymer scientist, as well as for other scientists interested in polymer science - as an introduction to a neighboring field, or as a compilation of detailed information for the specialist.

Review articles for the individual volumes are invited by the volume editors. Single contributions can be specially commissioned.

Readership: Polymer scientists, or scientists in related fields interested in polymer and biopolymer science, at universities or in industry, graduate students.

More information about this series at http://www.springer.com/series/12

R. Jayakumar • M. Prabaharan
Editors

Chitosan for Biomaterials IV

Biomedical Applications

With contributions by

R. Abhinandan · S. P. Adithya · A. Almeida · S. Anil ·
K. Balagangadharan · C.-S. Cho · K.-H. Cho ·
G. M. Guebitz · L.-F. Hu · R. Jayakumar · H.-L. Jiang ·
R. Kandasamy · V. Krishnaswami · V. S. Kumar ·
F. Laffleur · C. Li · V. P. Murali · S. S. Murugan ·
S. Natesan · G. S. Nyanhongo · R. Panonnummal ·
S. Parmaksız · D. Pathayappurakkal Mohanan · A. Pellis ·
C. Ponnusamy · M. Prabaharan · A. Pradeep · L.-Y. Qi ·
A. K. Rajendran · M. Sabitha · E. Sanchez Armengol ·
B. Sarmento · N. Selvamurugan · S. Şenel · M. S. Shim ·
D. S. Sidharthan · P. Sivakumar · A. S. Soubhagya ·
W. Sun · M. N. Sundaram · P. Sundararaju ·
S. Thekkilaveedu · P. K. Varma · R. Vellayutham ·
J. Venkatesan · Y. Wang · L. Xing · P.-S. Zhao · T.-J. Zhou

Springer

Editors
R. Jayakumar
Centre for Nanosciences and Molecular Medicine
Amrita Vishwa Vidyapeetham (University)
Kochi, India

M. Prabaharan
Department of Chemistry
Hindustan Institute of Technology and Science
Padur, Chennai, India

ISSN 0065-3195 ISSN 1436-5030 (electronic)
Advances in Polymer Science
ISBN 978-3-030-83020-5 ISBN 978-3-030-83021-2 (eBook)
https://doi.org/10.1007/978-3-030-83021-2

This Springer imprint is published by the registered company Springer Nature Switzerland AG.
The registered company address is: Gewerbestrasse 11, 6330 Cham, Switzerland

Preface

Chitosan, a derivative of chitin, is a versatile biopolymer that is considered a promising biomaterial due to its unique physicochemical and biological characteristics. Hence, considerable efforts have been made to develop chitosan-based materials for biomedical applications over the past few years. The reassuring outcomes of these efforts have established the importance of chitosan and its derivatives in biomedical fields. This volume summarizes recent advancements of chitosan-based materials in the forms of nanomaterials, 3D-printed materials, gels, hydrogels, membranes, films, and fibers as hemostatic agents, tissue sealants, tissue engineering scaffolds, drug/gene/vaccine delivery carriers, and wound dressings. Some chapters place emphasis on the fabrication techniques of chitosan-based materials for hemostasis, periodontal and bone tissue engineering as well as in regenerative medicine, while others focus on the construction of chitosan-based theranostics and anticancer drug delivery carriers for advanced cancer therapy. The chapters of this volume also summarize the recent progress of chitosan-based topical and transdermal drug/growth factor delivery systems. This volume will be of interest to material/clinical scientists, chemists, biologists, and researchers in the fields of biopolymer science and biomaterials by providing a deep insight of structure–property relationship of chitosan-based materials and through knowledge to develop novel chitosan-based systems suitable for advanced biomedical and various other applications.

Kochi, India R. Jayakumar
Chennai, India M. Prabaharan

Contents

Adv Polym Sci (2021) 288: 1–28
https://doi.org/10.1007/12_2021_98

Published online: 21 May 2021

Different Forms of Chitosan and Its Derivatives as Hemostatic Agent and Tissue Sealants

M. Nivedhitha Sundaram, Aathira Pradeep, Praveen Kerala Varma, and R. Jayakumar

Contents

Abstract Hemorrhage is the leading cause of death in combat settings, trauma injury, and during surgical procedures. Hemostatic agents and tissue sealants are required when body's coagulation cascade is not able to control hemorrhage effectively. Chitosan because of its hemostatic and tissue adhesive property has gained popularity as a topical hemostatic agent. Chitosan and derivatives of chitosan in different forms such as sponge, hydrogel, dressing, and particles have been extensively studied as hemostatic agent. This chapter provides an overview of the basic hemostatic property of chitosan and also reviews the recent progress on different forms of chitosan and its derivatives as hemostatic agents and tissue sealants.

M. N. Sundaram, A. Pradeep, and R. Jayakumar (✉)
Centre for Nanosciences and Molecular Medicine, Amrita Vishwa Vidyapeetham, Kochi, Kerala, India
e-mail: rjayakumar@aims.amrita.edu

P. K. Varma
Department of Cardiovascular and Thoracic Surgery, Amrita Institute of Medical Sciences, and Research Centre, Amrita Vishwa Vidyapeetham, Kochi, India

Keywords Chitosan · Chitosan derivatives · Hemostatic agent · Tissue sealant

1 Introduction

Hemostasis is the process of stopping bleeding from the site of injury by clot formation [1]. Hemostasis involves the interaction of endothelium, platelets, and coagulation factors to form a stable hemostatic plug as shown in Fig. 1. The three main steps involved in hemostasis are vasoconstriction, platelet plug formation, and activation of coagulation cascade [2].

Primary and secondary hemostasis are the two main phases of hemostasis. Damage to blood vessels initiates primary hemostasis, in which the damaged vessel wall constricts to reduce blood loss, platelets adhere to exposed collagen, platelets get activated and aggregation of platelets forms a soft platelet plug [3]. In the secondary hemostasis, dense granules of platelets release factors that activate the coagulation cascade. Activation of coagulation cascade leads to fibrin mesh formation over the platelet plug. The fibrin mesh is then cross-linked to form stable clot at the site of injury [4]. Steps involved in primary and secondary hemostasis are shown in Fig. 2.

Uncontrolled bleeding is a challenge during civilian/combat trauma and surgical procedures. In these situations, the body's natural coagulation system would not be sufficient to control bleeding. Topical hemostatic agents and tissue sealants are used as adjunct techniques to control bleeding during these critical situations [5]. Hemostatic agents are categorized as passive and active hemostats based on their mechanism of action as shown in Fig. 3. Passive hemostats available in the market are gelatin, collagen, oxidized cellulose, starch, and chitosan-based. These hemostats act as a physical barrier at the injury site and help clot formation by aggregating RBCs/platelets. Active hemostats like thrombin and a combination of fibrinogen and thrombin directly act on the coagulation cascade to form clot at the bleeding site. Thrombin converts fibrinogen to fibrin monomers which are then cross-linked to form a stable fibrin clot. Thrombin has also been used in combination with passive hemostats to improve the efficiency of the hemostat [6]. Tissue sealants are developed from natural or synthetic materials for variety of applications such as hemostatic agents, wound closure, and healing agents [7]. Although most of the developed bioadhesives exhibit poor adhesion in presence of body fluids, extensive research is carried out in developing wet resistant adhesives. The development of an ideal tissue sealant for sutureless wound closure is also a rising field of research.

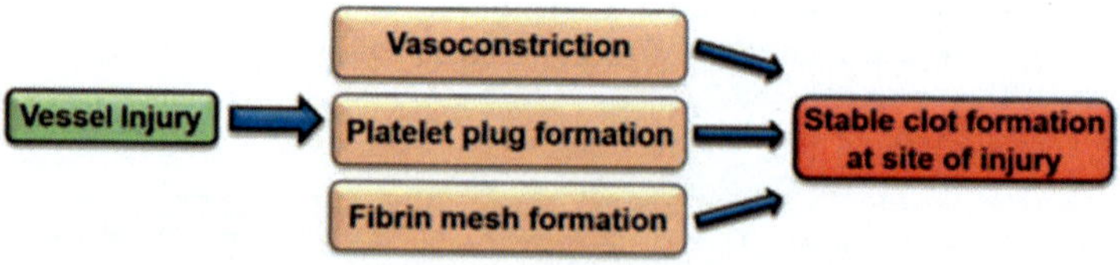

Fig. 1 Schematic diagram showing the factors involved in hemostasis

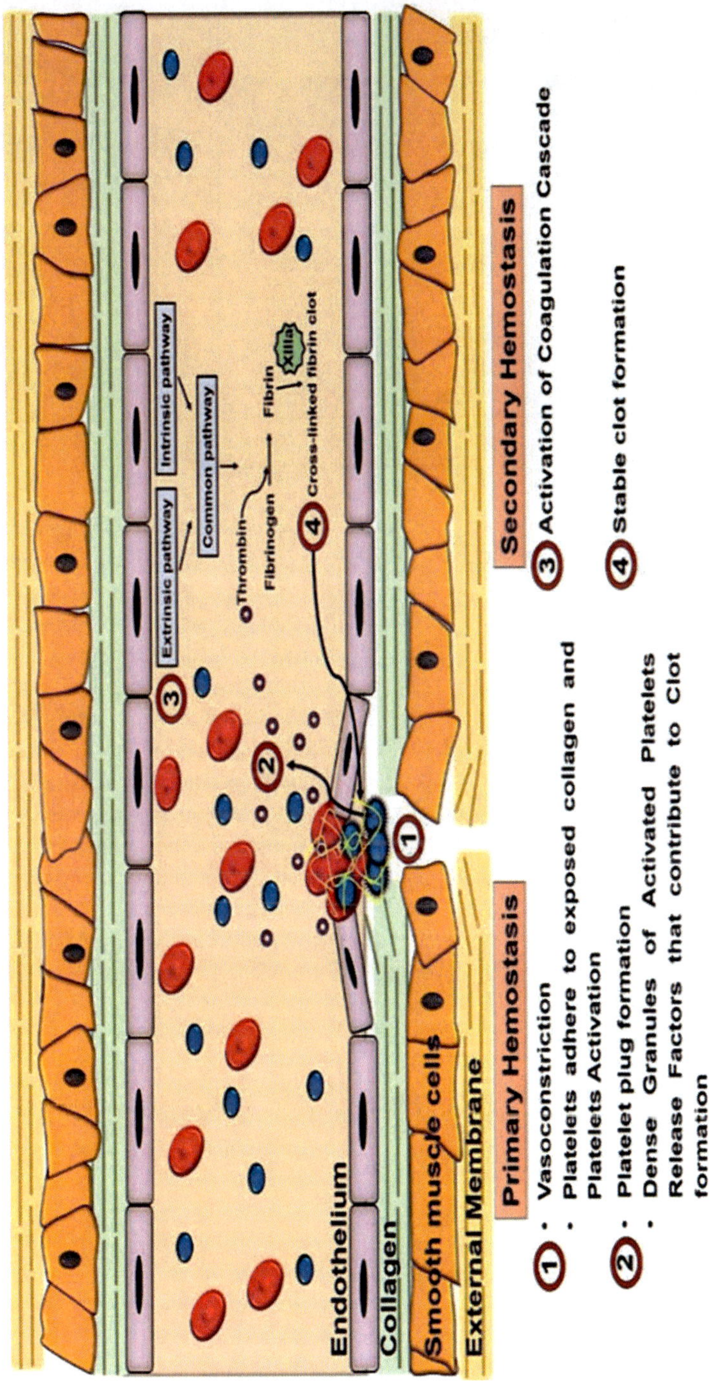

Fig. 2 Pictorial representation of steps involved in primary and secondary hemostasis

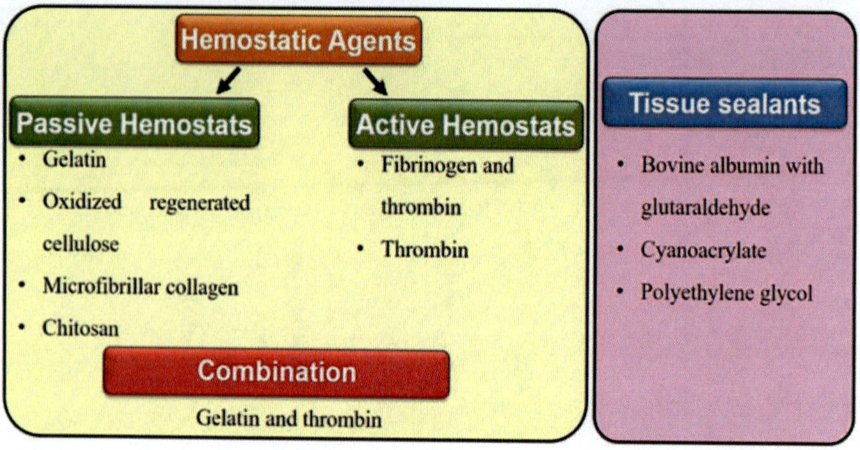

Fig. 3 Commercially available hemostatic agents and tissue sealants

Among the commercially available hemostatic agents, chitosan has gained popularity as a topical hemostatic agent [8]. Chitosan is a polycationic polysaccharide consisting of repeating units of glucosamine and N-acetyl glucosamine formed by partial deacetylation of chitin [9, 10]. Chitosan possesses hemostatic [11], antibacterial [12–14], antimicrobial [15, 16], and mucoadhesive properties. Chitosan in the form of composite sponge [17], nanoparticle for drug delivery [18], and hydrogel as a carrier for angiogenic agents [19–22] deliver growth factors for bone [23] regeneration, for wound healing [24] and also act as a matrix for tendon [25, 26]/ligament [27, 28] and cartilage regeneration [29]. The hemostatic properties of chitosan arise from the polycationic nature of the amino groups present in its molecular chain. Several studies have proved the hemostatic mechanism of chitosan [30, 31]. The RBCs and platelets with negatively charged cell membranes are attracted to positively charged chitosan causing its entrapment and hemostatic plug formation [32]. Degree of deacetylation (DDA) and molecular weight (MW) play an important role in the hemostatic property of chitosan. It has been reported that chitosan with higher DDA and MW showed improved aggregation of RBCs and platelets [33]. Commercially available FDA approved chitosan-based hemostatic agents are Hemcon®, Chitoflex®, Chitogauze®, Celox®, Omnistat®, Traumastat®, and Axiostat®. These hemostatic agents are in the form of gauze, granules, and sponge [34].

2 Different Forms of Chitosan and Its Derivatives as Hemostatic Agent

There are several reports on the hemostatic property of chitosan and functionally modified chitosan processed in different forms such as gauze, dressings, film, scaffold, sponge, foam, granules, particles, powder, spray, hydrogel, and glue (Fig. 4). These different forms of chitosan could be directly applied to the bleeding site to bring about effective hemostasis.

2.1 Gauze/Dressing Based Hemostatic Agent

There are many FDA approved chitosan-based hemostatic dressing that has widespread application as topical hemostatic dressings [35]. The first generation dressing is a Hemcon® bandage that is a chitosan acetate-based wafer that has been reported to control moderate to severe arterial and venous bleeding during military and civilian trauma [36]. Hemcon® dental dressing has been widely used to control

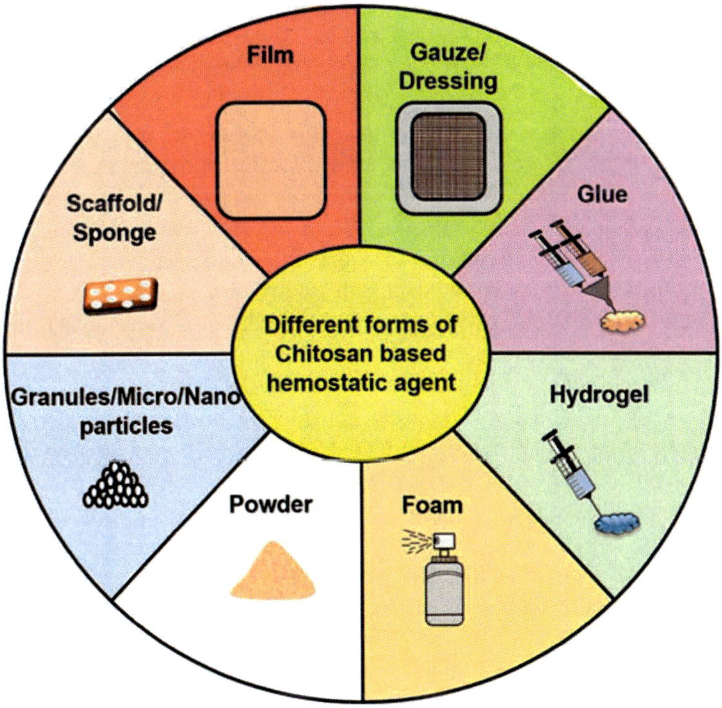

Fig. 4 Diagram showing the different forms of chitosan and chitosan derivatives based hemostatic agents

bleeding during dental procedures [37, 38]. Hemcon bandages have also been used in the sutureless repair of rupture of the left ventricle caused by myocardial infarction [39]. Apart from Hemcon®, there is Chitoflex pro® which is a double-sided flexible roll that is FDA approved as a topical hemostatic agent. This can be easily stuffed into the wound site and control bleeding effectively [40]. Then comes the third-generation chitosan dressing such as Chitogauze® XR PRO, Celox® gauze, Celox® rapid, and Mini sponge dressings. The Chitogauze® XR PRO is a Z folded hemostatic dressing of chitosan-coated polyester-rayon blended nonwoven gauze [41]. This dressing is highly flexible and can conform to the injured area and stop severe hemorrhage. The hemostatic dressing Celox® gauze was used by UK military forces to control hemorrhage in combat causalities. Celox® gauze controls bleeding by swelling and forming a gel-like plug at the site of application. Celox® rapid is an improved form of Celox® gauze that has been reported to control severe arterial bleeding with 60 s of compression compared to other chitosan-based hemostatic dressings that require 3 min compression at the bleeding site. Celox® rapid gauze speeds up packing time and reduces compression time [42]. At present, this chitosan-based gauze is the product of choice to control severe hemorrhage caused by combat wounds. The mini sponge dressing is a syringe like device loaded with chitosan-coated small cylindrical cellulose sponges. At the site of injury, the sponges rapidly absorb blood, expand filling the site, and aid in immediate control of blood loss [43].

There are also several reports on the development of chitosan-based dressings with improved hemostatic property. Chan et al. prepared a chitosan gauze with improved hemostatic property by coating the gauze with PolySTAT [32]. Plasma treatment was done on the chitosan gauze to improve the absorption of PolySTAT solution to form a uniform coating over the gauze. PolySTAT coated chitosan gauze in contact with blood would help in rapid blood absorption. In femoral artery hemorrhage created in rat, PolySTAT coated chitosan gauze when applied to the site of injury exhibited decreased blood loss compared to commercially chitosan gauze (Celox® Rapid). This could be because of the high absorbing capacity, cross-linking of fibrin monomers, entrapment, and aggregation of RBCs in the PolySTAT coated chitosan gauze. A bilayered hemostatic dressing was developed using nanofibrous silk fibroin as the upper layer and chitosan and bacterial cellulose as porous sub-layer [44]. Several coagulation cascade activators such as vitamin K, protamine sulfate, and kaolin were incorporated in the sub-layer of the dressing. The in vitro and in vivo blood clotting potential of the developed nano-micro layered dressing was evaluated. In in vivo rat femoral artery model, the protamine sulfate doped silk fibroin/chitosan/bacterial cellulose bilayer dressing showed reduced blood clotting time. This could be because of the simultaneous action of mucoadhesive nature of chitosan, absorption ability of cellulose, and platelet aggregation caused by silk fibroin as it has a similar structure to collagen, and the nanofibrous layer exhibits a high surface area to volume ratio. An electrospun composite dressing-based hemostatic agent was prepared from silk fibroin, chitosan, and halloysite nanotubes [45]. Incorporation of halloysite led to improved thermal stability, mechanical and hemostatic property of the composite dressing. In in vitro whole blood clotting study, the synthesized composite dressing could form a stable blood clot in 0.46 min. Increased concentration of halloysite decreased the blood clotting time. Halloysite could absorb blood and concentrate coagulation factors and at the same time the negative charge on halloysite could activate the coagulation pathway.

Gu et al. prepared gelatin blended chitosan nanofibrous mats by electrospinning technique [46]. The prepared mats were then sonicated to increase its pore size. The hemostatic potential of the prepared nanofibrous mats was evaluated in vitro. The sonicated gelatin blended chitosan nanofibrous mats exhibited better hemostatic potential than chitosan nanofibrous mats. The hemostatic property of the developed mats could be because of the improved blood absorption property and increased platelet aggregation due to the blending of chitosan and gelatin. The sonicated mats showed better fibroblast cell proliferation and infiltration which could be because of its increased pore size. Chitosan dressing was chemically modified with succinyl, carboxymethyl, and quaternary ammonium groups, and the hemostatic ability of the chitosan nonwoven dressing was evaluated [47]. The hemostatic potential was evaluated in the rabbit ear artery model in which among the three chemically modified groups, succinyl modified chitosan showed the least clotting time of 147 ± 3.7 s. This could be because of the ability of the negatively charged group of succinyl modified chitosan to activate the intrinsic coagulation pathway. Chitosan dressing with inorganic additives and levofloxacin was prepared and the developed dressing was reported to act as hemostatic agent with wound healing properties [48]. Commonly used inorganic additives are aluminium sulfate, aluminium chloride, and iron sulfate. These when incorporated in the chitosan dressing would aid in the activation of coagulation pathways. In the in vivo mice tail amputation study, the chitosan-aluminium sulfate-levofloxacin dressing exhibited good blood absorption property because of which this dressing could achieve effective hemostasis.

2.2 *Scaffold/Sponge/Foam/Film Based Hemostatic Agent*

Hemostatic agents in the form of sponge have been commonly used even in clinics to control blood loss. An absorbable chitosan sponge with different DDA (40% and 73%) was prepared and Huang *et al* evaluated its hemostatic potential in vivo in liver hemorrhage model [49]. It was found that the absorbable hemostatic dressing showed smaller pores, high swelling ratio and therefore in in vivo hemorrhage model it exhibited faster blood clot formation with less blood loss compared to the gelatin sponge that was used as control. The absorbable chitosan sponge with 40% DDA showed better hemostatic property and biodegradability. Wang et al. prepared phosphorylated chitosan with different degree of phosphorylation and evaluated its hemostatic property [50]. Soluble and insoluble phosphorylated chitosan were prepared and its hemostatic potential was evaluated in the mouse liver injury model. In the in vivo evaluation among the soluble phosphorylated chitosan, chitosan with 74% phosphate group showed blood clotting time of 39 s and among the insoluble group, chitosan with 56% showed a blood clotting time of

56 s. Comparing the soluble and insoluble chitosan groups with similar phosphate groups the insoluble chitosan phosphate could achieve hemostasis in reduced time. The synergistic effect of chitosan on RBCs and phosphate group on activation of intrinsic coagulation pathway could have attributed to its hemostatic property.

A hydrophobically modified chitosan foam that could achieve hemostasis even without compression was prepared [51]. When applied at the bleeding site the prepared foam can expand and fill the wound. As the prepared foam is hydrophobic it attaches its hydrophobic chains into the cell membrane of blood cells leading to clustering of the cells. In the in vivo rat liver hemorrhage model, compared to unmodified chitosan and no-treatment groups, the hydrophobically modified chitosan foam could achieve effective hemostasis without compression and therefore the animals treated with the prepared foam showed good animal survival rate. Dowling et al. further evaluated the efficiency of the sprayable foam in swine liver hemorrhage model. When foams with different percentages of hydrophobic modification were applied at the bleeding site, results showed a decrease in the amount of blood loss with an increase in the percentage of hydrophobic modification [52]. Therefore, the modified chitosan could exhibit better hemostatic property even without compression because of the stability, integrity of the prepared foam and its interaction with blood cells. Hydrophobically modified chitosan with zeolite has been developed and its hemostatic effect was evaluated in rat femoral artery injury model [53]. Hydrophobically modified chitosan sponge with zeolite showed the shortest blood clotting time because zeolite could absorb blood and concentrate coagulation factors while chitosan caused RBC aggregation. Chen et al. reported the preparation of a polyelectrolyte using carboxymethyl starch and chitosan oligosaccharide as hemostatic agent [54]. The polyelectrolyte could be formed by the ionic interaction between negatively charged carboxymethyl starch and positively charged chitosan oligosaccharide. In the in vivo study, the prepared polyelectrolyte complex could achieve effective hemostasis in rabbit liver hemorrhage by simultaneous activation of both the coagulation pathways.

A tannic acid cross-linked chitosan-gelatin composite sponge was prepared by gradual base extraction and lyophilization technique [55]. The hemostatic efficiency of the prepared composite sponge was evaluated by performing in vivo hemostasis studies (ear artery and liver hemorrhage) created in rabbit. In the in vivo studies, chitosan-gelatin composite sponge showed better hemostatic property than chitosan or gelatin sponge alone. This could be because of the good liquid absorption, enhanced RBC, and platelet aggregation property of the composite sponge that would have led to quick blood clot formation. A chitosan-cellulose composite sponge was developed by following the alkali-urea method [56]. The developed composite sponge showed good absorption, better mechanical and shape recovery property. In vivo hemostatic potential of the prepared composite sponge was evaluation in mouse tail amputation, rat liver and leg artery injury models and the time taken by composite sponge to control blood loss was found to be 29 s, 20 s, and 34 s, respectively. It was also reported that spherical-shaped composite sponge showed good expansion property and when used in irregular or narrow wounds, the spherical compressed sponge would expand by absorbing blood and also exerts pressure on the surrounding injured tissue. A composite hemostatic sponge made of carboxymethyl chitosan and sodium carboxymethyl cellulose was developed [57]. The cross-linker used in the preparation process (γ-(2, 3-epoxypropoxy) propyltrimethoxysilane) caused the formation of capillary structure in the prepared sponge. The prepared sponge possessed good absorption property and better mechanical strength. The composite sponge with 1% each of carboxymethyl chitosan, carboxymethyl cellulose and the cross-linker was chosen for further studies. In in vivo rat femoral artery injury model, compared to the controls, the developed composite sponge treated group exhibited the shortest time for hemostasis (73 ± 12 s) and least blood loss of 0.57 ± 0.04 g. The unique capillary mimicking structure of the composite sponge could absorb plasma rapidly and concentrate the clotting factors; the presence of carboxyl groups could stimulate adhesion and aggregation of platelets and also cause activation of the intrinsic coagulation cascade.

A nanocomposite sponge was developed using oxidized bacterial cellulose (OBC), chitosan, and collagen [58]. The nanocomposite sponge was formed by the electrostatic interaction between anionic oxidized bacterial cellulose and cationic chitosan. In the in vitro blood clotting study the OBC-Collagen-Chitosan nanocomposite showed significantly higher blood clotting ability compared to OBC alone and Surgicel a commercial hemostatic agent. The faster blood clotting ability of OBC-Collagen-Chitosan than OBC could be due to the addition of collagen to the nanocomposite sponge. The OBC-Collagen-Chitosan nanocomposite sponge also exhibited better erythrocyte adhesion which could be one of the mechanisms by which the composite works. In the in vivo rat liver injury model, OBC-Collagen-Chitosan sponge exhibited least blood loss (58 mg) and the time for hemostasis was recorded to be 86 s. The hemostatic action of the prepared nanocomposite sponge could be due to the synergistic effect of the porous structure of sponge that could absorb blood and concentrate plasma, Chitosan could aggregate RBCs and platelets, and Collagen and OBC could trigger platelet adhesion and activation followed by its aggregation. A cellulose-chitosan sponge was prepared for control of blood loss from deep wounds [59]. The cellulose sponge was prepared from Barca wood pulp with 2 wt% sodium dodecyl sulfate as a surfactant and 60 wt % sodium sulfate as the pore foaming agent. The prepared cellulose sponge with high porosity was then immersed in chitosan solution and freeze-dried to form the cellulose chitosan sponge. The cellulose sponge immersed in 1% chitosan solution was used for in vivo studies. In the in vivo mouse tail amputation, rat liver and leg artery injury studies were performed to evaluate the hemostatic potential of the prepared 1% cellulose-chitosan sponge. The hemostasis time taken by the prepared sponge in mouse tail amputation study was 67 s, in rat liver injury the time for blood clot formation was 89 s, and in rat leg artery injury the time for hemostasis was 105 s. Compared to the untreated group, gauze and gelatin sponge treated groups the cellulose-chitosan sponge showed reduced hemostasis time. Therefore, cellulose-chitosan sponge when placed on the wound site would rapidly absorb blood and concentrate coagulation factors, it could expand by absorbing blood because of its rapid shape recovery property and compress the wound site, chitosan in the sponge would aid in aggregation of platelets and erythrocytes thus helping in stable clot formation.

A gallium-containing mesoporous bioglass incorporated chitosan sponge was developed and its hemostatic potential was evaluated in vitro [60]. The prepared composite scaffold exhibited better water absorption property as its percentage porosity was found to be greater than 79%. In the in vitro blood clotting study, a chitosan sponge containing 50% gallium-containing mesoporous bioglass showed the least blood clotting time. The hemostatic effect of composite scaffold was by the interaction of amine groups of chitosan with blood cells causing RBC and platelet aggregation at the same time the scaffold could stimulate the release of Ca^{2+} ions that would contribute to thrombin mediated fibrin clot formation. A collagen sponge was prepared with the incorporation of chitosan-calcium pyrophosphate nanoflowers as a hemostatic agent [61]. The prepared nanoflowers were grafted in the collagen sponge by amide bonds formed between the carboxyl and amino group of collagen and chitosan. The hemostatic ability of the prepared sponge was evaluated by performing in vivo rabbit ear artery model and hepatic injury model. The collagen sponge with chitosan-calcium pyrophosphate nanoflowers showed the shortest time for bleeding control and the least blood loss. The hemostatic mechanism of each of the components of the prepared sponge is as follows: collagen would cause platelet adhesion and further initiation of coagulation, chitosan would interact electrostatically causing aggregation of RBCs, and calcium would activate coagulation. The prepared sponge also showed good absorption property and degraded in vivo in 21 days. A chitosan-collagen scaffold was developed with the addition of recombinant batroxobin that could show a synergistic effect in controlling blood loss [62]. Batroxobin is a snake venom that in contact with blood would cleave the fibrinopeptide A of fibrinogen and enhances fibrin mesh formation. In vivo evaluation of the developed sponge was performed in rat femoral artery and liver hemorrhage model. In the rat femoral artery injury model the developed sponge with collagen: chitosan (1:1) and 5BU/well of recombinant batroxobin stopped bleeding in 240 s and in the liver injury model the sponge could control bleeding in 180 s with the least blood loss. The hemostatic effect of the sponge could be because of the synergistic effect of chitosan, collagen, and recombinant batroxobin. N-alkylated chitosan/graphene oxide sponge was developed as a hemostatic agent [63]. The addition of graphene oxide improved the mechanical strength of the sponge. Incorporation of 20% ratio of graphene oxide to N-alkylated chitosan could aid in achieving shorter hemostasis time (134.64 ± 17.10 s) compared to commercial chitosan hemostatic product Celox in rabbit femoral artery injury model. The prepared porous sponge in contact with blood could cause RBCs/platelets aggregation, promote the release of Ca^{2+} ions from adhered platelets, and also lead to platelet activation.

Mesoporous silica nanoparticles incorporated N-alkylated chitosan sponge (MSN/NACS) was synthesized [64]. The hydrophilicity of chitosan was improved by modification of N-alkylated chitosan sponge with glycerol. Mesoporous silica with a particle size of 60 nm and pore size of 15 nm was synthesized. The negative charge of mesoporous silica and the positive charge of chitosan would act simultaneously in contact with blood and aid in rapid clotting of blood. In the in vivo rabbit femoral artery injury model, MSN/NACS could achieve hemostasis in 69 ± 5.57 s with the least blood loss (9.5 ± 0.85 mL/kg). In rat liver injury the MSN-NACS could clot blood completely in 64.33 ± 3.06 s with least blood loss of 2.160 ± 0.208 mL. The hemostatic efficiency of MSN/NACS was observed to be better than commercial mineral-based hemostatic sponge combat gauze. A composite sponge made of hydroxybutyl chitosan-diatom biosilica was developed for effective control of blood loss [65]. To make chitosan water-soluble and also impart thermosensitivity, chemically modified with hydroxybutyl group was performed in the hydroxyl (C-6) and amino (C-2) of chitosan. In the in vitro clotting study, the prepared composite sponge showed the least clotting time of 383 ± 20 s. The composite sponge could absorb blood, concentrate clotting factors, activate intrinsic coagulation pathway, and also cause erythrocytes and platelets aggregation.

An antibacterial and hemostatic sponge made of thiol-modified chitosan incorporated with silver nanoparticles (TMC-Ag NPs) has been reported [66]. The thiol modification of chitosan was carried out by the addition of mercaptosuccinic acid. The percentage porosity observed for the prepared sponge was found to be 99.42%. Therefore, the prepared hemostatic sponge would absorb blood rapidly and promote RBCs and platelet aggregation. In the in vivo rat liver injury and rabbit vein bleeding model, the TMC-Ag NPs exhibited a blood clotting time of 50 s and 15 s, respectively. The hemostatic time of the composite sponge was less compared to CS, TMC, and commercial poly (vinyl alcohol) dimethyl-formal (PVF) sponges. The thiol groups of the composite sponge could have activated the tissue factor pathway of the coagulation cascade and caused platelet activation and aggregation. The difference in the hemostatic time observed between developed sponges (TMC and TMC-Ag NPs) could be attributed to the Ag NPs added to prepare the composite sponge. The incorporated Ag NPs also exhibited good antibacterial property. A composite sponge made of chitosan-tilapia peptides microspheres incorporated in the chitosan matrix was synthesized [67]. The tilapia is a good source of aqua collagen that could substitute the use of animal collagen. Tilapia peptide was obtained from tilapia collagen by chemical and enzymatic hydrolysis. First chitosan-tilapia peptide microsphere with an average size of 15 μm was prepared. In the in vivo evaluation of the hemostatic potential of prepared composite sponge, the time for hemostasis observed for composite sponge was reduced by 6 and 15 s compared to chitosan alone, chitosan-tilapia collagen microparticles, respectively. The composite sponge could absorb blood and concentrate clotting factors and also facilitate RBC and platelet aggregation due to which the synthesized composite sponge could achieve rapid hemostasis. Chitosan-konjac glucomannan composite sponge with the incorporation of thrombin coated microporous starch particles for rapid hemostasis was studied [68]. The konjac glucomannan is a soluble polysaccharide that has good absorption property. In in vivo rabbit ear injury the prepared chitosan-konjac glucomannan composite sponge with 5% thrombin coated microporous starch particles could control bleeding in 50 s and in liver injury it could control bleeding in 40 s. The hemostatic actin of the composite sponge is as follows:

(1) the positive charge of the sponge interacts electrostatically and causes aggregation of blood cells, (2) the porous structure of the composite sponge exhibits good absorption property, and (3) thrombin would initiate intrinsic coagulation pathway and aid in fibrin mesh formation trapping the aggregated RBCs and platelets. The carboxymethyl chitosan grafted with marine collagen peptide was developed using EDC chemistry [69]. The marine collagen peptide has gained advantage over the other sources of collagen as its source is abundant, low immunogenicity, and easy to obtain high yield. In the rabbit ear artery and liver hemorrhage model, the prepared sponge could achieve hemostasis in 46.49 ± 3.58 s and 65.19 ± 5.47 s, respectively. Therefore, the reduction in blood clotting time and amount of blood loss observed for developed sponge was due to the simultaneous action of collagen that aids platelet adhesion/aggregation and chitosan that would contribute to RBCs and platelet aggregation.

2.3 Granules/Particles/Powder Based Hemostatic Agent

The second generation chitosan-based hemostatic agents in granular form (Celox granules) were reported to be more effective than chitosan-based hemostatic bandages [70]. Celox granules are easy to use and it fills even irregular-shaped wounds. Celox has been reported to control moderate to severe bleeding [71] and could also clot heparinized blood. Celox has been reported to control bleeding effectively in sheep groin injury [72]. Following are the research articles that report the development of chitosan-based hemostatic agents in the form of granules, particles, or powder with the improved hemostatic property.

Diatom is a single-celled microalga that has a cell wall made of silica and it has a 3D architecture with an ordered porous structure. It can be obtained from diatom culture or diatomite. Feng et al. reported the coating of extracted diatom with chitosan solution to prepare chitosan-coated diatom [73]. In vitro characterization, hemocompatibility, and blood clotting potential of different concentrations of prepared chitosan-coated diatom were evaluated. The 1% chitosan-coated diatom showed better absorption ability, hemocompatibility, and blood clotting potential. Further in vivo rat tail amputation study showed the efficiency of the chitosan-coated diatom to clot blood in least time with reduced blood loss compared to diatomite, diatom, gauze, and commercial zeolite-based product Quick clot. The hemostatic mechanism by which chitosan-coated diatom acts was by RBC aggregation by chitosan along with activation of coagulation cascade by the silica present in the diatom. Further Li et al. synthesized aerogel with chitosan-diatom biosilica as hemostatic agent [74]. The *tert*-butyl alcohol was introduced to create porous aerogel with better absorption property. The composite aerogel made of 2% chitosan-dopamine-diatom biosilica and 30% *tert*-butyl alcohol exhibited the shortest blood clotting time of 60.80 ± 4.38 s and 64.67 ± 4.93 s in in vivo rat tail amputation and femoral artery and vein injury model, respectively. The hemostatic action of the prepared aerogel would be by the porous structure of aerogel that would concentrate clotting factors, negatively charged silanol group of biosilica would cause activation of the intrinsic coagulation cascade, and the amine group of chitosan could aid in erythrocytes aggregation. A hemostatic composite beads made of chitosan, dopamine, and diatom biosilica was developed [75]. Micro-meter sized diatom-biosilica with nano-meter sized pores was used in this study. In the in vitro evaluation of the blood clotting time, the prepared composite beads could clot blood in 1.35 min, which was reduced because of the addition of diatom-biosilica to chitosan-dopamine beads. Chen et al. developed mesoporous silica nanoparticles modified with chitosan and hydrocaffeic acid [76] (MSN@CS-HCA). The hydrodynamic diameter of the developed MSN@CS-HCA was found to be 211 nm with a PDI of 0.085. The average pore distribution observed on MSN@CS-HCA was 7.437 nm. In in vivo rat liver and femoral artery hemorrhage model, the developed MSN@CS-HCA reduced blood clotting time by 62.6% and 67.7% compared to the control group, respectively. Comparing the time for hemostasis observed for MSN, MSN@CS, and MSN@CS-HCA in the in vivo hemorrhage models, the prepared MSN@CS-HCA exhibited superior hemostatic performance due to the synergistic effect of negative charge of silica nanoparticles that would activate coagulation cascade, chitosan would aggregate RBCs and platelets and at the same time, the catechol-quinones could help adhere the applied nanoparticles at the site of application.

Chitosan particles in the form of aerogel were prepared by following dripping and supercritical dying process [77]. The prepared aerogel possessed both micro and mesopores with large surface area and better porosity. In the in vivo pig femoral artery injury, the aerogel particles when applied to the site of injury adhered to the wound site and formed a stable clot. The chitosan aerogel particles could induce platelet adhesion and aggregation thereby it could form a stable platelet plug at the site of injury. The preparation of chitosan microspheres by microemulsion followed by thermally induced phase separation technique was reported [78]. The hemostatic potential of the prepared microspheres was tested in vivo in rat liver injury and tail amputation model. The pore size of the microspheres was found to be 2 μm. Chitosan microspheres exhibited improved hemostatic potential in in vitro and in vivo studies. In the rat liver injury model, the prepared microsphere could clot blood in 70 s which was less compared to compact chitosan nanoparticles. The prepared porous chitosan microspheres could absorb blood and concentrate the coagulation factors and at the same time, chitosan could cause RBC aggregation by electrostatic interaction. Chitosan PVA spheres were prepared by electrospraying technique followed by ionotropic gelation [79]. Spheres made of 50% each of chitosan and PVA with a particle size of 1.2 mm reduced blood clotting time to 11.2 min in in vitro study. In vivo study conducted in rat liver hemorrhage with 50% each of chitosan-PVA with a particle size of 1.19 ± 0.32 mm exhibited clotting time of 5.33 ± 0.34 min which was significantly lower than control gauze group. This could be because the spheres could agglutinate blood cells due to the positively charged amine groups and the large surface area of spheres could absorb plasma coagulation factors thereby enhancing the clotting. A hemostatic microparticles composed of carboxymethyl chitosan, tannic acid, starch, and hyaluronic acid was prepared [80]. Tannic acid was used as a cross-linker in the microparticle preparation. The microparticles were prepared by inverse emulsion and cross-linking of the surface of the particles by the use of tannic acid. In the in vitro evaluation of the hemostatic property of the prepared microparticles using thromboelastogram the prepared microparticles significantly shortened the clotting time to 126 s. Increasing tannic acid concentration shortened the clotting time but beyond a certain level increase in tannic acid concentration prolonged the blood clotting time. Therefore, the synergistic effect of carboxymethyl chitosan, hyaluronic acid, and corn starch contributed to the hemostatic, antibacterial, and wound healing property of the prepared microparticles.

Chitosan-gelatin microspheres were synthesized using genipin as cross-linker [81]. The chitosan-gelatin microparticles prepared with 0.75 mL (0.1 mmol/L) genipin in contact with blood exhibited the shortest clotting time and in in vivo rat skin injury, the prepared microspheres could stop bleeding in 12 s. The hemostatic potential of the microparticles could be due to its blood-absorbing property because of the high surface area that would have contributed to RBC and platelet aggregation. Microsphere composed of carboxymethyl chitosan, sodium alginate, and collagen was developed and evaluated its hemostatic potential in vivo [82]. The average size of the prepared microspheres is 39.33 μm. Rat tail amputation study was conducted in which the prepared microparticles exhibited the shortest clotting time of 249.2 ± 44.7 s. The decrease in clotting time would be because of the action of the microparticle on multiple hemostasis mechanism when in contact with blood. Hemostatic property of the prepared microparticles arises due to its good absorption property, RBC aggregating property of carboxymethyl chitosan, activation of coagulation cascade by the Ca ions of alginate and activation leading to aggregation of platelets by collagen. Porous chitosan microspheres with zinc ion were synthesized by microemulsion followed by thermal phase inversion and the average particle size was found to be 70 μm [83]. In in vivo rat liver injury and tail amputation study, the prepared microparticle showed reduced clotting time of 73 ± 5 s and 134 ± 5 s which was significantly less compared to its control groups. The hemostatic potential of the microparticles arises from the synergistic effect of electrostatic interaction of chitosan with RBCs, activation of coagulation pathways by Zn^{2+} ions, and plugging of the bleeding site by applied microparticles. A composite microsphere of carboxymethyl chitosan-sodium alginate and collagen was prepared and evaluated for its blood clotting property in vitro [84]. The microspheres exhibited higher absorption property, higher platelet adhesion, and aggregation compared to the commercial polysaccharide hemostatic powder. Prepared microspheres in contact with blood could swell and because of its large internal structure, the microspheres can adsorb plasma proteins and activate the coagulation cascade that would result in stable clot formation. Liu et al. compared the hemostatic potential of chitosan microspheres with carboxymethyl chitosan microspheres [85]. The particle size of the prepared 3% chitosan and 3% carboxymethyl chitosan microspheres were 3,561.3 and 2,375 nm. The hemostatic potential of the microspheres was evaluated in rat liver injury model wherein the prepared 3% carboxymethyl chitosan microspheres showed better swelling property and accelerated blood clotting potential than chitosan microspheres in both in vitro and in vivo studies. The swelling property and large surface area of carboxymethyl chitosan microspheres could have contributed to better blood clotting potential.

Chitosan powder from the shell of Antarctic krill was prepared and the in vitro and in vivo hemostatic potential of the prepared chitosan was examined in comparison with commercially available chitosan powders with the same MW and DDA [86]. In in vivo mice tail amputation study, the prepared chitosan powder could reduce clotting time by 38%. The hemostatic property exhibited by prepared chitosan powder was better than the commercial products. The prepared chitosan powder showed better absorption property and the DDA of chitosan could have contributed to platelet adhesion and RBC aggregation on the Antarctic krill chitosan. A composite hemostatic material of carboxymethyl chitosan-Ag^{+}-nano TiO_2 was synthesized and evaluated [87]. In the in vivo mice tail amputation study the synthesized composite powder shortened the blood clotting time which could be because of the combined effect of the hemostatic effect of carboxymethyl chitosan, antimicrobial activity of Ag^{+} in the form of silver nitrate and antibacterial effect of $nTiO_2$. Similarly, Wang et al. synthesized composite hemostatic powder by combining chitosan nanoparticles with silver nitrate and calcium chloride [88]. Even this nano Chitosan-Ag^{+}-Ca^{2+} composite powder could act simultaneously and reduce the blood clotting time observed during mice tail amputation study. Porous chitosan-starch composite particles were synthesized by cross-linking using sodium-trimetaphosphate [89]. Chitosan with a MW of 100 kDa was chosen to prepare the particles and ratio of starch: chitosan selected was 2:1 because it showed good absorption property. The hemostatic property of prepared chitosan-starch particles was evaluated by performing rat tail amputation and femoral artery injury model. In rat tail amputation and liver hemorrhage study, the chitosan-starch composite showed shortened time for hemostasis, 164.42 ± 8.67 s and 145.2 ± 19.2 s, respectively, compared to control. This reduction in blood clotting time observed for the prepared composite particle was because of the combined effect of porous starch and chitosan in contact with blood.

2.4 *Hydrogel-Based Hemostatic Agent*

Hydrogel is a 3D structure of hydrophilic polymer that can absorb and hold large amount of water in its interstitial space. Surgi shield is made of carboxymethyl chitosan (8%) and water (92%) and is available in the form of a hydrogel from D. med. Chung YJ et al. evaluated the hemostatic efficiency and wound healing ability of this chitosan hydrogel applied after endoscopic sinus surgery in refractory chronic rhinosinusitis patients [90]. The carboxymethyl chitosan showed reduced time to attain hemostasis, reduced postoperative adhesion and also there was no adverse side effect caused by the applied gel. This study enfaces on the effective use of carboxymethyl chitosan hydrogel to control blood loss and for better wound healing in patients after endoscopic sinus surgery. Tranquilan-Aranilla et al. developed carboxymethyl κ-carrageenan and chitosan with different degrees of substitution and evaluated the blood clotting potential of the prepared hydrogel in vitro [91]. Carboxymethyl κ-carrageenan and carboxymethyl chitosan with increasing degree of substitution and increase in the number of cross-linking steps showed accelerated clot formation and more platelet aggregation. The hemostatic potential of the developed hydrogels was found to be comparable to commercial product Celox®. A hemostatic hydrogel-based poly(vinyl alcohol)/human-like-collagen/carboxymethyl chitosan-based dressing with tween-80 as a pore-forming agent has been reported [92]. The prepared hydrogel showed good absorption and hemostatic property. The hemostatic property of the lyophilized hydrogels was studied in the rabbit ear artery bleeding model. The poly (vinyl alcohol)/human-like-collagen/tween-80 and poly(vinyl alcohol)/human-like-collagen/carboxymethyl chitosan/tween −80 based dressing showed the least time for blood clot formation. The poly(vinylalcohol)/human-like-collagen/tween-80 dressing could control bleeding in 55.25 s. This reduction in blood clotting time could be due to the porous structure of these hydrogels that could have improved its absorption property. Therefore, these hydrogels when applied on the bleeding site could absorb blood, concentrate clotting factors, and contribute to rapid blood clot formation. An injectable viscous nanocomposite of chitosan and rectorite was prepared by intercalating quaternized carboxymethyl chitosan into rectorite (layer silica) [93]. The developed nanocomposite showed a human blood clotting time of 4 min in in vitro blood clotting study. The combined effect of chitosan and rectorite in contact with blood could have led to a decreased time for clot formation. An injectable hydrogel was made from Schiff base reaction between carboxymethyl chitosan, oxidized sodium alginate, and gelatin [94]. In in vivo rat liver hemorrhage model, the prepared hydrogel could bring about effective hemostasis in 62 s and reduce clotting time by 82.2% compared to the untreated wound. This could be because of the good tissue adhesive property and synergistic effect of carboxymethyl chitosan, oxidized sodium alginate, and gelatin present in the hydrogel.

A short peptide of I_3QGK that can change from sol to gel in the presence of transglutaminase occurs by the formation of isopeptide bonds that promote gelation. Hao et al. reported the preparation of *O*-Carboxymethyl chitosan (*O*-CMCS) nanostructured hydrogel with I_3QGK peptides [95]. In the in vitro blood clotting study, the concentration of *O*-CMCS and I_3QGK optimized to show the shortest clotting time of 10 ± 2 s are 0.44 and 5.03 mg/mL, respectively. In in vivo mice tail amputation study, the prepared *O*-CMCS (1.32 mg/mL) and I_3QGK (7.55 mg/mL) hydrogel decrease the hemostasis time to 15 ± 3 s and blood loss to 64 ± 10 mg. In the rat liver injury model, the prepared hydrogel exhibited blood clotting time of 8.7 ± 2.7 s and blood loss of 92 ± 50 mg. The reduction in hemostasis time could be because of the synergistic effect of *O*-CMCS and I_3QGK. The peptide I_3QGK interacted with *O*-CMCS by enzymatic (γ-carboxamide and amine group) and electrostatic interactions and formed a gel in contact with blood. An injectable hydrogel made of tannic acid-*O*-carboxymethyl chitosan in presence of 1,4-benzenediboronic acid was developed [96]. The addition of 1, 4-benzenediboronic acid was to reduce the gelation time and also improve the mechanical properties of the prepared hydrogel. The injectable hydrogel is formed by a hydrogel bond between TA and *O*-carboxymethyl chitosan and boronate ester bonds between TA and 1, 4-benzenediboronic acid. The prepared hydrogel displayed a gelation time of 10 s mainly because of its interconnected porous structure. In in vivo mouse liver injury model, the prepared injectable hydrogel reduced blood loss by about 77% compared to the control group. The wet adhesion property of tannic acid and *O*-carboxymethyl chitosan could have contributed to the firm adhesion of the prepared hydrogel at the site of injury. A thermo-responsive chitosan hydrogel was developed by addition of hydroxybutyl group (1, 2-butene oxide) to its backbone by etherification reaction [97]. The hydroxybutyl chitosan hydrogel was found to be hydrophilic with good porous structure and also could form a gel at body temperature (37°C). In the in vivo study conducted in the rat renal artery model, the 5% hydroxybutyl chitosan hydrogel aggregated RBCs and adhered to the vessel wall thereby occluded blood flow in the applied renal artery. A catechol modified chitosan/β-glycerol phosphate/oyster peptides based thermosensitive hydrogel was developed as hemostatic agent [98]. The main source of oyster peptides are marine organisms and the peptides possess good absorption property and are also biocompatible. The hemostatic performance of the developed hydrogel was evaluated in vivo in mouse liver injury and tail amputation study. The time taken by the thermosensitive hydrogel to clot blood in the mouse liver injury model was reduced by 19.5% and the amount of blood loss during tail amputation study was decreased by 18.9%. Compared to the gelatin sponge the prepared hydrogel exhibited 95.87% higher RBC aggregation. The developed thermosensitive hydrogel could achieve hemostasis in shorter duration because of the action of the hydrogel on different mechanisms of hemostasis. In addition to this, the porous structure of the hydrogel enhanced its hemostatic property.

An injectable chitosan hydrogel with whitlockite nanoparticles was developed [99]. The average size of the prepared whitlockite nanoparticles was 75 ± 5 nm. The composite hydrogel was then prepared by physical mixing of whitlockite nanoparticles added to chitosan hydrogel. In in vivo rat liver and femoral artery injury model the prepared injectable hydrogel showed reduced time for hemostasis 62 ± 3 s and 229 ± 9 s, respectively. The time taken by the prepared hydrogel was even found to be less compared to the commercial hemostatic agent Floseal®. The hemostatic effect of the injectable composite hydrogel was because of the simultaneous action of amine groups of chitosan and magnesium, calcium, and phosphate ions released from whitlockite nanoparticles that would have caused RBCs/platelet aggregation and activation of the coagulation cascade (Fig. 5). Similarly, an injectable composite hydrogel consisting of chitosan and bioglass nanoparticles was also reported [100]. In the in vivo rat liver and femoral artery injury model, the developed chitosan-nano bioglass hydrogel exhibited clotting time of 54 ± 3 s and 185 ± 9 s, respectively. Time taken by chitosan-nano bioglass hydrogel was found to be less compared to control chitosan hydrogel. This could be due to the action of the amine group of chitosan along with silica, phosphorus, and calcium ions released from nano bioglass.

Leonhardt et al. synthesized a hemostatic hydrogel that was formed on β-cyclodextrin polyester that acts as a sacrificial template into which

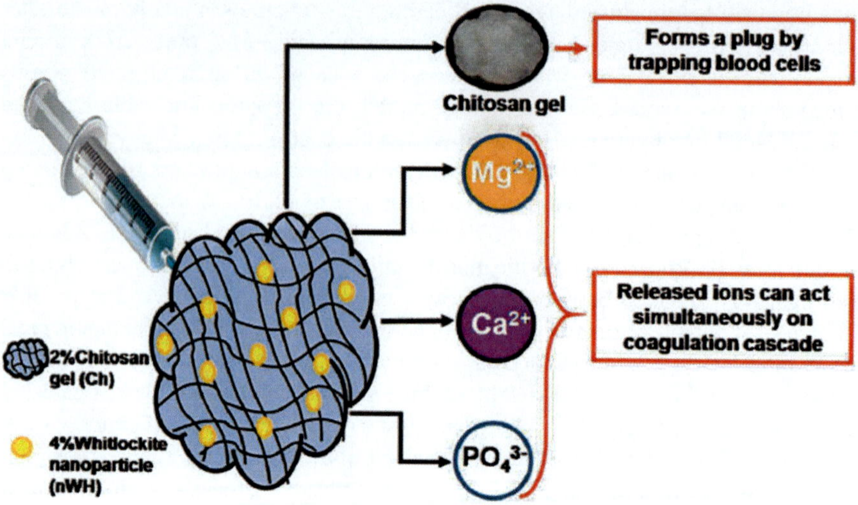

Fig. 5 Schematic representation of mechanism of hemostasis by Chitosan-Whitlockite hydrogel [99]. Reproduced with permission, Copyright American Chemical Society

honeycomb-like nanofibers of chitosan (9.2 ± 3.7 nm) were imprinted (CDPE-Chitosan) [101]. Dianhydride was used as a cross-linker to obtain the prepared cyclodextrin polyester so that the carboxylate groups would react with chitosan and form the hydrogel. The hemostatic potential of the prepared hydrogel was tested on injuries created in rats, rabbits, and pig models. In the in vivo studies developed CDPE-Chitosan hydrogel exhibited faster hemostasis with lower blood loss compared to blank CDPE hydrogel and commercial products Surgicel, Curaspon. A tough hydrogel for hemostasis and wound healing was developed by doping calcium carbonate into acetate chitosan [102]. The acidic nature of acetate chitosan could trigger the release of Ca^{2+} ions from calcium carbonate and would cross-link chitosan to form a tough hydrogel. In in vitro blood clotting study, the prepared hydrogel exhibited the least clotting time which could be because of the combined effect of cationic groups of chitosan and Ca^{2+} ions of calcium carbonate that would have aggregated blood cells and also activated the coagulation cascade. An injectable composite hydrogel was synthesized by the incorporation of vasoconstriction and coagulation activator [103]. The vasoconstrictor and coagulation activator used in the study were potassium aluminium sulfate and calcium chloride, respectively. In the in vivo rat femoral and liver injury model the developed hydrogel could stop bleeding in 105 ± 31 s and 20 ± 10 s, respectively. This could be because of the synergistic effect of components present in the hydrogel that would have caused the reduction in blood clotting time (Fig. 6).

Cryogels are supermacroporous gels prepared by polymerization of monomers at sub-zero temperature [104]. Cryogels based hemostatic agents have gained attention nowadays as it exhibits outstanding water absorption property and flexibility than hydrogels. Zhao et al. synthesized quaternized chitosan-based cryogel functionalized

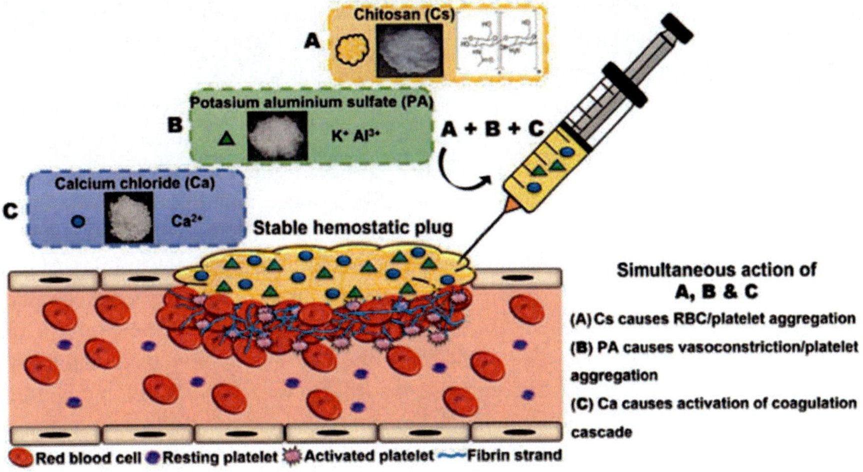

Fig. 6 Schematic representation of mechanism of hemostasis by Chitosan-Potash alum and Calcium chloride hydrogel [103]. Reproduced with permission, Copyright Elsevier

with carbon nanotube and glycidyl methacrylate [105] (QCSG-CNT). The porous nature of cryogel would absorb and concentrate blood cells, meanwhile the positively charged amino and quaternary ammonium groups would aggregate negatively charged plasma proteins and the carbon nanotubes would activate platelets. The synergistic action of all the above-mentioned factors contributed to the hemostatic performance of the developed cryogel in mouse liver injury and tail amputation study. The shape memory property of the carbon nanotubes reinforced cryogel was studied in a rabbit liver injury model. Developed QCSG-CNT4 showed reduced amount of blood loss because of its robust mechanical property. Therefore, the QCSG-CNT cryogel could fit into irregular-shaped injury sites and absorb blood thereby concentrating coagulation factors and aid in effective bleeding control. An injectable chitosan-polydopamine (CS-PDA) cryogel with expansion and blood clotting potential to control blood loss from lethal deep penetrating wounds was reported [106]. The cryogel prepared from 20 mg chitosan and 4.5 mg dopamine was chosen for in vivo studies. In the mouse liver injury model and rat liver injury model, the CS20-PDA4.5 showed the least blood loss which could be because of the rapid blood absorption and concentration of clotting factors leading to clot formation. In rabbit liver injury model, the application of CS20-PDA4.5 cryogel showed a significant reduction in blood clotting time (0.7 min) and blood loss (0.25 g). Further thrombin was incorporated into the CS20-PDA4.5 cryogel and the hemostasis potential of the cryogel was evaluated in coagulopathy hemorrhage created in swine. Incorporating thrombin in the cryogel significantly reduced blood loss (12.2 mg) during mouse liver injury and in rabbit liver hole injury, thrombin incorporated cryogel showed reduced blood loss of 0.13 g and hemostasis time of 23 s. The prepared cryogel could be easily injected into the irregular-shaped wound, where it could expand by absorbing blood forming a physical barrier to stop bleeding and at the same time, it could also concentrate clotting factors and aid in rapid blood clot formation. A semi-interpenetrating cryogel with a pore size of 124.57 ± 20.31 μm was synthesized by cross-linking oxidized dextran and thiolated chitosan where in locust bean gum was introduced to impart hydrophilicity to the prepared cryogel [107]. The addition of locust bean gum increased the swelling rate and also the compressive modulus of the cryogel. The prepared cryogel was cytocompatible to fibroblast cells and was also hemocompatible. In the in vitro blood clotting study the prepared semi-IPN cryogel showed the least blood clotting index which signifies its better blood clotting potential. The hemostatic ability of the cryogel could be due to its absorption property that would have concentrated coagulation factors. Along with the absorption property, the cationic group of chitosan could aggregate platelets and RBCs thereby aiding in stable blood clot formation. A polysaccharide-peptide-based cryogel for bleeding control and wound healing was synthesized by copolymerization of glycol chitosan methacrylate and ε-polylysine acrylamide (GC-EPL). In the in vivo evaluation of the hemostatic potential in rat liver injury model, prepared cryogel exhibited decreased blood loss (more than 90%) compared to control groups (Combat gauze and gelatin sponge). Blood loss in the glycol chitosan and GC-EPL hydrogel group was found to be more compared to the prepared cryogel. This could be because of the microporous architecture of the prepared cryogel. The enhanced hemostatic effect of the GC-EPL cryogel is by the addition of ε-polylysine acrylamide that would have imparted more positive groups to glycol chitosan and the microstructure that could absorb blood and concentrate coagulation factors.

3 Chitosan and Its Derivatives as Tissue Sealants

Tissue sealants/adhesives can be used for different applications including hemostasis, wound closure, and tissue repair. The natural polysaccharides like chitosan, gelatin, alginate and its derivatives are widely used for such applications. Numerous studies have been reported on chitosan-based tissue sealants and their applications due to its biocompatibility. The use of tissue adhesives as hemostatic agents is advantageous as it provides better contact between the bleeding site and hemostatic material.

An in situ forming adhesive hydrogel was prepared using chitosan hydrochloride and oxidized dextran [108]. The Schiff base formation occurs instantaneously between aldehyde group of oxidized dextran and amine group of chitosan and shows good tissue adhesive properties and better adhesive strength than commercially used fibrin glue. The hemostatic effects were assessed in rabbit liver injury model and show good hemostatic effects [108]. Similar Schiff base cross-linking was adopted by Liu et al., to develop an in situ forming adhesive hydrogel using aldehyde hydroxyethyl starch and aminocarboxymethyl chitosan. The material was found to be biocompatible with the vital organs in vivo and exhibits hemostatic effects thereby reducing the blood loss [109]. Since no external cross-linkers are required for Schiff base reaction, it was a widely accepted cross-linking method to develop hydrogels. Another Schiff base cross-linked hydrogel was developed using carboxymethyl chitosan and oxidized dextran. The mechanical strength and adhesive strength of the gel was enhanced with the incorporation of chitin nanowhiskers. The developed material has hemostatic, tissue adhesive, and antibacterial activity [110]. A self-healing adhesive hydrogel with dodecyl-modified chitosan (DCS) and four-armed benzaldehyde-terminated poly(ethylene glycol) (BAPEG) was developed through the Schiff base formation between the aldehyde and amino groups in the polymers. The material exhibits excellent tissue adhesion as the dodecyl tails can be anchored onto the lipid bilayer of the cell membrane and also shows effective hemostasis and antibacterial properties. Incorporation of VEGF into the hydrogel could aid in proliferation of cells and tissue remodeling in the wound site [111]. To combat weakening of adhesive strength under wet condition, hydrogels are developed with hydrophobic surfaces which can react spontaneously with the tissue surface. Inspired from biofilms produced by bacteria and mussel adhesives, a hydrogel based on chitosan grafted with methacrylate (CS-MA), dopamine (DA), and N-hydroxymethyl acrylamide (NMA) was developed via a facile radical polymerization process. CS-MA mimics the structure of Staphylococcal polysaccharide intercellular adhesin (PIA) with hydrophobic residues and a cationic free amine group. The developed hydrogel integrates well with the wet tissue surfaces and has effects in sealing the wound and rapid hemostasis in wet and beating hearts. The hydrogel also shows good biocompatibility and excellent wound healing in *S. aureus* infected cutaneous tissue [112]. An in situ gelling multifunctional chitosan-based adhesive was developed [113]. The glycol chitosan derivative modified with hydroxypropionic acid was cross-linked via Ru-catalyzed photo cross-linking mechanism and was incorporated with the drug amoxicillin. The developed material exhibits good tissue adhesive property in egg membrane and shows hemostatic effects after 10 s in mice liver injury model. The drug loaded in the hydrogel shows an initial burst release for 10 min followed by a slow release for 10 h and shows antibacterial activity against *E. coli* and *S. epidermidis* [113]. Inspired from mussel adhesives that are rich in 3,4-dihydroxyphenyalanine and lysine residues, different chitosan-g-polypeptide polymers were prepared via ring opening polymerization of 3, 4-di-hydroxyphenylalanine-N-carboxyanhydride (DOPA-NCA), cysteine-NCA (Cys-NCA), and arginine-NCA (Arg-NCA). Chitosan is used as an initiator with partially protected amino group. These chitosan graft polypeptides exhibit better adhesive property with tissues and metal. They also show low cytotoxicity and good biodegradability with good hemostatic potential and wound healing properties [114].

Mussel adhesion mechanism is used in developing a pH dependent adhesive hydrogel with hydrocaffeic acid conjugated glycol chitosan. Higher percentage of glycol chitosan conjugate was required to exhibit similar hemostatic effects as that of chitosan-catechol conjugate as glycol units hinder the catechol mediated blood coagulation. However it results in negligible undesired immune responses than chitosan-catechol conjugate confirming it as a promising alternative [115]. A wet adhesive based on chitosan was developed by grafting catechol and lysine groups into chitosan. The catechol group in chitosan provides the wet adhesion and lysine group enhances the solubility of chitosan and antibacterial effects. When dry polymer comes in contact with the blood, it gets hydrated and adheres to the site without causing any adverse effects to the cells. The hemostatic property of the wet adhesive was studied in Wistar rats by applying the adhesive on gauze followed by application at the bleeding site and the adhesive coated gauze shows less bleeding compared to the normal ones. The adhesive was also tested by coating on needles to make functional hemostatic needles and shows excellent blood clotting effects by preventing oozing blood from femoral artery [116]. Sanandiya et al. [117] also acquired inspiration from nature to develop tissue adhesive hydrogels where they incorporated pyrogallol groups to chitosan to mimic the self-healing properties of tunichrome in tunicates. The presence of pyrogallol groups enhanced the solubility and adhesiveness. They also show excellent platelet adhesion and blood clotting effects. Natural structure of the tunicates can be mimicked by fabricating this chitosan derivative as electrospun mats [117]. A multifunctional tissue adhesive cryogel in the form of a hemostatic tape was developed with rapid blood clotting effects. It is composed of quarternized chitosan cross-linked with polydopamine. The cryogel exhibited better hemostatic effects than gauze and gelatin sponge in in vivo bleeding models and can be used as a non-pressing hemostatic dressing [118].

4 Summary

Bleeding during surgeries and other situations has always been a concern and the necessity of an effective hemostatic technique is always important. To address the limitations due to toxic effects of hemostatic agents, different natural polymers are utilized in developing hemostatic agents with multifunctional properties. The natural polymers like chitosan endow excellent properties for application in the field of biomaterials. As discussed in the chapter, different types of hemostatic materials are developed from chitosan with good potential in effective bleeding control.

Acknowledgements M. Nivedhitha Sundaram is grateful to the Council of Scientific and Industrial Research (CSIR), Government of India for the financial support through Senior Research Fellowship [Award No: 09/963(0043) 2K18-EMR-I]. Aathira Pradeep is grateful to the Department of Science and Technology, Government of India for the financial support through DST-Inspire Fellowship (Grant no. IF170659) and the authors express sincere gratitude to Amrita Centre for Nanoscience and Molecular Medicine, Amrita Vishwa Vidyapeetham, Kochi 682041, India for providing infrastructural support.

References

1. Palta S, Saroa R, Palta A (2014) Overview of the coagulation system. Indian J Anaesth 58:515–523
2. Tanaka KA, Key NS, Levy JH (2009) Blood coagulation: hemostasis and thrombin regulation. Anesth Analg 108:1433–1446
3. Versteeg HH, Heemskerk JWM, Levi M, Reitsma PH (2013) New fundamentals in coagulation. Physiol Rev 93:327–358
4. Wolberg AS, Campbell RA (2008) Thrombin generation, fibrin clot formation and hemostasis. Transfus Apher Sci 38:15–23
5. Behrens AM, Sikorski MJ, Kofinas P (2014) Hemostatic strategies for traumatic and surgical bleeding. J Biomed Mater Res Part A 102:4182–4194
6. Peng T (2010) Biomaterials for hemorrhage control. Trends Biomater Artif Organs 24:27–68
7. Ryu JH, Hong S, Lee H (2015) Bio-inspired adhesive catechol-conjugated chitosan for biomedical applications: a mini review. Acta Biomater 27:101–115
8. Nivedhitha Sundaram M, Mony U, Jayakumar R (2016) Chitin and chitosan as hemostatic agents. In: Encyclopedia of polymer science and technology. Wiley, Hoboken, pp 1–12
9. Anitha A, Sowmya S, Kumar PTS et al (2014) Chitin and chitosan in selected biomedical applications. Prog Polym Sci 39:1644–1667
10. Jayakumar R, Menon D, Manzoor K et al (2010) Biomedical applications of chitin and chitosan based nanomaterials – a short review. Carbohydr Polym 82:227–232
11. Mohandas A, Nimal TR, Das V et al (2015) Drug loaded bi-layered sponge for wound management in hyperfibrinolytic conditions. J Mater Chem 3:5795–5805
12. Anjana J, Mohandasa A, Seethalakshmya S, Suresha MK et al (2018) Bi-layered nanocomposite bandages for controlling microbial infections and overproduction of matrix metalloproteinase activity. Int J Biol Macromol 110:124–132
13. Nithya S, Nimal TR, Baranwal G et al (2018) Preparation, characterization and efficacy of lysostaphin-chitosan gel against *Staphylococcus aureus*. Int J Biol Macromol 110:157–166
14. Elbi S, Nimal TR, Rajan VK et al (2017) Fucoidan coated ciprofloxacin loaded chitosan nanoparticles for the treatment of intracellular and biofilm infections of salmonella. Colloids Surf B Biointerfaces 160:40–47
15. Raveendran NT, Mohandas A, Ramachandran R et al (2019) Ciprofloxacin and fluconazole containing fibrin nanoparticles incorporated chitosan bandages for the treatment of polymicrobial wound infections ciprofloxacin and fluconazole containing fibrin nanoparticles incorporated chitosan bandages for the treatment. ACS Appl Bio Mater 2:243–254
16. Mohandas A, Deepthi S, Biswas R, Jayakumar R (2018) Chitosan based metallic nanocomposite scaffolds as antimicrobial wound dressings. Bioact Mater 3:267–277
17. Anisha BS, Sankar D, Mohandas A et al (2013) Chitosan-hyaluronan/nano chondroitin sulfate ternary composite sponges for medical use. Carbohydr Polym 92:1470–1476
18. Rajitha P, Gopinath D, Biswas R et al (2016) Chitosan nanoparticles in drug therapy of infectious and inflammatory diseases. Expert Opin Drug Deliv 13:1177–1194
19. Mohandas A, Rangasamy J (2021) Nanocurcumin and arginine entrapped injectable chitosan hydrogel for restoration of hypoxia induced endothelial dysfunction. Int J Biol Macromol 166:471–482
20. Mohandas A, Sun W, Nimal T et al (2018) Injectable chitosan-fibrin/nanocurcumin composite hydrogel for the enhancement of angiogenesis. Res Chem Intermed 44:4873–4887
21. Vignesh S, Sivashanmugam A, Mohandas A et al (2018) Injectable deferoxamine nanoparticles loaded chitosan-hyaluronic acid coacervate hydrogel for therapeutic angiogenesis. Colloids Surf B Biointerfaces 161:129–138
22. Mohandas A, Anisha BS, Chennazhi KP, Jayakumar R (2015) Chitosan-hyaluronic acid/VEGF loaded fibrin nanoparticles composite sponges for enhancing angiogenesis in wounds. Colloids Surf B Biointerfaces 127:105–113

23. Deepthi S, Venkatesan J, Kim S-K et al (2016) An overview of chitin or chitosan/nano ceramic composite scaffolds for bone tissue engineering. Int J Biol Macromol 93:1338–1353
24. Nimal TR, Baranwal G, Bavya MC et al (2016) Anti-staphylococcal activity of injectable nano tigecycline/chitosan-PRP composite hydrogel using *Drosophila melanogaster* model for infectious wounds. ACS Appl Mater Interfaces 8:22074–22083
25. Sundaram MN, Deepthi S, Mony U et al (2019) Chitosan hydrogel scaffold reinforced with twisted poly (L lactic acid) aligned micro fibrous bundle to mimic tendon extracellular matrix. Int J Biol Macromol 122:37–44
26. Deepthi S, Sundaram MN, Kadavan JD, Jayakumar R (2016) Layered chitosan-collagen hydrogel/aligned PLLA nanofiber construct for flexor tendon regeneration. Carbohydr Polym 153:492–500
27. Sundaram MN, Sowmya S, Deepthi S et al (2016) Bilayered construct for simultaneous regeneration of alveolar bone and periodontal ligament. J Biomed Mater Res Part B Appl Biomater 104:761–770
28. Deepthi S, Jeevitha K, Nivedhitha Sundaram M et al (2015) Chitosan-hyaluronic acid hydrogel coated poly(caprolactone) multiscale bilayer scaffold for ligament regeneration. Chem Eng J 260:478–485
29. Deepthi S, Abdul Gafoor AA, Sivaahanmugam A et al (2016) Nanostrontium ranelate incorporated injectable hydrogel enhanced matrix production supporting chondrogenesis in vitro. J Mater Chem B 4:4092–4103
30. Whang HS, Kirsch W, Zhu YH et al (2005) Hemostatic agents derived from chitin and chitosan. J Macromol Sci Polym Rev 45:309–323
31. Maksym PV (2015) Chitosan as a hemostatic agent: current state. Eur J Med Ser B 2:24–33
32. Chan LW, Kim CH, Wang X et al (2016) PolySTAT-modified chitosan gauzes for improved hemostasis in external hemorrhage. Acta Biomater 31:178–185
33. Hattori H, Ishihara M (2015) Changes in blood aggregation with differences in molecular weight and degree of deacetylation of chitosan. Biomed Mater 10:015014
34. Khan MA, Mujahid M (2019) A review on recent advances in chitosan based composite for hemostatic dressings. Int J Biol Macromol 124:138–147
35. Erpaçal B, Adigüzel Ö, Cangül S, Acartürk M (2019) A general overview of chitosan and its use in dentistry. Int Biol Biomed J 5:1–11
36. Wedmore I, Mcmanus JG, Pusateri AE, Holcomb JB (2006) A special report on the chitosan-based hemostatic dressing: experience in current combat operations. J Trauma 60:655–658
37. Gadde P (2016) Hemostasis and post-operative care of oral surgical wounds by Hemcon dental dressing in patients on oral anticoagulant therapy: a Split mouth randomized controlled clinical trial. J Clin Diagn Res 10:ZC37–ZC40
38. Pippi R, Santoro M, Cafolla A (2017) The use of a chitosan-derived haemostatic agent for post-extraction bleeding control in patients on anti-platelet treatment. J Oral Maxillofac Surg 75:1118–1123
39. Morimoto Y, Sugimoto T, Haba F, Sakahira H (2016) A new hybrid sutureless patch repair utilizing chitosan for left ventricle rupture after myocardial infarction: a case report. Int J Surg Case Rep 26:131–133
40. Littlejohn LF, Devlin JJ, Kircher SS et al (2011) Comparison of Celox-A, ChitoFlex, WoundStat, and combat gauze hemostatic agents versus standard gauze dressing in control of hemorrhage in a swine model of penetrating trauma. Acad Emerg Med 18:340–350
41. Jastrzębski P, Adamiak Z, Gesek M et al (2016) Safety of the long-term application of QuikClot Combat Gauze, ChitoGauze PRO and Celox Gauze in a femoral artery injury model in swine – a preliminary study. Pol J Vet Sci 19:337–343
42. Bennett BL (2017) Bleeding control using hemostatic dressings: lessons learned. Wilderness Environ Med 28:S39–S49
43. Rodriguez MI, Jensen JT, Gregory K et al (2017) A novel tamponade agent for management of post partum hemorrhage: adaptation of the Xstat mini-sponge applicator for obstetric use. BMC Pregnancy Childbirth 17:187

44. Karahaliloğlu Z, Demirbilek M, Ulusoy I et al (2016) Hemostatic activities of nano/microporous bilayer dressings in a femoral artery bleeding rat model. J Appl Polym Sci 133:43657
45. Ren X, Xu Z, Wang L, Zhao H (2019) Silk fibroin/chitosan/halloysite composite medical dressing with antibacterial and rapid haemostatic properties. Mater Res Express 6:125409
46. Kang B, Jun S, Sup M et al (2016) Gelatin blending and sonication of chitosan nanofiber mats produce synergistic effects on hemostatic functions. Int J Biol Macromol 82:89–96
47. Yan D, Hu S, Zhou Z et al (2018) Different chemical groups modification on the surface of chitosan nonwoven dressing and the hemostatic properties. Int J Biol Macromol 107:463–469
48. Koumentakou I, Terzopoulou Z, Michopoulou A et al (2020) Chitosan dressings containing inorganic additives and levofloxacin as potential wound care products with enhanced hemostatic properties. Int J Biol Macromol 162:693–703
49. Huang X, Sun Y, Nie J et al (2015) Using absorbable chitosan hemostatic sponges as a promising surgical dressing. Int J Biol Macromol 75:322–329
50. Wang Y, Kim K, Lee MS, Lee H (2019) Hemostatic ability of chitosan-phosphate inspired by coagulation mechanisms of platelet polyphosphates. Macromol Biosci 18:e1700378
51. Dowling MB, Smith W, Balogh P et al (2015) Hydrophobically-modified chitosan foam: description and hemostatic efficacy. J Surg Res 193:316–323
52. Dowling MB, Macintire IC, White JC et al (2015) Sprayable foams based on an amphiphilic biopolymer for control of hemorrhage without compression. ACS Biomater Sci Eng 1:440–447
53. Huang Y, Feng L, Zhang Y et al (2017) Hemostasis mechanism and applications of N-alkylated chitosan sponge. Polym Adv Technol 28:1107–1114
54. Chen X, Yan YY, Li H et al (2018) Evaluation of absorbable hemostatic agents ofpolyelectrolyte complexes using carboxymethylstarch and chitosan oligosaccharide bothin vitroandin vivo. Biomater Sci 6:3332–3344
55. Lan G, Lu B, Wang T et al (2015) Chitosan/gelatin composite sponge is an absorbable surgical hemostatic agent. Colloids Surf B Biointerfaces 136:1026–1034
56. Fan X, Li Y, Li N et al (2020) Rapid hemostatic chitosan/cellulose composite sponge by alkali/urea method for massive haemorrhage. Int J Biol Macromol 164:2769–2778
57. Wang L, Zhong Y, Qian C et al (2020) A natural polymer-based porous sponge with capillary-mimicking microchannels for rapid hemostasis. Acta Biomater 114:193–205
58. Yuan H, Chen L, Hong FF (2020) A biodegradable antibacterial nanocomposite based on oxidized bacterial nanocellulose for rapid hemostasis and wound healing. ACS Appl Mater Interfaces 12:3382–3392
59. Fan X, Li M, Yang Q et al (2021) Morphology-controllable cellulose/chitosan sponge for deep wound hemostasis with surfactant and pore-foaming agent. Mater Sci Eng C 118:111408
60. Pourshahrestani S, Zeimaran E, Kadri NA et al (2017) Potency and cytotoxicity of a novel gallium-containing mesoporous bioactive glass/chitosan composite scaffold as hemostatic agents. ACS Appl Mater Interfaces 9:31381–31392
61. Yan T, Cheng F, Wei X et al (2017) Biodegradable collagen sponge reinforced with chitosan/calcium pyrophosphate nanoflowers for rapid hemostasis. Carbohydr Polym 170:271–280
62. Mi G, Hee M, Kwon B et al (2017) Functional improvement of hemostatic dressing by addition of recombinant batroxobin. Acta Biomater 48:175–185
63. Zhang Y, Guan J, Wu J et al (2019) N-alkylated chitosan/graphene oxide porous sponge for rapid and effective hemostasis in emergency situations. Carbohydr Polym 219:405–413
64. Chen Z, Han L, Liu C et al (2018) A rapid hemostatic sponge based on large, mesoporous silica nanoparticles and N-alkylated chitosan. Nanoscale 10:20234–20245
65. Zhang K, Li J, Wang Y et al (2020) Hydroxybutyl chitosan/diatom-biosilica composite sponge for hemorrhage control. Carbohydr Polym 236:116051
66. Wu Z, Zhou W, Deng W et al (2020) Antibacterial and hemostatic thiol-modified chitosan-immobilized AgNPs composite sponges. ACS Appl Mater Interfaces 12:20307–20320

67. Ouyang Q, Hou T, Li C et al (2019) Construction of a composite sponge containing tilapia peptides and chitosan with improved hemostatic performance. Int J Biol Macromol 139:719–729
68. Shi Z, Lan G, Hu E et al (2020) Puff pastry-like chitosan/konjac glucomannan matrix with thrombin- occupied microporous starch particles as a composite for hemostasis. Carbohydr Polym 232:115814
69. Cheng Y, Lu S, Hu Z et al (2020) Marine collagen peptide grafted carboxymethyl chitosan: optimization preparation and coagulation evaluation. Int J Biol Macromol 164:3953–3964
70. Kozen BG, Kircher SJ, Henao J et al (2008) An alternative hemostatic dressing: comparison of CELOX, HemCon, and QuikClot. Acad Emerg Med 15:74–81
71. Kheirabadi BS, Edens JW, Terrazas IB et al (2009) Comparison of new hemostatic granules/ powders with currently deployed hemostatic products in a lethal model of extremity arterial hemorrhage in swine. J Trauma 66:316–326
72. Ersoy G, Ülkümen R, Yilmaz O et al (2016) Hemostatic efficacy of local chitosan linear polymer granule in an experimental sheep model with severe bleeding of arteria and vena femoralis. Ulus Travma Acil Cerrahi Derg 22:215–223
73. Feng C, Li J, Wu GS et al (2016) Chitosan-coated diatom silica as hemostatic agent for hemorrhage control. ACS Appl Mater Interfaces 8:34234–34243
74. Li J, Sun X, Zhang K et al (2020) Chitosan/diatom-biosilica aerogel with controlled porous structure for rapid hemostasis. Adv Healthc Mater 9:2000951
75. Wang Y, Fu Y, Li J et al (2018) Multifunctional chitosan/dopamine/diatom-biosilica composite beads for rapid blood coagulation. Carbohydr Polym 200:6–14
76. Chen J, Ai J, Chen S et al (2019) Synergistic enhancement of hemostatic performance of mesoporous silica by hydrocaffeic acid and chitosan. Int J Biol Macromol 139:1203–1211
77. Lovskaya D, Menshutina N, Mochalova M et al (2020) Chitosan-based aerogel particles as highly effective local hemostatic agents. Production process and in vivo evaluations. Polymers (Basel) 12:2055
78. Li J, Wu X, Wu Y et al (2017) Porous chitosan microspheres for application as quick in vitro and in vivo hemostat. Mater Sci Eng C 77:411–419
79. Chen Q, Liu Y, Wang T et al (2017) Chitosan-PVA monodisperse millimeter-sized spheres prepared by electrospraying reduce the thromboembolic risk in hemorrhage control. J Mater Chem B 5:3686–3696
80. Li N, Yang X, Liu W et al (2018) Tannic acid cross-linked polysaccharide-based multifunctional hemostatic microparticles for the regulation of rapid wound healing. Macromol Biosci 18:e1800209
81. Liu W, Wang M, Liang B, Li W (2018) Genipin crosslinked microspheres as an effective hemostatic agent. Polym Adv Technol 29:2632–2642
82. Jin J, Ji Z, Xu M et al (2018) Microspheres of carboxymethyl chitosan, sodium alginate and collagen as a hemostatic agent in vivo. ACS Biomater Sci Eng 4:2541–2551
83. Pan M, Tang Z, Tu J et al (2018) Porous chitosan microspheres containing zinc ion for enhanced thrombosis and hemostasis. Mater Sci Eng C 85:27–36
84. Shi X, Fang Q, Ding M et al (2015) Microspheres of carboxymethyl chitosan, sodium alginate and collagen for a novel hemostatic in vitro study. J Biomater Appl 30:1092–1102
85. Liu L, Lv Q, Zhang Q et al (2015) Preparation of carboxymethyl chitosan microspheres and their application in hemostas. Disaster Med Public Health Prep 11:660–667
86. Wu S, Huang Z, Yue J et al (2015) The efficient hemostatic effect of Antarctic krill chitosan is related to its hydration property. Carbohydr Polym 132:295–303
87. Wang F, Wei C, Li J et al (2016) Preparation and properties characterization of modified chitosan composited hemostatic materials. Mater Sci Forum 852:1271–1276
88. Wang F, Wei C, Li J et al (2016) Preparation and performance characterization of hemostatic materials. Mater Sci Forum 852:1282–1287
89. Chen Y, Qian J, Zhao C et al (2019) Preparation and evaluation of porous starch/chitosan composite cross-linking hemostatic. Eur Polym J 118:17–26

90. An YCS, Woo JY, Shim S (2016) Effect of a chitosan gel on hemostasis and prevention of adhesion after endoscopic sinus surgery. Clin Exp Otorhinolaryngol 9:143–149
91. Tranquilan-aranilla C, Barba BJD, Vista JRM, Abad LV (2016) Hemostatic efficacy evaluation of radiation crosslinked carboxymethylkappa-carrageenan and chitosan with varying degrees of substitution. Radiat Phys Chem 124:124–129
92. Pan H, Fan D, Cao W et al (2017) Preparation and characterization of breathable hemostatic hydrogel dressings and determination of their effects on full-thickness defects. Polymers (Basel) 9:727
93. Li X, Li Y-C, Chen M et al (2018) Chitosan/rectorite nanocomposite with injectablefunctionality for skin hemostasis. J Mater Chem B 6:6544–6549
94. Cao J, Xiao L, Shi X (2019) Injectable drug-loaded polysaccharide hybrid hydrogels for hemostasis. RSC Adv 9:36858–36866
95. Hao R, Peng X, Zhang Y et al (2020) Rapid hemostasis resulting from the synergism of self-assembling short peptide and O-carboxymethyl chitosan. ACS Appl Mater Interfaces 12:55574–55583
96. Geng H, Dai Q, Sun H et al (2020) Injectable and sprayable polyphenol-based hydrogels for controlling hemostasis. ACS Appl Mater Interfaces 3:1258–1266
97. Bao Z, Zuo Y, Kong M, Cheng X (2017) Thermo-responsive hydroxybutyl chitosan hydrogel as artery intervention embolic agent for hemorrhage control. Int J Biol Macromol 105:566–574
98. Zhang D, Hu Z, Zhang L et al (2020) Chitosan-based thermo-sensitive hydrogel loading oyster peptides for hemostasis application. Materials (Basel) 13:5038
99. Sundaram N, Pillai M, Eswar K et al (2019) Injectable nano Whitlockite incorporated chitosan hydrogel for effective hemostasis. ACS Appl Bio Mater 2:865–873
100. Sundaram MN, Amirthalingam S, Mony U et al (2019) Injectable chitosan-nano bioglass composite hemostatic hydrogel for effective bleeding control. Int J Biol Macromol 129:936–943
101. Leonhardt EE, Kang N, Hamad MA et al (2019) Absorbable hemostatic hydrogels comprising composites of sacrificial templates and honeycomb-like nanofibrous mats of chitosan. Nat Commun 10:1–9
102. Zhou Y, Wang Y, Li X (2018) Acetate chitosan with CaCO3 doping form tough hydrogel for hemostasis and wound healing. Polym Adv Technol:1–10
103. Sundaram MN, Mony U, Kerala P, Jayakumar R (2021) Vasoconstrictor and coagulation activator entrapped chitosan based composite hydrogel for rapid bleeding control. Carbohydr Polym 258:117634
104. Saylan Y, Denizli A (2019) Supermacroporous composite cryogels in biomedical applications. Gels 5:20
105. Zhao X, Guo B, Wu H et al (2018) Injectable antibacterial conductive nanocomposite cryogels with rapid shape recovery for noncompressible hemorrhage and wound healing. Nat Commun 9:2784
106. Zhao X, Liang Y, Guo B et al (2021) Injectable dry cryogels with excellent blood-sucking expansion and blood clotting to cease hemorrhage for lethal deep-wounds, coagulopathy and tissue regeneration. Chem Eng J 403:126329
107. Kumar L, Raval P, Kedaria D, Vasita R (2018) Study of locust bean gum reinforced cyst-chitosan and oxidized dextran based semi-IPN cryogel dressing for hemostatic application. Bioact Mater 3:370–384
108. Balakrishnan B, Soman D, Payanam U et al (2017) A novel injectable tissue adhesive based on oxidized dextran and chitosan. Acta Biomater 53:343–354
109. Liu J, Li J, Yu F et al (2020) In situ forming hydrogel of natural polysaccharides through Schiff base reaction for soft tissue adhesive and hemostasis. Int J Biol Macromol 147:653–666
110. Pang J, Bi S, Kong T et al (2020) Mechanically and functionally strengthened tissue adhesive of chitin whisker complexed chitosan/dextran derivatives based hydrogel. Carbohydr Polym 237:116138

111. Chen G, Yu Y, Wu X et al (2018) Bioinspired multifunctional hybrid hydrogel promotes wound healing. Adv Funct Mater 28:1801386
112. Han W, Zhou B, Yang K et al (2020) Biofilm-inspired adhesive and antibacterial hydrogel with tough tissue integration performance for sealing hemostasis and wound healing. Bioact Mater 5:768–778
113. Lu M, Liu Y, Huang YC et al (2018) Fabrication of photo-crosslinkable glycol chitosan hydrogel as a tissue adhesive. Carbohydr Polym 181:668–674
114. Lu D, Wang H, Wang X et al (2019) Biomimetic chitosan-graft-polypeptides for improved adhesion in tissue and metal. Carbohydr Polym 215:20–28
115. Park E, Lee J, Huh KM et al (2019) Toxicity-attenuated glycol chitosan adhesive inspired by mussel adhesion mechanisms. Adv Healthc Mater 8:1900275
116. Gao L, Ma S, Luo J et al (2019) Synthesizing functional biomacromolecular wet adhesives with typical gel-sol transition and shear-thinning features. ACS Biomater Sci Eng 5:4293–4301
117. Sanandiya ND, Lee S, Rho S et al (2019) Tunichrome-inspired pyrogallol functionalized chitosan for tissue adhesion and hemostasis. Carbohydr Polym 208:77–85
118. Li M, Zhang Z, Liang Y et al (2020) Multifunctional tissue-adhesive cryogel wound dressing for rapid nonpressing surface hemorrhage and wound repair. ACS Appl Mater Interfaces 12:35856–35872

Adv Polym Sci (2021) 288: 29–86
https://doi.org/10.1007/12_2021_91

Published online: 25 May 2021

Chitosan Nanofibers in Regenerative Medicine

Vishnu Priya Murali and Priyadarshan Sundararaju

Contents

Abstract Chitosan, a natural biopolymer, has been widely explored in the field of biomedical engineering. Because of its chemical similarity to the extracellular matrix (ECM) components, chitosan is known to be extremely biocompatible. It has been fabricated into various scaffold forms such as membranes, hydrogels, films, beads, particles, etc. for a large number of regenerative applications. Chitosan nanofibrous membranes are quite popular in the field of tissue regeneration because of their structural and chemical similarity to the natural ECM. The nanostructure very closely mimics the ECM and also provides increased surface to deliver biotherapeutics. This chapter attempts to summarize and highlight some of the recent research efforts that have been carried out to explore the tissue engineering, drug delivery, and wound healing applications of chitosan nanofibrous membranes. It also discusses some of the human trials where chitosan membranes have been effectively used to treat certain conditions, and also mentions issues which need to be addressed

V. P. Murali (✉)
Department of Chemical and Biomedical Engineering, FAMU-FSU College of Engineering, Tallahassee, FL, USA
e-mail: vmurali@fsu.edu

P. Sundararaju
Embry-Riddle Aeronautical University, Daytona Beach, FL, USA

before chitosan nanofibrous membranes can be translated to the clinics for human use.

Keywords Chitosan · Drug delivery · Electrospinning · Nanofibers · Tissue engineering · Wound healing

1 Introduction

Chitosan nanofibrous membranes have been investigated for a wide range of biomedical applications. Because of their biocompatibility with various types of cells, these membranes have been evaluated for a wide range of tissue regenerative purposes. Chitosan membranes very closely mimic the extracellular matrix (ECM), both structurally and chemically. Chitosan, a natural biopolymer, is chemically similar to the glycosaminoglycans that make up the ECM, and the nanofibrous structure of the membranes closely represents the fibrous structure of the ECM, thereby making these membranes ideal surfaces for cells to attach and grow [1–3]. Further, the nanostructure of these membranes makes them an attractive option for drug delivery applications because of their increased surface area [1]. Chitosan nanofibrous membranes have been investigated for regenerating tissues like bone, tendon, ligament, liver, skin, nerve, etc. and to deliver drugs for antimicrobial and wound healing applications [4–8]. Figure 1 lists the different applications for which chitosan nanofibrous membranes have been used in the past few years.

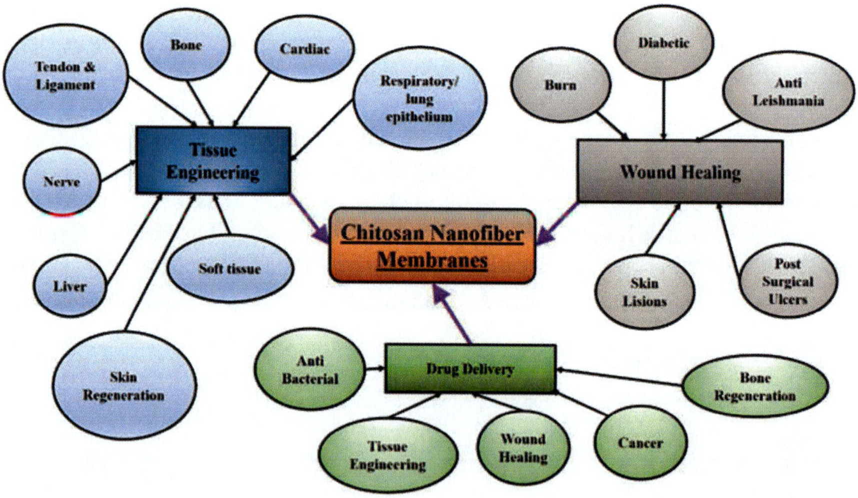

Fig. 1 List of some of the recent tissue engineering, drug delivery, and wound healing applications using chitosan nanofibrous membranes

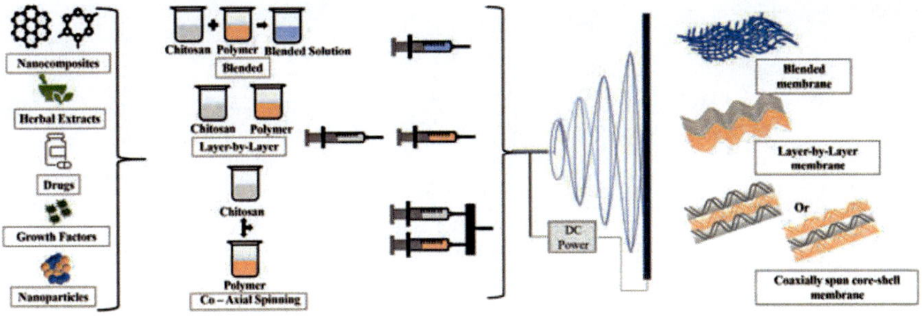

Fig. 2 A schematic of different types of electrospinning techniques used to make chitosan membranes for delivering various biotherapeutics like nanocomposites, drugs, herbal extracts, growth factors, and nanoparticles

Electrospinning is extensively used for fabricating these nanofibrous membranes. In this technique, the polymer solution is subjected to a very high electric field, which draws the polymer into thin fibrous structures and deposits them on to a collector plate [9]. The diameter of the fibers can be manipulated by changing the electric field potential, molecular weight of the polymer, viscosity and flow rate of the polymer solution, and the distance between the solution and collector plate [10]. For example, while spinning chitosan and poly-caprolactone (PCL) dissolved in a mixture of formic acid and acetic acid, Sharifi et al. obtained fibers in the range of 400 nm, whereas membranes fabricated by Jhala et al. had fiber diameter of around 75 nm [11, 12]. The concentrations of polymers used, the ratio of the solvents, applied voltage and the distance between the solution and the collector plate were slightly different in both these cases, which were thought to cause the variations in the fiber diameters. Electrospinning is a versatile process, as it enables spinning a wide variety of polymers and polymeric solutions loaded with drugs, nanoparticles (NPs), ceramics, and so on (as long as they are compatible with the solvent used to dissolve the polymer) [2, 13–15]. Chitosan blended membranes are fabricated using co-spinning (blended solution), layer-by-layer spinning, or co-axial spinning process. Figure 2 represents the schematic of different electrospinning techniques and lists the classes of biotherapeutics that have been incorporated into the electrospun membranes.

Despite the advantages of chitosan membranes, electrospinning chitosan has mainly been restricted by its high crystallinity, poor solubility in a regular solvent, and increased surface tension of the chitosan solution, which makes it difficult to form uniform fibers [16]. As an attempt to overcome this issue, several solvent systems have been investigated. Table 1 lists some of the commonly used solvents for making chitosan solution which include acetic acid, trifluoroacetic acid (TFA), and formic acid to name a few. However, using acidic solvents makes the chitosan membranes hydrophilic, causing them to swell and lose their nanofibrous structure when placed in aqueous environment [17]. To protect the membranes from swelling, they are often neutralized in an alkaline solution, or crosslinked using agents like glutaraldehyde, genipin, or UV exposure [18–20]. However, these techniques, more

Table 1 Solvent systems commonly used to dissolve and form chitosan and/or chitosan-polymer solution

TFA- DCM
Acetic acid ± DMSO
TFE-water
Acetic acid-formic acid
Chloroform-acetic acid-water-ethanol
Succinic acid, water
Chloroform-acetic acid
HFIP-TFA-DCM
Acetic acid-water
Formic acid-acetone

TFA trifluoroacetic acid, *DCM* dichloromethane, *DMSO* dimethylsulfoxide, *HFIP* hexafluoroisopropanol, *TFE* trifluoroethanol

often than not, decrease the cytocompatibility of the membranes, make the membranes brittle, and eventually lead to either partial or complete loss of the fibrous structure [21].

As an alternative to the neutralizing or crosslinking techniques, Su et al. developed a novel chemical treatment procedure to stabilize the nanofibers. In this technique, the acidic salts on the nanofibers are removed using triethylamine in an acetone solution and the amino groups of chitosan are protected using tert-butyloxycarbonyl (tBOC) groups [4]. In another technique developed by Wu et al., the hydroxyl groups of chitosan were acylated with fatty acids to form a hydrophobic protective wrap around the chitosan polymeric chains [22]. Membranes treated with these two techniques, maintained their nanofibrous structure for over 2 weeks in an aqueous environment and promoted bone formation in rat calvaria comparable to commercially available BioMend Extend™ Collagen membranes [4, 22, 23].

To improve the solubility of chitosan, the polymer has been modified or functionalized with some chemical groups, to make it water soluble. For example, Chen et al. grafted water soluble N-maleoyl functional groups on to the amino groups of chitosan [24]. This modified chitosan produced nanofibrous membranes with a diameter of 300 nm and was successfully tested as a drug delivery carrier [24]. In another work, Mahoney et al. depolymerized chitosan by subjecting high molecular weight chitosan to oxidative degradation [6]. The depolymerized chitosan was soluble in water, and the nanofibrous membranes of this chitosan effectively promoted the growth and attachment of epithelial cells [6].

To improve the spinnability of chitosan solutions, a fiber forming agent, polyethylene oxide (PEO), is often dissolved with chitosan, to improve the fiber quality. However, this solution undergoes phase separation and has to be electrospun within 24 h of blending [16]. Numerous other synthetic as well natural biopolymers have been electrospun with chitosan to improve the spinnability of the polymer solution, to improve the mechanical properties of the membranes and/or to make it compatible with specific cell types. In case of blended polymer spinning, it was interesting to note that, as the concentration of chitosan increased, the diameter of the fibers tended to decrease [6, 25]. This decrease in the fiber diameter could be thought to have increased the cytocompatibility of the membranes.

This chapter aims to discuss some of the recent advances in the field of tissue engineering, drug delivery, and wound healing, using chitosan nanofibrous membranes and certain drawbacks and challenges that need to be overcome before these membranes can be translated to clinical use.

2 Chitosan Nanofibrous Membranes in Regenerative Medicine

2.1 *Tissue Engineering*

2.1.1 Guided Bone Regeneration and Bone Tissue Engineering

Electrospun chitosan nanofibrous membranes have been extensively evaluated for their application as guided bone regeneration membranes in a bone grafting setting. Table 2 summarizes some of the recent studies using chitosan nanofibrous membranes for bone regeneration applications. The currently used membranes either require a second surgery for their removal, as in the case of non-resorbable polytetrafluoroethylene, degrade into acidic by-products, like poly-lactic and polyglycolic acid (PLA, PGA), or are susceptible to premature degradation, like collagen membranes [34–36]. Chitosan membranes treated using the tBOC protection technique and the acylation technique described above demonstrated a controlled degradation rate. These membranes maintained good barrier function and promoted better bone density than a commercial collagen membrane, when evaluated in rat calvarial defects for 8 weeks (Fig. 3) [26]. These membranes were fabricated by dissolving chitosan in a mixture of TFA and dichloromethane (DCM) and had a fiber diameter of 200 nm [26]. Both fibroblasts and bone cells showed improved proliferation in the presence of these membranes [26].

Chitosan and chitosan-PEO membranes crosslinked with genipin and glutaraldehyde, having a fiber diameter of 150 nm, also supported proliferation of bone cells, similar to tissue culture plastic (control) [18, 19]. However, when these membranes were neutralized using sodium hydroxide (NaOH), instead of being crosslinked, the fiber diameter increased to around 300 nm and the membranes supported better differentiation of mesenchymal stem cells (MSCs) than the controls [20]. Also, the random fiber membranes with 300 nm diameter promoted better mineralization than aligned fiber membranes with more than 400 nm diameter [20].

When comparing electrospun polycaprolactone (PCL) membranes with chitosan-PCL membranes, osteoblasts were more viable and proliferated better on chitosan-PCL membranes than PCL membranes [12]. The blended membranes also supported better mineralization of these osteoblasts with increased calcium deposition, bone nodule formation, and alkaline phosphatase (ALP) activity [12]. Jhala et al. observed that MC3T3 osteoblasts spread better and confirmed to their polygonal morphology

Table 2 Studies investigating chitosan nanofibrous membranes for bone tissue engineering

Construct material	Solvent (s) used	Fiber diameter	In vitro studies		In vivo studies				Ref
			Cells used	Main findings	Animal model	Control groups	Time points	Main findings	
Acylated chitosan	TFA, DCM	99.3 ± 33.7 nm	NIH 3T3	Cells adhered and proliferated well on acylated chitosan membranes	8 mm rat calvarial defect	BioMend Extend™ collagen membranes	3 and 12 weeks	Defects treated with chitosan membranes have similar BV, better bone quality and no cell penetration in the defects as compared to control membranes	[22]
tBOC modified chitosan; Acylated chitosan	TFA, DCM	tBOC (231.2 ± 93.3 nm); Acylated (283.9 ± 158.3 nm)	N/A	N/A	8 mm rat calvarial defect	BioMend Extend™ collagen membranes	3 and 8 weeks	Defects treated with chitosan membranes have similar BV but better BMD than control membranes	[26]
tBOC modified chitosan	TFA, DCM	330 ± 130 nm	Saos-2	Cells proliferated well on chitosan membranes	8 mm rat calvarial defect	BioMend Extend™ collagen membranes	3 and 12 weeks	Defects treated with chitosan membranes have similar mild inflammatory response and bone formation as in control membranes	[4]
Chitosan	TFA, DCM	68-241 nm	Saos-2	Cells proliferated well on chitosan membranes	N/A	N/A	N/A	N/A	[18]

Chitosan-PEO	Acetic acid,	149 nm	MG63	Cells proliferated and mineralized on chitosan membranes similar to tissue culture plastic control	N/A	N/A	N/A	N/A	[19]
Chitosan-PEO	Acetic acid-DMSO	410 ± 163 nm (aligned); 288 ± 107 nm (random)	MG63; human embryonic MSCs	Cells viable on both types of membranes. Stem cells deposited more calcium and penetrated more on random membranes	N/A	N/A	N/A	N/A	[20]
Chitosan-PCL	HFIP, acetone	Not mentioned	Rat BMSCs, NIH 3T3	Cells attached and proliferated better on (30-70) chitosan-PCL membranes than (50-50) or control PCL membranes. Cells differentiated more in blended membranes than PCL alone. All membranes prevented fibroblast penetration	N/A	N/A	N/A	N/A	[27]
Chitosan-PCL	Acetic acid, formic acid	439 nm	MG63	Cells proliferated and were more viable on blended	N/A	N/A	N/A	N/A	[12]

(continued)

Table 2 (continued)

Construct material	Solvent (s) used	Fiber diameter	In vitro studies		In vivo studies				Ref
			Cells used	Main findings	Animal model	Control groups	Time points	Main findings	
				membranes than on PCL membranes					
Chitosan-PCL	Acetic acid, formic acid	75 nm	MC3T3	Cell adhered, spread, and differentiated better on the membranes than on tissue culture plastic with increased calcium deposition, bone nodule formation, and ALP activity	N/A	N/A	N/A	N/A	[11]
Chitosan-PCL-CaP (Sr doped)	TFA	474 ± 94 nm	Rat MSCs	Cells adhered, proliferated, and differentiated better on the blended composite membranes than on non-CaP blended membranes. Cells expressed more VEGF in Sr-CaP blended membranes than on blended membranes	N/A	N/A	N/A	N/A	[28]
Chitosan-PVA with	Succinic acid, water	200-70 nm	MC3T3	Cells proliferated better in HaP blended	N/A	N/A	N/A	N/A	[29]

chitin nanowhiskers				membranes, and the proliferation increased with increase in chitin nanowhiskers' concentration					
Chitosan-polyamide-6,6	HFIP, acetic acid	80-250 nm (polyamide - 6,6); 20-50 nm (chitosan)	MC3T3	Cell viability and spreading increased with increasing chitosan weight proportions	N/A	N/A	N/A	N/A	[30]
Chitosan-silk fibroin-(intrafibrillar/extrafibrillar) HaP	TFA, DCM	447 ± 168 nm	Human BMSCs	Cells adhered and proliferated well on both types of membranes. Cells differentiated more on membranes with increasing HaP proportion	SQ nude mice	Intrafibrillar composite membranes without hMSCs	4 and 8 weeks	Defects treated with intrafibrillar composite membranes with hMSCs have more ectopic bone formation and increased vascularization with increased BMD and OCN and col 1 production than control membranes	[31]
Chitosan-collagen-HaP (PEO)	Acetic acid, DMSO	190-230 nm	iPSC-MSCs	Cells proliferated, attached, and spread better on composite blended membranes than on blended membranes. Cells on composite blended membranes expressed higher	3 mm mouse calvarial defect	Blank defect, chitosan membranes, chitosan-HaP membranes	4,6 and 8 weeks	Defects treated with iPSC-MSC seeded composite blended membranes healed completely and had the highest BMD in 6 weeks as compared to other	[32]

(continued)

Table 2 (continued)

Construct material	Solvent (s) used	Fiber diameter	In vitro studies: Cells used	In vitro studies: Main findings	In vivo studies: Animal model	In vivo studies: Control groups	In vivo studies: Time points	In vivo studies: Main findings	Ref
				levels of osteogenic gene expression and ALP activity				control and non-cell seeded membranes	
Chitosan-collagen, trace amounts of PCL	HFIP, formic acid	384 ± 5 nm	Human PDLSCs	Cells proliferated more on the blended membranes than on collagen membranes	5 mm rat calvarial defect	Blank defect, collagen electrospun membranes	4 and 8 weeks	Defects treated with blended membranes had higher serum ALP and OCN (but not statistically significant) and produced higher BV and complete defect closure than control groups	[33]

TFA trifluoroacetic acid, *DCM* dichloromethane, *MicroCT* micro-computed tomography, *tBOC* tert-butyloxycarbonyl, *PEO* polyethylene oxide, *DMSO* dimethyl sulfoxide, *PCL* polycaprolactone, *HFIP* hexafluoro isopropanol, *MSC* mesenchymal stem cells, *ALP* alkaline phosphate, *CaP* calcium phosphate, *Sr-CaP* strontium doped CaP, *PVA* polyvinyl alcohol, *HaP* hydroxyapatite, *SQ* subcutaneous, *OCN* osteocalcin, *Col1* collagen-1, *iPSC* induced pluripotent stem cells, *ELISA* enzyme linked immunosorbent assay, *BMD* bone mineral density, *BV* bone volume, *VEGF* vascular endothelial growth factor, *IHC* immunohistochemistry, *PDLSCs* periodontal ligament stem cells, *BMSC* bone marrow derived MSC

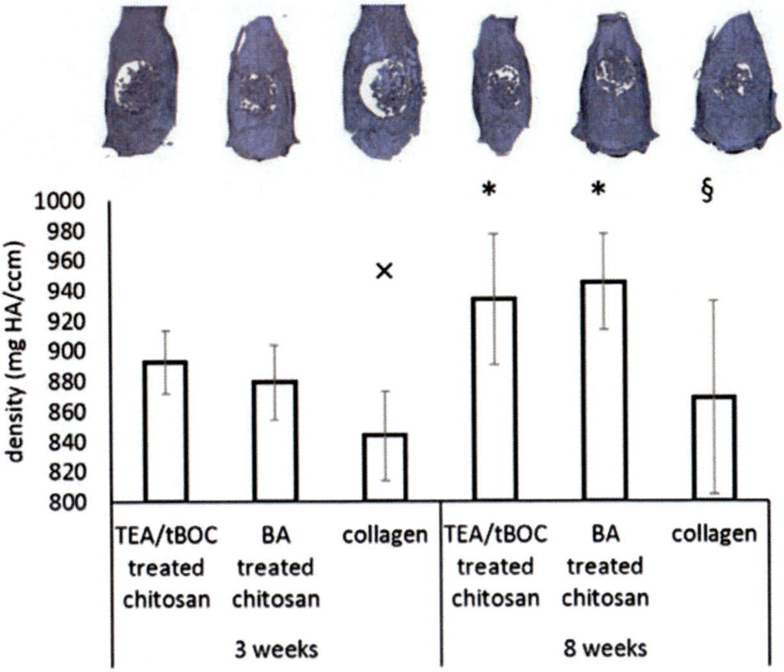

Fig. 3 Micro computed tomography analysis of bone density in rat calvarial defects treated with tBOC protected membrane (TEA/tBOC treated chitosan), acylated membrane (BA treated chitosan), and a commercial collagen membrane (collagen) at 3 and 8 weeks. The defects treated with the tBOC protected and acylated membranes had significantly higher bone density than the commercial collagen membrane at both time points. [26] Figure reproduced with kind permission from Elsevier

when seeded on chitosan-PCL nanofibrous membranes as opposed to tissue culture plastic, where the cells displayed an elongated fibroblast-like structure (Fig. 4) [11]. It has been reported previously that cell spreading and attachment was essential for precursor bone cells to undergo osteogenic differentiation, in the absence of which they tend to differentiate into adipose cells [11, 37].

In another study by He et al., it was observed that increasing chitosan percentage in the blend from 30 to 50% decreased cell viability and proliferation, but did not affect the differentiation of bone marrow MSCs [27]. The decrease in cell viability was thought to be because of the presence of nodules in the membrane structure which increased with increasing chitosan concentration. Interestingly, when co-spinning PEO with chitosan, increasing chitosan concentration from 10 to 20% increased the viability of bone cells seeded on them [30]. These membranes however did not have any nodule formation due to the presence of chitosan. Natural polymers like collagen have also been cospun with chitosan and evaluated for their osteogenic potential [33]. Membranes made of a blend of chitosan, collagen, and PCL were able to promote better proliferation of periodontal ligament cells and stimulated higher

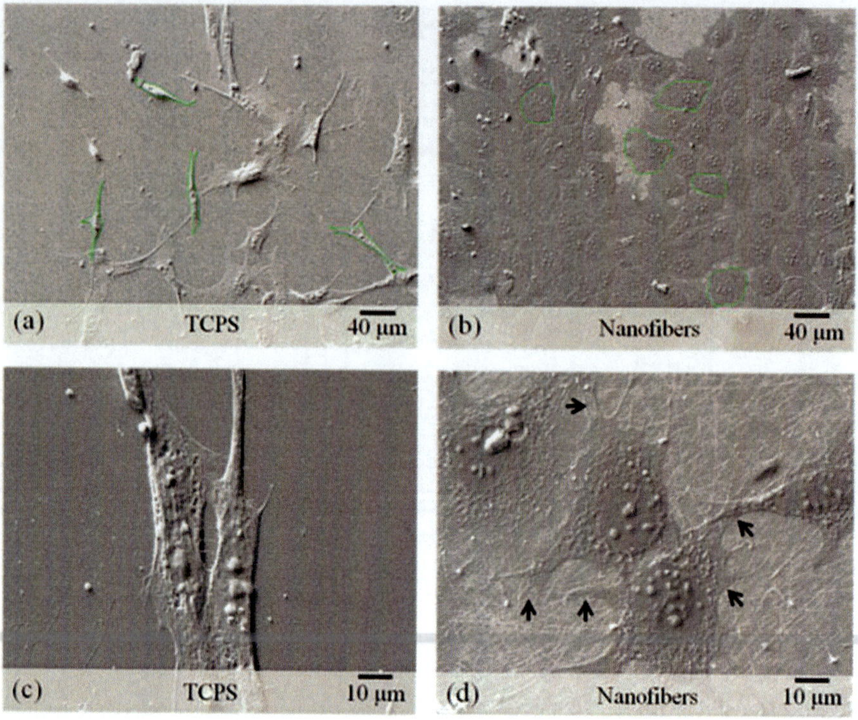

Fig. 4 Morphology of MC3T3 cells grown on tissue culture plastic (TCPS) (**a** and **c**) and chitosan-PCL membranes (**b** and **d**) for 7 days reported by Jhala et al. The black arrows show cell attachment with the fibrous membranes. Cells grown on the membranes displayed filopodia indicating good cell attachment which is critical for osteogenic differentiation [11]. Figure reproduced with kind permission from Elsevier

bone volume and complete defect healing by 8 weeks in 5 mm rat calvarial defects as compared to collagen membranes [33].

To further improve the bone healing potential of chitosan and chitosan-blended membranes, nanocomposites like hydroxyapatite (HaP) and calcium phosphate (CaP) have been incorporated into them [28, 31, 32]. As expected, membranes with the nanocomposites promoted better osteogenic differentiation of cells in vitro and promoted better bone formation in animal models [28, 31, 32]. For example, MSCs seeded on chitosan-PCL membranes incorporated with strontium doped CaP promoted enhanced osteogenic differentiation and vascular endothelial growth factor (VEGF) expression than control membranes [28]. In another study by Lai et al., MSC-seeded, chitosan-silk fibroin membranes incorporated with HaP, stimulated ectopic bone formation in nude mice with significant vascularization in 8 weeks [31].

2.1.2 Soft Tissue Regeneration

Chitosan membranes have been investigated to regenerate non-calcified tissues like tendon and ligament, muscle, skin, nerve, liver, cardiac and the lung epithelium [6–8, 25, 38–44] (Table 3). Other than for bone regeneration, chitosan-PCL membranes incorporated with HaP have been evaluated for tendon and ligament regeneration using osteoblasts [43]. Chitosan-PEO membranes with aligned and random fiber orientation, crosslinked with dibasic sodium phosphate, were evaluated with myoblasts to assess their biocompatibility for soft tissue regeneration [38]. Though initially cells adhered better on control surfaces than the membranes, with time, they attached and proliferated more on the aligned membranes than the random ones [38].

Skin Chitosan-PCL blended membranes have been explored for their skin regeneration potential using fibroblasts [39–41]. The cells seemed to adhere, proliferate, and integrate well with the blended membranes as compared to PCL control membranes [39], and adding silk fibroin to the chitosan-PCL blend further seemed to improve the cytocompatibility of the membranes [41]. However, when chitosan-PCL-gelatin membranes were studied for regenerating skin tissue, it was interesting to note that the cells grow less on chitosan membranes as compared to the blended membranes [40]. The cells adhered in the order of PCL/gelatin > chitosan/gelatin > chitosan/PCL/gelatin > chitosan/PCL, whereas the fiber diameter increased in the order of PCL/gelatin (424 ± 177 nm) < chitosan/gelatin (919 ± 230 nm) < chitosan/PCL/gelatin ($1{,}236 \pm 254$ nm) [40]. This indicated that fibroblasts might prefer adhering to smaller nanofibrous structures than on larger fibrous structures in the micrometer range. It was also interesting to note that the membranes without chitosan had a higher cell attachment, indicating that cells prefer surfaces that are moderately hydrophilic.

Nerve Chitosan-PCL membranes have also been investigated to regenerate nerve tissue [8, 42]. Schwann cells seemed to attach and proliferate better on blended membranes than control membranes [8]. To improve membrane conductivity, they have been doped with gold NPs [42]. It was observed that increasing chitosan concentration led to greater gold ion reduction, which in turn increased the electrical conductivity of the membranes [42]. Also, the Schwann cells proliferated better on gold doped membranes. However, there was no difference in the extent of attachment between gold doped membranes and non-doped membranes [42].

Liver Semnani et al. demonstrated the cytocompatibility of chitosan-PCL membranes with liver epithelial cells [44]. Rajendran et al. evaluated the long-term liver functions of hepatocytes on electrospun chitosan membranes coated with fibronectin [7]. Hepatocytes, fibroblasts, hepatocellular liver carcinoma cells, and liver sinusoidal endothelial cells, all showed excellent viability in the presence of fibronectin coated chitosan membranes [7]. Endothelial cells attached and spread well on fibronectin coated chitosan membranes. Hepatocytes co-cultured with fibroblasts

Table 3 Studies investigated chitosan nanofibrous membranes for non-calcified tissue regeneration

Application	Construct material	Solvent(s) used	Fiber diameter	In vitro studies		Ref
				Cells used	Main findings	
Tendon and ligament regeneration	Chitosan-PCL with HaP nanoparticles	Acetic acid, DMSO	~200 nm	Human osteoblasts	Cells adhered well with distinct morphology on the blended membranes	[43]
Soft tissue engineering	Chitosan-PEO	Acetic acid, DMSO	128 ± 19 nm (random) 140 ± 41 nm (aligned)	C2C12 myoblasts	Cells adhered more on the control cover slips than the membranes, but the cells proliferated and attached better on aligned membranes than random membranes with time	[38]
Skin regeneration	Chitosan-PCL	Formic acid, acetone	Not mentioned	L929 & HaCaT keratinocytes	Cells attached, proliferated, and integrated better with chitosan-PCL membranes than PCL membranes	[39]
Skin regeneration	Chitosan-PCL-gelatin	Acetic acid, water	1,236 ± 254 nm (PCL/CS/gel); 919 ± 230 nm (CS/gel); 424 ± 177 nm (PCL/gel)	Human fetal fibroblasts	Cells adhered in the order of PCL/gelatin> CS/gelatin > CS/PCL/gelatin > CS/PCL > tissue culture plastic. Cells on PCL/gelatin/CS tended to become more aligned with culture time	[40]
Skin regeneration	Chitosan-PCL-silk fibroin	Formic acid	170 ± 55.76 nm	L929	Cells were more viable on blended membranes than control membranes	[41]
Nerve tissue engineering	Chitosan-PCL	HFIP, DCM, TFA	240-110 nm	L929, rat Schwann cells	Cells grew, attached, and spread more on blended membranes than control membranes	[8]

Nerve tissue engineering	Chitosan-PCL with Au NPs	Acetic acid, formic acid	250-95 nm	Rat Schwann cells	Cells grew better on Au NPs doped blended membranes than blended membranes, but the attachment was similar in both membrane groups	[42]
Liver tissue engineering	Chitosan with fibronectin coatings	TFA, DCM	156 ± 48 nm	Rat hepatocytes, 3T3-J2 fibroblasts, human hepatocellular liver carcinoma cell, rat liver sinusoidal endothelial cells	Endothelial cells attached and spread better on fibronectin coated chitosan membranes. Co-culture of hepatocytes and fibroblasts on coated chitosan membranes promoted better cell spreading, attachment, migration, albumin secretion and cytochrome P-450 activity by hepatocytes than monoculture of hepatocytes	[7]
Liver tissue engineering	Chitosan-PCL	TFA, Trifluoroethanol	500-200 nm	Mouse epithelial liver cells	Cells were more viable when grown with media extracted from soaking the blended membranes for 7 days	[44]
Cardiac tissue engineering	Chitosan-PLA	TFA, DCM	Random fibers: 210-350 nm; aligned fibers: 120-200 nm	Rat cardiomyocytes	Cells adhered and were more viable on aligned fibers than random fibers. Cells on aligned fibers expressed higher levels of α-actinin and Tn-I	[25]
Respiratory tissue engineering	Depolymerized chitosan-PCL	Water, trifluoroethanol	800-300 nm	Porcine tracheobronchial epithelial cells	Cells grew and attached well on the blended membranes	[6]

TFA trifluoroacetic acid, *DCM* dichloromethane, *PEO* polyethylene oxide, *DMSO* dimethyl sulfoxide, *PCL* polycaprolactone, *HFIP* hexafluoroisopropanol, *HaP* hydroxyapatite, *CS* chitosan, *Gel* gelatin, *PLA* polylactic acid, *Au NPs* gold nanoparticles on coated chitosan membranes secreted higher levels of albumin and demonstrated higher cytochrome P-450 activity than monoculture of hepatocytes [7].

Respiratory and Cardiac Chitosan-PCL membranes have also supported the growth and attachment of porcine tracheobronchial epithelial cells demonstrating their potential in respiratory tissue engineering [6]. Apart from that, chitosan-PLA membranes have exhibited improved cell viability of cardiomyocytes [25]. The cells adhered and maintained higher viability on aligned fibrous membranes than on random fibrous membranes with increased expression of α-actinin and Tn-I [25].

Though a large number of studies have evaluated the regenerative potential of chitosan membranes and have extensively characterized them in vitro, very few studies have tested these membranes in in vivo animal models. In fact, in the papers discussed in this section, only studies evaluating chitosan membranes for bone tissue engineering tested the membranes in animal models [4, 22, 26, 32, 33]. Though cell culture provides credible insight into the cytocompatibility of the membranes, their true potential needs to be tested in a physiological environment. The static nature of cell culture might influence the membranes' behavior very differently than a dynamic biological environment. Therefore, it is essential to develop efficient and reliable in vivo models to test the regeneration potential of chitosan nanofibrous in regenerating tissues like liver, heart, nerve, and skin.

Further, it was noticed that the fiber orientation of the membranes had different effects on different cell types. In case of MG63 osteosarcoma cells grown on chitosan-PEO membranes, the cells attached and differentiated better on random fibrous membranes than on aligned fibrous membranes [20]. However, when being evaluated for potential soft and cardiac tissue regeneration, myoblasts and cardiomyocytes attached and proliferated better on aligned chitosan-PEO and chitosan-PLA membranes, respectively, as compared to their respective random counterparts [25, 38]. Also, the cardiomyocytes grown on aligned chitosan-PLA membranes expressed higher levels of α-actinin [25]. These studies indicate that cells grow better on surfaces that mimic their native tissue structure. Since muscle and cardiac tissue are naturally aligned, their respective cells grew better on membranes that resembled the aligned structure of their native ECM. Thus, the orientation of the membrane fibers needs to be carefully considered during the fabrication process, as it significantly influences the cell proliferation and growth. This orientation would change based on the specific application of the membrane.

2.2 *Drug Delivery*

Because of the increased surface area and biocompatibility of chitosan nanofibrous membranes, they have been investigated for several drug delivery applications in in vitro, in vivo, and even human models [45–48]. The increased surface area of the nanofibrous membranes provides increased sites for biotherapeutics to bind and incorporate into the nanostructure. The binding can be simple physical adsorption or chemical bonding by means of hydrogen bonding, covalent bonding, electrostatic interactions, hydrophobic/hydrophilic interactions, or even ionic bonding [5, 49–52]. Some of the recent studies where chitosan nanofibrous membranes have been evaluated for drug delivery applications have been discussed in the following sections.

2.2.1 Antibacterial

Since chitosan inherently possesses antibacterial property, chitosan nanofibrous membranes have been investigated for their antibacterial potential [53–55]. The protonated amino groups of chitosan interact with negatively charged bacterial surfaces to damage membrane permeability, cause cell leakage and finally cell death. Chitosan-PCL membranes showed antibacterial activity against *Escherichia coli (E. coli)* and the activity increased with increasing chitosan concentration [53].

In order to further improve the antibacterial potential of chitosan nanofibrous membranes, they are often incorporated with bioactive agents like antibacterial molecules, drugs, or NPs [56–59]. Table 4 summarizes some of the recent studies where chitosan membranes have been investigated as antibacterial drug delivery scaffolds. The controlled release of these agents from the membranes allows them to be present at the target site for prolonged duration and be more effective in inhibiting bacterial growth. For example, Monteiro et al. demonstrated that grafting gentamicin loaded liposomes on chitosan membranes promoted a very gradual and sustained gentamicin release for 24 h as compared to a burst release of gentamicin from drug loaded liposomes within the first hour followed by a plateau in 10 h [60]. The membranes with sustained gentamycin release were more effective in reducing bacterial growth [60]. In another study, while comparing chitosan-PEO nanofibers and sponges for delivering tea-tree oil (known to have wound healing and antimicrobial properties) loaded liposomes, the membranes released 60% of the tea-treeoil in about 7 days, whereas the sponges had a small burst release in 1 h and plateaued out in 12 h [56]. The tea-tree oil loaded liposome incorporated membranes effectively inhibited the growth of *Staphylococcus aureus (S. aureus)*, *E. coli*, and *Candida albicans (C. albicans)* [56]. Chitosan-PEO membranes have also been used to deliver colocasia tuber protein, which is known to have antibacterial property [57]. These membranes not only reduced bacterial growth, but also supported fibroblast proliferation for 7 days [57].

Because of their well-established antimicrobial properties, silver NPs (Ag NPs) have also been incorporated into chitosan membranes. Chitosan-PEO membranes incorporated with Ag NPs and chlorhexidine demonstrated a 100% chlorhexidine release in 2 days, whereas a sustained release of silver for more than 28 days [58]. The burst release is thought to be crucial to eliminate any initial infection at the bacterial wound site and the prolonged release is expected to prevent further infection while the wound was healing. In this study, Ag NPs were formed by reducing $AgNO_3$ in the chitosan solution and chlorhexidine was added to this solution before electrospinning [58]. Interestingly, when $AgNO_3$ was reduced by

Table 4 Studies investigating chitosan nanofibrous membranes for drug delivery applications

		Deliverable			In vitro studies		
Application	Construct material	Delivery agent	Conc	Loading efficiency/ Release profile	Bacteria/Cells used	Main findings	Ref
Antibacterial/ wound dressing	Chitosan	Gentamicin-loaded liposome	140 mM gentamicin in liposomes	EE of gentamicin in liposomes: 17.14 ± 1.09%, sustained release for 7 h, and a slower release for 24 h	*E. coli, P. aeruginosa* and *S. aureus*	Drug loaded membranes effectively inhibited bacterial growth	[60]
Antibacterial/ wound healing	Chitosan-PEO	Tea tree oil liposomes	2 wt%	Sustained release of 60% in 7 days	*S. aureus, E. coli* and *C. albicans*	Drug loaded blended membranes effectively inhibited bacterial growth	[56]
Antibacterial	Chitosan-PEO	Colocasia tuber protein	1, 2 wt%	Not mentioned	*S. aureus, E. coli,* and human skin fibroblasts	Protein loaded blended membranes reduced *S. aureus* viability by 98% and *E. coli* viability by 90-95% within 1 h of exposure. All membranes supported cell proliferation for 7 days	[57]
Antibacterial	Chitosan and PEO	Ag NPs ($AgNO_3$), chlorhexidine (CHX)	$AgNO_3$: 0.1, 1, and 5 wt%.	CHX: ~80% burst release in 8 h and 100% release in 2 days. $AgNO_3$: No burst release, relatively fast release for 4 days followed by a sustained release for 28 days	Human fibroblast, and *S. aureus*	Membranes with >60 μg CHX showed severe toxicity. No concentration of $AgNO_3$ showed any toxicity	[58]

Antibacterial/ wound dressing	Chitosan and PEO	Ag NPs, *Falcaria vulgaris*	0.25% and 0.50% (wt/wt) of *F. vulgaris*. Reduced AgNPs	Ag: Sharp increase in first 8 h, followed by sustained release till 96 h	*S. aureus* and *E. coli*	Ag NP loaded blended membranes effectively inhibited bacterial growth at both Ag NP concentrations	[61]
Antibacterial	Chitosan and PEO	ZIF-8	3, 5, and 10% w/w	Sustained release for 7 h, much slower for 24 h.	*S. aureus* and *E. coli*	3% ZIF-8 loaded membranes effectively inhibited bacterial growth	[62]
Anti-inflammatory and antibacterial	Chitosan-PEO	Hydrocortisone-imipenem/ cilastatin-loaded ZnO NPs	2% hydrocortisone and 3% imipenem/ cilastatin	EE of imipenem/ cilastatin in ZnO particles: 82%. Release: 20% in 12 h and only ~50% in 15 days. Hydrocortisone: >80% release in 12 h, 91% in 24 h and 96% in 15 days	*S. aureus* and *E. coli*	All drug loaded composite membranes inhibited bacterial growth even after 15 days of drug release	[63]
Antibacterial	Chitosan-PVA	Mafenide acetate	20, 40 wt/wt%	Burst release of 50% in 30 min. Release rate increased with increase in initial drug concentration, PVA membranes released faster than blended membranes	*S. aureus* and *P. aeruginosa*	All drug loaded membranes inhibited *S. aureus* growth in 8 h. The effect was less pronounced with *P. aeruginosa* with drug loaded blended membranes with higher chitosan concentration showing better antibacterial effect	[64]
Antibacterial	Chitosan-PVA	Honey	20,30,40%	N/A	*S. aureus*, *E. coli*, and neonatal mice	Honey loaded blended membranes showed good antibacterial activity against *S. aureus* and	[59]

(continued)

Table 4 (continued)

Application	Construct material	Deliverable			In vitro studies		Ref
		Delivery agent	Conc	Loading efficiency/ Release profile	Bacteria/Cells used	Main findings	
					skin fibroblasts	weak activity against *E. coli*. The activity increased with increase in chitosan concentration. All membranes showed good cell viability.	
Antibacterial	Chitosan-PVA	Honey	10, 20, 30%	N/A	*E. coli*, and *S. aureus*	Honey loaded blended membranes effectively inhibited bacterial growth	[65]
Antibacterial	Chitosan-PVA	Graphene oxide	1 wt%	N/A	*E. coli*, and *S. aureus*	Membranes showed enhanced antibacterial activity against both bacteria	[55]
Antibacterial	Chitosan-PVA	Ag NP ($AgNO_3$)	$AgNO_3$: 0.25, 0.5, and 1 wt%.	0.25% loaded membranes released 40% Ag, 0.5% loaded membranes released 56% and 1% loaded membranes released 36% in 16 days	*E. coli* and *S. aureus*	0.5% Ag membranes were most effective in inhibiting bacterial with slightly higher effectiveness against *S. aureus*	[66]
Drug delivery	Chitosan-PVA	Lidocaine hydrochloride, erythromycin loaded gelatin nanoparticles	700 mg lidocaine, 10% drug loaded nanoparticles	Erythromycin EE:89.34%. Sustained release of 70% till 72 h. Lidocaine: Burst release of 30% at 2 h followed	*S. aureus* and *P. aeruginosa*	Drug loaded membranes effectively inhibited bacterial growth	[67]

				by a sustained release of >80% in 72 h.			
Antibiotic drug delivery	N-maleoyl functional chitosan-PVA	Tetracycline hydrochloride	6.25 wt%	EE: 67%. 20% burst release in 5 h, followed by sustained release till 72 h and 95% release in 120 h	*E. coli*, and L929	N-maleoyl chitosan membranes inhibited bacterial growth. Extracts from drug loaded blended membranes supported good cell viability.	[24]
Antibacterial drug delivery	Chitosan(shell)-PLA (core)	Curcumin	1 wt% of polymer	Drug loading efficiency: Core-82.4% ± 1.3% and shell- 66.7% ± 0.9% Release from core: 20% in 4 h, followed by sustained release of ~80% in 5 days. Release from shell: ~45% in 4h, 80% release in <10 h.	N/A	N/A	[68]
Antimicrobial	Chitosan-PCL	Tetracycline hydrochloride	3, 5, 10, 20, and 30%, wt/wt	~10-15% release by 1 h followed by sustained release of 60-80% in 5 days. Release increased with increase in drug concentration	*S. aureus*	Drug loaded blended membranes inhibited bacterial growth from 3-20% drug concentration. After a certain point (of 20% drug loading), increasing drug concentration did not influence the antibacterial activity.	[69]
Antibacterial	Chitosan-PCL with bioactive glass	N/A	N/A		*E. coli*	Only silver bioglass incorporated membranes formed zone of	[54]

(continued)

Table 4 (continued)

Application	Construct material	Deliverable			In vitro studies		Ref
		Delivery agent	Conc	Loading efficiency/ Release profile	Bacteria/Cells used	Main findings	
	(doped with cerium, copper, or silver)					inhibition against *E. coli* in 24 h	
Tissue engineering	Chitosan-PCL	MgO	10, 25, 50 wt% of PCL		L292	MgO loaded blended membranes showed slight toxicity on day 1 but soon recovered and supported good cell proliferation in 3 days	[70]

EE encapsulation efficiency, *E. coli Escherichia coli, P. aeruginosa Pseudomonas aeruginosa, S. aureus Staphylococcus aureus, NPs* nanoparticles, *Ag NPs* silver nanoparticles, *PEO* polyethylene oxide, *CHX* chlorhexidine, *ZIF-8* zeolitic imidazolate framework-8 nanoparticles, *ZnO NP* zinc oxide nanoparticles, *PLA* polylactic acid, *PVA* polyvinyl alcohol, *PCL* polycaprolactone, Conc concentration

Falcaria vulgaris herbal extract, known to possess antioxidant and antimicrobial property, and added to chitosan-PEO solution for electrospinning, there seemed to be a burst release of silver within 9 h, followed by a plateau in 48 h [61]. Apart from Ag NPs, zeolitic imidazolate NPs and drug loaded zinc oxide NPs have also been incorporated into chitosan-PEO nanofibrous membranes and proven to have good antibacterial property [62, 63]. Drug levels releasing out of hydrocortisone-imipenem/cilastatin-loaded ZnO NPs showed enhanced antibacterial properties against *S. aureus* and *E. coli* even after 15 days of release [63].

Other than PEO, membranes formed by co-spinning poly(vinyl alcohol) (PVA) and chitosan have also been explored for delivering antibacterial agents [59, 64, 66, 67]. While comparing PVA and chitosan-PVA membranes for delivering mafenide acetate, an antibacterial agent used for treating burn scars, Abbaspour et al. showed that the drug release from blended membrane was more sustained than the PVA membranes [64]. Further, the drug loaded membranes inhibited bacterial growth within 4 h of exposure and the antibacterial effect of the membranes increased with increase in chitosan concentration [64]. The antibacterial and wound healing property of honey motivated Sarhan et al. to evaluate its controlled release from chitosan-PVA membranes [59, 65, 71]. Interestingly, though membranes loaded with 20 and 30 wt% honey were able to effectively inhibit the growth of gram positive *S. aureus* in a dose-dependent manner, no such effect was seen against gram negative *E. coli* [59]. Further, the antibacterial effect of 30 wt% honey loaded membranes increased with increase in chitosan concentration, with a more pronounced effect on *S. aureus* [59].

In another study, graphene oxide incorporated chitosan-PVA membranes demonstrated effective antibacterial activity against gram negative *E. coli* and gram positive *S. aureus* [55]. Chitosan-PVA membranes have also been incorporated with Ag NPs and have shown to be slightly more effective in inhibiting the growth of *S. aureus* than *E. coli* [66]. Chitosan-PVA membranes have also been evaluated as dual drug delivery systems by delivering lidocaine hydrochloride and erythromycin loaded gelatin NPs, simultaneously [67]. Erythromycin showed a sustained drug delivery of about 70% in 72 h, whereas lidocaine hydrochloride from the membranes showed a short burst release of 30% in 2 h followed by a sustained release of over 80% in 72 h (after which the study was terminated) [67]. The dual drug loaded membranes showed good antibacterial activity against *S. aureus* and *P. aeruginosa*, while lidocaine alone loaded membranes were not antibacterial [67].

In an attempt to develop stimuli responsive chitosan membranes, Chen et al. fabricated tetracycline hydrochloride loaded N-maleoyl functional chitosan-PVA membranes, which were responsive to glutathione [24]. These modified chitosan-PVA membranes were able to crosslink in the presence of UV and allyl sulfide and released 20% of the drug in 5 h, followed by a sustained release till 72 h and ~95% release in 120 h [24]. However, the release rate of the drug was effectively increased by exposing these crosslinked membranes to glutathione, which cleaved the disulfide crosslinking and released 40% of the drug in 6 h, followed by 90% in 24 h and 97% in 120 h [24]. This type of stimuli response would be effective in treating acute wounds, which require larger doses of the drug to control wound site infections.

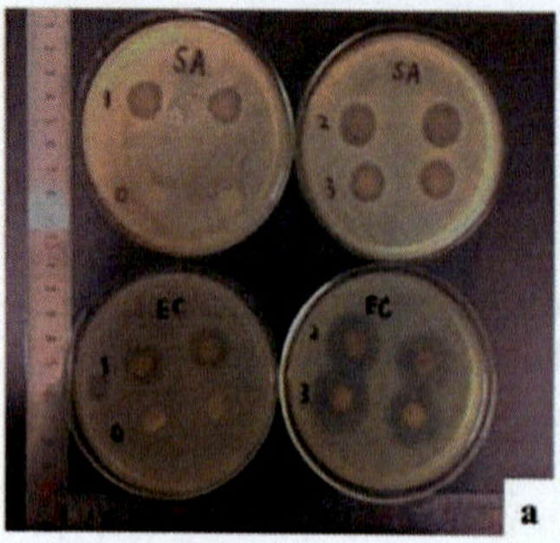

a

Sample	Antibacterial potency against *Staphylococcus aureus* (ATCC 25923)	Antibacterial potency against *Escherichia coli* (ATCC 25922)
CS-PEO	79.56%	76.11%
CS-PEO-*F. vulgaris*	85.41%	81.38%
CS-PEO-*F. vulgaris*-0.25% Ag NPs	Not growth (>99%)	Not growth (>99%)
CS-PEO-*F. vulgaris*-0.50% Ag NPs	Not growth (>99%)	Not growth (>99%)

b

Fig. 5 **(a)** Inhibition zone of Ag-chitosan/PEO membranes against *S. aureus* (SA) and *E. coli* (EC). The membranes labelled 0 are blank chitosan membranes, not showing any antibacterial activity [72]. **(b)** A table summarizing the antibacterial effect of different chitosan membranes. The blank (CS-PEO) chitosan membranes showed close to 80% bacterial inhibition [61]. Figure and table reproduced with kind permission from Sage and Elsevier

These drugs loaded membranes effectively reduced *E. coli* growth and supported fibroblast viability.

Tetracycline hydrochloride has also been incorporated with chitosan-PCL membranes [69]. These membranes showed 10–15% release on day one followed by 60–80% release in five days [69]. Interestingly, the drug loaded membranes were able to inhibit bacterial growth when the drug concentration in the membranes increased from 3 to 20 w%, but beyond 20 wt%, no significant reduction in bacterial activity was observed [69]. Curcumin, another natural antibacterial agent with established antibacterial and anticancer properties has also been incorporated into a core-shell PCL-chitosan membrane [68]. The loading efficiency and the release profile were investigated by loading the drug in the core and shell components of the fiber. When loaded in the core, the loading efficiency was 82.4% ± 1.3% and 20% of the drug released in 4 h, followed by a sustained release of ~80% in 5 days [68]. On the other hand, when loaded in the shell, the loading efficiency dropped to 66.7% ± 0.9% with ~45% release in 4 h, and 80% release in <10 h [68].

It was interesting to note that while in some studies the blank chitosan membranes showed antibacterial effect while in some others they did not (Fig. 5). For example, in a study conducted by Wang et al., where silver ions and calcium phosphate loaded chitosan membranes were being evaluated for their antibacterial effects, the membranes without silver ions did not inhibit bacterial growth [72]. However, in another study conducted by Kohsari et al., blank chitosan membranes did inhibit bacterial growth to certain extent [61]. The antibacterial effect of chitosan was thought to be influenced by the source, molecular weight, and degree of deacetylation of chitosan and also the type of bacteria being tested [73]. Thus, to make accurate decisions on the best type of chitosan to be used for making antibacterial drug delivery scaffolds, it is important to characterize the chitosan thoroughly and test the scaffold with a wide a variety of microbial species.

2.2.2 Dental and Bone Regeneration

Chitosan nanofibrous membranes have also been used to deliver and control the release of drugs that can fight infection and also improve bone healing. These studies have been summarized in Table 5. Calcium phosphate (CaP) and silver doped CaP were incorporated into chitosan membranes and investigated for antibacterial and stem cell proliferation potential [73]. The membranes showed a burst release of 76% of silver in 24 h, followed by a sustained release till 28 days [73]. When tested with bacteria, the silver doped CaP loaded membranes effectively inhibited their growth. However, when tested with rat bone marrow derived mesenchymal stem cells (BMSCs), the non-Ag CaP loaded membranes stimulated better cellular proliferation and attachment than Ag containing membranes, but all the composite membranes performed better than blank chitosan membranes [73]. In another study, chitosan membranes were subjected to different hydrophobic fatty acid treatments to improve their stability in aqueous environment [5]. These membranes were then loaded with an anti-cholesterol hydrophobic drug called simvastatin that is known to have osteogenic potential [5]. It was noted that as the hydrophobicity of the membranes increased, the drug release decreased. All the membranes showed a sustained drug release with no burst release [5]. The non-toxic drug loaded membranes when tested with W20-17 mouse stromal cells, stimulated their osteogenic differentiation with and without the presence of bone morphogenetic protein-2 (BMP-2) growth factor [5].

Chitosan blended with other natural and synthetic polymers have also been used to deliver agents to improve bone regeneration. Chitosan-gelatin membranes loaded with an antibacterial peptide and nano HaP have been investigated for their antibacterial and osteogenic potential [74]. The membranes released 35% of the peptide in 24 h followed by a cumulative release of 65% in 4 days [74]. Further, these membranes showed antibacterial activity even after 4 weeks of release and promoted rat BMSC proliferation, attachment, and osteogenic differentiation by stimulating increased ALP activity [74]. In another study, HaP and titania were incorporated in chitosan-gelatin membranes which stimulated osteogenic differentiation of mouse embryonic fibroblasts within 7 days [75].

In an attempt to incorporate both osteoinductive and osteoconductive agents in nanofibrous membranes to support faster bone regeneration, Shalumon et al. fabricated BMP-2 loaded chitosan-silk fibroin-HaP membranes [47]. Two different amounts of BMP-2 were loaded in the membranes –17 μg BMP-2/g of membrane (thick membranes) and 28 μg BMP-2/g of membrane (thin membranes) [47]. It was interesting to note that the thickness of the membranes did not seem to affect the drug release profile of BMP 2 as both the membranes released around 20% of BMP-2 in 24 h followed by a sustained release of 80% in 2 weeks [47]. Interestingly, though all the BMP-2 membranes supported osteogenic differentiation, the thin BMP-2 loaded membranes stimulated higher ALP expression in MSCs than the thick membranes on days 7 and 14 [47]. Further, subcutaneous implantation of MSC-seeded thin BMP-2 membranes stimulated ectopic bone formation with increased expression of

Table 5 Studies investigating chitosan nanofibrous membranes as drug delivery platforms to deliver therapeutics for bone (/dental) regeneration and/or antibacterial effect

Application	Construct material	Deliverable			In vitro studies		Ref
		Delivery agent	Conc	Loading efficiency/Release profile	Bacteria/Cells used	Main findings	
Bone regeneration, antibacterial	Chitosan	CaP or Ag-CaP	0.075, 0.144%	Burst release of ~76% in 24 h, and gradual cumulative release till 28 days	*S. mutans*, *P. gingivalis* and rat BMSCs	Ag CaP membranes inhibited bacterial growth in 24 h. All membranes promoted better cell growth, attachment, and proliferation than chitosan membrane. CaP membranes promoted better cell proliferation than Ag CaP membrane	[73]
Bone regeneration	Differently modified chitosan	SMV	50, 100, 250, 500 μg	All membranes showed a sustained drug release till 28 days. Drug release decreased with increasing hydrophobicity of the membranes.	W20-17	50 μg loaded most hydrophobic membranes were non-toxic to the cells and promoted osteogenic differentiation with increased ALP activity and calcium deposition	[5]
Bone regeneration, antibacterial	Chitosan-gelatin	nHaP and Pac-525	Pac-525: 2.5 mg/ml of PLGA sol. PLGA – 60 mg/ml.	Pac-525: 35% drug release in 24 h, 65% cumulative release in 4 days	*S. aureus*, *E. coli* and Rat BMSCs	All membranes supported good cell biocompatibility, proliferation, and cell attachment. nHaP membranes promoted good osteogenic differentiation (increased ALP activity). Pac-525 membranes showed good bacterial inhibition even after 4 weeks of release	[74]

Bone regeneration	Chitosan-gelatin	HaP, titania NPs	1% titania, 1% HaP	N/A	Mouse embryonic fibroblasts	Composite membranes promoted cell proliferation and osteogenic differentiation with increased ALP activity	[75]
Bone regeneration	Chitosan-silk fibroin-HaP	BMP-2	Thick membranes: 17 μg BMP-2/g of membrane. Thin membranes: 28 μg BMP-2/g of membrane	~20% release by day 1 followed by sustained release of 80% in 2 weeks	Human MSCs	All membranes supported cell proliferation. BMP-2 loaded thin membranes stimulated higher ALP expression and calcium deposition than thick membranes	[47]
Periodontal regeneration	Chitosan-PVA-HaP	Meloxicam	0.03%, 0.06%, 0.09%, and 0.12%	>60% burst release in 1 h followed by 100% release in 24 h	VERO cells	All drug loaded blended membranes supported good cell viability	[76]
Periodontal regeneration	Chitosan-PVA-HaP	Piroxicam	0.12% (wt/wt)	~50% release by 1 h followed by 100% release in 24 h	VERO cells	Drug loaded blended composite membranes supported cell viability	[77]

CaP Calcium phosphate, *Ag CaP* silver doped calcium phosphate, *S. mutans Streptococcus mutans, P. gingivalis Porphyromonas gingivalis, S. aureus Staphylococcus aureus, E. coli Escherichia coli, NP* nanoparticles, *BMSC* bone marrow derived mesenchymal stem cells, *SMV* simvastatin, *ALP* alkaline phosphate, *nHAP* nanohydroxyapatite, *Pac 525* antibacterial peptide, *PLGA* polylactic-glycolic acid, *BMP-2* bone morphogenetic protein-2, *PVA* polyvinyl alcohol, *VERO* monkey kidney epithelial cells, Conc Concentration osteocalcin and collagen1 [47]. Interestingly, the acellular membranes promoted significant tissue ingrowth within 4 weeks of implantation [47].

Instead of using natural polymers, Rehman et al. used PVA to fabricate chitosan composite scaffold to deliver non-steroidal anti-inflammatory drugs that are commonly administered to reduce periodontal destruction [76, 77]. Meloxicam and piroxicam were incorporated into chitosan-PVA-HaP membranes and their drug release profile and cytocompatibility was evaluated [76, 77]. Meloxicam and piroxicam had a burst release of >60% and 50%, respectively, in 1 h followed by a 100% release in 24 h [76, 77].

There have been a few human trials where drug loaded chitosan-PVA and chitosan-PCL membranes have been tested for their potential in treating dental caries and periodontitis [45, 46, 78]. Samprasit et al. fabricated chitosan-PVA and thiolated chitosan-PVA membranes loaded with a natural extract, mangostina, that is known to have antioxidant and antimicrobial properties [46]. Both the membranes released 80% of the drug in 1 h and 100% of the drug in 4 h [46]. When tested with *Streptococcus mutans* and *Streptococcus sanguinis*, both the drug loaded membranes inhibited bacterial growth within 30 min [46]. However, higher drug loaded membranes (3 and 5%) were toxic to keratinocytes and fibroblasts when grown for more than 1 or 2 h [46]. Further, when these drug loaded membranes were placed in the buccal mucosa of human volunteers for 15, 30 and 60 min, a significant reduction in the oral bacteria was seen within 15 min [46]. Similarly, Khan et al. investigated tinidazole (another antimicrobial drug) loaded chitosan-PCL membranes for treating periodontitis [45]. The drug loaded membranes showed a 30% release in 8 h followed by a 90% release in 18 days [45]. These membranes inhibited bacterial growth but supported fibroblast proliferation for 21 days [45]. When placed in periodontal pockets of human volunteers after the standard scaling and probing treatment, the drug loaded membrane group showed improved healing within 4 weeks as compared to the control and non-drug loaded membrane groups [45].

2.2.3 Others

Other than regeneration and antibacterial applications, drug loaded blended chitosan membranes have also been investigated for delivering anticancer agents [51, 79]. Two composite membranes made of chitosan-PLA-graphene oxide-TiO_2 and chitosan-cobalt ferrite-TiO_2 loaded with doxorubicin showed significant toxicity towards cancer cell lines without inhibiting the growth of fibroblasts [51, 79]. The chitosan-PLA membranes showed a sustained release of the drug for over 14 days, with the encapsulation efficiency decreasing with increase in drug concentration [79]. The drug release from chitosan-cobalt ferrite-TiO_2 membranes were controlled by subjecting them to alternating magnetic field as an external stimulus [51]. In the absence of a magnetic field, the membranes released ~35% (pH 7.4) and 60% (pH 5.3) of the drug in 3 days, whereas in the presence of a magnetic field, the release increased to 50 and 80%, respectively [51]. It was interesting to note that the membranes released higher drug levels at acidic environments (typical to cancer cells) than at physiological conditions, which explained their toxicity towards the cancer cells and cytocompatibility with fibroblasts [51, 79].

Other than the applications, discussed here, drug loaded chitosan membranes have been investigated for a few other wound healing and tissue engineering applications. These studies are summarized in Table 6.

2.3 *Wound Healing*

Chitosan membranes have played a major role in wound healing application because of the well-established biocompatibility, bioresorbability, and antimicrobial properties of chitosan. Membranes fabricated using different synthetic and natural polymer blends have been tested in several in vitro and in vivo animal models. Many of such recent studies have been summarized in Table 7. Synthetic polymers like PCL, PVA, polyaniline, PLA, PEO, etc. and natural polymers like gelatin, silk, and hyaluronic acid (HA) have been investigated for this application.

Mahmoudi et al. showed that chitosan-Poly(vinyl pyrrolidone)-PEO membranes incorporated with 1.5 w/w% graphene oxide nanosheets promoted in vitro cell proliferation and significantly improved wound healing in rat full thickness wounds within two weeks [84]. The improved biological property of the membranes was thought to be because of the graphene oxide nanosheets which increased the specific surface area of the membranes, in turn promoting increased cell attachment and viability. The graphene oxide incorporated membranes also had improved mechanical properties with a controllable water permeability similar to the natural skin [84].

In another attempt to mimic the skin ECM, Bazmandeh et al. fabricated a two-layered chitosan/gelatin-chitosan/HA membranes [88]. When placed in an aqueous environment, the chitosan/HA component formed a gel, whereas chitosan/gelatin component remained as a nanofibrous membrane, forming a gel/fibrous structure. This structure significantly promoted cell proliferation and supported maximum wound healing in an animal model, with reduced inflammation, better organized collagen fibers and regeneration of skin adnexal elements within 14 days [88]. The presence of gelatin and HA in the membranes seemed to support enhanced wound healing. In another study, membranes fabricated using oleoyl grafted chitosan and gelatin, with and without human amniotic membrane-derived stem cells, promoted significantly better wound healing in an animal model than a commercially used occlusive wound dressing material within two weeks of implantation [86]. This enhanced wound healing was attributed to the presence of the oleoyl groups, which reduced the inflammation mediators, raft formation, and fluidity of cell membrane, which are vital for wound repair [92].

Chitosan-PEO-PCL membranes have been investigated for treating acute skin lesions [89, 90]. Skin lesions on mice backs treated with these blended membranes healed faster with greater presence of smooth muscle actin and collagen deposition than the control PCL membranes [89]. Chitosan membranes blended with poly (vinylidene fluoride) and poly(hydroxybutyrate) have been investigated to treat

Table 6 Studies investigating chitosan nanofibrous membranes for other drug delivery applications (non-antibacterial wound healing and tissue engineering and cancer drug delivery)

		Deliverable			In vitro studies		In vivo studies			
Application	Construct material	Delivery agent	Conc	Loading efficiency/ Release profile	Bacteria/ Cells used	Main findings	Animal model	Control groups	Main findings	Ref
Wound healing	Chitosan and PEO	*Lawsonia inermis* (henna)	1 and 2% wt/wt	N/A	Normal human fore-skin fibroblasts	Henna loaded membranes supported better cell viability and proliferation	Rat dorsal skin full thickness wound	Commercial henna extract oint-ment, blank blended membranes	Wounds treated with henna loaded membranes showed maximum wound clo-sure in 14 days, followed by blank blended membranes and henna ointment groups	[49]
Wound healing	Chitosan-PEO	Ag NP ($AgNO_3$)	12 µg, 60 µg	Sustained release for 28 days	Human PDLSCs, *P. gingivalis* and *F. nucleatum*	Ag loaded membranes effectively inhibited bacterial growth even after 28 days, in a	Rabbit SQ implantation	Sham	Wounds treated with Ag and non-Ag membranes had similar healing response	[58, 80]

						dose-dependent manner. All membranes supported cell viability				
Wound healing	Chitosan-PEO	Cefazolin, fumed silica (F. silica) and cefazolin-loaded fumed silica NPs	1%	F. Silica: Burst release of 80% in 24 h. NPs: 30% release in 6 h, followed by sustained release of ~60% in 15 days	*S. aureus*, and *E. coli.*	NP and F. silica loaded membranes effectively inhibited bacterial growth	Rat dorsal skin full thickness wound	Comfeel glue	Wounds treated with NP loaded membranes showed complete wound healing with new hair formation in 10 days	[81]
Wound therapeutics	Chitosan-PCL	Ferulic acid, resveratrol	1, 2, 5, and 10 mg/ml of the polymer solution	Ferulic acid: 28% release by 1 h, 42% release by 24 h, sustained ~55% release for 120 h. Resveratrol: 15% release at 1 h, 38% release in 24 h, ~45% release in 120 h	Human epidermal keratinocytes	All blended membranes supported cell viability. Drug loaded blended membranes supported better cell migration	Rat dorsal full thickness wound	Saline treated wounds	Wounds treated with drug loaded membranes healed faster, with more collagen deposition, sebaceous gland, and hair follicle formation by day 16	[48]

(continued)

Table 6 (continued)

		Deliverable			In vitro studies		In vivo studies			
Application	Construct material	Delivery agent	Conc	Loading efficiency/ Release profile	Bacteria/ Cells used	Main findings	Animal model	Control groups	Main findings	Ref
Tissue engineering	Chitosan-PEO	Au NPs, Si NPs	10 and 30%	N/A	Human keratinocytes	NP loaded blended membranes supported cell viability	N/A	N/A	N/A	[50]
Lung cancer	Chitosan-PLA-GO-TiO_2	DOX	10, 25, 50 mg/g	LE: 10 mg/g- 97%, 25 mg/g- 94%, 50 mg/g- 92%. Initial release of ~10-15% followed by sustained release of ~80% in 14 days	A549-human lung carcinoma cells and L-132 normal human lung alveolar cells	DOX loaded GO-TiO_2 membranes showed toxicity towards A549 in a dose-dependent manner. None of the membranes showed toxicity towards L-132	N/A	N/A	N/A	[79]

Cancer treatment	Chitosan-cobalt ferrite-TiO_2	DOX	2% wt/wt	LE: 96.5 ± 1%.	B16F10 melanoma cells	Drug loaded composite membranes decreased the cell viability from 70% to 20% in 3 days	N/A	N/A	N/A	[51]

NPs nanoparticles, *Ag NPs* silver nanoparticles, *Au NPs* gold nanoparticles, *Si NPs* silica nanoparticles, *PEO* polyethylene oxide, *PCL* polycaprolactone, *PLA* polylactic acid, *PDLSCs* periodontal ligament stem cells, *SQ* subcutaneous, *LE* loading efficiency, *GO* graphene oxide, *DOX* doxorubicin, *P. gingivalis Porphyromonas gingivalis, F. nucleatum Fusobacterium nucleatum, S. aureus Staphylococcus aureus, E. coli Escherichia coli*, Conc Concentration

Table 7 Studies investigating chitosan nanofibrous membranes for wound healing and skin regeneration applications

Application	Construct material	Deliverable			In vitro studies		In vivo studies			Ref
		Delivery agent	Conc	Loading efficiency/ Release profile	Cells used	Main findings	Animal model	Control groups	Main findings	
Wound dressing	Chitosan-PCL	N/A	N/A	N/A	N/A	N/A	Mouse cutaneous excisional skin wound	Tegaderm™	Wounds treated with Chitosan-PCL membranes healed faster with faster re-epithelialization, increased collagen density and complete wound closure by day 14	[82]
Wound healing	Chitosan-PANI	N/A	N/A	N/A	Host-T85-osteoblasts, neonatal human dermal fibroblasts	Blended membranes supported good cell attachment and proliferation for both cell types	N/A	N/A	N/A	[83]
Wound dressing	Chitosan-PVP, PEO	GO nanosheets	0.5, 1, 1.5, 2%wt/wt	N/A	Rat bone marrow MSCs	GO loaded blended membranes supported increasing cell proliferation with increasing GO concentration till 1.5%wt/wt	Rat dorsal skin wound	Sterile gauze	Defects treated with 1.5% wt/wt GO loaded blended membranes showed more wound closure	[84]

Skin regeneration	Chitosan-PVA-silk	N/A	N/A	N/A	L929	Blended membranes supported better cell attachment and growth than on chitosan-PVA and PVA membranes	Rat full thickness dorsal wound	Paraffin sterile gauze	Defects treated with cell loaded blended membranes showed same amount of wound closure but higher epithelialization than control membranes	[85]
Wound healing	Oleoyl grafted chitosan-gelatin	N/A	N/A	N/A	HAMSCs	Blended membranes supported increasing cell proliferation, with increasing amount of chitosan	Rat dorsal skin full thickness wound	Tegaderm™	Wounds treated with HAMSC seeded and acellular membranes had significantly enhanced wound healing. By day 15, the cellular membranes had complete epithelium formation and increased collagen density	[86]
Wound dressing	Methacrylated gelatin/ chitosan-PCL/ PLA	N/A	N/A	N/A	Normal human dermal fibroblasts	Blended membranes promoted good cell viability, attachment, and proliferation. The membranes also displayed good thrombogenicity and	N/A	N/A	N/A	[87]

(continued)

Table 7 (continued)

Application	Construct material	Deliverable			In vitro studies		In vivo studies			Ref
		Delivery agent	Conc	Loading efficiency/ Release profile	Cells used	Main findings	Animal model	Control groups	Main findings	
						non-hemolytic activity				
Wound dressing	Chitosan/ PEO-gelatin membrane+ chitosan-HA-PEO electrospun together	N/A	N/A	N/A	Normal human dermal fibroblasts	All membranes supported cell viability. Blended hybrid membranes promoted better cell attachment	Rat dorsal skin full thickness wound	Wounds treated with medical gauze	Wounds treated with blended hybrid membranes showed more wound closure with dense aligned collagen fibers, complete reepithelialization, and no inflammation	[88]
Skin lesions	Chitosan/ PEO-PCL	N/A	N/A	N/A	N/A	N/A	Mice back skin lesion	PCL membranes	Wounds treated with blended membranes showed greater inflammatory infiltrate, greater expression of proliferating cell nuclear antigen, and smooth muscle actin and collagen deposition	[89]
Skin lesions	Chitosan-PEO and	N/A	N/A	N/A	L929	All membranes supported cell	N/A	N/A	N/A	[90]

	PCL-cellulose acetate					growth and proliferation				
Surgical wound dressing	chitosan-PHB-PVDF	Gentamycin	1,2, 3%	30% release in 2 h, and sustained release till 6 days, followed by a plateau	N/A	N/A	N/A	N/A	N/A	[91]

PCL polycaprolactone, *PANI* polyaniline, *PVP* polyvinylpyrrolidone, *PEO* polyethylene oxide, *GO* graphene oxide, *MSC* mesenchymal stem cells, *PVA* polyvinyl alcohol, *HAMSC* human amniotic membrane derives stem cells, *PLA* polylactic acid, *PHB* polyhydroxybutyrate, *PVDF* polyvinylidene fluoride, Conc Concentration post-surgical ulcers [91]. These membranes loaded with gentamycin showed a 30% release in 2 h, followed by a sustained release for 6 days, and the release increased with increase in chitosan concentration [91]. No further in vitro and in vivo studies were carried out to test their biological response.

2.3.1 Antibacterial Wound Healing

Apart from the antibacterial studies mentioned in the previous drug delivery section, chitosan membranes have been incorporated with antibacterial therapeutics specifically to promote wound healing. In an attempt to develop an antibacterial wound dressing, Woo et al. fabricated a bilayer composite membrane made of TiO_2 incorporated chitosan nanofibrous membrane and a human adipose-derived ECM sheet [93]. The TiO_2 layer served as the antibacterial barrier and the ECM layer promoted tissue regeneration. When tested in full thickness rat wounds, though the control and test membranes showed similar levels of wound healing, wounds treated with test membranes promoted higher vascularization, which is crucial in the regeneration process [93]. To enhance the antibacterial property of chitosan, $Ti_3C_2T_z$ flakes have been incorporated and electrospun with chitosan [94]. These membranes inhibited bacterial growth but were non-toxic to HeLa cells [94]. $Ti_3C_2T_z$ or MXenes are a relatively new family of 2D transition metal layered hexagonal ternary carbide and/or nitrides, whose application in various fields has just started to be investigated [95, 96].

In a study conducted by Antunes et al., chitosan membranes and arginine grafted chitosan membranes were evaluated for their wound healing potential [97]. The arginine grafted chitosan membranes inhibited bacterial growth to a greater extent while supporting fibroblast attachment and proliferation on them [97]. The grafted membranes also promoted enhanced wound healing in rat full thickness wound model with increased re-epithelialization and tissue reorganization [97]. Other than these, several blended chitosan membranes loaded with drugs like tetracycline, doxycycline, ciprofloxacin, lidocaine, mupirocin, etc., natural wound healing extracts from *Aloe vera*, *Cleome droserifolia*, and *Allium sativum*, growth factors like EGF and bFGF and NPs made of silver and iron oxide have been investigated for antibacterial wound healing purposes [71, 98–101]. Some of such recent studies have been listed in Table 8.

2.3.2 Anti-Leishmania Wound Dressing

Apart from antibacterial treatments, chitosan membranes have also been investigated for anti-parasitic treatments as well (Table 9). These dressings were fabricated using a blend of chitosan, PEO and PCL and loaded with drugs like Glucantime and Berberine [111–113]. Glucantime is the standard drug used to treat the parasitic infection – leishmaniasis, caused by the organism of the genus *Leishmania,* and Berberine is a natural alkaloid with anti-inflammatory properties and is known to be

Table 8 Studies investigating chitosan nanofibrous membranes for wound healing applications with a focus on reducing antibacterial infections

	Deliverable			In vitro studies		In vivo studies			
Construct material	Delivery agent	Conc	Loading efficiency/ Release profile	Bacteria/Cells used	Main findings	Animal model	Control groups	Main findings	Ref
Chitosan	$Ti_3C_2T_z$ flakes	0, 0.05, 0.10, 0.25, 0.50, 0.75 wt%	Not mentioned	*E. coli* and *S. aureus*, Hela	0.75% $Ti_3C_2T_z$ membranes reduced *E. coli* and *S. aureus* growth by 95% and 62%, respectively. Membranes were non-toxic to cells	N/A	N/A	N/A	[94]
Arginine grafted chitosan	N/A	N/A	N/A	*E. coli* and *S. aureus*, primary human dermal fibroblasts	Grafted membranes inhibited bacterial growth but promoted the attachment and proliferation of fibroblasts on them	Rat dorsal skin full thickness wound	Untreated wounds	Wounds treated with grafted membranes showed better healing with higher re-epithelialization and tissue reorganization	[97]
Chitosan-human adipose-derived extracellular matrix sheet	TiO_2	1% w/v	N/A	*E. coli* and *S. aureus*	Composite blended membranes prevented bacterial penetration	Rat dorsal skin full thickness wound	Untreated wounds	Untreated wounds and wounds treated with membranes showed similar healing, but the membrane treated	[93]

(continued)

Table 8 (continued)

Construct material	Deliverable			In vitro studies		In vivo studies			Ref
	Delivery agent	Conc	Loading efficiency/ Release profile	Bacteria/Cells used	Main findings	Animal model	Control groups	Main findings	
								wounds showed higher number of CD31 positive cells and micro vessel density	
Chitosan-PEO	Silver nanoparticles	0, 2, 4, 8 wt %, agno3 (precursor) to polymer	Burst release in 10 h, followed by sustained release for 72 h	*E. coli* and *S. aureus*, pig iliac endothelial cells	Composite blended membranes inhibited bacterial growth. Membranes were non-toxic to cells	N/A	N/A	N/A	[72]
Chitosan-PEO	Aloe vera	N/A	N/A	*E. coli* and *S. aureus*, NIH 3 T3	Blended membranes inhibited bacterial growth, but promoted cell proliferation and attachment	Mice dorsal full thickness skin wound	Untreated wounds	Wounds treated with aloe vera loaded membranes had better wound closure rate with increased blood vessel and full thickened epidermal regeneration	[102]

Chitosan/PEO-bacterial cellulose	N/A	N/A	N/A	*E. coli*	Chitosan-PEO and cospun chitosan-PEO-bacterial cellulose membranes inhibited bacterial growth	N/A	N/A	N/A	[103]
Chitosan-PCL	Aloe vera	N/A	N/A	Normal human dermal fibroblasts, *S. aureus* and *E. coli*	Aloe vera loaded membranes promoted better cells attachment, proliferation, and growth, and prevented bacterial penetration and biofilm formation on them	N/A	N/A	N/A	[104]
Chitosan-PCL	Mupirocin, lidocaine hydrochloride	Mupirocin: 2% w/v; lidocaine: 1% w/v	Lidocaine: 66% release in 1 h, cumulative 85% release in 6 h; mupirocin: 57% release in 6 h; cumulative ~85% release in 114 h	*S. aureus*, *E. coli*, and *P. aeruginosa*, human dermal fibroblasts	Blended drug loaded membranes inhibited bacterial growth but promoted cell attachment and growth	N/A	N/A	N/A	[105]

(continued)

Table 8 (continued)

Construct material	Deliverable			In vitro studies		In vivo studies			Ref
	Delivery agent	Conc	Loading efficiency/ Release profile	Bacteria/Cells used	Main findings	Animal model	Control groups	Main findings	
Chitosan/ PEO-PCL	DDAC	1 and 8% wt/wt of PCL	N/A	*E. coli*, *S. aureus* and Normal human dermal fibroblasts	DDAC membranes inhibited *S. aureus* growth in a dose dependent manner but had no effect on *E. coli*. All membranes showed good cell viability and attachment	Mice dorsal full thickness skin wound	Untreated wounds, Tegaderm™	Wounds treated with DDAC membranes showed improved wound healing in a dose-dependent manner. All the drug loaded membranes showed increased vascularization, epithelialization, and hair follicle formation	[106]
Chitosan/ PEO-PCL collagen	EGF, bFGF, SSD	GF: 25 µg/ml, SSD: 3 mg/ml	35% SSD released on day1, followed by a sustained release for 20 days. 50% of GF released on day 1 followed by a rapid	*P. aeruginosa*, *S. aureus*, and human dermal fibroblasts	SSD loaded membranes prevented bacterial growth for 20 days. GF loaded membranes supported higher cell viability attachment and proliferation. SSD did not	Rat full-thickness dorsal wound	Open wound	Defects treated with GF loaded membranes had reduced granulation tissue, increased presence of fibroblasts and collagen fibers, reduced inflammatory cells, increased vascularization and no detectable signs of	[98]

			release for 11 days.		seem to influence the cell viability			scar formation as compared to defects treated with membrane alone and open wound control.	
Chitosan-PCL-HA	N/A	N/A	N/A	*E. coli*, and monkey kidney epithelial cells	Blended membranes prevented bacterial growth but promoted cell growth, proliferation, and migration	N/A	N/A	N/A	[107]
Chitosan-PCL-HA-Zein	Salicylic acid	8%	16% release in 5 days followed by a sustained release	*S. aureus*. and Normal human dermal fibroblasts	Drug loaded composite membranes supported cell viability and exhibited antibacterial activity till at least 5 days.	N/A	N/A	N/A	[108]
Chitosan-PVA	Tetracycline hydrochloride	5 mg/ml (5 wt%)	~80% burst release in 2 h	*E. coli*, *S. epidermidis*, *S. aureus* and rabbit aortic smooth muscle cells	Drug loaded membrane effectively inhibited antibacterial growth (more effective against *S. aureus* and	N/A	N/A	N/A	[99]

(continued)

Table 8 (continued)

Construct material	Deliverable			In vitro studies		In vivo studies			Ref
	Delivery agent	Conc	Loading efficiency/ Release profile	Bacteria/Cells used	Main findings	Animal model	Control groups	Main findings	
					S. epidermidis than *E. coli*), but promoted cell viability and attachment				
Chitosan-PVA (films vs nanofibers)	Doxycycline	1% wt/wt	85% burst release in 48 h from nanofibers; 75% burst release in 48 h from films	*P. aeruginosa*, *S. aureus*, and *A. baumannii*	Drug loaded films had better antibacterial activity than membranes. The membranes had no effect on *P. aeruginosa*.	N/A	N/A	N/A	[100]
Chitosan-starch-PVA	N/A	N/A	N/A	L 929, *S. aureus* and *E. coli*	All blended membranes supported cell viability, with the starch membrane supporting more. Membranes with more chitosan and starch promoted better	N/A	N/A	N/A	[109]

					in vitro wound healing (scratch assay). Antibacterial activity increased with increase in chitosan concentration. More effective against *S. aureus* than *E. coli*.				
Chitosan-PVA-GO	Cipro and Cipro HCl	5 wt%	Cipro release: 18.7% in 4 h and 96.5% in 7 days; Cipro HCl: 30% in 4 h and 62.1% in 7 days	*E. coli*, *S. aureus*, *B. subtilis*, and melanoma cells	Chitosan-PVA-GO- Cipro membranes greatly inhibited bacterial growth. All blended composite membranes with and without the drugs were non-toxic to the cells	N/A	N/A	N/A	[110]
Chitosan-PVA-honey	AS, CE	CE: 10 wt/v AS: 50% of the solvent	N/A	*S. aureus*, *E. coli*, MRSA, multidrug resistant *P. aeruginosa*,	AS and SE-AS loaded membranes completely inhibited *S. aureus*.	Mice dorsal back wound	Untreated wound with cotton gauze, Aquacel® Ag	Wounds treated with AS loaded membranes promoted enhanced wound healing as compared to the	[71]

(continued)

Table 8 (continued)

Construct material	Deliverable			In vitro studies		In vivo studies			Ref
	Delivery agent	Conc	Loading efficiency/ Release profile	Bacteria/Cells used	Main findings	Animal model	Control groups	Main findings	
				and Human fibroblasts	SE-AS showed mild inhibitory effect against MRSA. Membranes were cytocompatible as compared to Aquacel®Ag which showed toxicity			commercial Aquacel®Ag	
Chitosan-gelatin	Fe_3O_4 nanoparticles	0.5, 1, 2 and 4 wt%	N/A	*E. coli* and *S. aureus*	Control and the particle loaded membranes inhibited bacterial growth. Particle loaded membranes displayed a dose-dependent antibacterial activity	N/A	N/A	N/A	[101]

DDAC N-decyl-N, N-dimethyl-1-decanaminium chloride, *SSD* silver sulfadiazine, *PEO* polyethylene oxide, *PCL* polycaprolactone, *GF* growth factor, *EGF* epidermal growth factor, *bFGF* basic fibroblast growth factor, *HA* hyaluronic acid, *PVA* polyvinyl alcohol, *GO* graphene oxide, *CE* cleome droserifolia, *AS* allium sativum, *MRSA* methicillin resistant *S. aureus*, *Cipro* Ciprofloxacin, *Cipro HC* Ciprofloxacin hydrochloride, *E. coli Escherichia coli, S. aureus Staphylococcus aureus, P. aeruginosa Pseudomonas aeruginosa, S. epidermidis Staphylococcus epidermidis, A. baumannii Acinetobacter baumannii, B. subtilis Bacillus subtilis*, Conc Concentration

Table 9 Studies investigating chitosan nanofibrous membranes for parasitic, burn and diabetic wound healing

		Deliverable		In vitro studies			In vivo studies			
Application	Construct material	Delivery agent	Conc	Loading efficiency/ Release profile	Cells used	Main findings	Animal model	Control groups	Main findings	Ref
Leishmania wound dressing	Chitosan-PEO	Berberine	0.5-50%	50% release in 18 h, followed by sustained~80% release till day 14	Leishmania major-promastigote stage and amastigote stage, human fibroblasts	Drug loaded blended membranes reduced parasitic growth with increasing drug concentrations. 30% and 50% drug loaded membranes showed toxicity towards fibroblasts.	N/A	N/A	N/A	[111]
Leishmania wound dressing	Chitosan-PEO	Berberine	0.5-20%	50% release in 12 h, 70.7% on day 1, 77.1% on day 2, and 80.8% on day 3	N/A	N/A	Leishmania infected wound in Balb/c mice	Untreated wounds, wounds treated with berberine or Glucantime (positive control)	Wounds treated with drug loaded membranes showed comparable results as with those treated with the positive controls	[112]
Leishmania wound dressing	Chitosan-PEO-gelatin-PVA	Glucantime	10, 20% wt/v	10% drug loaded membranes released 84.6% of the drug in 9 h, 20% drug loaded membranes released 83.7% in 9 h	Promastigotes of Leishmaniasis major, and NIH 3T3	Drug loaded blended membranes reduced parasitic growth, but did not show toxicity towards fibroblasts	N/A	N/A	N/A	[113]

(continued)

Table 9 (continued)

Application	Construct material	Deliverable		In vitro studies			In vivo studies			Ref
		Delivery agent	Conc	Loading efficiency/ Release profile	Cells used	Main findings	Animal model	Control groups	Main findings	
Burn wound healing	Chitosan-PEO	Bromelain	2 and 4% wt/v	2% drug loaded membrane released >90% drug in 1 day; 4% drug loaded membranes released ~90% in 6 days.	Human dermal fibroblasts	Drug loaded blended membranes were cytocompatible	Rat back burn wound	Wounds treated with chitosan-PEO membranes	Wounds treated with drug loaded membranes healed completely by day 21 with more collagen fiber formation and complete epithelialization	[114]
Burn infection management	Chitosan-PEO	ZnO NP and Cipro	ZnO: 5 μg/ml; Cipro: 1 μg/ml	LE of ZnO: 92%. LE of Cipro: 98%. The drug release was faster in acidic skin conditions	*S. aureus* and *E. coli*, human dermal fibroblasts and keratinocytes	Cipro blended membranes showed higher antibacterial activity than ZnO membranes. Drug loaded membranes were not cytotoxic	N/A	N/A	N/A	[115]
Burn wound healing	Chitosan-PVA-alginate (core-shell)	Asiaticoside	1.5, 2.5, and 5%	20% burst release in 1 h followed by ~85% release in 12 h	N/A	N/A	Rat dorsal burn wound	Untreated wounds, wounds treated with centella triterpenes cream (positive control)	Wounds treated with drug loaded membranes and positive control showed enhanced wound healing with increased blood vessel formation, re-epithelialization and collagen formation and decreased inflammatory cytokine expression	[116]

Diabetic wound dressing	Chitosan-PVA	N/A	N/A	N/A	N/A	N/A	Wounds in streptozotocin-induced diabetic rats	Non-diabetic wound and untreated diabetic wound	Wounds treated with blended membranes showed better wound healing with more wound closure, epidermal layer restoration and blood vessel formation	[117]
Diabetic wound dressing	Chitosan-PVA	ZnO NP	N/A	N/A	*E. coli*, P. aeruginosa, *B. subtilis*, and *S. aureus*	ZnO loaded membranes showed better bacterial inhibition in 24 h	Diabetic rabbit SQ wound (dressing replaced every 3 days)	Untreated wounds	Wounds treated with ZnO membranes showed enhanced wound healing. By day 12, the ZnO membrane treated groups had almost completely healed with complete epidermal regeneration and increased collagen formation	[52]

PEO polyethylene oxide, *PVA* polyvinyl alcohol, *Cipro* ciprofloxacin, *NP* nanoparticle, *LE* loading efficiency, *SQ* subcutaneous, *S. aureus Staphylococcus aureus, E. coli Escherichia coli, P. aeruginosa Pseudomonas aeruginosa, B. subtilis Bacillus subtilis*, Conc Concentration effective against the parasite [112, 118]. The blended membranes released 50% of the loaded berberine in 12–18 h, followed by a sustained release of 80% for 8 to 14 days depending on the initial drug loading [111]. In the case of glucantime, the membranes released more than 80% of the drug in 9 h [113]. All these drug loaded membranes were effective in reducing parasitic growth in vitro while not showing any toxicity toward fibroblasts [111, 113]. When tested in Balb/c mice with parasitic infection, the berberine loaded blended chitosan membranes showed wound healing similar to the positive controls of berberine or glucantime [112].

2.3.3 Burn and Diabetic Wound Healing

Burn Wound Healing Chitosan-PEO membranes loaded with bromelain were investigated to treat burn wounds in rat back burn wounds (Table 9) [114–116]. In vitro, 2% drug loaded membranes seemed to release >90% of the drug within 24 h, whereas when the loading increased to 4%, the drug release continued for 6 days (90% cumulative) [114]. In vivo, the 2% drug loaded membranes completely healed the wound in 21 days with increased collagen fiber formation and complete re-epithelialization (Fig. 6) [114]. In order to treat infectious burn wounds, Abid et al. fabricated ZnO NPs and ciprofloxacin loaded chitosan-PEO membranes [115]. These membranes were effective in reducing bacterial growth while not being cytotoxic to the cells. Interestingly, the membranes showed a larger drug release in an acidic environment [115].

Chitosan-PVA-alginate core-shell membranes loaded with asiaticoside have also been evaluated for burn wound management. These membranes released 20% of

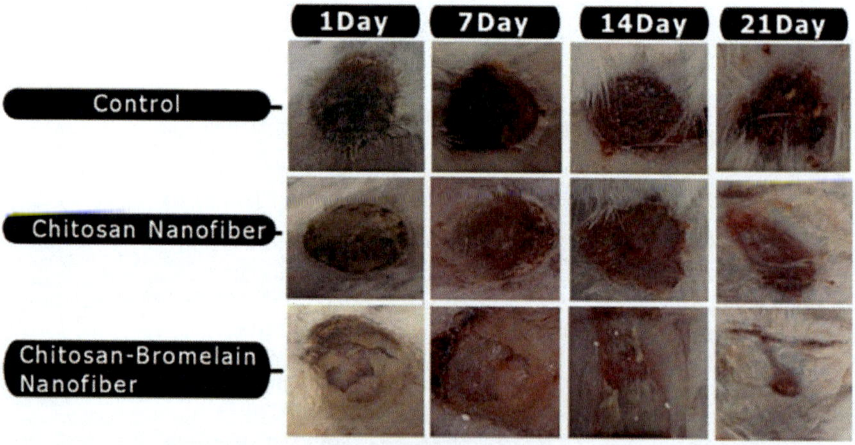

Fig. 6 Images of a second degree burn on rat's dorsal skin healing over 21 days. Chitosan nanofibrous membranes loaded with Bromelain (therapeutic extract from pineapple) showed almost complete wound healing with complete re-epithelialization in 21 days [114]. Figure reproduced with kind permission from Elsevier

the drug in 1 h and 85% in 12 h [116]. Despite a burst release, when used to treat a burn wound in rat, these drug loaded membranes promoted wound healing similar to positive control in terms of increased blood vessel formation, re-epithelialization and collagen formation and decreased inflammatory cytokine expression [116]. However, observational studies showed that the drug loaded membranes promoted better wound healing than the positive control with better wound closure and hair growth [116].

Diabetic Wound Healing Since diabetic wounds take extended time to heal, investigators have attempted to develop chitosan-PVA membranes that can promote diabetic wound healing (Table 9) [52, 117]. When used to treat wounds in streptozotocin-induced diabetic rats, there was a significant improvement in wound closure, epidermal layer restoration and blood vessel formation as compared to the untreated diabetic or non-diabetic wounds [117]. To improve the effectiveness of these membranes, ZnO NPs were incorporated into them to provide antibacterial properties [52]. ZnO loaded membranes effectively reduced bacterial growth in 24 h and promoted complete wound healing in diabetic rabbit subcutaneous wounds within 12 days of implantation [52].

3 Future Direction and Conclusion

Though several improvements are being made in the field of regenerative medicine, not a lot of electrospun biomaterial-based scaffolds are currently being used in the clinics. Though chitosan nanofibrous membranes have proved effective in a wide range of biomedical, regenerative applications, they still need to be evaluated in appropriate (large) animal models in order to establish their efficiency in treating damaged tissue or promoting wound healing. Further, since chitosan is obtained from a wide variety of natural sources, it is critical to thoroughly characterize the polymer and make efforts to reduce batch-to-batch variability. In order to further improve chitosan's popularity in the field of tissue regeneration, newer and more reliable efforts need to be made to modify chitosan to improve its solubility in non-harsh solvents. While fabricating nanofibrous membranes, the target tissue and application need to be kept in mind, since one type of chitosan membrane will not be able to effectively regenerate all types of tissues or deliver all types of biotherapeutics efficiently. Therefore, thorough evaluation of chitosan nanofibrous membranes in large animal models with appropriate characterization is crucial for the future translation of these membranes into clinical use.

References

1. Jayakumar R, Prabaharan M, Nair S, Tamura H (2010) Novel chitin and chitosan nanofibers in biomedical applications. Biotechnol Adv 28:142–150
2. Bhattarai N, Edmondson D, Veiseh O, Matsen FA, Zhang M (2005) Electrospun chitosan-based nanofibers and their cellular compatibility. Biomaterials 26:6176–6184
3. Chen Z, Wang P, Wei B, Mo X, Cui F (2010) Electrospun collagen–chitosan nanofiber: a biomimetic extracellular matrix for endothelial cell and smooth muscle cell. Acta Biomater 6:372–382
4. Su H, Liu K-Y, Karydis A, Abebe DG, Wu C, Anderson KM, Ghadri N, Adatrow P, Fujiwara T, Bumgardner JD (2016) *In vitro* and *in vivo* evaluations of a novel post-electrospinning treatment to improve the fibrous structure of chitosan membranes for guided bone regeneration. Biomed Mater 12:015003
5. Murali VP, Fujiwara T, Gallop C, Wang Y, Wilson JA, Atwill MT, Kurakula M, Bumgardner JD (2020) Modified electrospun chitosan membranes for controlled release of simvastatin. Int J Pharm:119438
6. Mahoney C, Conklin D, Waterman J, Sankar J, Bhattarai N (2016) Electrospun nanofibers of poly (ε-caprolactone)/depolymerized chitosan for respiratory tissue engineering applications. J Biomater Sci Polym Ed 27:611–625
7. Rajendran D, Hussain A, Yip D, Parekh A, Shrirao A, Cho CH (2017) Long-term liver-specific functions of hepatocytes in electrospun chitosan nanofiber scaffolds coated with fibronectin. J Biomed Mater Res A 105:2119–2128
8. Bolaina-Lorenzo E, Martínez-Ramos C, Monleón-Pradas M, Herrera-Kao W, Cauich-Rodríguez JV, Cervantes-Uc JM (2016) Electrospun polycaprolactone/chitosan scaffolds for nerve tissue engineering: physicochemical characterization and schwann cell biocompatibility. Biomed Mater 12:015008
9. Subbiah T, Bhat GS, Tock RW, Parameswaran S, Ramkumar SS (2005) Electrospinning of nanofibers. J Appl Polym Sci 96:557–569
10. Beachley V, Wen X (2009) Effect of electrospinning parameters on the nanofiber diameter and length. Mater Sci Eng C 29:663–668
11. Jhala D, Rather H, Vasita R (2016) Polycaprolactone–chitosan nanofibers influence cell morphology to induce early osteogenic differentiation. Biomater Sci 4:1584–1595
12. Sharifi F, Atyabi SM, Norouzian D, Zandi M, Irani S, Bakhshi H (2018) Polycaprolactone/carboxymethyl chitosan nanofibrous scaffolds for bone tissue engineering application. Int J Biol Macromol 115:243–248
13. Demir MM, Yilgor I, Yilgor E, Erman B (2002) Electrospinning of polyurethane fibers. Polymer 43:3303–3309
14. Duan B, Yuan X, Zhu Y, Zhang Y, Li X, Zhang Y, Yao K (2006) A nanofibrous composite membrane of plga–chitosan/pva prepared by electrospinning. Eur Polym J 42:2013–2022
15. Yang E, Qin X, Wang S (2008) Electrospun crosslinked polyvinyl alcohol membrane. Mater Lett 62:3555–3557
16. Qasim SB, Zafar MS, Najeeb S, Khurshid Z, Shah AH, Husain S, Rehman IU (2018) Electrospinning of chitosan-based solutions for tissue engineering and regenerative medicine. Int J Mol Sci 19:407
17. Sangsanoh P, Supaphol P (2006) Stability improvement of electrospun chitosan nanofibrous membranes in neutral or weak basic aqueous solutions. Biomacromolecules 7:2710–2714
18. Norowski Jr PA, Fujiwara T, Clem WC, Adatrow PC, Eckstein EC, Haggard WO, Bumgardner JD (2015) Novel naturally crosslinked electrospun nanofibrous chitosan mats for guided bone regeneration membranes: material characterization and cytocompatibility. J Tissue Eng Regen Med 9:577–583
19. Singh YP, Dasgupta S, Nayar S, Bhaskar R (2020) Optimization of electrospinning process and parameters for producing defect-free chitosan/polyethylene oxide nanofibers for bone tissue engineering. J Biomater Sci Polym Ed 31:781–803

20. Qasim SB, Najeeb S, Delaine-Smith RM, Rawlinson A, Rehman IU (2017) Potential of electrospun chitosan fibers as a surface layer in functionally graded gtr membrane for periodontal regeneration. Dent Mater 33:71–83
21. Muzzarelli RA (2011) New techniques for optimization of surface area and porosity in nanochitins and nanochitosans. Chitosan Biomater II:167–186
22. Wu C, Su H, Karydis A, Anderson KM, Ghadri N, Tang S, Wang Y, Bumgardner JD (2017) Mechanically stable surface-hydrophobilized chitosan nanofibrous barrier membranes for guided bone regeneration. Biomed Mater 13:015004
23. Ghadri N, Anderson KM, Adatrow P, Stein SH, Su H, Garcia-Godoy F, Karydis A, Bumgardner JD (2018) Evaluation of bone regeneration of simvastatin loaded chitosan nanofiber membranes in rodent calvarial defects. J Biomater Nanobiotechnol 9:210
24. Chen C-K, Huang S-C (2016) Preparation of reductant–responsive n-maleoyl-functional chitosan/poly (vinyl alcohol) nanofibers for drug delivery. Mol Pharm 13:4152–4167
25. Liu Y, Wang S, Zhang R (2017) Composite poly (lactic acid)/chitosan nanofibrous scaffolds for cardiac tissue engineering. Int J Biol Macromol 103:1130–1137
26. Su H, Fujiwara T, Anderson KM, Karydis A, Ghadri MN, Bumgardner JD (2021) A comparison of two types of electrospun chitosan membranes and a collagen membrane *in vivo*. Dent Mater 37:60–70
27. He Y, Wang W, Tang X, Liu X (2017) Osteogenic induction of bone marrow mesenchymal cells on electrospun polycaprolactone/chitosan nanofibrous membrane. Dent Mater J 36:325–332
28. Ye H, Zhu J, Deng D, Jin S, Li J, Man Y (2019) Enhanced osteogenesis and angiogenesis by pcl/chitosan/sr-doped calcium phosphate electrospun nanocomposite membrane for guided bone regeneration. J Biomater Sci Polym Ed 30:1505–1522
29. Pangon A, Saesoo S, Saengkrit N, Ruktanonchai U, Intasanta V (2016) Hydroxyapatite-hybridized chitosan/chitin whisker bionanocomposite fibers for bone tissue engineering applications. Carbohydr Polym 144:419–427
30. Shrestha BK, Mousa HM, Tiwari AP, Ko SW, Park CH, Kim CS (2016) Development of polyamide-6, 6/chitosan electrospun hybrid nanofibrous scaffolds for tissue engineering application. Carbohydr Polym 148:107–114
31. Lai G-J, Shalumon K, Chen J-P (2015) Response of human mesenchymal stem cells to intrafibrillar nanohydroxyapatite content and extrafibrillar nanohydroxyapatite in biomimetic chitosan/silk fibroin/nanohydroxyapatite nanofibrous membrane scaffolds. Int J Nanomedicine 10:567
32. Xie J, Peng C, Zhao Q, Wang X, Yuan H, Yang L, Li K, Lou X, Zhang Y (2016) Osteogenic differentiation and bone regeneration of ipsc-mscs supported by a biomimetic nanofibrous scaffold. Acta Biomater 29:365–379
33. Guo S, He L, Yang R, Chen B, Xie X, Jiang B, Weidong T, Ding Y (2020) Enhanced effects of electrospun collagen-chitosan nanofiber membranes on guided bone regeneration. J Biomater Sci Polym Ed 31:155–168
34. Aurer A, Jorgić-Srdjak K (2005) Membranes for periodontal regeneration. Acta Stomatol Croat 39:107–112
35. Meinig RP (2010) Clinical use of resorbable polymeric membranes in the treatment of bone defects. Orthopedic Clin 41:39–47
36. Turri A, Elgali I, Vazirisani F, Johansson A, Emanuelsson L, Dahlin C, Thomsen P, Omar O (2016) Guided bone regeneration is promoted by the molecular events in the membrane compartment. Biomaterials 84:167–183
37. McBeath R, Pirone DM, Nelson CM, Bhadriraju K, Chen CS (2004) Cell shape, cytoskeletal tension, and rhoa regulate stem cell lineage commitment. Dev Cell 6:483–495
38. Tonda-Turo C, Ruini F, Ramella M, Boccafoschi F, Gentile P, Gioffredi E, Labate GFDU, Ciardelli G (2017) Non-covalently crosslinked chitosan nanofibrous mats prepared by electrospinning as substrates for soft tissue regeneration. Carbohydr Polym 162:82–92

39. Prasad T, Shabeena E, Vinod D, Kumary T, Kumar PA (2015) Characterization and *in vitro* evaluation of electrospun chitosan/polycaprolactone blend fibrous mat for skin tissue engineering. J Mater Sci Mater Med 26:28
40. Gomes S, Rodrigues G, Martins G, Henriques C, Silva JC (2017) Evaluation of nanofibrous scaffolds obtained from blends of chitosan, gelatin and polycaprolactone for skin tissue engineering. Int J Biol Macromol 102:1174–1185
41. Chen H-W, Lin M-F (2020) Characterization, biocompatibility, and optimization of electrospun sf/pcl/cs composite nanofibers. Polymers 12:1439
42. Saderi N, Rajabi M, Akbari B, Firouzi M, Hassannejad Z (2018) Fabrication and characterization of gold nanoparticle-doped electrospun pcl/chitosan nanofibrous scaffolds for nerve tissue engineering. J Mater Sci Mater Med 29:1–10
43. Wu G, Deng X, Song J, Chen F (2018) Enhanced biological properties of biomimetic apatite fabricated polycaprolactone/chitosan nanofibrous bio-composite for tendon and ligament regeneration. J Photochem Photobiol B Biol 178:27–32
44. Semnani D, Naghashzargar E, Hadjianfar M, Dehghan Manshadi F, Mohammadi S, Karbasi S, Effaty F (2017) Evaluation of pcl/chitosan electrospun nanofibers for liver tissue engineering. Int J Polym Mater Polym Biomater 66:149–157
45. Khan G, Yadav SK, Patel RR, Kumar N, Bansal M, Mishra B (2017) Tinidazole functionalized homogeneous electrospun chitosan/poly (ε-caprolactone) hybrid nanofiber membrane: development, optimization and its clinical implications. Int J Biol Macromol 103:1311–1326
46. Samprasit W, Kaomongkolgit R, Sukma M, Rojanarata T, Ngawhirunpat T, Opanasopit P (2015) Mucoadhesive electrospun chitosan-based nanofibre mats for dental caries prevention. Carbohydr Polym 117:933–940
47. Shalumon K, Lai G-J, Chen C-H, Chen J-P (2015) Modulation of bone-specific tissue regeneration by incorporating bone morphogenetic protein and controlling the shell thickness of silk fibroin/chitosan/nanohydroxyapatite core–shell nanofibrous membranes. ACS Appl Mater Interfaces 7:21170–21181
48. Poornima B, Korrapati PS (2017) Fabrication of chitosan-polycaprolactone composite nanofibrous scaffold for simultaneous delivery of ferulic acid and resveratrol. Carbohydr Polym 157:1741–1749
49. Yousefi I, Pakravan M, Rahimi H, Bahador A, Farshadzadeh Z, Haririan I (2017) An investigation of electrospun henna leaves extract-loaded chitosan based nanofibrous mats for skin tissue engineering. Mater Sci Eng C 75:433–444
50. Al-Kattan A, Nirwan VP, Munnier E, Chourpa I, Fahmi A, Kabashin AV (2017) Toward multifunctional hybrid platforms for tissue engineering based on chitosan (peo) nanofibers functionalized by bare laser-synthesized au and si nanoparticles. RSC Adv 7:31759–31766
51. Radmansouri M, Bahmani E, Sarikhani E, Rahmani K, Sharifianjazi F, Irani M (2018) Doxorubicin hydrochloride-loaded electrospun chitosan/cobalt ferrite/titanium oxide nanofibers for hyperthermic tumor cell treatment and controlled drug release. Int J Biol Macromol 116:378–384
52. Ahmed R, Tariq M, Ali I, Asghar R, Khanam PN, Augustine R, Hasan A (2018) Novel electrospun chitosan/polyvinyl alcohol/zinc oxide nanofibrous mats with antibacterial and antioxidant properties for diabetic wound healing. Int J Biol Macromol 120:385–393
53. Hardiansyah A, Tanadi H, Yang M-C, Liu T-Y (2015) Electrospinning and antibacterial activity of chitosan-blended poly (lactic acid) nanofibers. J Polym Res 22:1–10
54. Goh Y-F, Akram M, Alshemary A, Hussain R (2016) Antibacterial polylactic acid/chitosan nanofibers decorated with bioactive glass. Appl Surf Sci 387:1–7
55. Yang S, Lei P, Shan Y, Zhang D (2018) Preparation and characterization of antibacterial electrospun chitosan/poly (vinyl alcohol)/graphene oxide composite nanofibrous membrane. Appl Surf Sci 435:832–840

56. Ge Y, Tang J, Fu H, Fu Y, Wu Y (2019) Characteristics, controlled-release and antimicrobial properties of tea tree oil liposomes-incorporated chitosan-based electrospun nanofiber mats. Fibers Polym 20:698–708
57. Wardhani RAK, Asri LA, Rachmawati H, Khairurrijal K, Purwasasmita BS (2020) Physical–chemical crosslinked electrospun colocasia esculenta tuber protein–chitosan–poly (ethylene oxide) nanofibers with antibacterial activity and cytocompatibility. Int J Nanomedicine 15:6433
58. Song J, Remmers SJ, Shao J, Kolwijck E, Walboomers XF, Jansen JA, Leeuwenburgh SC, Yang F (2016) Antibacterial effects of electrospun chitosan/poly (ethylene oxide) nanofibrous membranes loaded with chlorhexidine and silver. Nanomedicine 12:1357–1364
59. Sarhan WA, Azzazy HM (2015) High concentration honey chitosan electrospun nanofibers: biocompatibility and antibacterial effects. Carbohydr Polym 122:135–143
60. Monteiro N, Martins M, Martins A, Fonseca NA, Moreira JN, Reis RL, Neves NM (2015) Antibacterial activity of chitosan nanofiber meshes with liposomes immobilized releasing gentamicin. Acta Biomater 18:196–205
61. Kohsari I, Shariatinia Z, Pourmortazavi SM (2016) Antibacterial electrospun chitosan–polyethylene oxide nanocomposite mats containing bioactive silver nanoparticles. Carbohydr Polym 140:287–298
62. Kohsari I, Shariatinia Z, Pourmortazavi SM (2016) Antibacterial electrospun chitosan-polyethylene oxide nanocomposite mats containing zif-8 nanoparticles. Int J Biol Macromol 91:778–788
63. Fazli Y, Shariatinia Z, Kohsari I, Azadmehr A, Pourmortazavi SM (2016) A novel chitosan-polyethylene oxide nanofibrous mat designed for controlled co-release of hydrocortisone and imipenem/cilastatin drugs. Int J Pharm 513:636–647
64. Abbaspour M, Makhmalzadeh BS, Rezaee B, Shoja S, Ahangari Z (2015) Evaluation of the antimicrobial effect of chitosan/polyvinyl alcohol electrospun nanofibers containing mafenide acetate. Jundishapur J Microbiol 8
65. Sarhan WA, Azzazy HM, El-Sherbiny IM (2016) The effect of increasing honey concentration on the properties of the honey/polyvinyl alcohol/chitosan nanofibers. Mater Sci Eng C 67:276–284
66. Liu Y, Liu Y, Liao N, Cui F, Park M, Kim H-Y (2015) Fabrication and durable antibacterial properties of electrospun chitosan nanofibers with silver nanoparticles. Int J Biol Macromol 79:638–643
67. Fathollahipour S, Abouei Mehrizi A, Ghaee A, Koosha M (2015) Electrospinning of pva/chitosan nanocomposite nanofibers containing gelatin nanoparticles as a dual drug delivery system. J Biomed Mater Res A 103:3852–3862
68. Afshar S, Rashedi S, Nazockdast H, Ghazalian M (2019) Preparation and characterization of electrospun poly (lactic acid)-chitosan core-shell nanofibers with a new solvent system. Int J Biol Macromol 138:1130–1137
69. Jiang S, Lv J, Ding M, Li Y, Wang H, Jiang S (2016) Release behavior of tetracycline hydrochloride loaded chitosan/poly (lactic acid) antimicrobial nanofibrous membranes. Mater Sci Eng C 59:86–91
70. Rijal NP, Adhikari U, Khanal S, Pai D, Sankar J, Bhattarai N (2018) Magnesium oxide-poly (ε-caprolactone)-chitosan-based composite nanofiber for tissue engineering applications. Mater Sci Eng B 228:18–27
71. Sarhan WA, Azzazy HM, El-Sherbiny IM (2016) Honey/chitosan nanofiber wound dressing enriched with allium sativum and cleome droserifolia: enhanced antimicrobial and wound healing activity. ACS Appl Mater Interfaces 8:6379–6390
72. Wang X, Cheng F, Gao J, Wang L (2015) Antibacterial wound dressing from chitosan/polyethylene oxide nanofibers mats embedded with silver nanoparticles. J Biomater Appl 29:1086–1095

73. Jin S, Li J, Wang J, Jiang J, Zuo Y, Li Y, Yang F (2018) Electrospun silver ion-loaded calcium phosphate/chitosan antibacterial composite fibrous membranes for guided bone regeneration. Int J Nanomedicine 13:4591
74. He Y, Jin Y, Wang X, Yao S, Li Y, Wu Q, Ma G, Cui F, Liu H (2018) An antimicrobial peptide-loaded gelatin/chitosan nanofibrous membrane fabricated by sequential layer-by-layer electrospinning and electrospraying techniques. Nano 8:327
75. Acevedo CA, Olguín Y, Briceño M, Forero JC, Osses N, Díaz-Calderón P, Jaques A, Ortiz R (2019) Design of a biodegradable uv-irradiated gelatin-chitosan/nanocomposed membrane with osteogenic ability for application in bone regeneration. Mater Sci Eng C 99:875–886
76. Yar M, Farooq A, Shahzadi L, Khan AS, Mahmood N, Rauf A, Chaudhry AA, Ur Rehman I (2016) Novel meloxicam releasing electrospun polymer/ceramic reinforced biodegradable membranes for periodontal regeneration applications. Mater Sci Eng C 64:148–156
77. Farooq A, Yar M, Khan AS, Shahzadi L, Siddiqi SA, Mahmood N, Rauf A, Manzoor F, Chaudhry AA, Ur Rehman I (2015) Synthesis of piroxicam loaded novel electrospun biodegradable nanocomposite scaffolds for periodontal regeneration. Mater Sci Eng C 56:104–113
78. Samprasit W, Rojanarata T, Akkaramongkolporn P, Ngawhirunpat T, Kaomongkolgit R, Opanasopit P (2015) Fabrication and *in vitro/in vivo* performance of mucoadhesive electrospun nanofiber mats containing α-mangostin. AAPS PharmSciTech 16:1140–1152
79. Samadi S, Moradkhani M, Beheshti H, Irani M, Aliabadi M (2018) Fabrication of chitosan/poly (lactic acid)/graphene oxide/tio2 composite nanofibrous scaffolds for sustained delivery of doxorubicin and treatment of lung cancer. Int J Biol Macromol 110:416–424
80. Shao J, Yu N, Kolwijck E, Wang B, Tan KW, Jansen JA, Walboomers XF, Yang F (2017) Biological evaluation of silver nanoparticles incorporated into chitosan-based membranes. Nanomedicine 12:2771–2785
81. Fazli Y, Shariatinia Z (2017) Controlled release of cefazolin sodium antibiotic drug from electrospun chitosan-polyethylene oxide nanofibrous mats. Mater Sci Eng C 71:641–652
82. Levengood SL, Erickson AE, F-c C, Zhang M (2017) Chitosan–poly (caprolactone) nanofibers for skin repair. J Mater Chem B 5:1822–1833
83. Moutsatsou P, Coopman K, Georgiadou S (2017) Biocompatibility assessment of conducting pani/chitosan nanofibers for wound healing applications. Polymers 9:687
84. Mahmoudi N, Simchi A (2017) On the biological performance of graphene oxide-modified chitosan/polyvinyl pyrrolidone nanocomposite membranes: *In vitro* and *in vivo* effects of graphene oxide. Mater Sci Eng C 70:121–131
85. Fathi A, Khanmohammadi M, Goodarzi A, Foroutani L, Mobarakeh ZT, Saremi J, Arabpour Z, Ai J (2020) Fabrication of chitosan-polyvinyl alcohol and silk electrospun fiber seeded with differentiated keratinocyte for skin tissue regeneration in animal wound model. J Biol Eng 14:1–14
86. Datta S, Rameshbabu AP, Bankoti K, Maity PP, Das D, Pal S, Roy S, Sen R, Dhara S (2017) Oleoyl-chitosan-based nanofiber mats impregnated with amniotic membrane derived stem cells for accelerated full-thickness excisional wound healing. ACS Biomater Sci Eng 3:1738–1749
87. Alves P, Santos M, Mendes S, Miguel PS, de Sá DK, Cabral SDC, Correia JI, Ferreira P (2019) Photocrosslinkable nanofibrous asymmetric membrane designed for wound dressing. Polymers 11:653
88. Bazmandeh AZ, Mirzaei E, Fadaie M, Shirian S, Ghasemi Y (2020) Dual spinneret electrospun nanofibrous/gel structure of chitosan-gelatin/chitosan-hyaluronic acid as a wound dressing: in-vitro and in-vivo studies. Int J Biol Macromol 162:359–373
89. Zanchetta FC, Trinca RB, Gomes Silva JL, Breder JSC, Cantarutti TA, Consonni SR, Moraes ÂM, Pereira de Araújo E, Saad MJA, Adams GG (2020) Effects of electrospun fibrous membranes of polycaprolactone and chitosan/poly (ethylene oxide) on mouse acute skin lesions. Polymers 12:1580
90. Trinca RB, Westin CB, da Silva JAF, Moraes ÂM (2017) Electrospun multilayer chitosan scaffolds as potential wound dressings for skin lesions. Eur Polym J 88:161–170

91. Amini F, Semnani D, Karbasi S, Banitaba SN (2019) A novel bilayer drug-loaded wound dressing of pvdf and phb/chitosan nanofibers applicable for post-surgical ulcers. Int J Polym Mater Polym Biomater 68:772–777
92. Cardoso C, Favoreto Jr S, Oliveira L, Vancim J, Barban G, Ferraz D, Silva J (2011) Oleic acid modulation of the immune response in wound healing: a new approach for skin repair. Immunobiology 216:409–415
93. Woo CH, Choi YC, Choi JS, Lee HY, Cho YW (2015) A bilayer composite composed of tio2-incorporated electrospun chitosan membrane and human extracellular matrix sheet as a wound dressing. J Biomater Sci Polym Ed 26:841–854
94. Mayerberger EA, Street RM, McDaniel RM, Barsoum MW, Schauer CL (2018) Antibacterial properties of electrospun ti 3 c 2 t z (mxene)/chitosan nanofibers. RSC Adv 8:35386–35394
95. Rasool K, Helal M, Ali A, Ren CE, Gogotsi Y, Mahmoud KA (2016) Antibacterial activity of ti3c2t x mxene. ACS Nano 10:3674–3684
96. Ghidiu M, Halim J, Kota S, Bish D, Gogotsi Y, Barsoum MW (2016) Ion-exchange and cation solvation reactions in ti3c2 mxene. Chem Mater 28:3507–3514
97. Antunes BP, Moreira AF, Gaspar V, Correia I (2015) Chitosan/arginine–chitosan polymer blends for assembly of nanofibrous membranes for wound regeneration. Carbohydr Polym 130:104–112
98. Nejaddehbashi F, Hashemitabar M, Bayati V, Abbaspour M, Moghimipour E, Orazizadeh M (2019) Application of polycaprolactone, chitosan, and collagen composite as a nanofibrous mat loaded with silver sulfadiazine and growth factors for wound dressing. Artif Organs 43:413–423
99. Alavarse AC, de Oliveira Silva FW, Colque JT, da Silva VM, Prieto T, Venancio EC, Bonvent J-J (2017) Tetracycline hydrochloride-loaded electrospun nanofibers mats based on pva and chitosan for wound dressing. Mater Sci Eng C 77:271–281
100. Hedayatyanfard K, Bagheri-Khoulenjani S, Hashemi A, Ziai SA (2019) Semi-ipn films and electrospun nanofibers based on chitosan/pva as an antibacterial wound dressing. Iran J Pharm Res IJPR 18:1156
101. Cai N, Li C, Han C, Luo X, Shen L, Xue Y, Yu F (2016) Tailoring mechanical and antibacterial properties of chitosan/gelatin nanofiber membranes with fe3o4 nanoparticles for potential wound dressing application. Appl Surf Sci 369:492–500
102. Pathalamuthu P, Siddharthan A, Giridev V, Victoria V, Thangam R, Sivasubramanian S, Savariar V, Hemamalini T (2019) Enhanced performance of aloe vera incorporated chitosan-polyethylene oxide electrospun wound scaffold produced using novel spirograph based collector assembly. Int J Biol Macromol 140:808–824
103. Ardila N, Medina N, Arkoun M, Heuzey M-C, Ajji A, Panchal CJ (2016) Chitosan–bacterial nanocellulose nanofibrous structures for potential wound dressing applications. Cellulose 23:3089–3104
104. Miguel SP, Ribeiro MP, Coutinho P, Correia IJ (2017) Electrospun polycaprolactone/aloe vera_chitosan nanofibrous asymmetric membranes aimed for wound healing applications. Polymers 9:183
105. Li X, Wang C, Yang S, Liu P, Zhang B (2018) Electrospun pcl/mupirocin and chitosan/lidocaine hydrochloride multifunctional double layer nanofibrous scaffolds for wound dressing applications. Int J Nanomedicine 13:5287–5299
106. Zafari M, Mansouri M, Omidghaemi S, Yazdani A, Pourmotabed S, Hasanpour Dehkordi A, Nosrati H, Validi M, Sharifi E (2020) Physical and biological properties of blend-electrospun polycaprolactone/chitosan-based wound dressings loaded with n-decyl-n, n-dimethyl-1-decanaminium chloride: an *in vitro* and *in vivo* study. J Biomed Mater Res B Appl Biomater 108:3084–3098
107. Chanda A, Adhikari J, Ghosh A, Chowdhury SR, Thomas S, Datta P, Saha P (2018) Electrospun chitosan/polycaprolactone-hyaluronic acid bilayered scaffold for potential wound healing applications. Int J Biol Macromol 116:774–785

108. Figueira DR, Miguel SP, de Sá KD, Correia IJ (2016) Production and characterization of polycaprolactone-hyaluronic acid/chitosan-zein electrospun bilayer nanofibrous membrane for tissue regeneration. Int J Biol Macromol 93:1100–1110
109. Adeli H, Khorasani MT, Parvazinia M (2019) Wound dressing based on electrospun pva/chitosan/starch nanofibrous mats: fabrication, antibacterial and cytocompatibility evaluation and *in vitro* healing assay. Int J Biol Macromol 122:238–254
110. Yang S, Zhang X, Zhang D (2019) Electrospun chitosan/poly (vinyl alcohol)/graphene oxide nanofibrous membrane with ciprofloxacin antibiotic drug for potential wound dressing application. Int J Mol Sci 20:4395
111. Rahimi M, Tabaei SJS, Ziai SA, Sadri M (2020) Anti-leishmanial effects of chitosan-polyethylene oxide nanofibers containing berberine: an applied model for leishmania wound dressing. Iran J Med Sci 45:286
112. Tabaei SJS, Rahimi M, Akbaribazm M, Ziai SA, Sadri M, Shahrokhi SR, Rezaei MS (2020) Chitosan-based nano-scaffolds as antileishmanial wound dressing in balb/c mice treatment: characterization and design of tissue regeneration. Iran J Basic Med Sci 23:788
113. Alishahi M, Khorram M, Asgari Q, Davani F, Goudarzi F, Emami A, Arastehfar A, Zomorodian K (2020) Glucantime-loaded electrospun core-shell nanofibers composed of poly (ethylene oxide)/gelatin-poly (vinyl alcohol)/chitosan as dressing for cutaneous leishmaniasis. Int J Biol Macromol 163:288–297
114. Bayat S, Amiri N, Pishavar E, Kalalinia F, Movaffagh J, Hashemi M (2019) Bromelain-loaded chitosan nanofibers prepared by electrospinning method for burn wound healing in animal models. Life Sci 229:57–66
115. Abid S, Hussain T, Nazir A, Zahir A, Ramakrishna S, Hameed M, Khenoussi N (2019) Enhanced antibacterial activity of peo-chitosan nanofibers with potential application in burn infection management. Int J Biol Macromol 135:1222–1236
116. Zhu L, Liu X, Du L, Jin Y (2016) Preparation of asiaticoside-loaded coaxially electrospinning nanofibers and their effect on deep partial-thickness burn injury. Biomed Pharmacother 83:33–40
117. Majd SA, Khorasgani MR, Moshtaghian SJ, Talebi A, Khezri M (2016) Application of chitosan/pva nano fiber as a potential wound dressing for streptozotocin-induced diabetic rats. Int J Biol Macromol 92:1162–1168
118. Santos DO, Coutinho CE, Madeira MF, Bottino CG, Vieira RT, Nascimento SB, Bernardino A, Bourguignon SC, Corte-Real S, Pinho RT (2008) Leishmaniasis treatment—a challenge that remains: a review. Parasitol Res 103:1–10

Adv Polym Sci (2021) 288: 87–116
https://doi.org/10.1007/12_2021_101

Published online: 18 June 2021

3D-Printed Chitosan Composites for Biomedical Applications

Sesha Subramanian Murugan, Sukumaran Anil, Padmanaban Sivakumar, Min Suk Shim, and Jayachandran Venkatesan

Contents

S. S. Murugan and J. Venkatesan (✉)
Biomaterials Research Laboratory, Yenepoya Research Centre, Yenepoya (Deemed to be University), Deralakatte, Mangalore, Karnataka, India
e-mail: jvenkatesan@yenepoya.edu.in

S. Anil
Department of Dentistry, Oral Health Institute, Hamad Medical Corporation, College of Dental Medicine, Qatar University, Doha, Qatar

P. Sivakumar and M. S. Shim (✉)
Division of Bioengineering, Incheon National University, Incheon, Republic of Korea
e-mail: msshim@inu.ac.kr

Abstract Regeneration of defective or diseased tissue by 3D-printed biomaterials is an emerging area of research, and 3D printing technology will meet the shortage of organ transplantation and therapeutic clinical applications. The development of novel bio-inks for 3D printing has challenges, including the rheological, physical, chemical, and biological properties of materials, the risk of an immune response, cytotoxicity, and regeneration rate. In recent years, chitosan and its composites as bio-inks for 3D bioprinting to develop artificial organs have been studied. The results infer that the regenerative capacity of the 3D printed chitosan composites varies depending on size, porosity, stimulating effect, cell interaction, cell adhesion, and the differentiation potential of stem cells. In this review, the types of 3D printing technology for the fabrication system and their role in tissue engineering applications are studied in detail.

Keywords 3D printing · Bio-ink · Growth factors · Stimulators · Tissue engineering

1 Introduction

Organ/tissue replacement is the state in which the damaged or defective organ/tissue is replaced by clinical surgery. It requires external clinical procedures to maintain homeostatic conditions. Autograft or allograft transplants are the most common organ transplant techniques. For instance, bone is the most necessary tissue to replace tissue loss from trauma, cancer, and accidents. Traditionally, clinicians follow autograft, as they are immunologically safe and tissue-compatible. It is a conventional method in which bone tissue can be collected from the iliac crest of the same patient [1]. Chronic pain, scar formation, bleeding, inflammation, and infection at the injury site are the inconvenience of autograft, in addition to the second surgery. There is no autograft treatment option, in case the need for bone is high at the defective site [2]. Allograft transplantation is the second bone grafting technique that uses the donor graft. It poses a high risk for infection and immune rejection. By this allograft method, the bone tissue from the donor has been demineralized as well as treated with radiation and kept for freeze drying before transplantation. This leads to limitations in osteo-inductive properties and disease transmission like HIV and hepatitis [3]. As well, available clinical inventions of bone cement and bone morphogenic proteins fail to show angiogenesis and osteoinductions. As a result, the alternative synthetic bone graft material is needed to replace the traditional bone graft technique [4]. 3D printing is the future imminent technology that will make the native tissues by using printed cells in the biomaterials. In regenerative medicine and tissue engineering, 3D culture scaffolds have been used to imitate tissue anatomy. These scaffolds can be individual or blended with two to more biomaterials (composites). Pluronic, polyethylene glycol, and alginate are renowned for their rheological behavior. Collagen, gelatin, chitosan, fibrin, extracellular matrix decellularized scaffolds have poor rheology and mechanical resistance. But mixing both in a form of composite blend promotes bio-ink production with reliable mechanical strength and biological attributes [5]. The 3D bioprinting cell culture naturally imitating the biology is further effective toward the therapeutic strategies to repairing and restoring the diseased organs. Printed 3D culture scaffolds and their composites are expected to have porosity for cell growth, proliferation, and differentiation while the traditional scaffold system has inability to recapitulate cell growth. In tissue engineering, chitosan is the most commonly used biomaterial for fabricating the scaffolds for both soft and hard tissues. In this review, we discussed the different types of 3D printing method for the development of chitosan composites. Further, preparation of bio-inks, stem cell interaction, tissue engineering, drug delivery, and biosensor applications of 3D printed chitosan composites are discussed (Table 1).

2 Tissue Engineering

Tissue engineering is a promising discipline in the medical sector that holds multidisciplinary applications in technology, life sciences, and biotechnology to regenerate diseased organs and maintain homeostatic conditions. Successful regeneration will require scaffolds with niche factors and cells. The scaffolds can be fabricated by many conventional methods but they also have their disadvantages including improper pore structure, customized scaffold size and volume, and mechanical strength. As a remedy for these limitations, additive manufacturing techniques have been introduced. Subsequently, the researchers and clinicians believed that the scaffolds that have been fabricated by the 3D printing can provide a cell support device for the cells to adhere to the surface. The paracrine cell signaling, endocrine signaling, epigenetic signaling, cytokines, oxygenation, and transcription factors are the primary cues required for cells to grow, repair, restore, and regenerate. The extracellular matrix of the cell provides the essential appropriate niche factors needed.

In addition to that, the cells behave according to the type of biomaterial. For instance, the production of glycol-amino acid varies according to the scaffolds in cartilage tissue engineering. Selecting the cell type source is the critical background in tissue engineering [6, 7]. The innovative scaffold architecture and high-quality scaffold techniques have been highly important for tissue engineering applications. 3D-printed scaffolds have stimulated cell development and transformation into tissue regeneration. The behavior of the cells within the scaffolds depends on the type of target cell and the structure of the scaffolds. The mineralization has to deposit on the surface of the scaffolds. The nano-scale topography of scaffolds influences the cellular habit of proliferation capacity. To construct the scaffold architecture for 3D printing, magnetic resonance imaging (MRI) and computed tomography (CT)

Table 1 3D-printed chitosan composites and their applications

Composites	Methods	Cell line used/ animal model	Applications	Reference
Chitosan	Pneumatic	–	Wound healing	[38]
Chitosan	Extrusion	Chondrocyte cells/ATDC5 cells	Cartilage tissue engineering	[22]
Chitosan ducts	Extrusion	–	Tissue engineering	[61]
Chitosan	Extrusion	–	Tissue engineering and drug delivery	[62]
Chitosan	3D bio plotter	Osteoblast cells	Drug release and bone tissue engineering	[63]
Chitosan	Extrusion	Normal dermal human fibroblast cells/Aneuploid immortal keratinocyte cells	Skin restoration and soft tissue healing	[64]
Chitosan	Fused deposition model	C84 fibroblast cells	Tissue engineering	[65]
Chitosan	Pneumatic	L929 fibroblast cells	Tissue engineering	[66]
Chitosan hydrogels	Extrusion	–	Skin restoration and soft tissue regeneration	[67]
Carboxymethyl chitosan	Extrusion	L929 fibroblast cells	Soft tissue regeneration	[68]
Methacrylated chitosan	Extrusion	SH-SY5Y cells	Tissue engineering	[69]
Methyl-furan functionalized chitosan	Extrusion	Human glioma U87- MG cells	Tissue engineering	[70]
Chitosan (CHI-MA)	Digital light processing printing	Human umbilical vein endothelial cells	Tissue engineering	[71]
Hydroxyl butyl chitosan polymer gel	Dispenser type 3D printer for thermoresponsive polymer	Human-induced pluripotent stem cell-derived cardiomyocytes and normal human cardiac fibroblasts (NHCF) cells	Cardiac tissue regeneration	[72]
Hydroxybutyl chitosan	Robotic dispensing 3D printer	Normal human cardiac fibroblastic cells and human cardiac microvascular endothelial cells	Regenerative medicine and tissue engineering	[73]

(continued)

Table 1 (continued)

Composites	Methods	Cell line used/ animal model	Applications	Reference
Carboxymethyl chitosan	Pneumatic and piston-driven	Rabbit chondrocytes cells	Cartilage tissue engineering	[14]
Chitosan–guar gum	Extrusion	–	Biomedical applications	[74]
Chitosan–hydroxyapatite	3D printed pore-forming mold	MC3T3E1 cells	Tissue engineering	[39]
Chitosan–graphene oxide	Extrusion	–	Biomedical applications	[75]
Chitosan–silk	Pneumatic	Human skin fibroblast cells	Soft tissue engineering	[44]
Chitosan–beta-tricalcium phosphate	3D printing	New Zealand rabbit/bone marrow mesenchymal stem cells	Anatomic radial head shape	[76]
Chitosan–polylactic acid	A solid-freeform fabrication of 3D printing	Human fibroblast cells	Drug delivery	[77]
Chitosan–polylactic acid	Fused deposition model	–	Tissue engineering	[78]
Chitosan–polylactic acid–alginate	Extrusion	Human adipose tissue-derived mesenchymal stem cells	Repair tympanic membrane perforation	[79]
Chitosan–1-ethyl-3 methylimidazolium acetate	Flow injection	NIH3T3 fibroblast cells	Hemostatic dressings	[80]
Chitosan–poly (gamma glutamic acid)	Pneumatic 3D bio plotter	Human dermal fibroblasts	Tissue regeneration	[81]
Chitosan–hydroxyapatite	Fused deposition model	Human osteosarcoma cells	Bone regeneration	[82]
Hydroxybutyl chitosan- oxidized chondroitin sulfate hydrogel	Hydrogel fabrication by 3D bioprinting system	Mesenchymal stem cells	Cartilage-subchondral bone regeneration	[83]
Chitosan–gelatin methacryloyl	Extrusion	L929 fibroblast cells	Soft tissue engineering	[84]
Chitosan–collagen	Extrusion	–	Tissue engineering	[85]
Chitosan–starch	Temperature controlled 3D printing	CCL-131 mouse neuroblastoma cells	Neural cell growth	[86]
Chitosan–polycaprolactone	Electrospinning and rapid prototyping	–	Fabrication of artificial vascular graft	[45]

(continued)

Table 1 (continued)

Composites	Methods	Cell line used/ animal model	Applications	Reference
Chitosan–Genipin bio-ink	Pneumatic	Human epidermal keratinocyte cells and human dermal fibroblast cells	Skin regeneration	[87]
Chitosan–gelatin hydrogel	3D bio plotter	–	Tissue engineering	[40]
Chitosan–poly(ethylene glycol) diacrylate	Extrusion	Mouse bone marrow-derived mesenchymal stem cells	Cartilage regeneration and wearable device applications	[88]
Hydroxybutyl chitosan hydrogels–poly(lactic-co-glycolic acid)	Electrospinning	Human mesenchymal stem cells	Cartilage restoration	[89]
Chitosan–collagen	Thermoplastic extrusion	Neural stem cells	Axon regeneration and neuronal recovery	[90]
Chitosan–polymethacrylate	Stereolithography 3D printing	–	Wound healing	[91]
Chitosan–carbon microtubes	3D bio plotter	–	Strain sensor applications	[92]
Chitosan – polylactic acid	Extrusion	–	Filament production for biomedical tools or implants	[93]
Chitosan–pectin hydrogel	Fused deposition model	–	Wound dressing application	[94]
Chitosan–Genipin	Extrusion	Human skin fibroblast cells	Chronic wound healing	[95]
Chitosan–poly (vinyl alcohol)	Extrusion	–	Hard tissue regeneration	[96]
Chitosan–silk particle	3D bio plotter	Human skin fibroblast cells	Soft tissue repair	[97]
Chitosan–multiwall carbon nanotubes	Extrusion	–	Stretchable sensors	[98]
Poly (L-lactide)/ chitosan-quercetin-polydopamine	Fused deposition model	MC3T3-E1 cells	Bone tissue engineering	[20]
Chitosan–catechol	Extrusion	L929 fibroblast cells	Writable bio-ink preparation	[99]
Chitosan polyion–alginate	3D bio plotter	Human adipose-derived stem cells	Tissue engineering	[100]

(continued)

Table 1 (continued)

Composites	Methods	Cell line used/ animal model	Applications	Reference
Chitosan–sodium alginate	3D inkjet printer	–	Bone tissue engineering	[46]
Chitosan–polycaprolactone	Electrohydrodynamic jet 3D printing	Human embryonic stem cell-derived fibroblast cells	Tissue engineering	[101]
Chitosan–polylactic acid	Temperature based 3D printer	Normal human foreskin fibroblast cells	Biomedical applications	[102]
Peptide chitosan–dextran core/shell	Pneumatic	Human bone marrow-derived mesenchymal stem cells	Wound healing	[103]
Chitosan–polyvinyl-alcohol	Extrusion	Human adipose tissue-derived mesenchymal stem cells	Corneal stromal transplantation	[104]
Chitosan–alginate	Extrusion	L929 fibroblast cells	Biomedical applications	[105]
Chitosan–poly(lactic acid)–hydroxyapatite	Fused deposition model	Human mesenchymal stem cells	Bone tissue engineering	[106]
Hydroxybutyl chitosan–hydrogel	3D nanoplotter	Chondrocytes cells/neonatal Sprague-Dawley rats	Biomedical applications	[107]
n chitosan–dextran sulfate microparticles	3D bio plotter	Mesenchymal stem cells	Tissue engineering	[108]
Chitosan hydrochloride–Hydroxypropyl methylcellulose	Rapid prototyping	C2C12 cells	Drug delivery	[109]
Chitosan–polyethylene glycol diacrylate resin	Stereolithography 3D printing	Human mesenchymal stem cells	Cartilage tissue engineering Applications	[110]
Methacrylate chitosan–Laponite nano silicate composite	Extrusion	MC3T3-E1 cells	Bone tissue engineering	[111]
Chitosan–egg shell chitosan	Extrusion	Human mesenchymal stem cells	Bone graft applications	[112]
Chitosan hydrogels–Lactoferrin-loaded carboxymethyl cellulose glycol	Thermoplastic extrusion	MC3T3-E1 cells	Tissue engineering	[113]

(continued)

Table 1 (continued)

Composites	Methods	Cell line used/ animal model	Applications	Reference
Glycol chitosan-oxidized hyaluronate–superparamagnetic iron oxide nanoparticles	Extrusion	ATDC5 cells	Wound healing	[48]
Chitosan–beta cyclodextrin–propolis extract	Pneumatic	Dermal human fibroblast cells	Ulcers and wound Healing	[114]
Carboxymethyl chitosan–hydroxy-apatite–polydopamine	Pneumatic	MC3T3-E1 cells	Bone tissue engineering	[115]
Chitosan–polylactic acid–keratin	Fused deposition model	Human gingival fibroblast cells	Tissue engineering	[116]
Chitosan–silk fibroin–collagen	Extrusion	Bone marrow mesenchymal stem cells/Wistar rat	Tissue engineering	[117]
Polycaprolactone–chitosan–hydrogel biocomposites	Thermoplastic printing	Human umbilical vein endothelial cells	Cardiovascular regeneration	[118]
Carboxymethyl chitosan hydrogel–gelatin–sodium alginate	Extrusion	Bone mesenchy-mal stem cells	Tissue engi-neering and regenerative medicine	[41]
Gelatin methacrylamide–N-maleyl chitosan methacrylate	Multiphoton polymerization	Human dental pulp stem cells	Orofacial bone tissue engineering	[119]
Methacrylated chitosan–chemi-cally converted graphene–graphene oxide	Extrusion	L929 murine fibroblasts cells	Tissue engineering	[120]
Chitosan–gelatin–hydroxyapatite	Fused deposition model	–	Bone tissue engineering	[121]
Hydroxybutyl chitosan–oxidized chondroitin sulfate hydrogels	Pneumatic	Human adipose-derived mesen-chymal stem cells	Cartilage tissue engineering	[122]
Glycol chitosan–oxidized hyaluronate–adipic acid dihydrazide	Extrusion	ATDC5 cell line	Wound healing and cartilage regeneration	[123]
Chitosan–gelatin–hyaluronic acid	Extrusion	Bone marrow stem cells	Osteochondral Healing	[124]

(continued)

Table 1 (continued)

Composites	Methods	Cell line used/ animal model	Applications	Reference
Chitosan–poly (caprolactone) diacrylate–poly (ethylene glycol) diacrylate	Digital light processing printing	L929 cells	Tissue engineering	[125]
Polyethylene glycol (PEG) grafted chitosan–alpha cyclodextrin–gelatin	Extrusion	Mesenchymal stem cells/ C57BL/6 mice	Bio-ink preparation for tissue engineering application	[126]
Chitosan–graphene–gelatin–tricalcium phosphate	Pneumatic	Human osteoblast (hOB) cells	Bone fracture healing	[50]
Chitosan–halloysite nanotubes–tea polyphenol	Extrusion	–	As an antioxidant	[127]
Chitosan–polylactic co glycolic acid–Nano hydroxyapatite–recombinant human bone morphogenetic protein-2	Thermoplastic printing	New Zealand white rabbits	Mandibular bone regeneration	[128]
Chitosan–polylactic acid–polycaprolactone–silk fibroin	Fused deposition model	Human chondrosarcoma cells	Bone regeneration	[129]
Chitosan–hydroxyapatite–collagen–recombinant human bone morphogenetic protein-2	Rapid prototyping	Human mesenchymal stem cells	Drug delivery and bone tissue repair	[130]
Carboxymethyl–chitosan–alginate–agarose	Extrusion	Human neural stem cells	Neuronal regeneration	[131]
Chitosan–Hyaluronan–silk fibroin–polyurethane	Inkjet printing	Mouse embryonic fibroblasts, and SNLP 76/7-4 feeder cells induced pluripotent stem cells/ ICR mouse	Tissue engineering	[132]
Chitosan–alginate–gelatin–Gellan gum or collagen type I hydrogels	Extrusion	Human dermal microvascular endothelial cells	Tissue engineering	[47]

(continued)

Table 1 (continued)

Composites	Methods	Cell line used/ animal model	Applications	Reference
Chitosan–calcium phosphate–poly (lactic acid)–gelatin	Thermoplastic printing	–	Bone tissue engineering	[133]
Chitosan-induced pluripotent stem cells–alginate–carboxymethyl–agarose–calcium chloride	3D bio plotter	Induced pluripotent stem cells	Drug development and regenerative medicine	[49]

pictures were translated to CAD files. Subsequently, the CAD files are sliced and converted to ASC II STL (Standard Tessellation Language) files, which are used to build the implant scaffolds [8, 9] (Fig. 1).

3 Chitosan

Chitosan (Fig. 2) is a naturally occurring derivative polymer widely used for biomedical and agricultural purposes. It is found in natural sources such as fungi, bacteria, crabs, and shrimps. Chitosan is an easily soluble chitin polymer with a greater availability on the market. The chitosan and its composites have been extensively used in tissue engineering applications, especially in wound dressing and bone tissue engineering because of its biocompatibility, biodegradability, nontoxicity, and excellent antimicrobial activity. The strength of chitosan is very low, so it has always been mixed with other hydroxyapatite composites, zirconia, and other polymers to negotiate effects [11]. Certain nosocomial pathogenic organisms acquired from the hospital, such as *S. aureus* and *S. epidermidis*, cause harmful effects during titanium implantation. Antibacterial activity of these bacterial species has been explained by the use of N-carboxymethyl chitosan (CMC), O-CMC, or N, O-CMC [12, 13].

4 Chitosan Bio-Ink Production

The development of bio-ink is important for the biomedical engineering areas. It can print the scaffolds, which can be provided to the hospitalized users for the immediate requirements of tissue repair. Bio-ink has the capability to provide appropriate mechanical strength and can print and store the scaffolds for long-term use with unique physical and biochemical properties. Chitosan can be formulated into

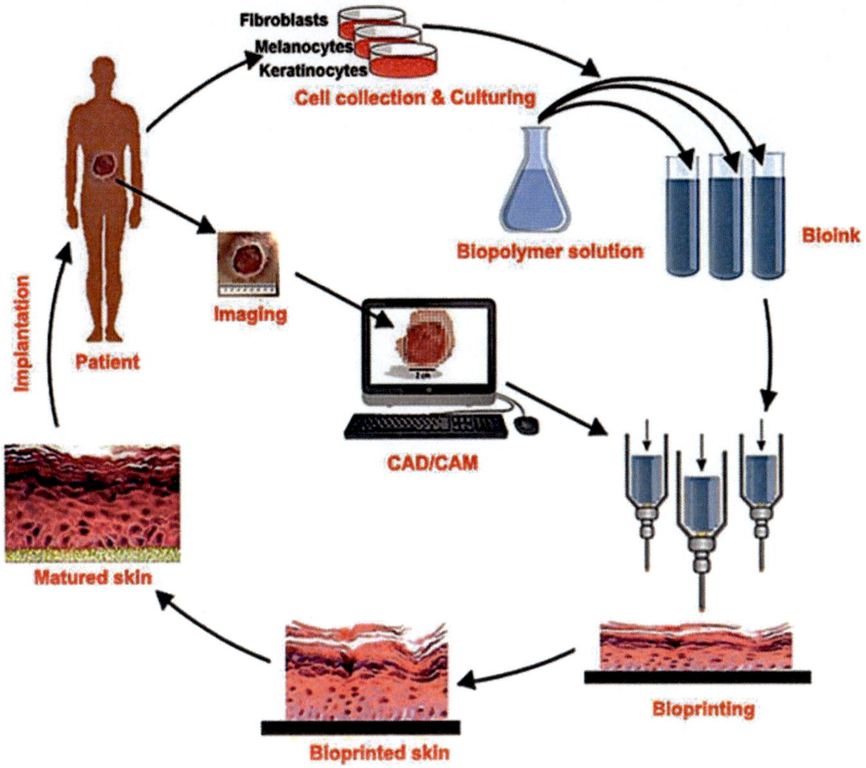

Fig. 1 Diagram of the fundamental stages of 3D printing – based on skin tissue engineering approach. Figure reproduced with permission from [10]

hydrogel bio-ink where their mechanical ability has been improved by the count of ethylene diamine tetraacetic acid (EDTA). The calcium has been used as the physical cross-linker in a neutral pH environment. This bio-chitosan hydrogel ink has been widely applied in cartilage tissue engineering applications [14]. GelMA is a widely used bio-ink which contains both anionic (alginate, xanthan, and κ-carrageenan) and cationic hydrogel (chitosan, gelatin, and gelatin methacrylate). The printing hydrogel should meet the time-dependent printing conditions, shape fidelity, and should have structural integrity and biocompatibility with living cells and tissues. Besides, printing bio-ink must have interfacial strength, before and after printing to avoid layer defects [15]. Printability, print fidelity, viscosity recovery, long-term shape fidelity are the perfect features of bio-ink production. Several methods were adapted to change the viscosity behavior. Viscosity varies according to temperature, molecular weight, polymer concentration, and molecular interactions [16]. During cell deposition during printing, minimal force is required (Yield strain) for the homogeneity of the encapsulated cells. The yield stress is measured by a stress ramp experiment [17].

Cellink skin, Cellink fibrin, Cellink bone, Gelxa skin, Gelxa fibrin, Gelxa bone, Bio conductink, Gelma fibrin, Gelxa cartilage, Cellink a-RGD, Cellink rgd, Gelxa

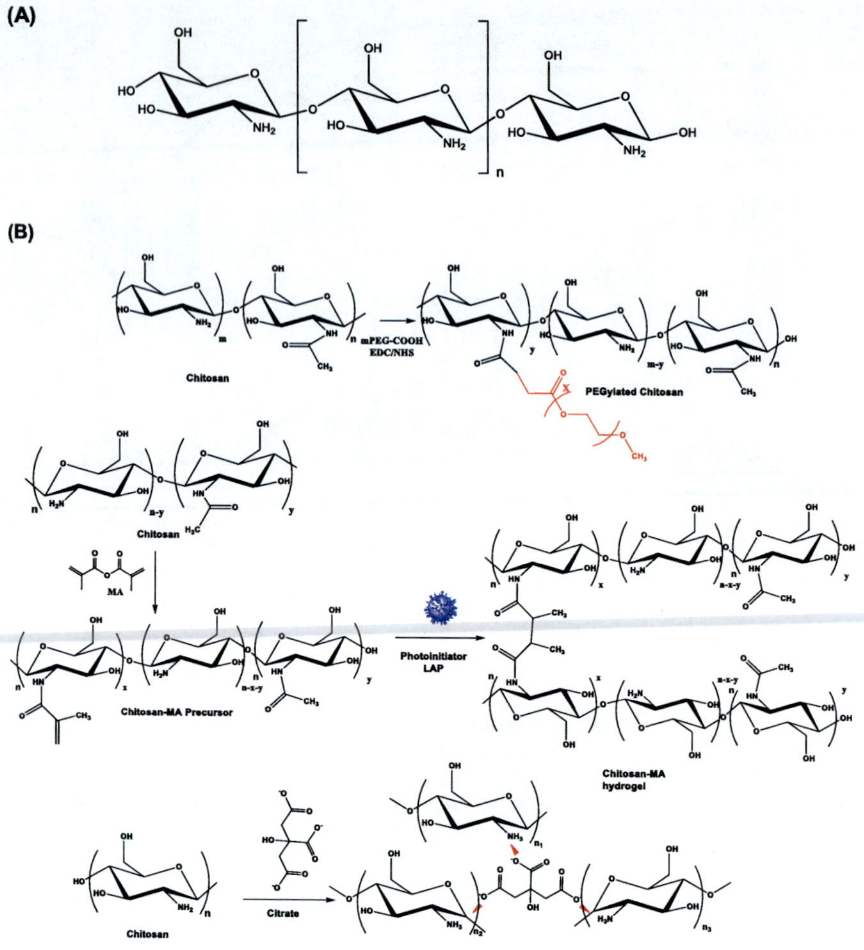

Fig. 2 Structure of (**a**) chitosan and (**b**) chitosan derivatives

lamininks, Cellink lamin inks, Cellink bio-ink, Gelma bio-ink, Gelma A, Gelma C, Gelx series, Gelxg, Chitoink are the commercially available bio-inks supplied by the bio-ink exporter cell ink [18].

5 3D-Printed Chitosan Composites

In tissue engineering, tissue restoration will depend upon the composition of the materials. Each material has only its function in restoring tissues. Polymers are the source of biomaterials, either organically or inorganically to mimic the exact composition of the desired tissue regeneration. For example, the bone matrix has an

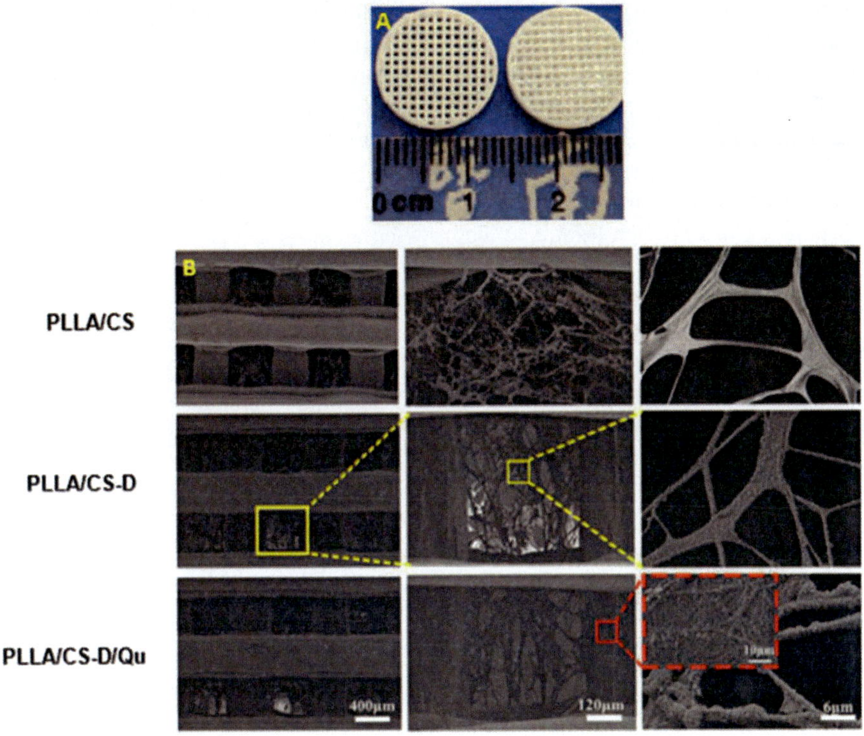

Fig. 3 Chitosan and its composite scaffolds (**a**) SEM Image of the chitosan composites (**b**) FESEM image of chitosan composites and its internal architecture. Figure reproduced with permission from [20]

organic matrix of 60%. Polymers can be synthetic or naturally occurring. Polymers could blend with any type of substrate material for adhesion, proliferation, and differentiation of specific stem cells. In most tissue engineering applications, chitosan has been used as a mixture material (Fig. 3). Moreover, in some studies, the biomaterials are functionalized to enhance their fast healing activity as well as to maintain the regeneration and degeneration kinetics [19]. The following table indicates the type of 3D printed chitosan composites and their target tissue engineering applications.

6 Different Methods Used in 3D Printing

6.1 *Additive Manufacturing*

3D printing is well commonly known for the techniques of solid-freeform fabrication, additive manufacturing, and rapid prototyping with the aid of computer-based software. Additive manufacturing (AM) is a 3D printing technology extensively used in the biomedical sector for tissue engineering applications. AM is the computer-aided software (CAD) based fabrication unit that controls the internal infill density and outer perimeters with multiple geometries required for suitable scaffolds fabrication. In medical implanting scaffolds, it overcomes the drawbacks and limitations of such conventional material scaffold preparation methods. 3D bioprinting is the blooming emerging technologies that can print the cells with materials. It is one such technique that uses bio-inks for the 3D organ culture. The 3D bioprinter has its sterilization unit, clean chamber options, and printing surface area for well plates, glass slides, or the petri dishes with suitable size for fabricating the scaffolds. The highest challenges for 3D printing are the print resolution, print speed, temperature, and rheology of the biomaterial ink. The specific automatic program of the 3D printer controls the geometric shape and spatial pattern of the cells to be printed [21]. The main drawback of bioprinting is the blockage of materials at the ends of the cartridge and nozzle.

6.2 Extruder

Extrusion-based bioprinting is the most common additive manufacturing unit. In general, polymers are most often used in such extruders. Particularly the chitosan scaffolds are studied for the cartilage tissue engineering application. It follows a fluid-based dispensing system where external pressure is required to push the bio-ink down for printing (Fig. 4) [22]. It contains the movable print head with software aided to control the pressure, temperature including bed, nozzle, and extruder. Multi-head extruders are used to print two materials simultaneously. For instance, the bioprinting of 3D organs accumulates the combination of stem cells, growth factors, and biomaterials layer by layer. This scaffolding system mimics as much as possible the niche conditions of the artificial biological organs of the physiological conditions [23–25]. Pneumatic and piston-based extruders are used in multi-nozzle bioprinting [26]. The reports are published based on a mixture of biomaterials and cells (bio-ink) used in extrusion-based bioprinting. This organic ink must have physiological, mechanical, biological, and rheological characteristics. High viscous biomaterials with cells need high compressed air for easy extrusion (prevent the nozzle clogging) to print the structure of the 3D cell but that may harm the cells. On the other hand, low-viscous biological inks reduce the blockage of the nozzle and allow cells in a viable state, but cannot retain the structural capability of the 3D cell. To illustrate, high-viscosity bio-inks are readily extruded and can retain their shape after extrusion. However, they require strong deformation forces during extrusion, which can affect the encapsulated cells. Various control methods have appeared to manage the settings of the bio 3D printer. The use of these technologies to produce tissue engineering scaffolds is widely explored because they can overcome the limitations of traditional porous material fabrication methods (such as solvent casting/salt leaching, and phase separation) in terms of geometry and process coherence [27–30].

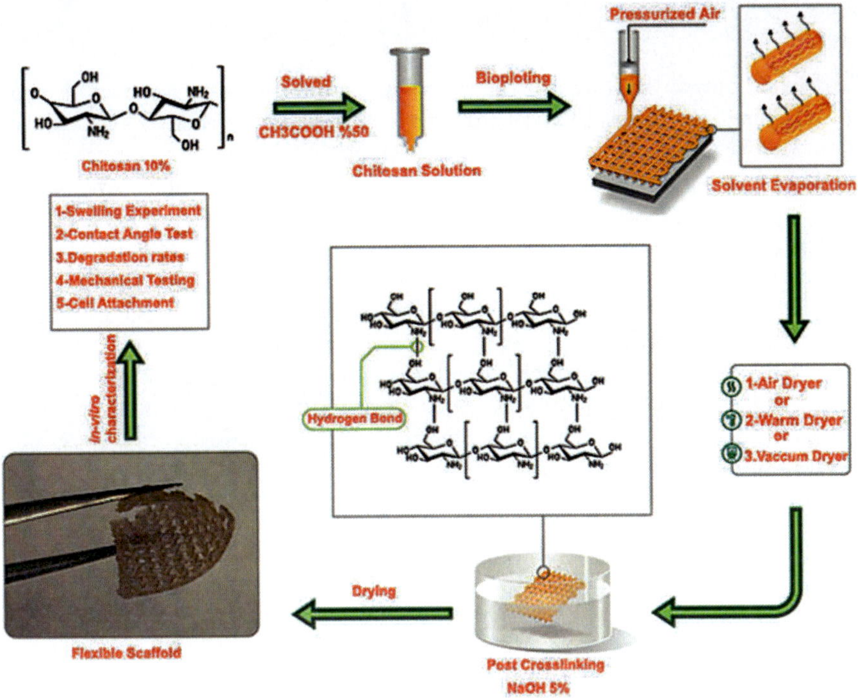

Fig. 4 Extrusion-based bioprinting of chitosan. Figure reproduced with permission from [22]

6.3 *Fused Deposition Model (FDM)*

FDM is the first extruded additive manufacturing system that uses the polymer to melt for the fabrication system. It can provide architectural adaptation of micro-level and high-strength functional biomaterial parts for tissues to develop. Many 3D printers are available with unique features to print the desired tissue from scaffolds in a morphological kind. Some printers use hard or soft materials for fabrication. Selective laser sintering (SLS) and fusion depository modeling (FDM) are selected 3D printing technologies that have been used for fabrication [31]. This is a widely available and economically feasible technology. Thermoplastics materials and temperature maintenance in an extruder nozzle are the basic adjustments of FDM printing. The FDM needs the desired materials in an unwound filament coil. The applying temperature heats the extrusion nozzle to melt the desired materials and makes it possible to print the scaffolds. Occasionally, the printed scaffolds are coated with other materials to enhance antibacterial activity, in particular PLA scaffold with chitosan coated for antibacterial action in bone implants.

6.4 Selective Laser Sintering (SLS)

This is the powder melting process where the high-power laser has been used to manufacture 3D models. They have their pros and cons. Complicated 3D parts can be successfully manufactured with this type of printing. It is structurally stronger as the laser falls where the other side gives way. In particular, the particle size and the viscosity of the powder are the key criteria in the laser sintering [32]. It fails to fabricate the heat-sensitive biological molecules of proteins, drugs, and cells in scaffolding. In particular, it can print even the tissue with complex anatomy structures, including craniofacial bone or cartilage. The powdered materials are used as the support to fabricate layer by layer structures through a laser beam. The bioceramic materials, and organic synthetic polymers of polylactic acid (PLA), polycaprolactone (PCL), poly ethyl ether ketone (PEEK), and polyether ketone (PEKK) are the suitable temperature-dependent materials used in the SLS. The hydroxyapatite and the growth factors are the processing, stimulating factors used to integrate within the scaffolds after printing [33, 34].

6.5 Inkjet Printing

The piezoelectric actuator is the inkjet-based multi-nozzle bioprinter. Inkjet printing has been used to print the living cells directly with high-resolution in a 3D form. Aqueous gelatin is used as a bio-ink to produce a 3D scaffold with cells. The droplet form of printing has been patterned as the desired structure in a substrate. The inkjet heating extruder generates the air bubbles that help the ink drops eject from the nozzle which does not affect biological molecules such as amino acids, proteins, and peptides but cell viability was poor [35]. The print head has run in the three axes: X, Y, and Z. The literature highlighted that the piezoelectric inkjet has been used for the electronic application, but recently it is applied to biomedical engineering applications. Even though many complications are in printing the scaffolds for the biological end users, the inkjet is used to directly print the cells on the glass slides or in its substrates (Fig. 5) [36].

6.6 Thermoplastic Printing

In a thermoplastic printer, an extruder is used to melt materials, including polymer composite, at a high temperature of up to 240°C. The bottom heating coil makes the materials melt continuously. The optimum printing speed is the important parameter, enables the materials to melt during printing. The rheology of the blended composite is crucial for good resolution and impression. Materials inside the cartridge should be packed uniformly to prevent the formation of air bubbles during heating. The

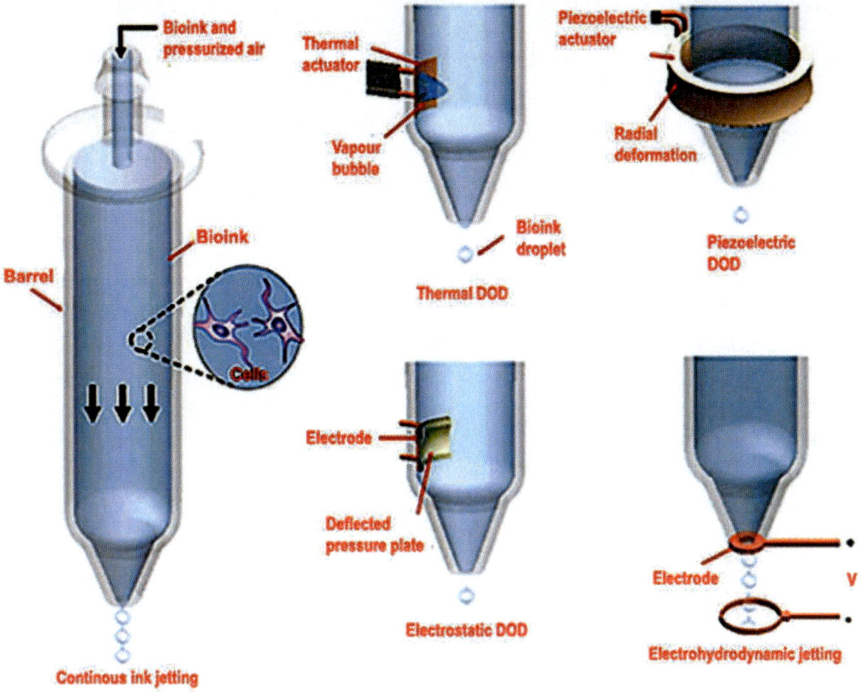

Fig. 5 Schematic diagram of Ink Jet bioprinting [37]

nozzle of variable sizes 0.2, 0.3, 0.4, 0.6, and 0.8 mm are recommended to achieve the scaffolds. Some printers use both types of a thermoplastic extruder, including screw extruder and direct dispenser based on a specific type; for example: regenHU bioprinter.

6.7 *Pneumatic Printing*

The pneumatic 3D printer is the method of press printing. It is capable of extruding both high and low-viscous materials. This pneumatic type of bioprinter has its compressor unit which has a maximum pressure of 700 kPa; for example: Cell ink pneumatic print head. The pneumatic printer controlled the print head temperature from 30 to 65°C. The print bed temperature was maintained in the range of 4–60°C. In some systems, the two pneumatic print heads are used simultaneously for layer-by-layer printing. The printer calibration system has been attached to the central automated system of the software. Different types of the conical nozzle and blend needles of various sizes including 18G, 20G, 22G, 25G, and 27G have been used for the pneumatic type of the printer; for example: Cell ink pneumatic 3D printer (Cellink).

7 Biomedical Applications

7.1 Tissue Engineering

3D printed composites have been the subject of footprints in the main field of tissue engineering. Chitosan has been widely applied in hard and soft tissue engineering applications. Even though the chitosan has poor mechanical strength, due to its self-assembling gelation properties under alkali conditions, it has been used as the printing ink for the direct ink writing (DIW) to enhance the biocompatibility properties [38]. Besides, the 10% (w/v) concentration and air drying of chitosan hydrogels increase mechanical strength compared to 8%, 10%, and 12% [22]. In some cases, combining the hydroxyapatite material with chitosan shows improved mechanical strength [39]. To enhance water retention, nutrient transfer, cell adhesion, and migration, additional gelatin materials are desirable. Subsequently, gelatin improves osteoblast proliferation when mixed with chitosan, graphene, and tricalcium phosphate [40]. Later the biomaterials were developed along with the cells for fabricating the tissues by using the 3D printing technology. For instance, the bone marrow stem cells are encapsulated along with the carboxymethyl chitosan, gelatin, and sodium alginate which shows the homogenous cell distribution on the scaffolds with good mechanical properties and antibacterial activity [41]. Furthermore, the natural silk protein produced by silk fibers has been extensively used in the manufacture of biomedical textiles [42]. Silk has crystalline nature which has been used with chitosan to enhance biomaterial rheology and scaffold printing fidelity. Literature has been reported that silk fibroin with collagen shows better cellular growth and proliferation compared to silk fibroin/chitosan scaffolds [43]. Later, research was conducted to combine silk fibroin with two polymers of polylactic acid and polycaprolactone which gives additional mechanical strength. Besides, the blending of hydroxyapatite gives additional support to silk fibroin/polylactic acid/polycaprolactone scaffolds. These scaffolds mimic natural bone minerals where human chondrosarcoma cells show improved adherence and proliferation growth [44]. In fabricating the artificial vascular graft, compared to the other biocompatible materials, the use of chitosan can induce vascular regeneration but it has poor mechanical strength. Polycaprolactone polymers are used to address these limitations [45]. Further, the developing approach of the bio compatible materials also focused on the alginate. Alginate has its biomedical applications toward wound dressings, tissue healing, and pathogenic infections. It shows promising treatment toward repairing bone tissue when combined with chitosan [46]. The combination of more biomaterials with the chitosan draws much attention to fabricating the bioconstructs. In such cases, the alginate, chitosan, gellan gum, gelatin, and collagen hydrogel are combined with vascular endothelial growth factor and bone morphogenetic protein-2 for tissue engineering applications [47]. Besides, the graphene oxide and carbon nanotubes are required and blended with chitosan to improve the rheological behavior of the materials extensively used in biomedical and sensor applications [48]. The bio-ink of carboxymethyl chitosan, agarose, and calcium chloride with induced pluripotent stem cells has been applied in pharmaceutical developments [49]. Chitosan can be blended with graphene, gelatin, and tricalcium phosphate, which are used in the bone healing process. The proliferation rate has been investigated by using human osteoblast (hOB) cells showing excellent biocompatibility [50]. Hydroxyapatite-chitosan-L-arginine with crosslinking agent genipin has been developed through an additive manufacturing process along with the freeze-dried techniques to form a precise geometry and microstructure. The porosity and mechanical strength of the scaffolds are equivalent to the cancellous bone. Further, the biocompatibility of the scaffold has been assessed through MG-63 cells [51]. It is reported that the Poly (vinyl alcohol)/N, O-Carboxymethyl Chitosan composite scaffolds have enhanced bone formation and proved by the in vivo model of rats [52] (Fig. 6).

7.2 *Drug Delivery*

Chitosan can be modulated into various polymeric scaffold forms, adapted to target drug delivery systems. Chitosan's free amino acid groups can easily encapsulate negatively charged proteins and peptides. Various composite mixtures are coated on chitosan scaffolds to enhance the drug delivery system and antimicrobial activity. For instance, α-helical antimicrobial peptide (AMP) shows excellent bactericidal

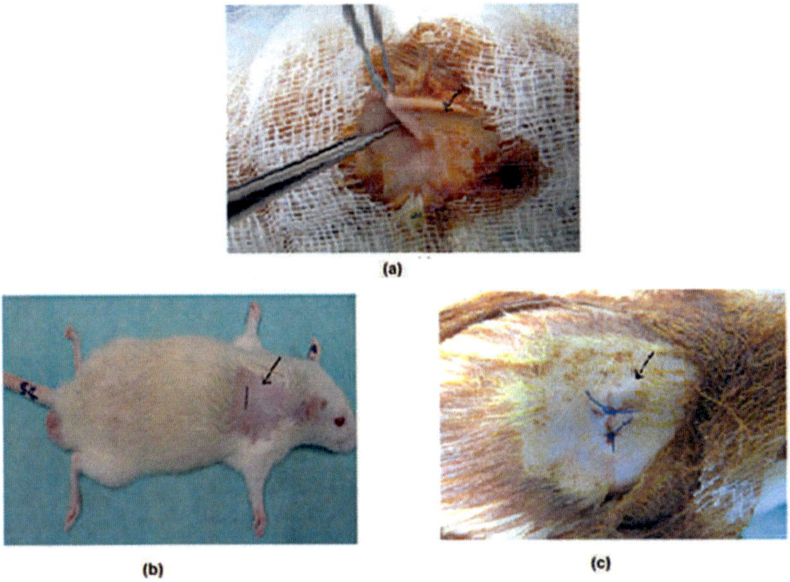

Fig. 6 (**a**) The black arrow mark indicates the fur is cleared before implantation. (**b**) Shows site where the materials were implanted. (**c**) The Implanted poly (vinyl alcohol)/N,O-carboxymethyl chitosan scaffold to the Sprague-Dawley male rat. Figures reproduced with permission from [52]

activity and also confers the cascade signaling system to interact bone cells and immune cells of healthy tissues to repair damaged tissue wound healing and minimize infections. Kartogenin drugs are used as regenerative medicine in several aspects, including bone healing, limb development, and able to solve problems associated with cartilage. These drugs have to release in a sustained manner to achieve the best results of regeneration. The main need is extended drug release, where chitosan resolves. Oral medications are used to administer insulin in the treatment of diabetes. Quaternized thermal-sensitive hydrogel based on chitosan and insulin improves the sustained release of insulin. This increases the efficacy of diabetes treatment and provides hope for the alternative to islets of pancreatic cells for insulin secretion. The acidic pH favors the growth of the cancerous cells and triggers the metastasis conditions. Clinicians are hopeful that targeted chemotherapy is close to preventing metastasis. This depends primarily on the pH maintained within a body system [13]. Chitosan is a pH-sensitive polymer that enhances the effectiveness of encapsulation and drug release [53]. In addition, vascular endothelial growth factors, fibroblast growth factors, and bone morphogenetic proteins are essential for tissue engineering applications. The endothelial cell line requires VEGF as an angiogenic protein whereas the fibroblast growth factors are critical for inducing angiogenesis through the specific signaling pathways. In addition, the bone morphogenic proteins, induce the differentiation pathways of the mesenchymal stem cells and osteoprogenitor cells to mature [54].

7.3 Biosensor

Biosensor is an electrochemical application that has been used in biomedical tissue engineering. It is used to recognize analytes from complicated samples to detect biomolecules [55]. The ligands and receptors of the target biomolecules on the sensor receive the signaling systems and detect the results in a form displayed by serious electrochemical reactions. The cellular biosensor has been used to directly detect biochemical effects and carry out sensitive analyses between biology and electronics [55, 56]. Piezoelectric inkjet printing has been used in the manufacture of biochip inks. In tissue engineering, cell adhesion, morphology, motility, and proliferation of adherent cells could be detected using electrical cell-substrate impedance. Light-addressable potentiometric sensors (LAPS) cell-based biosensor systems are used to record the electrophysiological properties of the differentiated state of embryonic stem cells [57]. Hybrid microchips with bioelectronic nose sensors can detect extracellular activities in olfactory cells. The microelectrode array sensors (MEAs), electrical cell–substrate impedance sensors (ECIS), field-effect transistor sensors (FETs), light-addressable potentiometric sensors (LAPS), and wearable sensors are the type of bio-application sensors used in tissue engineering applications.

7.4 *Stem Cell Interaction with 3D-Printed Chitosan Composites*

Human fat derivative stem cells have better communication with 3D-printed hydrogels of the polyionic complex alginate chitosan. A cell density of 2×10^5 per mL is suitable for the 10 mm × 10 mm × 2 mm scaffolds. The proliferation rate implies that stem cells replenish themselves (Fig. 7). The scaffolds give the appropriate temperature, pH, and niche factors to make stem cells in a dormant state and trigger for self-renew or constrained to differentiate the state of osteoblast. The composite scaffolds activate the extrinsic signals of stem cells to interact and integrate with the hosting tissue. The stem cells grow in the composite scaffolds and can easily adapt and communicate with the other cells (cell cross talk) for sustained metabolic pathways. The cellular pathways of Wnt, Notch, ANG1, OPN, and BMP are recognized for differentiating the hematopoietic stem cells. As biomaterials are implanted, the proteins around the scaffolds adhered to the surface. Following, the immune cells bound to the adsorbed protein release the inflammatory kinases that attract the stem cells to scaffolds. Subsequently, ribosomal proteins adhered to the surface of the composite biomaterials form metalloproteinase to re-modulate the extracellular matrix. The extracellular matrix connects the implanted composite biomaterials to endogenous cells [58, 59]. The rigidity and topography of the materials and the scaffold system have been shown to determine the rate of proliferation and differentiation (Fig. 8).

8 Future Approaches

The printing pressure, printing speed, extruder and bed temperature, the viscosity of the materials, the nozzle size, the infill density, and the layer of the printing are the common parameters for all types of 3D printers. The optimizing parameters are

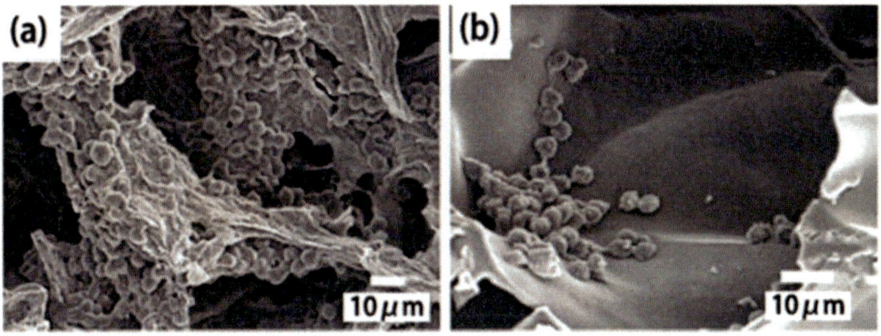

Fig. 7 SEM images chondrocyte cells with (**a**) chitosan–alginate and (**b**) chitosan. Figures reproduced with permission from [60]

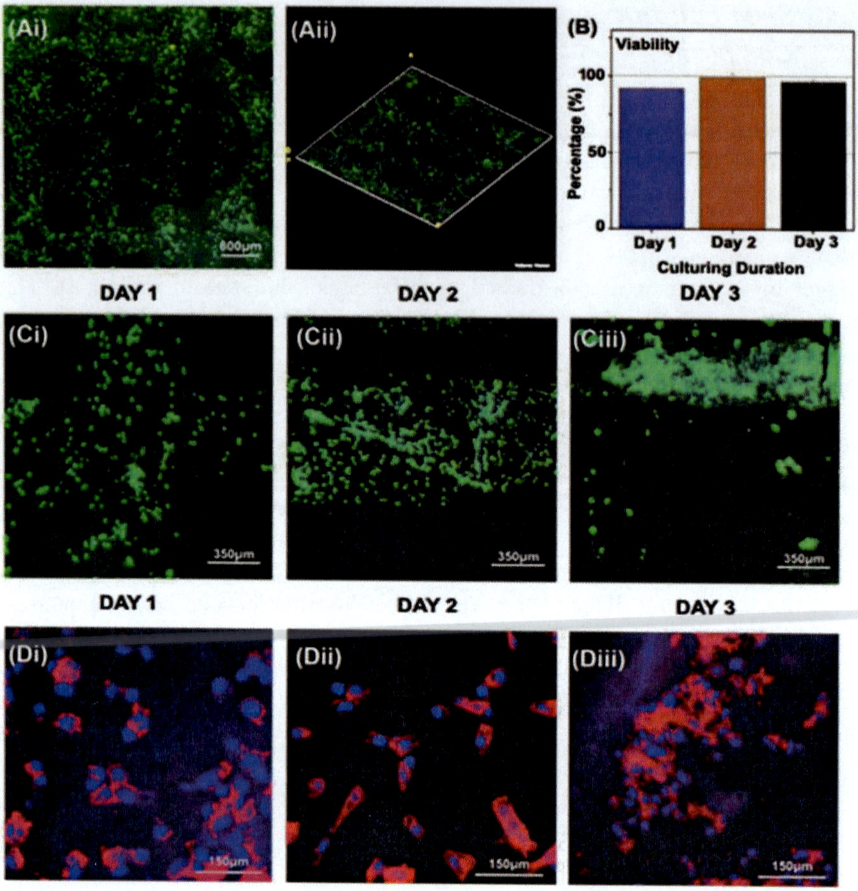

Fig. 8 (**a**) Merge images of the cultured chitosan hydrogel scaffold with human umbilical vein endothelial cells. (Green: live cells, red: dead cells) (**b**) Live cells on 3D printed scaffolds (**c**) The Live-dead staining (**d**) The F-actin and nucleus morphology of the seeded HUVECs on the scaffold (Red: F-actin, blue: nucleus). Figures reproduced from [38] with permission

desirable, as they may vary between material to material preparations. The desired composite scaffolds have satisfied the levels of suitable in vitro and in vivo studies for technological application to end users. The physiological properties of the composite scaffold need to be investigated, as it does not offer the appropriate mechanical strength and tensile strength. Vascularization is a critical challenge and must be studied in detail in all aspects of tissue engineering applications. Consequently, the ideal bio-ink based on the approved qualitative and quantitative materials has been demanded in the future to resolve tissue engineering needs. The success rate of the translation of the implant implies the regeneration and the degeneration kinetics of the biomaterials. So the preparing bio-ink should possess the regeneration and degeneration kinetics.

9 Conclusion

Compared to other fabrication techniques, 3D printing offers advantages for the fabrication of scaffolds with desired pore size, struts size, and surface area. The 3D bioprinter was thought to be useful for organ-on-a-chip, drug delivery systems, and regenerative medicine. Chitosan's flexibility makes an ideal material for 3D printing biomedical applications. It has the properties to acclimate all types of body cells to grow and to restore the defective portions of both hard and soft tissues. Several types of research of the chitosan composites are in the level of bench side which has not been transferred to the technology process due to several limitations and challenges. So the biomedical 3D printing chitosan products have to be studied in detail to translate the research from bench to bedside.

Acknowledgments This work was supported by the Post-Doctor Research Program (2017) through Incheon National University (INU), Incheon, Republic of Korea.

References

1. Younger EM, Chapman MW (1989) Morbidity at bone graft donor sites. J Orthop Trauma 3 (3):192–195
2. Banwart JC, Asher MA, Ruth S (1995) Iliac crest bone graft harvest donor site morbidity. A statistical evaluation. Spine 20(9):1055–1060
3. Delloye C, Olivier C, Druez V, Barbier O (2007) Bone allografts: what they can offer and what they cannot. J Bone Joint Surg 89(5):574–580
4. Goldberg VM, Danek MS (2002) Bone 400(500):600
5. Ramesh S, Harrysson OLA, Rao PK, Tamayol A, Cormier DR, Zhang Y, Rivero IV (2021) Extrusion bioprinting: recent progress, challenges, and future opportunities. Bioprinting 21
6. Kim K, Lee JW, Shin KS (2012) Polyethylenimine-capped Ag nanoparticle film as a platform for detecting charged dye molecules by surface-enhanced raman scattering and metal-enhanced fluorescence. ACS Appl Mater Interfaces 4(10):5498–5504
7. Francioli SE, Martin I, Sie CP, Hagg R, Tommasini R, Candrian C, Heberer M, Barbero A (2007) Growth factors for clinical-scale expansion of human articular chondrocytes: relevance for automated bioreactor systems. Tissue Eng 13(6):1227–1234
8. Labbaf S, Ghanbar H, Stride E, Edirisinghe M (2014) Preparation of multilayered polymeric structures using a novel four-needle coaxial electrohydrodynamic device. Macromol Rapid Commun 35(6):618–623
9. Singh S, Afara IO, Tehrani AH, Oloyede A (2015) Effect of decellularization on the load-bearing characteristics of articular cartilage matrix. Tissue Eng Regen Med 12(5):294–305
10. Augustine R (2018) Skin bioprinting: a novel approach for creating artificial skin from synthetic and natural building blocks. Prog Biomater 7(2):77–92
11. Balagangadharan K, Dhivya S, Selvamurugan N (2017) Chitosan based nanofibers in bone tissue engineering. Int J Biol Macromol 104(Pt B):1372–1382
12. LogithKumar R, KeshavNarayan A, Dhivya S, Chawla A, Saravanan S, Selvamurugan N (2016) A review of chitosan and its derivatives in bone tissue engineering. Carbohydr Polym 151:172–188

13. Chen Y, Liu JM, Xiong XX, Qiu XY, Pan F, Liu D, Lan SJ, Jin S, Yu SB, Chen XQ (2015) Piperlongumine selectively kills hepatocellular carcinoma cells and preferentially inhibits their invasion via ROS-ER-MAPKs-CHOP. Oncotarget 6(8):6406–6421
14. He Y, Derakhshanfar S, Zhong W, Li B, Lu F, Xing M, Li X (2020) Characterization and application of carboxymethyl chitosan-based bioink in cartilage tissue engineering. J Nanomater 2020
15. Li H, Tan YJ, Liu S, Li L (2018) Three-dimensional bioprinting of oppositely charged hydrogels with super strong Interface bonding. ACS Appl Mater Interfaces 10 (13):11164–11174
16. Holzl K, Lin S, Tytgat L, Van Vlierberghe S, Gu L, Ovsianikov A (2016) Bioink properties before, during and after 3D bioprinting. Biofabrication 8(3):032002
17. Pereira RF, Sousa A, Barrias CC, Bártolo PJ, Granja PL (2018) A single-component hydrogel bioink for bioprinting of bioengineered 3D constructs for dermal tissue engineering. Mater Horizons 5(6):1100–1111
18. Cellink https://www.cellink.com/
19. Feng X (2009) Chemical and biochemical basis of cell-bone matrix interaction in health and disease. Curr Chem Biol 3(2):189–196
20. Zhu L, Chen S, Liu K, Wen W, Lu L, Ding S, Zhou C, Luo B (2020) 3D poly (L-lactide)/ chitosan micro/nano fibrous scaffolds functionalized with quercetin-polydopamine for enhanced osteogenic and anti-inflammatory activities. Chem Eng J 391
21. Xie Z, Gao M, Lobo AO, Webster TJ (2020) 3D bioprinting in tissue engineering for medical applications: the classic and the hybrid. Polymers (Basel) 12(8)
22. Sadeghianmaryan A, Naghieh S, Alizadeh Sardroud H, Yazdanpanah Z, Afzal Soltani Y, Sernaglia J, Chen X (2020) Extrusion-based printing of chitosan scaffolds and their in vitro characterization for cartilage tissue engineering. Int J Biol Macromol 164:3179–3192
23. Wang X, Yan Y, Zhang R (2007) Rapid prototyping as a tool for manufacturing bioartificial livers. Trends Biotechnol 25(11):505–513
24. Wang X (2012) Intelligent freeform manufacturing of complex organs. Artif Organs 36 (11):951–961
25. Zhang S, Wang G, Lin X, Chatzinikolaidou M, Jennissen HP, Laub M, Uludağ H (2008) Polyethylenimine-coated albumin nanoparticles for BMP-2 delivery. Biotechnol Prog 24 (4):945–956
26. Li S, Tian X, Fan J, Tong H, Ao Q, Wang X (2019) Chitosans for tissue repair and organ three-dimensional (3D) bioprinting. Micromachines 10(11)
27. Skeldon G, Lucendo-Villarin B, Shu W (2018) Three-dimensional bioprinting of stem-cell derived tissues for human regenerative medicine. Philos Trans R Soc Lond B Biol Sci 373 (1750)
28. Ozbolat IT, Hospodiuk M (2016) Current advances and future perspectives in extrusion-based bioprinting. Biomaterials 76:321–343
29. Tabriz AG, Hermida MA, Leslie NR, Shu W (2015) Three-dimensional bioprinting of complex cell laden alginate hydrogel structures. Biofabrication 7(4):045012
30. Aguado BA, Mulyasasmita W, Su J, Lampe KJ, Heilshorn SC (2012) Improving viability of stem cells during syringe needle flow through the design of hydrogel cell carriers. Tissue Eng Part A 18(7–8):806–815
31. Sarkar SD, Farrugia BL, Dargaville TR, Dhara S (2013) Chitosan–collagen scaffolds with nano/microfibrous architecture for skin tissue engineering. J Biomed Mater Res 101 (12):3482–3492
32. Mazzoli A (2013) Selective laser sintering in biomedical engineering. Med Biol Eng Comput 51(3):245–256
33. Tao O, Kort-Mascort J, Lin Y, Pham HM, Charbonneau AM, ElKashty OA, Kinsella JM, Tran SD (2019) The applications of 3D printing for craniofacial tissue engineering. Micromachines (Basel) 10(7)

34. Preethi Soundarya S, Haritha Menon A, Viji Chandran S, Selvamurugan N (2018) Bone tissue engineering: scaffold preparation using chitosan and other biomaterials with different design and fabrication techniques. Int J Biol Macromol 119:1228–1239
35. Dhillon H, Chikara S, Reindl KM (2014) Piperlongumine induces pancreatic cancer cell death by enhancing reactive oxygen species and DNA damage. Toxicol Rep 1:309–318
36. Nakamura M, Nishiyama Y, Henmi C, Iwanaga S, Nakagawa H, Yamaguchi K, Akita K, Mochizuki S, Takiura K (2008) Ink jet three-dimensional digital fabrication for biological tissue manufacturing: analysis of alginate microgel beads produced by ink jet droplets for three dimensional tissue fabrication. J Imag Sci Tech 52(6)
37. Gudapati H, Dey M, Ozbolat I (2016) A comprehensive review on droplet-based bioprinting: past, present and future. Biomaterials 102:20–42
38. Zhou L, Ramezani H, Sun M, Xie M, Nie J, Lv S, Cai J, Fu J, He Y (2020) 3D printing of high-strength chitosan hydrogel scaffolds without any organic solvents. Biomater Sci 8 (18):5020–5028
39. Zhao H, Liao J, Wu F, Shi J (2020) Mechanical strength improvement of chitosan/hydroxy-apatite scaffolds by coating and cross-linking. J Mech Behav Biomed Mater
40. Fischetti T, Celikkin N, Contessi Negrini N, Farè S, Swieszkowski W (2020) Tripolyphosphate-crosslinked chitosan/gelatin biocomposite ink for 3D printing of uniaxial scaffolds. Front Bioeng Biotechnol 8
41. Huang J, Fu H, Wang Z, Meng Q, Liu S, Wang H, Zheng X, Dai J, Zhang Z (2016) BMSCs-laden gelatin/sodium alginate/carboxymethyl chitosan hydrogel for 3D bioprinting. RSC Adv 6(110):108423–108430
42. Vepari C, Kaplan DL (2007) Silk as a biomaterial. Prog Polym Sci 32(8–9):991–1007
43. Sun K, Li H, Li R, Nian Z, Li D, Xu C (2015) Silk fibroin/collagen and silk fibroin/chitosan blended three-dimensional scaffolds for tissue engineering. Eur J Orthop Surg Traumatol 25 (2):243–249
44. Zhang J, Allardyce BJ, Rajkhowa R, Kalita S, Dilley RJ, Wang X, Liu X (2019) Silk particles, microfibres and nanofibres: a comparative study of their functions in 3D printing hydrogel scaffolds. Mater Sci Eng C 103
45. Lee SJ, Heo DN, Park JS, Kwon SK, Lee JH, Lee JH, Kim WD, Kwon IK, Park SA (2015) Characterization and preparation of bio-tubular scaffolds for fabricating artificial vascular grafts by combining electrospinning and a 3D printing system. Phys Chem Chem Phys 17 (5):2996–2999
46. Fedotov AY, Egorov AA, Zobkov YV, Mironov AV, Popov VK, Barinov SM, Komlev VS (2016) 3D printing of mineral-polymer structures based on calcium phosphate and poly-saccharides for tissue engineering. Inorg Mater Appl Res 7(2):240–243
47. Akkineni AR, Ahlfeld T, Lode A, Gelinsky M (2016) A versatile method for combining different biopolymers in a core/shell fashion by 3D plotting to achieve mechanically robust constructs. Biofabrication 8(4)
48. Ko ES, Kim C, Choi Y, Lee KY (2020) 3D printing of self-healing ferrogel prepared from glycol chitosan, oxidized hyaluronate, and iron oxide nanoparticles. Carbohydr Polym 245
49. Gu Q, Tomaskovic-Crook E, Wallace GG, Crook JM (2017) 3D bioprinting human induced pluripotent stem cell constructs for in situ cell proliferation and successive multilineage differentiation. Adv Healthc Mater 6(17)
50. Lu H, Pan X, Hu M, Zhang J, Yu Y, Hu X, Jiang K (2020) Fabrication of graphene/gelatin/chitosan/tricalcium phosphate 3D printed scaffolds for bone tissue regeneration applications. Appl Nanosci 11:335–346
51. Zafeiris K, Brasinika D, Karatza A, Koumoulos E, Karoussis IK, Kyriakidou K, Charitidis CA (2021) Additive manufacturing of hydroxyapatite–chitosan–genipin composite scaffolds for bone tissue engineering applications. Mater Sci Eng C 119
52. Kamarul T, Krishnamurithy G, Salih ND, Ibrahim NS, Raghavendran HR, Suhaeb AR, Choon DS (2014) Biocompatibility and toxicity of poly(vinyl alcohol)/N,O-carboxymethyl chitosan scaffold. ScientificWorldJournal 2014:905103

53. Babu A, Ramesh R (2017) Multifaceted applications of chitosan in cancer drug delivery and therapy. Mar Drugs 15(4)
54. Bose S, Vahabzadeh S, Bandyopadhyay A (2013) Bone tissue engineering using 3D printing. Mater Today 16(12):496–504
55. Li YE, Lee IC (2020) The current trends of biosensors in tissue engineering. Biosensors (Basel) 10(8)
56. Otero F, Magner E (2020) Biosensors-recent advances and future challenges in electrode materials. Sensors (Basel) 20(12)
57. Liu Y, Chang Y, Yang C, Sang Z, Yang T, Ang W, Ye W, Wei Y, Gong C, Luo Y (2014) Biodegradable nanoassemblies of piperlongumine display enhanced anti-angiogenesis and anti-tumor activities. Nanoscale 6(8):4325–4337
58. Ferraro F, Celso CL, Scadden D (2010) Adult stem cells and their niches. Adv Exp Med Biol 695:155–168
59. Brown BN, Valentin JE, Stewart-Akers AM, McCabe GP, Badylak SF (2009) Macrophage phenotype and remodeling outcomes in response to biologic scaffolds with and without a cellular component. Biomaterials 30(8):1482–1491
60. Li Z, Zhang M (2005) Chitosan-alginate as scaffolding material for cartilage tissue engineering. J Biomed Mater Res A 75(2):485–493
61. Zhao CQ, Liu WG, Xu ZY, Li JG, Huang TT, Lu YJ, Huang HG, Lin JX (2020) Chitosan ducts fabricated by extrusion-based 3D printing for soft-tissue engineering. Carbohydr Polym 236
62. Wu Q, Therriault D, Heuzey MC (2018) Processing and properties of chitosan inks for 3D printing of hydrogel microstructures. ACS Biomater Sci Eng 4(7):2643–2652
63. Lin HY, Chang TW, Peng TK (2018) Three-dimensional plotted alginate fibers embedded with diclofenac and bone cells coated with chitosan for bone regeneration during inflammation. J Biomed Mater Res A 106(6):1511–1521
64. Intini C, Elviri L, Cabral J, Mros S, Bergonzi C, Bianchera A, Flammini L, Govoni P, Barocelli E, Bettini R, McConnell M (2018) 3D-printed chitosan-based scaffolds: an in vitro study of human skin cell growth and an in-vivo wound healing evaluation in experimental diabetes in rats. Carbohydr Polym 199:593–602
65. Elviri L, Foresti R, Bergonzi C, Zimetti F, Marchi C, Bianchera A, Bernini F, Silvestri M, Bettini R (2017) Highly defined 3D printed chitosan scaffolds featuring improved cell growth. Biomed Mater 12(4)
66. Wu Q, Maire M, Lerouge S, Therriault D, Heuzey MC (2017) 3D printing of microstructured and stretchable chitosan hydrogel for guided cell growth. Adv Biosyst 1(6)
67. Darabi MA, Khosrozadeh A, Mbeleck R, Liu Y, Chang Q, Jiang J, Cai J, Wang Q, Luo G, Xing M (2017) Skin-inspired multifunctional autonomic-intrinsic conductive self-healing hydrogels with pressure sensitivity, stretchability, and 3D printability. Adv Mater 29(31)
68. Puertas-Bartolomé M, Włodarczyk-Biegun MK, Del Campo A, Vázquez-Lasa B, Román JS (2020) 3d printing of a reactive hydrogel bio-ink using a static mixing tool. Polymers 12 (9):1–17
69. Tonda-Turo C, Carmagnola I, Chiappone A, Feng Z, Ciardelli G, Hakkarainen M, Sangermano M (2020) Photocurable chitosan as bioink for cellularized therapies towards personalized scaffold architecture. Bioprinting 18
70. Magli S, Rossi GB, Risi G, Bertini S, Cosentino C, Crippa L, Ballarini E, Cavaletti G, Piazza L, Masseroni E, Nicotra F, Russo L (2020) Design and synthesis of chitosan – gelatin hybrid hydrogels for 3D printable in vitro models. Front Chem 8
71. Shen Y, Tang H, Huang X, Hang R, Zhang X, Wang Y, Yao X (2020) DLP printing photocurable chitosan to build bio-constructs for tissue engineering. Carbohydr Polym 235
72. Tsukamoto Y, Akagi T, Akashi M (2020) Vascularized cardiac tissue construction with orientation by layer-by-layer method and 3D printer. Sci Rep 10(1)

73. Tsukamoto Y, Akagi T, Shima F, Akashi M (2017) Fabrication of orientation-controlled 3D tissues using a layer-by-layer technique and 3D printed a thermoresponsive gel frame. Tissue Eng Part C Methods 23(6):357–366
74. Cleymand F, Poerio A, Mamanov A, Elkhoury K, Ikhelf L, Jehl JP, Kahn CJF, Ponçot M, Arab-Tehrany E, Mano JF (2021) Development of novel chitosan/guar gum inks for extrusion-based 3D bioprinting: process, printability and properties. Bioprinting 21
75. Ahmed J, Mulla M, Maniruzzaman M (2020) Rheological and dielectric behavior of 3D-printable chitosan/graphene oxide hydrogels. ACS Biomater Sci Eng 6(1):88–99
76. Wang JQ, Jiang BJ, Guo WJ, Zhao YM (2019) Indirect 3D printing technology for the fabrication of customised β-TCP/chitosan scaffold with the shape of rabbit radial head – an in vitro study. J Orthop Surg Res 14(1)
77. Wang J, Nor Hidayah Z, Razak SIA, Kadir MRA, Nayan NHM, Li Y, Amin KAM (2019) Surface entrapment of chitosan on 3D printed polylactic acid scaffold and its biomimetic growth of hydroxyapatite. Compos Interfaces 26(5):465–478
78. Singh S, Singh G, Prakash C, Ramakrishna S, Lamberti L, Pruncu CI (2020) 3D printed biodegradable composites: an insight into mechanical properties of PLA/chitosan scaffold. Polym Test 89
79. Ilhan E, Ulag S, Sahin A, Yilmaz BK, Ekren N, Kilic O, Sengor M, Kalaskar DM, Oktar FN, Gunduz O (2021) Fabrication of tissue-engineered tympanic membrane patches using 3D-printing technology. J Mech Behav Biomed Mater 114
80. Li B, Wang J, Gui Q, Yang H (2020) Continuous production of uniform chitosan beads as hemostatic dressings by a facile flow injection method. J Mater Chem B 8(35):7941–7946
81. Pisani S, Dorati R, Scocozza F, Mariotti C, Chiesa E, Bruni G, Genta I, Auricchio F, Conti M, Conti B (2020) Preliminary investigation on a new natural based poly(gamma-glutamic acid)/chitosan bioink. J Biomed Mater Res B Appl Biomater 108(7):2718–2732
82. Nazeer MA, Onder OC, Sevgili I, Yilgor E, Kavakli IH, Yilgor I (2020) 3D printed poly(lactic acid) scaffolds modified with chitosan and hydroxyapatite for bone repair applications. Mater Today Commun 25
83. Li C, Wang K, Li T, Zhou X, Ma Z, Deng C, He C, Wang B, Wang J (2020) Patient-specific scaffolds with a biomimetic gradient environment for articular cartilage-subchondral bone regeneration. ACS Appl Bio Mater 3(8):4820–4831
84. Mora-Boza A, Włodarczyk-Biegun MK, Del Campo A, Vázquez-Lasa B, Román JS (2020) Glycerylphytate as an ionic crosslinker for 3D printing of multi-layered scaffolds with improved shape fidelity and biological features. Biomater Sci 8(1):506–516
85. Heidenreich AC, Pérez-Recalde M, González Wusener A, Hermida ÉB (2020) Collagen and chitosan blends for 3D bioprinting: a rheological and printability approach. Polym Test 82
86. Butler HM, Naseri E, MacDonald DS, Andrew Tasker R, Ahmadi A (2020) Optimization of starch- and chitosan-based bio-inks for 3D bioprinting of scaffolds for neural cell growth. Materialia 12
87. Hafezi F, Shorter S, Tabriz AG, Hurt A, Elmes V, Boateng J, Douroumis D (2020) Bioprinting and preliminary testing of highly reproducible novel bioink for potential skin regeneration. Pharmaceutics 12(6):1–21
88. Deng Z, Qian T, Hang F (2020) Three-dimensional printed hydrogels with high elasticity, high toughness, and ionic conductivity for multifunctional applications. ACS Biomater Sci Eng 6 (12):7061–7070
89. Liu X, Song S, Huang J, Fu H, Ning X, He Y, Zhang Z (2020) HBC-nanofiber hydrogel scaffolds with 3D printed internal microchannels for enhanced cartilage differentiation. J Mater Chem B 8(28):6115–6127
90. Sun Y, Yang C, Zhu X, Wang JJ, Liu XY, Yang XP, An XW, Liang J, Dong HJ, Jiang W, Chen C, Wang ZG, Sun HT, Tu Y, Zhang S, Chen F, Li XH (2019) 3D printing collagen/chitosan scaffold ameliorated axon regeneration and neurological recovery after spinal cord injury. J Biomed Mater Res A 107(9):1898–1908

91. Maalihan RD, Chen Q, Agueda JRHS, Pajarito BB, Tamura H, Advincula RC (2020) On the use of surfactant-complexed chitosan for toughening 3D printed polymethacrylate composites. Macromol Mater Eng 306(1):2000448
92. Nasri-Nasrabadi B, Kaynak A, Adams SD, Heidarian P, Kouzani AZ (2019) Fabrication of a conductive composite structure with enhanced stretchability using direct-write 3D printing. Mater Res Express 6(8)
93. Mania S, Ryl J, Jinn JR, Wang YJ, Michałowska A, Tylingo R (2019) The production possibility of the antimicrobial filaments by co-extrusion of the pla pellet with chitosan powder for FDM 3D printing technology. Polymers 11(11)
94. Long J, Etxeberria AE, Nand AV, Bunt CR, Ray S, Seyfoddin A (2019) A 3D printed chitosan-pectin hydrogel wound dressing for lidocaine hydrochloride delivery. Mater Sci Eng C 104
95. Hafezi F, Scoutaris N, Douroumis D, Boateng J (2019) 3D printed chitosan dressing crosslinked with genipin for potential healing of chronic wounds. Int J Pharm 560:406–415
96. Ergul NM, Unal S, Kartal I, Kalkandelen C, Ekren N, Kilic O, Chi-Chang L, Gunduz O (2019) 3D printing of chitosan/poly(vinyl alcohol) hydrogel containing synthesized hydroxyapatite scaffolds for hard-tissue engineering. Polym Test 79
97. Zhang J, Allardyce BJ, Rajkhowa R, Zhao Y, Dilley RJ, Redmond SL, Wang X, Liu X (2018) 3D printing of silk particle-reinforced chitosan hydrogel structures and their properties. ACS Biomater Sci Eng 4(8):3036–3046
98. Wu Q, Zou S, Gosselin FP, Therriault D, Heuzey MC (2018) 3D printing of a self-healing nanocomposite for stretchable sensors. J Mater Chem C 6(45):12180–12186
99. Lee D, Park JP, Koh MY, Kim P, Lee J, Shin M, Lee H (2018) Chitosan-catechol: a writable bioink under serum culture media. Biomater Sci 6(5):1040–1047
100. Liu Q, Li Q, Xu S, Zheng Q, Cao X (2018) Preparation and properties of 3D printed alginate-chitosan polyion complex hydrogels for tissue engineering. Polymers 10(6)
101. Wu Y, Sriram G, Fawzy AS, Fuh JYH, Rosa V, Cao T, Wong YS (2016) Fabrication and evaluation of electrohydrodynamic jet 3D printed polycaprolactone/chitosan cell carriers using human embryonic stem cell-derived fibroblasts. J Biomater Appl 31(2):181–192
102. Wu CS (2016) Modulation, functionality, and cytocompatibility of three-dimensional printing materials made from chitosan-based polysaccharide composites. Mater Sci Eng C 69:27–36
103. Turner PR, Murray E, McAdam CJ, McConnell MA, Cabral JD (2020) Peptide chitosan/dextran Core/Shell vascularized 3D constructs for wound healing. ACS Appl Mater Interfaces 12(29):32328–32339
104. Ulag S, Ilhan E, Sahin A, Karademir Yilmaz B, Kalaskar DM, Ekren N, Kilic O, Nuzhet Oktar F, Gunduz O (2020) 3D printed artificial cornea for corneal stromal transplantation. Eur Polym J 133
105. Ren Y, Lou R, Liu X, Gao M, Zheng H, Yang T, Xie H, Yu W, Ma X (2016) A self-healing hydrogel formation strategy: via exploiting endothermic interactions between polyelectrolytes. Chem Commun 52(37):6273–6276
106. Rogina A, Pribolšan L, Hanžek A, Gómez-Estrada L, Gallego Ferrer G, Marijanović I, Ivanković M, Ivanković H (2016) Macroporous poly(lactic acid) construct supporting the osteoinductive porous chitosan-based hydrogel for bone tissue engineering. Polymer 98:172–181
107. Wang X, Wei C, Cao B, Jiang L, Hou Y, Chang J (2018) Fabrication of multiple-layered hydrogel scaffolds with elaborate structure and good mechanical properties via 3D printing and ionic reinforcement. ACS Appl Mater Interfaces 10(21):18338–18350
108. Akkineni AR, Luo Y, Schumacher M, Nies B, Lode A, Gelinsky M (2015) 3D plotting of growth factor loaded calcium phosphate cement scaffolds. Acta Biomater 27:264–274
109. Vorndran E, Klammert U, Ewald A, Barralet JE, Gbureck U (2010) Simultaneous immobilization of bioactives during 3D powder printing of bioceramic drug-release matrices. Adv Funct Mater 20(10):1585–1591

110. Morris VB, Nimbalkar S, Younesi M, McClellan P, Akkus O (2017) Mechanical properties, cytocompatibility and manufacturability of chitosan:PEGDA hybrid-gel scaffolds by stereolithography. Ann Biomed Eng 45(1):286–296
111. Cebe T, Ahuja N, Monte F, Awad K, Vyavhare K, Aswath P, Huang J, Brotto M, Varanasi V (2020) Novel 3D-printed methacrylated chitosan-laponite nanosilicate composite scaffolds enhance cell growth and biomineral formation in MC3T3 pre-osteoblasts. J Mater Res 35 (1):58–75
112. Dadhich P, Das B, Pal P, Srivas PK, Dutta J, Ray S, Dhara S (2016) A simple approach for an eggshell-based 3D-printed osteoinductive multiphasic calcium phosphate scaffold. ACS Appl Mater Interfaces 8(19):11910–11924
113. Janarthanan G, Tran HN, Cha E, Lee C, Das D, Noh I (2020) 3D printable and injectable lactoferrin-loaded carboxymethyl cellulose-glycol chitosan hydrogels for tissue engineering applications. Mater Sci Eng C 113
114. Andriotis EG, Eleftheriadis GK, Karavasili C, Fatouros DG (2020) Development of bio-active patches based on pectin for the treatment of ulcers and wounds using 3D-bioprinting technology. Pharmaceutics 12(1)
115. Chen T, Zou Q, Du C, Wang C, Li Y, Fu B (2020) Biodegradable 3D printed HA/CMCS/PDA scaffold for repairing lacunar bone defect. Mater Sci Eng C 116
116. Rojas-Martínez LE, Flores-Hernandez CG, López-Marín LM, Martinez-Hernandez AL, Thorat SB, Reyes Vasquez CD, Del Rio-Castillo AE, Velasco-Santos C (2020) 3D printing of PLA composites scaffolds reinforced with keratin and chitosan: effect of geometry and structure. Eur Polym J 141
117. Sun K, Li R, Li H, Li D, Jiang W (2018) Comparison of three-dimensional printing for fabricating silk fibroin-blended scaffolds. Int J Polym Mater Polym Biomater 67(8):480–486
118. Ulag S, Kalkandelen C, Oktar FN, Uzun M, Sahin YM, Karademir B, Arslan S, Ozbolat IT, Mahirogullari M, Gunduz O (2019) 3D printing artificial blood vessel constructs using PCL/chitosan/hydrogel biocomposites. ChemistrySelect 4(8):2387–2391
119. Parkatzidis K, Chatzinikolaidou M, Kaliva M, Bakopoulou A, Farsari M, Vamvakaki M (2019) Multiphoton 3D printing of biopolymer-based hydrogels. ACS Biomater Sci Eng 5 (11):6161–6170
120. Sayyar S, Gambhir S, Chung J, Officer DL, Wallace GG (2017) 3D printable conducting hydrogels containing chemically converted graphene. Nanoscale 9(5):2038–2050
121. Lee CM, Yang SW, Jung SC, Kim BH (2017) Oxygen plasma treatment on 3D-printed chitosan/gelatin/hydroxyapatite scaffolds for bone tissue engineering. J Nanosci Nanotechnol 17(4):2747–2750
122. Li C, Wang K, Zhou X, Li T, Xu Y, Qiang L, Peng M, Xu Y, Xie L, He C, Wang B, Wang J (2019) Controllable fabrication of hydroxybutyl chitosan/oxidized chondroitin sulfate hydrogels by 3D bioprinting technique for cartilage tissue engineering. Biomed Mater 14(2)
123. Kim SW, Kim DY, Roh HH, Kim HS, Lee JW, Lee KY (2019) Three-dimensional bioprinting of cell-laden constructs using polysaccharide-based self-healing hydrogels. Biomacromolecules 20(5):1860–1866
124. Hu X, Man Y, Li W, Li L, Xu J, Parungao R, Wang Y, Zheng S, Nie Y, Liu T, Song K (2019) 3D bio-printing of CS/Gel/HA/Gr hybrid osteochondral scaffolds. Polymers 11(10)
125. Cheng YL, Chen F (2017) Preparation and characterization of photocured poly (-ε-caprolactone) diacrylate/poly (ethylene glycol) diacrylate/chitosan for photopolymerization-type 3D printing tissue engineering scaffold application. Mater Sci Eng C 81:66–73
126. Hu T, Cui X, Zhu M, Wu M, Tian Y, Yao B, Song W, Niu Z, Huang S, Fu X (2020) 3D-printable supramolecular hydrogels with shear-thinning property: fabricating strength tunable bioink via dual crosslinking. Bioact Mater 5(4):808–818
127. Wang Y, Yi S, Lu R, Sameen DE, Ahmed S, Dai J, Qin W, Li S, Liu Y (2021) Preparation, characterization, and 3D printing verification of chitosan/halloysite nanotubes/tea polyphenol nanocomposite films. Int J Biol Macromol 166:32–44

128. Deng N, Sun J, Li Y, Chen L, Chen C, Wu Y, Wang Z, Li L (2019) Experimental study of rhBMP-2 chitosan nano-sustained release carrier-loaded PLGA/nHA scaffolds to construct mandibular tissue-engineered bone. Arch Oral Biol 102:16–25
129. Thunsiri K, Pitjamit S, Pothacharoen P, Pruksakorn D, Nakkiew W, Wattanutchariya W (2020) The 3D-printed bilayer's bioactive-biomaterials scaffold for full-thickness articular cartilage defects treatment. Materials (Basel) 13(15):1–26
130. Wang H, Wu G, Zhang J, Zhou K, Yin B, Su X, Qiu G, Yang G, Zhang X, Zhou G, Wu Z (2016) Osteogenic effect of controlled released rhBMP-2 in 3D printed porous hydroxyapatite scaffold. Colloids Surf B Biointerfaces 141:491–498
131. Gu Q, Tomaskovic-Crook E, Lozano R, Chen Y, Kapsa RM, Zhou Q, Wallace GG, Crook JM (2016) Functional 3D neural mini-tissues from printed gel-based bioink and human neural stem cells. Adv Healthc Mater 5(12):1429–1438
132. Wong CW, Chen YT, Chien CL, Yu TY, Rwei SP, Hsu SH (2018) A simple and efficient feeder-free culture system to up-scale iPSCs on polymeric material surface for use in 3D bioprinting. Mater Sci Eng C 82:69–79
133. Schneider M, Günter C, Taubert A (2018) Co-deposition of a hydrogel/calcium phosphate hybrid layer on 3D printed poly(lactic acid) scaffolds via dip coating: towards automated biomaterials fabrication. Polymers 10(3)

Adv Polym Sci (2021) 288: 117–162
https://doi.org/10.1007/12_2021_99

Published online: 4 June 2021

Chitosan and Its Potential Use for the Delivery of Bioactive Molecules in Bone Tissue Engineering

D. Saleth Sidharthan, R. Abhinandan, S. Pranav Adithya, K. Balagangadharan, and N. Selvamurugan

Contents

D. S. Sidharthan, R. Abhinandan, S. P. Adithya, K. Balagangadharan, and N. Selvamurugan (✉)
Department of Biotechnology, College of Engineering and Technology, SRM Institute of Science and Technology, Kattankulathur, Tamil Nadu, India
e-mail: selvamun@srmist.edu.in

Abstract Bone tissue engineering (BTE) is a transitional research field that focuses on material science and biology to construct novel biocomposites proficient in treating impaired bone. Bioactive molecules are unique therapeutic agents that have prompted rapid advances in the field of tissue engineering. Polymers of natural sources play a crucial role in the fabrication of biocompatible delivery systems that facilitate bioactive agents' efficient delivery. Chitosan (CS) is known for its distinct pharmacological properties, and its function in regenerative medicine is well recorded. The versatility of CS enables the formulation of a wide range of drug carriers. This chapter highlights various scientific findings concerning different forms of nanomaterials produced from CS. CS-based composites' utility as effective delivery systems for the potent bioactive compounds such as growth factors, nucleic acids, and phytocompounds in treating bone defects is discussed.

Keywords Chitosan · Drug delivery system · Growth factors · Nucleic acids · Phytocompounds · Bone tissue engineering

1 Introduction

Bone is a highly organized and diverse connective tissue in the human body that constructs the skeleton. It is a rigid mineralized structure that plays a crucial function in mechanical stability and locomotion, protects the soft visceral organs, and maintains the mineral homeostasis [1]. Bone tissue majorly harbors hard organic matrix like collagen (Col) fibers in which hydroxyapatite (HAp) mineral salts are deposited. Bone stability and vasculature are endured by two important structures, such as compact and trabecular bone. Compact bone is a hard material, which provides mechanical stability and protects bone marrow, whereas trabecular bone offers compressive strength to the skeleton [2]. The development of bone is a complex mechanism accomplished by two significant modes involving the metamorphosis of pre-existing mesenchyme to a structural osseous tissue. The direct transformation of mesenchyme to the bone is known as intramembranous ossification, and the transformation via cartilage formation to bone tissue is endochondral ossification. In both these ossification, modeling and remodeling of bone are mediated by bone cells like osteoblasts and osteoclasts [3]. These constructive mechanisms are widely regulated by various factors like hormones, growth factors (GFs), and minerals. There are many possibilities of misregulation of these factors or injuries that can lead to anomalies in the bone formation, such as skeletal dysplasia, osteoarthritis, osteopenia, and Paget's disease [3, 4]. Although bone is highly vascularized and

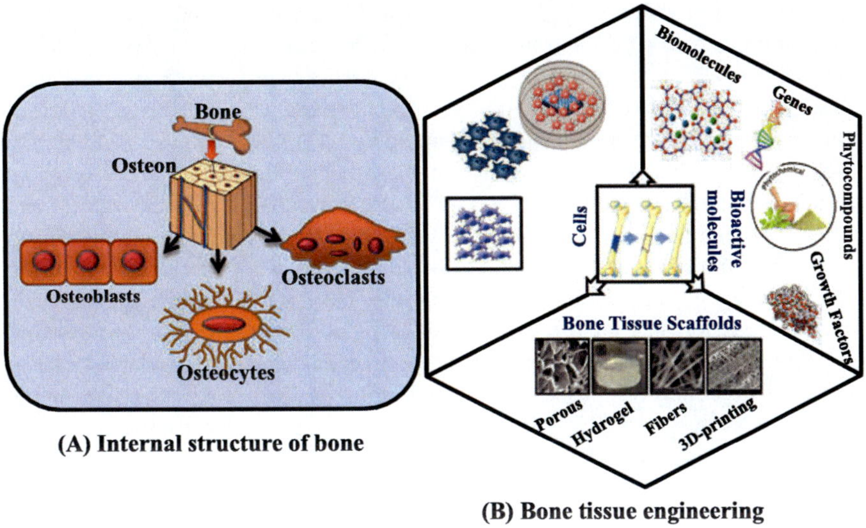

Fig. 1 A schematic diagram of bone remodeling, bone defect, and bone tissue engineering. (**a**) Bone undergoes remodeling by osteoblasts and osteoclasts. (**b**) A defect in the bone can get repaired by bone tissue engineering utilizing a combination of cells, bioactive compounds, and biomaterials. Scaffolds are prepared using different biomaterials with the help of various fabrication techniques

has the potential to support regeneration, it is a slow and prolonged process that has consequences of hematoma formation and inflammation. To reduce the risks and improve bone regeneration faster, BTE plays a significant role in healing and regenerative medicine [4].

BTE is a field that is comprised of three major parts, namely (1) GFs, (2) scaffold material, and (3) cultured cells, and they can effectively combine to repair a bone defect (Fig. 1). The primary goal of BTE is to regenerate an ideal bone that could efficiently integrate and function with the neighboring tissues as a native bone. This leads to an alternative treatment for bone defects clinically with a decline in obstacles such as immune rejection, transfer of pathogens, and limited availability [5]. To accomplish the objective of BTE, grafting materials employed for scaffold fabrication should possess some important properties like biocompatibility in the host environment that mimics native bone, effectively deliver bioactive substances to the injured site, and provide an excellent homing site that can support cell adhesion and differentiation [5, 6]. Delivering bioactive molecules such as GFs, nucleic acids, and phytocompounds resulted in tissue regeneration acceleration at the defective region (Fig. 1). The grafting materials are natural, synthetic polymers and ceramics employed to design an optimal scaffold that can successfully deliver a drug or a molecule to the targeted site [6, 7]. The natural polymers such as collagen (Col), CS, gelatin (Gel), and alginate (Alg) are widely used, and these polymers showed improved biocompatibility and biodegradability. Synthetic polymers like polyglycolic acid (PGA), polylactic acid (PLA), poly ε-caprolactone (PCL), and poly (lactic-co-glycolide) (PLGA) showed suitable mechanical property, and their degradation rate can be controlled. Similarly, bioceramics like HAp, tricalcium phosphate (TCP) exhibited chemical characteristics similar to bone, including enhanced osteoinductive effect, high mechanical strength, and improved degradation rate [7]. There are many limitations posed by a single polymer scaffold, and to improve its characteristics, copolymeric composites are fabricated. These composites showed enormous mechanical strength and cytocompatibility improvement and were very much similar to the natural feature of a bone extracellular matrix (ECM) [7, 8]. The combinations of different polymers proved as a potential candidate for BTE in therapeutic applications. To successfully fabricate a biomimetic composite, several parameters like biodegradation, porosity, mechanical strength, and biocompatibility are considered [9]. The fabrication methods commonly available to fabricate a 3D-structured composite are solvent casting, lyophilization, electrospinning, particulate leaching, and 3D-printing [9, 10]. These fabricating techniques satisfy mostly all the required parameters to construct an ideal scaffold. These methods have a very promising scope for scaffold synthesis for BTE (Fig. 1).

2 Bioactive Molecules in Bone Tissue Engineering

Bioactive molecules are a class of substances that can affect or alter the state of living tissue. For example, bioactive molecules such as GFs, hormones, and phytocompounds play a significant role in regulating cellular processes through various biochemical pathways [11, 12]. Hence, employing these bioactive molecules for enhancing bone regeneration has become a recent trend in BTE.

2.1 Growth Factors

GFs are organic polypeptides that can elicit the target cells involving in various metabolic activities. GFs act as signaling molecules that can bind to specific transmembrane receptors on a cell and initiate a cascade of biochemical events, which drastically affect the target cell's fate [13]. Bone regeneration is a complex mechanism, and its execution inevitably requires the spatiotemporal presence of certain GFs. Bone morphogenetic proteins (BMPs), fibroblast growth factors (FGFs), vascular endothelial growth factor (VEGF), and interleukins (ILs) are the main classes of GFs that are predominantly involved in bone formation. Numerous studies have been conducted by administering these GFs to the cells in simulated environments to assess bone regeneration effects [14]. BMPs are the key GFs involved in the initiation of bone formation by promoting the differentiation of mesenchymal stem cells (MSCs) towards osteogenic lineage. BMPs are widely used in BTE, and BMP-2 and BMP-7 have been clinically used to treat bone defects [15]. Angiogenesis is an essential step in bone formation as it supports the tissues in the vicinity by providing nutrients. GFs like FGF and VEGF, the well-known regulators of angiogenesis, have been utilized to engineer bone tissues with promising results [16, 17]. Inflammation being the first step in the bone healing process, inflammatory cytokines such as IL-1, IL-6, tumor necrosis factor-α (TNF-α), and macrophage colony-stimulating factor (MCSF) have also been used to elevate bone regeneration [18]. Thus, utilizing these bioactive GFs in BTE turns out to be a great way to treat bone defects.

2.2 *Endocrine Hormones*

Endocrine hormones are signaling complexes that can target and activate distant cells or tissues. They are secreted by specialized glands and reach the target cells via the circulatory system. Hormones are majorly involved in regulating cells' physiology, such as maturation, differentiation, and development. Parathyroid hormone (PTH) is an essential factor in the bone remodeling process as it plays a crucial role in bone formation and resorption mechanisms by regulating the action of osteoclasts. It is the only drug approved by the FDA to treat osteoporosis [19]. PTH is known for its regulatory activities in calcium homeostasis, and it indirectly induces bone regeneration. Systemic delivery of PTH for treating bone defects showed increased mineralization and bone regeneration [20]. PTH has also been delivered locally to the bone defect site using biocompatible scaffolds, which showed a dosage-dependent increase in bone formation [21]. Calcitonin is another endocrine secretion responsible for regulating the calcium and phosphate levels in the blood. Employing calcitonin to treat in vivo models with simulated osteopenia showed improved bone regeneration [22]. Calcitonin can inhibit osteoclast activity and prevents osteoporosis, hence proving its potential application in BTE [23]. Calcitriol, also known as bioactive vitamin D, is another hormone that can regulate bone formation. It up-regulated bone-specific gene markers such as alkaline phosphatase (ALP), osteocalcin (OCN), type 1 collagen (COL1), and osteopontin (OPN) [24, 25]. Hence, such endocrine hormones can be utilized as regeneration eliciting drugs to enhance bone formation.

2.3 *Other Bioactive Drugs*

Bioactive molecules obtained from plants, commonly known as phytochemicals, are exclusively being used in BTE. Phenolic compounds such as Sinapic acid (SA) [25], Veratric acid [26], Zingerone [27], and Anethole [28] have been reported to possess an increased potential to promote the differentiation of MSCs towards osteogenic lineage. Flavonoids of plant sources such as Diosmin [29], Chrysin [30], Valproic acid [31], and terpenoids like Phytol [32] have also been proven to be a promising drug for BTE. To avoid unwanted problems like inflammatory responses and bacterial infiltration while using an osteogenic scaffold, a wide range of anti-inflammatory drugs and antibiotics are also being used as deliverable payloads in BTE [33]. Nucleic acid delivery is another aspect of mitigating bone defects, which involves delivering DNA and mRNA coding for GFs such as BMP, FGF, VEGF, and IGF. Delivering non-coding genes like miRNAs and siRNAs are also being studied in BTE [34]. Apart from using these extrinsic bioactive molecules, indigenous bone-specific minerals such as HAp and calcium phosphate, which have been proven to induce osteoblast differentiation, are used to construct bone tissues. Using these natural minerals reduces the risk of eliciting immune responses and compatibility issues [35].

3 Chitosan

CS is a cationic polysaccharide that has excellent biocompatibility, biodegradability, and mucoadhesive properties [36]. Chitin (CT), a long chain nitrogenous glucose derivative, is the primary source of CS. CT is found in the exoskeleton of some insects, fungi and is predominantly present in the shells of crustaceans such as lobsters, crabs, krill, etc. CT is considered one of the most abundant polysaccharides available in nature [37, 38]. Relative to the source, CT exists in three allomorphs, namely α-CT, β-CT, and γ-CT, which differ in their polymer chain orientation. α-CT is the most commonly occurring form [40]. CS is a copolymer of D-glucosamine and N-acetyl-D-glucosamine residues, and the properties of CS vary relative to the configuration of these residues. Due to the presence of amino groups in its backbone, CS has a net positive surface charge, enabling the ability to coat them with biocompatible anionic compounds and serve as an efficient drug delivery system (DDS) [36, 40]. The process through which CS is extracted from CT is called deacetylation. By manipulating the degree of deacetylation, CS with varying molecular weights and drastically different properties can be obtained, which can be utilized for a wide range of applications. Studies showed that CS has inherent antimicrobial properties and the ability to promote osteoblast differentiation in MSCs, thus making CS a much sought out material for tissue engineering and wound healing applications [40–42]. CS has a malleable architecture. Its morphology can be manipulated through 3D-printing, electrospinning, lyophilization, solvent casting, etc., which pave the way to be used synergistically with other biomaterials as composites [43]. Due to its ample availability and exemplary properties, CS is the most sought out biopolymer in the field of regenerative medicine and drug delivery applications.

3.1 Sources

Although insects and some fungi naturally produce CS, there are only a limited number of sources for pristine CS. Hence CT, the substantially available natural polysaccharide, is utilized for CS extraction. CT is widely distributed in the exoskeleton of marine crustaceans. Since the marine environment makes up 90% of the earth's biosphere, many commercial CS is sourced from marine ecology [39]. Terrestrial arthropods, the most diverse and successful forms of life, become the second-largest CS source, followed by broadly available microbial diversity. CT is the most prevalent compound in marine arthropods' exoskeleton, namely shrimps, crabs, lobsters, krill, etc., making them the significant contributors of CS. CS isolated from prawn species such as *Fenneropenaeus indicus, Penaeus monodon, Penaeus indicus, and Fusarium sp.* has a great potential to be used in the pharmaceutical industry, wastewater treatment, and for food packaging [44]. CS from these sources also possesses excellent antimicrobial properties, making it a suitable applicant to be employed in drug delivery purposes [45]. Apart from yielding good CS quality, the marine ecosystem provides us with such CS in an enormous quantity, hence serving as an efficient source of CS [46, 47].

CT forms a major part of terrestrial arthropods' structural components, which act as a scaffold and provides physical structure and integrity. These insects have their biosynthetic pathway for CT production, making them a reliable source for CS extraction [48]. The wings of a roach species *Periplaneta americana* serve as an excellent source of α-CT, a similar type of CT commonly found in crustaceans' shells. Due to the ease of culturing and maintaining these organisms, this could be considered a potential source for CS production [49]. Capturing invasive arthropods like Grasshoppers, beetles, and bumblebees is an environmentally friendly and effective CS extraction method [50]. *Musca domestica*, also known as a housefly, has been proven to be an excellent source of fats, vitamins, minerals, and proteins. CS obtained from this species showed increased antibacterial, antioxidant, hypolipidemic properties, and good chelating abilities, enabling them to be of good use in food processing and biomedical industries [51, 52].

CT and CS are widely distributed in microbes like fungi, yeast, chrysophyte algae, and even bacteria like streptomycetes and prosthecate bacteria. Among these, the contribution of fungi to CT production is significant. CT is present in various fungi like Ascomycetes, Zygomycetes, Deuteromycetes, and Basidiomycetes and is an important component of cell walls and septa of fungi, which strengthens them and provides rigidity. Fungi are also involved in the biosynthesis of CT mediated by CT synthases, and α-CT is the most commonly found form in fungal cells [53]. Fungi are used in various industrial applications like baking, brewing, antibiotics production, enzyme production, etc. The industries excrete a large amount of fungal biomass, which can be a viable CS extraction source [54]. Fungus such as *Mucor roxii* has been used to obtain CT and CS, which showed high yield % [55, 56]. CS obtained from a fungal species *Aspergillus flavus* has been seen to increase certain antibiotics' antimicrobial activities. This stated the ability of fungal CS to be used as synergistic adjuvants with antibiotics for treatment against bacterial diseases [57]. The ability of fungal cultures to cherish and survive on simple substrates becomes an advantage for large-scale cultivation. *Cunninghamella bertholletiae*, a fungal species, could thrive only on sugarcane extract as the carbon source and produced industrial-grade CS from the substrate [58]. Fungi are also known to utilize food processing industrial wastes such as steep corn liquor and apple pomace as substrates and produce quality metabolites [59].

3.2 Extraction

The first step in CS extraction is the isolation of CT. As CT is found in a wide array of natural compounds, it is often associated with impurities such as minerals, proteins, and even coloring pigments in some cases. The CT derivation procedure consists of two main processes, namely demineralization and deproteination. Demineralization is the removal of unwanted minerals, and deproteination is the process of removing undesirable protein residues. After a successful retrieval of CT, CS can be synthesized by the deacetylation process. The most common method used by large-scale industries for extracting CS is chemical extraction, which uses synthetic chemicals to remove proteins and minerals [60, 61]. The chemical method of demineralization involves treating the raw material with acids, which mainly focuses on removing deposited minerals like calcium carbonate and calcium phosphate. To remove the unwanted proteins, the deproteination process involving alkali treatment is undergone. For this process, NaOH is preferred as it has efficient protein eliminating capabilities. If there are unwanted pigments associated with the raw material, an additional step of decoloration is required, and this makes use of organic solvents like acetone [62]. After these processes, the resulting CT is subjected to alkaline deacetylation. The alkaline treatment hydrolyzes the acetyl groups and exposes the NH2 groups, which lead to CS formation [63]. Biological extraction is another approach for extracting CS by utilizing microbes and their metabolites. For demineralization, lactic acid of the bacterial source is being used. Lactic acid can react with minerals like calcium carbonate to form calcium lactate, further washed off. The biological way of deproteination employs proteases derived from bacteria to remove proteins. For the deacetylation process, CT deacetylase enzymes obtained from fungi and bacteria are used. An advantage of the biological extraction procedure is that it is environmentally safe, and the protein and mineral excreta of these processes can be used as human and animal nutrients [64].

3.3 Structure and Physicochemical Properties

CS is a semi-crystalline linear polysaccharide composed of repeating monomers of β (1–4)-linked 2-amino-2-deoxy-D-glucose (D-glucosamine) and 2-acetamido-2-deoxy-D-glucose (N-acetyl-D-glucosamine). It is generally colorless and odorless and has a structure identical to cellulose except for the amine group (NH2) in its C-2 position rather than a hydroxyl group (OH). The presence of NH2 in CS makes it positively charged, allowing it to bind with negatively charged macromolecules like lipids, proteins, metal ions, etc. [65]. Due to both NH2 and OH groups' presence, CS can form strong covalent bonds through esterification, reductive amination, or etherification reactions. The properties of CS majorly depend on the proportion of D-glucosamine and N-acetyl-D-glucosamine present in them and their degree of deacetylation and molecular weight [66]. The degree of deacetylation is an essential property of CS that determines the percentage of NH2 groups present. The presence of exposed NH2 groups makes CS highly soluble in acidic solutions. Since CS is a weak base, it is readily soluble in acidic media pH of less than 6. CS solubilizes through protonation of the NH2 groups in the C-2 position, and it is converted into a polycation in aqueous acids, which enables them to form complexes with various anionic compounds. This property makes CS a useful material for drug delivery and tissue engineering applications [67]. Molecular weight is another eminent parameter that strongly influences other characteristics of the polymer, such as viscosity. CS of larger molecular mass shows high viscosity and mucoadhesion levels, thus making them an efficient drug carrier. The significant attributes of CS, such as degree of deacetylation and molecular weight, are highly manipulatable during the fabrication process. Hence, CS having different properties can be obtained and used in a broad range of applications [66, 67].

3.4 *Medicinal and Pharmaceutical Properties*

CS is one of the few polymers that are known for its exceptional biocompatibility. The term biocompatibility defines a substance's ability to integrate with the living cells and function cordially without provoking an unwanted response or having toxic effects on the cells [68]. As CS is a naturally occurring polymer, it is highly biocompatible with living tissues and has minimal toxicity. Nearly all the works done on CS boasts about its exclusive biocompatible nature. This makes CS a suitable biopolymer for a wide range of pharmaceutical applications and therapeutics. It is redundant that an implantable biomaterial should have a controlled and stable degradation mechanism for drug delivery. If the material fails to degrade, it will either become cytotoxic or lead to unsuccessful payload delivery. CS is susceptible to numerous indigenous enzymes readily available in a cellular ambiance [69]. CS is mainly degraded in vertebrates by lysozymes and bacterial enzymes present in the colon, making it an ideal ingestible material. CS is also degraded by various human chitinases such as acidic mammalian chitinase, chitotriosidase, and di-N-acetylchitobiase enabling it to be applied in various medicinal applications [70]. Another unique property of CS is its enhanced mucoadhesiveness. Mucoadhesion is the ability of a material to effectively adhere to the mucous membrane, which armors the vulnerable areas of the respiratory, gastrointestinal, and reproductive tract [71]. Its cationic nature owes this exquisite property of CS. This property enables it to establish an electrostatic force of attraction with the anionic mucosa, thus establishing strong adhesion. Therefore, employing a CS-based drug delivery system encourages drug efficient targeting and increases the target tissues' drug intake [72]. One of the practical applications of CS is its role in wastewater treatment. Heavy metal ions being the primary contaminant of water pose an imminent hazard to the life forms. Due to its cationic property, CS can effectively bind to metal contaminants such as molybdenum, cadmium, copper, arsenic, zinc, etc. [73, 74]. CS has been used to fabricate numerous materials like biofilms, beads, membranes, and flocculation systems that have been utilized to remove water contaminants [75]. Although CS is said to be non-toxic, it has an antagonistic effect on a wide range of microbes, making it a useful material for wound dressing. CS is also known to possess wound healing capabilities by inhibiting hemorrhage and inducing fibroblasts [76]. Composites of CS and various polymers and bioactive drugs have been utilized to enhance wound healing [77]. As CS possesses a broad array of biological properties, it is the widely preferred biopolymer in tissue engineering [78]. CS is a versatile material that has been used to fabricate a variety of bio-mimicking scaffolds and has been used for regenerative medicine by DDS [25, 79, 80].

4 Drug Delivery Systems

A drug is any chemical or biological substance that exhibits pharmacological activity, useful in treating diseases [81]. The drug release pattern can have a significant impact on the therapeutic efficiency of the drug upon administration. Hence it is essential to design and construct suitable drug delivery models to achieve the maximum therapeutic effect. Historical perspective on drug delivery provides us two crucial drug release systems, namely sustained release systems and controlled release systems. Sustained release systems provide spontaneous release of drugs upon administration and a gradual release over an extended period. Classic examples of sustained-release systems include commonly concocted suspensions, emulsions, compressed tablets, etc. Improvements to sustained release systems have yielded the novel controlled release systems that have spawned generations of highly efficient drug delivery models operating with increased efficacy. Controlled release systems release drugs in a predetermined manner, maintaining the loaded drug's bioavailability within the therapeutic range over a prolonged time [82]. This steady release of drug is achieved by the careful selection of polymers and suitable additive agents that aid in the tedious task of maintaining stable drug release.

Different mechanisms dictate drug release mode from the delivery systems, they are diffusion, chemical regulation, and solvent controlled [83]. A consortium of physical chemistry, polymeric sciences, and nanoengineering has led to the inception of novel drug carriers operating at both the micro and nanoscale, employing the mechanisms mentioned above in performing the designated function. These DDS have been instrumental in shaping the course of pharmacological sciences over the past five decades. A plethora of DDS exists, each with its limitations. The ingenuity of these DDS is tested when exposed to the complex physiology of the human body. This conglomerate of multiple systems working together in synchrony throws various obstacles in these miracle workers' path, which hinders their full potential.

The reticuloendothelial system (RES) is one such obstacle that needs to be conquered as a prerequisite for DDS. It protects our body from invading pathogens by facilitating their immediate digestion. However, the inability of RES to recognize therapeutic drug carriers as harmless structures causes it to be actively involved in clearing these pharmacological reinforcements taken up by the patient to treat the ailment concerned [84, 85]. This calls for engineering sophisticated DDS that could evade physiological barriers and provide maximum therapeutic effect at the tissue concerned.

Colloidal carriers appear to be one of the up-and-coming classes of drug carriers exhibiting the much-needed property. We usually expand our classification by colloidal carriers to include nanoparticles, liposomes, microspheres, and polymeric micelles [84]. Recent studies suggested constructing these drug carriers with hydrophilic polysaccharides. This can provide a steric barrier that effectively prevents opsonization, thereby evading phagocytic capture [85]. Nature, ever being generous, has given rise to numerous polysaccharides ready to fulfill this role. This has resulted in a peaked interest in polysaccharide-based colloidal drug carriers in terms of nanomedicine. The delivery of drugs using nanotechnology offers maintaining the drugs' bioavailability, thus enhancing therapeutic efficiency. Some of these nanoscale materials include carbon nanotubes (CNTs), nanosheets, nano horns (NHs), nanogels (NGs), nanoplexes, and nanofibers. Another radical approach of the twenty-first century is "biology-inspired engineering." This idea has given rise to two exemplary DDS, namely exosomes and dendrimers. These two systems have been engineered based on our understanding of the cellular trafficking mechanism that maintains multicellular organisms' integrity. These systems have the added advantage of resembling familiar cellular carriers and increasing the chance of being neglected by the intensely scrutinizing RES [85, 86]. The following section will provide newfound insight into some of the most intriguing DDS that has the potential to change the face of current therapeutics.

4.1 Colloidal Carriers

Liposomes are bilayer lipoidal vesicles capable of acting as carriers, ensuring cytosolic delivery of both hydrophobic and hydrophilic drugs [86]. Their chemical similarity to the dynamic mammalian cell membrane allows it to pass through the membrane achieving drug release intracellularly. They have a delivery spectrum unrivaled by any conventional DDS, making it extremely popular in clinical nanomedicine. They are produced by dispersing lipid monomers in an aqueous environment exploiting their innate amphiphilicity. A layer of such impulsively assembled phospholipids is called lamellae, and liposomes are classified by the number of lamellae each of them comprises in their molecular architecture and their size. Considering the two deterministic features of liposomes, they can be classified into unilamellar vesicles and multilamellar vesicles. The presence of multiple lamellae ensures compartmentalization of the drug carrier, facilitating multiple payloads of different polarities. Based on drug loading, liposome synthesis techniques can be classified into active loading and passive loading techniques. In passive loading (for example, solvent dispersion), the payload is encapsulated during the meticulously manipulated self-assembly of the lipid bilayers into functional liposomes, whereas active loading is achieved after the formulation of liposomes, brought on by appropriate manufacturing practices [87]. Liposomes exhibit the famous enhanced permeability and retention effect making them invaluable in cancer therapy [88].

Polymeric microspheres, a collection of monomers networked together, forming a casket surrounding the drug, have tremendous potential applications in the field of drug delivery. The microspheres possess excellent biocompatibility, large surface area, high stability, low immunogenicity, and drug's bioavailability. Microspheres have two distinct subdivisions, namely microcapsules and micromatrices. In microcapsules, the drug is located at the central core of the enclosing polymeric capsule with release triggered either by dissolution or diffusion. In contrast, in micromatrices, the drug is elegantly spread over its intrinsic architecture [89]. Various methods for the synthesis of drug encapsulating microspheres have been developed, taking advantage of the chemistry behind polymers and the finished product's microgeometry. Some of them include interfacial polymerization, ionic gelation, spray drying, and coacervation methods [90–92]. Recent trends in polymeric chemistry have given rise to the trending approach of employing emulsions in the production of biodegradable microspheres scaling up the production of these prominent drug carriers [93].

Nanoparticles are nanoengineered drug trafficking entities that retain their loaded drug in their core region, protecting it from the outside environment. Many types of nanoparticles are available mainly due to the collective knowledge obtained through extensive research in polymeric chemistry and nanotechnology. Some classical features of nanoparticles are their tiny size, high surface to volume ratio, relatively stable molecular arrangement, and their ability to be endocytosed upon interacting with the cell membrane. Their small size is extremely useful in avoiding the hypervigilant RES of the human physiology. A few popular types of nanoparticles are metallic nanoparticles, polymeric nanoparticles, solid lipid nanoparticles, and protein nanoparticles [94].

Iron oxide nanoparticles are typical metal oxide nanoparticles that have found a niche in biomedicine over recent decades. They are naturally magnetic since iron is the most popular ferromagnetic material globally, making it highly susceptible to the magnetic field. This property is beneficial in externally guiding the administered nanoparticles toward the anatomical locale of the target tissue [95]. Other examples of metallic nanoparticles include gold and silver nanoparticles, each embodying their elemental precursors' essential aspects. Polymeric nanoparticles have been wildly successful in bringing nanomedicine towards a new frontier. The similar characteristics shared by polymeric nanoparticles are their small size (100–400 nm), the ability to encapsulate a drug and hinder its degradation, and the capacity to be endocytosed [96]. The most common method of synthesizing polymeric nanoparticles happens to be ionic gelation. Ionic gelation occurs due to the difference in the polymers and their crosslinkers, resulting in fully functional drug-loaded nanoparticles. Many polymers have been fabricated as nanoparticles, including CS, Alg, PLGA, etc. [97–99]. Hybridized copolymeric nanoparticles display increased effectiveness as drug carriers, compared to homogenous polymeric nanoparticles [100]. Despite the many successes polymeric nanoparticles have met within recent years, large-scale production is still a prerequisite for such novel drug carriers, and research is being done to commercialize clinically accomplished nanoparticles [101]. Solid lipid nanoparticles are nanoscale lipid-based nanocarriers chemically composed of lipids, surfactants, and the drug to be delivered [102]. Their unique chemical morphology paves the way for better stability in comparison with other colloidal carriers. The inheritance of certain defining characteristics of nanoparticles without their corresponding limitations makes them a highly coveted hybrid DDS, exhibiting enhanced drug release in the wake of biological systems. Proteins have often been used to functionalize multiple DDS improving their bio-functionality. Interestingly enough, proteins themselves can act as excellent carriers of molecules inside the body and can be exploited when engineered appropriately. Examples of the proteins fabricated as nanoparticles include albumin, casein, gliadin, and Gel. The choice of proteins as nanocarriers is encouraged by their innate biocompatibility from their close interaction with biological systems. The macromolecular 3D-structure of proteins allows for the existence of small crevices filled by the loaded drug. Proteins are highly susceptible to catalytic and hydrolytic degradation in vivo upon administration, leading to ineffective drug release. This predicament can be overcome by suitable chemical modifications that can significantly increase the loaded drug's bioavailability restoring the utility of protein-based nanoparticles [103].

Micelle is usually associated with the superstructures, and free-floating surfactant unimers aggregate to form a solution [104]. The self-assembly results from the hydrophobic portions of the surfactant monomer aggregating together. Van der Waals forces stabilize the hydrophobic core while hydrogen bonds stabilize the outer layer [105]. Polymeric micelles are created by the same micellization process that gives rise to the surfactant micelles, involving amphiphilic copolymeric monomers. Their similarity in structure has helped researchers to utilize this novel nanostructure in the pharmacological, paint, and cosmetic industries [106]. Polymeric micelles are macromolecular structures existing at the nanoscale level with a complex geometry comprising a hydrophilic exterior protecting a hydrophobic core [107]. The chemically synthesized micelles are highly vulnerable to destabilization owing to their dynamic nature. Hence, traditional biopolymers are often associated with chemically inert substances in fabricating polymeric micelles. This results in copolymeric micelles that achieve enough stability to ensure proper drug uptake by the target cells. This approach has met with success in delivering a wide variety of bioactive molecules, including docetaxel, genexol, gemcitabine, etc. [108].

4.2 Nanomaterials for Drug Modulated Tissue Repair

Nanofibers are organized structures made up of a collection of polymeric tendrils, capable of supporting new tissue formation [109]. They are an important class of nanostructures useful in a wide array of biomedical applications, including drug delivery, tissue engineering, and wound healing. Their intrinsic architecture promotes cell adhesion, proliferation, and even differentiation when loaded with individual inductive molecules [110]. This makes nanofibers an excellent platform for drug modulated tissue repair. Besides enhancing tissue regeneration, nanofibers entrapped with bioactive molecules can serve as a repository for therapeutics upon implantation. The nanofibrous subunits' chemistry can be manipulated to adjust the degradation rate, maintaining a steady release of the drug in vivo. Many techniques have been developed for the synthesis of nanofibers. Some of the more popular methods include molecular self-assembly, electrospinning, and thermally induced phase separation [111]. Electrospinning is a versatile technique capable of producing highly uniform and stable nanostructures. Previous literature has shown that a wide range of polymers, including Gel, PCL, PLLA, PLA, CS, have been fabricated as nanofibers with this method [26, 78, 112]. Tissue repair depends on the bioactive nature of various biomolecules, including GFs, hormones, and phytocompounds. Apart from a steady release of the therapeutics mentioned above, a biomimetic environment appears to be a prerequisite for the influenced cells to flourish, yielding a functional tissue. These nanofibers are extremely useful in encapsulating target drugs and providing such an artificial environment accomplishing drug modulated tissue repair. Various mechanisms have been proposed for drug entrapment using nanofibers. The simplest of all appears to be physical adsorption encouraged by the nanofibers' large porous surface, discouraged by its increased chances of burst release. The encapsulating drugs in accomplished nanocarriers and entrapping them inside the nanofibrous matrix appear to alleviate the threat of burst release to a vast extent [113]. Thus, nanofibers could play a critical role in integrating drug delivery and tissue engineering [114].

4.3 Bio-inspired Nanocarriers

Dendrimers are nanoscale, hyperbranched polymeric constructs capable of serving as carriers of therapeutic biomolecules [114, 115]. They are highly biocompatible, non-toxic, and water-soluble, making them more appealing in the trending age of personalized medicine [116]. The dendrimer structure typically has three parts: a central core, multiple layers made up of repetitive chemical units referred to as generations, and peripherally located functional groups, crucial in determining the bio-functionality of the designed dendrimer. Drugs can be covalently linked to the dendrimer by their externally located chemical moieties. The mechanism of dendrimer mediated drug delivery is said to be the breakage of linking bonds between the payload and the dendrimer by the action of enzymes in vivo [117]. Another proposed mechanism is an alteration in dendrimers' molecular configuration in response to physicochemical parameters like pH, temperature, etc. [118]. Commercially available dendrimers are classified, based on their molecular constituents into poly amidoamine dendrimers, polyester dendrimers, peptide dendrimers, polyacetyl dendrimers, glycodendrimers, and hybrid dendrimers [119]. Dendrimer synthesis has been improved over time owing to the combined efforts of both polymeric and molecular chemistry. Some of the classical methods of synthesizing dendrimers are divergent synthesis and convergent synthesis [120]. The dendrimers obtained by these methods have been successfully employed in the ocular, oral, and transdermal delivery of drugs [121]. The clinical efficiency and scalability of dendrimers encourage vigorous research into these branched nanocarriers [122].

Extracellular vesicles (EVs) play an essential role in intracellular signaling and communication between cells [123]. The three distinct types of EVs are microvesicles, apoptotic bodies, and exosomes [124]. The exosomes are the products of the invagination of the multivesicular bodies, releasing intraluminal vesicles in the extracellular environment. The circumstances involved in its inception bring a certain sense of familiarity the recipient cells can relate to, making exosomes a handy personalized drug carrier. The common biomolecules decorating exosomal architecture include lipids, cholesterol, integrins, heat shock proteins, and tetraspanins [125]. Integrins are important in ensuring that the exosome adheres to the recipient cell membrane delivering its contents. In vivo studies employing exosomes to deliver anti-inflammatory and anti-cancer drugs have established the successful uptake of the drugs by target cells endorsing the credibility of these biological nanocarriers [126, 127]. The smaller sizes, non-immunogenicity, and inherent specificity make exosomes a physiologically abundant nanocarrier.

4.4 Miscellaneous Nanocarriers

CNTs are versatile nanomaterials that consist of a network of graphite entities in a cylinder configuration [128]. These nanomaterials have been studied extensively for their possible utility in the ever-growing field of drug delivery. The tube-like architecture accounts for its increased drug loading capacity, and efficient delivery of the payload is attributed to its needle-like construction. Most of the drugs are loaded into CNTs by physical adsorption or covalent bonds formed between functional groups of CNTs and the drug [129]. CNTs have been known to traverse through the lipid bilayer by the endocytosis pathway. Although CNTs, along with nanoparticles and liposomes, exhibit the enhanced permeability and retention effect, their specificity towards cancer cells can be significantly improved by conjugating with targeting moieties that enhance their innate bio-functionality. PEG, CS, peptides, and transferrins are examples of such biofunctional conjugating molecules [130]. These diffusion-controlled nanocarriers are also known to carry several therapeutic molecules, including chemotherapeutic agents, antibodies, antioxidants, and siRNAs [131, 132].

Recent advances in nanotechnology have given rise to numerous nanoscale materials, which could be potential DDS. One such novel drug carrier conceived by nanoscience happens to be nanosheets. Nanosheets are a class of nanomaterials characterized by their 2D-structure, with the entire entity's existence confined to the nanometer range [133]. The characteristic large surface to volume ratio of nanosheets is accompanied to a certain extent by a slew of exemplary physicochemical properties that could prove useful in applied biomedicine. Their large surface area is beneficial in sustaining the loaded drug by the nanosheets-mediated pi–pi interactions. Nanosheets have been classified into various classes, including metallic nanosheets (e.g., molybdenum disulfide), non-metallic nanosheets (e.g., black phosphorous), polymeric nanosheets (e.g., PLLA), and monolayered hydroxide nanosheets [134–136]. Nanosheets with appropriate functionalization have proven useful in the controlled delivery of potent therapeutics, including anti-cancer drugs, anti-inflammatory drugs, nucleic acids, etc. [133].

NGs are compact nanoscopic hydrogels versions embodying certain aspects of both a typical hydrogel and a nanoparticle [137]. The synergy of two traditional DDSs sets NGs apart as a hybridized controlled drug release system of biochemical mediators. These hydrogel derivatives are entitled to the hydrogels' classic fluid retention capacity, making them excellent carriers capable of accommodating multiple bioactive agents, attributed to the presence of hydrophilic groups [138]. These groups can significantly impact drug loading and release characteristics by forming covalent bonds with the payload. Additionally, they help achieve functionalization with homing molecules, including antibodies and receptor-specific ligands for targeted release. CS, dextran, pullulan, cellulose, etc. are some of the polymers commonly fabricated as NGs. The polymeric makeup of NGs ensures that these 3D-nanocarriers can be subjected to crosslinking, increasing the stability of drug-loaded NGs accompanied by physical entrapment of drugs [139]. Unlike other DDS, all three release mechanisms come into play when NGs interact with our body's aqueous environment. Their extremely small sizes and their hydrophilic chemistry serve them well in evading RES and avoiding unwanted immune reactions. NGs synthesized from established biopolymers have innate biocompatibility and biodegradability in releasing drugs safely and stably. Polymers can give rise to NGs by two methods: polymerization of individual monomers and size reduction of hydrogels. NGs can be administered orally, nasally, and other traditional routes of administration [140].

Nanoplexes are the resultant complexes of a drug interacting with an oppositely charged macromolecule in an aqueous environment. These nanoplexes are effortlessly synthesized nanocarriers, manufactured by electrostatic interaction directed self-assembly [141]. They are similar in size to conventional nanoparticles and more receptive to drug loading as the drug itself is heavily invested in nanoplex synthesis. Nanoplexes exhibit enhanced stability, exponentially increasing their shelf life. Despite their impressive drug loading capacity, the threat of burst release discourages the formulation of nanoplexes with potent therapeutics such as chemotherapeutic drugs [142]. This susceptibility towards burst release can be rectified by suitable crosslinkers achieving a controlled release of notable therapeutics [143]. NHs are a novel class of carbon-based nanoplatforms, distinguished by their conically terminating structures, derived by laser ablation of graphite [144]. The standard properties of NHs useful in the science of drug delivery include a large surface area facilitating selective adsorption of chemical compounds, a nano reservoir for drug entrapment, a desirable exertion of the enhanced permeation and retention effect, and their innate non-toxic nature. The adsorption is particularly useful in stably entrapping drugs, with release achieved by applying NIR (near-infrared radiation) [145, 146]. Recent years have witnessed a call for smart DDS, capable of deciding in vivo drug release in response to a predefined set of appropriate biochemical (pH, ATP) or biophysical stimuli (NIR, temperature). As mentioned above, of these DDS, polymeric micelles, NGs, nanosheets, and NHs appear to be worthy candidates for the futuristic "smart delivery systems." These nanocarriers displaying various desirable features show tremendous potential in the future of nanomedicine.

5 Chitosan Formulation in Drug Delivery Applications

CS is a natural alkaline polysaccharide that is non-toxic, biocompatible, and biodegradable [147]. Some peculiar properties that make CS the best option for delivery systems are (1) protects the drug molecules from the gastric acidic environment, (2) enhances tissue penetration by breaching tight epithelial junction, (3) facilitates transcellular and paracellular drug transport, (4) improves absorption of drugs, and (5) conjugates with anionic macromolecules via electrostatic attraction [148, 149]. CS formulation is very much necessary because it becomes hydrated in aqueous acidic surroundings, and gelation occurs. Due to gel formation, there is no sustained release of drugs at the target; hence, CS should be formulated to support and facilitate sustained release of macromolecules or drug at the targeted tissue [150]. CS has been widely formulated as various drug delivery structures like nanoparticles, nanofibers, hydrogels, microneedle, and microspheres (Fig. 2). Depending on the formulated CS structure, the drugs are adsorbed, chemically attached, or coated [151, 152]. Since there are various types of formulated CS used in drug delivery, it is necessary to explore the methods used in CS formulation. Several fabrication methods are available to formulate a specific structure that can effectively deliver a drug to the targeted tissue [153].

5.1 Chitosan Nanoparticles

Nanoparticles are a sub-microscopic colloidal structure of 1–100 nm in size [154]. Due to its sub-microscopic size, it exhibits enhanced targeting and tissue

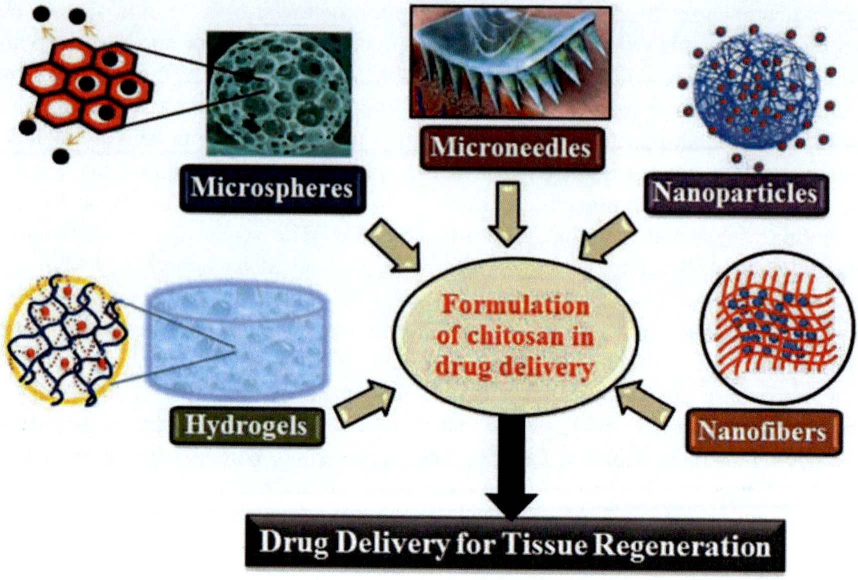

Fig. 2 Illustration of chitosan formulations for drug delivery in tissue engineering

penetration. CS possesses non-toxic and mucoadhesive properties, making CS an ideal nanoparticle formulating material for drug delivery [155]. As an agent of drug delivery, the nanoparticles should be encapsulated with the compound of interest. The properties of drugs, such as their molecular weight, short half-life, low melting point, and affinity for both lipophilic and hydrophilic phases, should be considered before their encapsulation into nanoparticles. The two commonly used nanoparticle fabrication methods are ionotropic gelation and micro emulsification [156].

The ionotropic gelation method in CS nanoparticle synthesis is done by using an anionic crosslinker like tripolyphosphate. This technique applies the electrostatic attraction between the negatively charged tripolyphosphate and the positive charge of the NH2 group in CS. The CS solution is wholly blended with anionic tripolyphosphate solution by a steady magnetic stirring in this process. The resulting suspension exhibits milky white like appearance as an indication of nanoparticle formation [157]. The concentrations of CS and crosslinker play an essential role in determining the size of the nanoparticle. Necessary parameters in this method are the stirring speed, the polymer solution's temperature, and the pH. Preferably at low ambient temperature, there is an increase in the cooling rate of the suspension, enhancement in hydrogen bonding between water and CS molecules, and there is an improvement in stiffness and stability of the synthesized nanoparticles. These synthesized nanoparticles are appropriate for drug delivery applications [158]. During the blending process, the drug of interest can be loaded into the nanoparticles by mixing it with the crosslinker solution. Nanoparticles can also be used as a dual drug carrier; the two drug compounds are taken for dual drug systems. These two drugs interact with each other either by augmenting a drug action by another drug or can exhibit a new effect combination. The dual drug system also expands the study of various hydrophilic with hydrophobic compounds [156].

Micro emulsification is another promising method adopted for efficient nanoparticle synthesis. It is a chemical crosslinking method, which involves the blending of CS solution with glutaraldehyde and hexane/surfactant. This technique takes place for an overnight period for efficient crosslinking between the glutaraldehyde and NH2 group of CS. The sizes of the nanoparticles are controlled by altering the concentration of glutaraldehyde [159]. The drugs can be encapsulated by mixing the drug solution with polymer solution and application of heat to produce a transparent delivery system. The formulated CS nanoparticle sizes are analyzed by various techniques such as TEM, AFM, and DLS. Several samples are measured quantitatively in DLS and measured more qualitatively by TEM or AFM [157]. Hence, CS nanoparticles have a great scope in drug delivery applications.

5.2 *Chitosan Nanofibers*

Nanofibers are solid polymeric fibers, and their diameters range from 300 to 1000 nm. CS-based nanofibers have been widely used in drug delivery. Among various useful DDS, nanofibers are very promising by exhibiting several advantages such as very high surface to volume ratio, porosity, immense drug loading capacity, multi-drug delivery system, controlled drug release, and mass transfer properties [158]. Nanofibers in the DDS, drugs could be administered on various routes such as topical, oral, transmucosal, and transdermal [159]. As a drug carrier, nanofibers can also protect the drug from degradation in the body, increase the drug concentration at the targeted sites, and decrease the drug dosage, therefore reducing the side effects [160].

Electrospinning is a commonly used technique for the formulation of nanofibers. It is employed for the continuous steady production of nanofibers at a large scale. It is a technique that involves the application of a strong electric field to the polymeric sample. The process involves converting an elastic polymeric solution to nanofibers at the influence of a strong electric field. The influence voltage causes the polymeric solution to be charged, and there is an electric field formed between the collector and nozzle. This high electric field causes a cone-shaped configuration of the liquid sample called the Taylor cone. When the voltage is strong enough to overwhelm the surface forces, the charged liquid stream elongates and rapidly evaporates. Finally, continuous CS nanofibers are formed [152]. In the context of drug delivery, the drug can be loaded into the nanofibers by directly dissolving the drug into the sample solution or can physically be bound to the nanofibers' surfaces [78]. Several parameters determine the architecture of electrospun nanofiber structures like polymeric solution concentration, a flow rate of solution into the syringe, screen to nozzle distance, temperature, and humidity [161]. When all the above parameters are appropriate, the properties such as diameter, thickness, and porosity of the fibers can be controlled. The remarkable features of CS nanofibers that prove as a potential candidate for drug release are (1) large surface area that favors dissolution, (2) support drug transformation from crystalline to amorphous, and (3) also possess high porosity [162]. Since CS is a swellable polymer, it swells and degrades gradually in the biological environment [163]. Therefore, CS nanofibers are excellent structures for their application in the DDS.

5.3 *Chitosan Hydrogels*

The hydrogels have a 3D-swellable polymer network that can absorb and hold a large water volume [164]. This swollen form of hydrogels possesses some important physical traits shared with living tissues, such as less surface tension with biological fluids, rubbery texture, and flexible consistency. The elasticity of swollen hydrogels minimizes irritability with surrounding tissues [165]. CS polymer is usually employed for the fabrication of hydrogels in DDS. Since CS possesses bioadhesive property, it improves tissue penetration and drug homing time. The adhesiveness is due to the polymer's interchain interaction with the mucosal glycoprotein at specific sites such as colon, oral, and nasal [166]. Hydrogels can also undergo deformation in shape depending on the target delivery site. Depending on hydrogels' formulation, the duration of drug release occurs from a few hours to days [167]. The fabrication of hydrogels is done by physical and chemical crosslinking.

The physical crosslinking formulation of hydrogels relies on significant physical interactions such as hydrophobic, ionic, and polyelectrolytic associations. The gels formed by these interactions are entirely physical and reversible. Physical CS-based gels are good drug delivery agents for short-term release [166]. In the hydrophobic interaction method, the process of gelation is based on sol-gel chemistry. The introduction of hydrophobic interactions in the formulation of CS hydrogels depends on N-acylation's degree with the hydrophobic segments. This technique involves the formation of amphiphile gel by coupling of hydrophobic solvent with the hydrophilic polymer. In amphiphile gels, the hydrophobic segments aggregate on the gels' inferior portion and minimize the water contact on the hydrophobic surface. This hydrophobic segment enhances the drug entrapment and the stability of gels [153, 167]. Another widely used physical crosslinking method is ionic complexation. The procedure involves using anionic agents to ionize with the cationic NH2 group of CS [168]. Metallic anions are often used to form crosslinking via covalent bonding with the polymer rather than electrostatic attraction. The gelation formed due to covalent crosslinking is much strong and stable. Another similar method employed in physical crosslinking is polyelectrolyte complexation [169]. This fabrication technique depends on the interaction of two opposite charged polymers. Due to the positive charge of CS, anionic polymers like PLA, alginate, and pectin are used to prepare hydrogels. The interaction between the two opposite charged polymers results in the formation of polyelectrolytic complexes. The anionic polymers chosen should exhibit high charge density so that the electrostatic interactions between the polymers are strong enough to maintain the appropriate physical characteristics of hydrogel [170].

Chemical crosslinking is a direct method to produce permanent and irreversible hydrogels. The irreversible hydrogels are produced by the formation of covalent bonding between polymer chains through crosslinkers. These crosslinkers molecules react with amino groups of CS and lead to irreversible intermolecular bridges between polymer chains [170]. Permanent covalent hydrogels can also be formed by photo crosslinking or in the presence of reactive enzyme molecules on CS. Photo crosslinking is carried out in the presence of a photoinitiator and subsequent exposure to UV light. A photoinitiator such as an azide or methacrylate coexists in the polymer. Upon UV exposure, azide conversion to a sensitive nitrene group binds with the NH2 groups of CS, causing rapid gelation [171]. The photochemically cross-linked hydrogels have the capability of prolonged drug release. The crosslinking efficiency depends on the UV exposure time and the UV source's distance [172]. Similarly, in the enzyme crosslinking technique, enzyme sensitive moieties such as hydroxyl phenyl groups of CS are covalently coupled with an NH2 group of another CS. This reaction is catalyzed by horseradish peroxidase. This leads to the formation of enzyme-catalyzed gel [173].

The efficiency of the hydrogels as a drug carrier system depends on the physico-chemical characteristics of them. The encapsulation of drugs depends on the porosity of the gels and the size of the drug. Suppose the drug of interest is a small molecule. In that case, the encapsulation is done by placing the hydrogels in a drug saturated medium, and the drug molecule will slowly travel inside the gel and gets entrapped [173, 174]. If the drug is a larger molecule, it should be added during the gelation of the polymer. The interaction between drugs and polymer is susceptible to the in vivo enzymes so that the drugs can be covalently linked to the polymer before gelation. The encapsulation also depends on the hydrophilicity and charge of the polymer. If the drugs used are hydrophobic, they are combined with amphiphilic molecules before loading [174].

5.4 *Chitosan Microneedles*

Microneedles are a micro-scale needle like design in a range of maximum 1 mm length and width in microns. They can breach a skin without the stimulation of sensory nerve and painlessly deliver drugs [154]. The materials used to formulate microneedles should improve mechanical strength, hardness, and toughness to ensure the needle piercing underwent without damage and fracture [175]. In the context of drug delivery, polymers like CS are used for the formulation of microneedles. CS-based drug carrier systems release drugs via biodegradation and swelling, thus ensuring a sustained and prolonged drug release. These properties clinch CS as a preferred microneedle fabricating material. A cost-effective and straightforward fabricating method is the cross-over lines laser engraving technique. This method can fabricate solid or hollow biodegradable microneedles for drug delivery [176]. In this method, the acrylic microneedle molds are prepared by the CO_2 laser cutter. The laser beam engraves patterns of lines that pass over each other at a point to form a sharp negative canonical midpoint in the acrylic sheet. In preparation for hollow CS microneedles, polydimethylsiloxane (PDMS) is cast on the acrylic mold, and it is plasma treated. The CS solution is then coated on PDMS microneedles, and they undergo solidification due to drying. A fragile layer of CS is accumulated on PDMS microneedles and forms a hollow shell [177]. Finally, rapid cooling causes shrinkage of PDMS and strengthens the solidified CS. These hollow microneedles allow the flow of drugs into the holes of the needle. Upon injection, the drug travels within the epidermis, which leads to absorption by dermal blood vessels. The major advantage of the use of a hollow needle in the DDS is that they grant continuous dispersion through the skin [178].

The solid microneedles' formulation is very much similar to hollow microneedles except for a few additional steps. The PDMS microneedles are silanized for casting PDMS on the microneedles [177, 178]. This silanization forms a barrier between two PDMS layers, making them easy to separate from each other. The casted PDMS mold is cured, and then it is severed from PDMS needles. Next, the CS solution is coated on the PDMS mold and is allotted under vacuum to enhance the density of CS. Finally, the sample undergoes drying to improve the stability and strength of the needle [179]. These needles invade the top skin layer and allow the journey of drugs into the lower skin regions. The drug gets entrapped by coating it onto the surface of the needle. Injection of a solid needle in the skin causes drug dissolution into the skin. Solid microneedles have superior strength and sharpness compared to hollow microneedles [180].

5.5 *Chitosan Microspheres*

Microspheres have tiny spheroidal particles, and their sizes range from 0.1 to 300 μm. Natural or synthetic polymers fabricate them. Due to their diminished structure, the particles improve drug absorption and bioavailability [181]. Microspheres are very beneficial for controlled and prolonged drug release. Prolonged drug release usually depends on the polymer's mucoadhesiveness, and therefore, the microspheres are formulated with a natural polymer such as CS. CS-based microspheres improve sustained drug release and benefit by avoiding frequent dosage. They are competent drug carriers as they could also cross the blood-brain barrier for drug release [182].

The abundantly used techniques of microsphere formulation are solvent evaporation, spray drying, and coacervation methods. The solvent extraction is a physical separation mechanism and is also called dry in the oil method. Initially, the CS/drug solution undergoes emulsification to separate the aqueous phase that contains the polymer and drug. The removal of excess organic solvent occurs by stable stirring or an increase in temperature and pressure. The solvent evaporation leads to precipitation of polymer on oil/water interphase, which forms a cavity and leads to microspheres [183]. Another commonly used method is spray drying. In this method, the first step involves the dispersion of CS drug in an organic solvent by homogenization. The homogenized solution is exposed to a hot air stream for atomization and forms tiny droplets. From these droplets, the solvent is evaporated, and microsphere formation takes place. The solvent trace in microspheres is removed by vacuum drying. The microspheres' size relies on atomization pressure, air temperature, exposure to crosslinking, and nozzle size [150].

There is a reduction of solubility of the polymer in an organic solvent in the coacervation method, forming a phase rich in a polymer called coacervates [184]. The drug compound is dissolved along with the polymer solution or embedded in an inappropriate polymer and added to the process, making the first polymer encompass the drug. After precipitation, the particles are isolated by centrifugation or filtration. Crosslinking agents are used to harden the particles and control the drug release. Other than the crosslinking agent, drug release also depends on parameters such as pH, enzymes, and polarity [153]. The release of drugs by CS microspheres occurs by three distinct steps, namely (1) release via biodegradation, (2) release from microsphere surface, and (3) release by swelling of the polymer. The microsphere drug release system undergoes a burst release. Minimization of burst release can be done by enhancing the crosslinker density [185]. The particle size of drug-loaded and unloaded microspheres can be estimated by optical microscopy, and the percentage of drug loading efficiency is evaluated by UV spectrophotometer [184].

6 Potential of Chitosan Delivery System in Bone Tissue Engineering

6.1 Growth Factor Delivery

The maintenance and regulation of cellular activities ultimately depend on numerous signaling molecules, which can initiate a torrent of downstream pathways. This causality is crucial for cellular processes like tissue regeneration and development. GFs are a class of soluble signaling proteins secreted by cells that play an extensive role in regulating cell proliferation, differentiation, migration, ECM synthesis, etc. GFs typically display short-range indigenous actions by diffusion through ECM [185, 186]. A wide variety of GFs are found to be involved in the bone regeneration process especially, GFs such as BMP-2, BMP-7, and FGF-2 have been used in preclinical studies for their bone-forming potential [187]. Delivering such bioactive signal molecules by employing biocompatible DDS has shown enhanced therapeutic effects. The delivery system must be designed so that the delivered GF should reach the desired dosage level at the target site [188]. CS-based delivery systems are widely popular in BTE due to their efficient and precise payload delivering potential. In this section, the deliveries of GFs by the CS-based systems for bone regeneration are discussed.

CS has a steady and controlled degradation rate, which enables it to be used in numerous sustained drug release systems for various proteins and peptides. Various forms of CS, such as nanoparticles, gels, nanofibers, etc., are used to achieve an effective DDS [189]. BMPs serve as the inevitable biochemical stimuli needed for MSCs to differentiate into functional osteoblasts [190, 191]. Hence, several studies have been made and suggested the delivery of BMPs through CS delivery systems for bone regeneration. CS gels possess unique properties such as improved viscosity and biocompatibility, making them a suitable biomaterial for implantation. In a study, CS gels were used to trap and deliver the osteoinductive cytokine rhBMP-2. The in vivo studies on mouse models suggested that the rhBMP-2 incorporated CS gels effectively adhered to the defective bone walls. The CS gels efficiently filled the area of bone defect, thus perpetuating rhBMP-2 at the site of the defect, which hastened bone formation [192] (Fig. 3). Another study reported a novel scaffold composed of electrospun PCL nanofibers coated with BMP-2 entrapped CS nanoarchitectures. This unparalleled structure was achieved by alternating layer by layer nano-immobilization of BMP-2 and CS. This contraption's osteogenic effects were analyzed in vitro on human primary osteoblast cells and in vivo on defective mouse calvaria. The results showed promising outcomes for bone regeneration. This approach also involved a substantially minimal amount of GF, thus eliminating the risk of BMP-2 overdosing and being highly cost-effective [193]. CS can also be used to immobilize GFs to achieve efficient delivery and prolonged dosage. A guided bone regenerative membrane composed of CS nanofibers was utilized to immobilize rhBMP-2 chemically. In vitro studies on MC3T3 cells (pre-mouse osteoblasts) showed increased proliferation and differentiation of the above membrane suggesting its potential application in BTE [194]. CS hybrids like CS-silica xerogels have been successfully employed to deliver GFs. Augmenting CS with silica greatly increased the scaffolds' porosity, contributing to their high drug bearing capacity. When these xerogels were used to deliver BMP-2, the hybrid scaffolds showed increased bone formation in both in vitro and in vivo studies [195, 196]. The chemically modified CS is a potent delivery system for GFs. Thiolated CS (thio-CS) is one of the widely used materials for drug delivery. The thio-CS was loaded with BMP-2 and checked for its bone-forming properties. The in vitro studies revealed that the modified CS showed a delay in release kinetics, hence exhibiting a prolonged drug delivery. The in vivo studies demonstrated the rapid formation of ectopic bone in mice compared to other scaffolds [197].

PDGF (platelet-derived growth factor) is another crucial GF that is involved in bone formation. It is responsible for activities like mitogenesis, angiogenesis, and macrophage activation, through which it indirectly induces bone regeneration [198]. A report demonstrated the delivery of PDGF-BB via CS-TCP sponge carrier. The drug release kinetics of the GF loaded CS-TCP showed that the porous scaffold was responsible for a consistent release of PDGF-BB. A similar study involving PDGF-BB delivery using a CS-TCP sponge for periodontal regeneration was reported. This study endorsed the drug delivery potential and osteogenic effect of the CS-based scaffolds [199, 200]. Another investigation reported the delivery of PDGF-BB through a PCL-CS hybrid matrix. The drug-loaded scaffolds initially

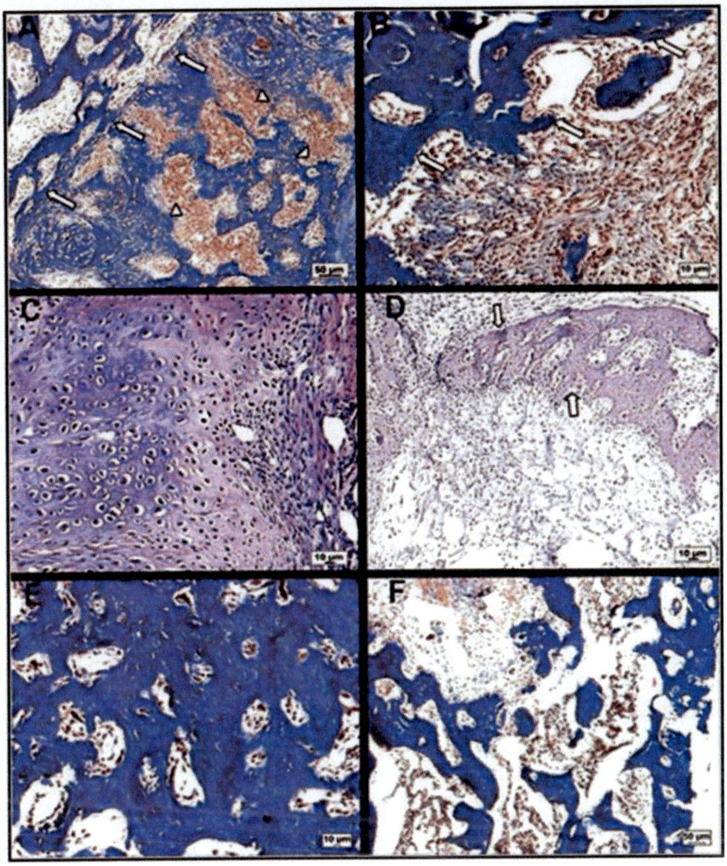

Fig. 3 (**a**) Micrograph showed the interface between old bone and new bone tissues, indicating a faster bone healing process via wide vessels and osteocytes in the middle area of the group treated with rhBMP-2 (MTS). (**b**) The group's occlusal region with the bone defect filled by coagulum and the interface between the old bone tissue and a large amount of fibrous connective tissue was observed (MTS). (**c**) rhBMP-2/monoolein gel showed a large quantity of chondrocytes (H&E stain). (**d**) Representative photomicrograph of the middle area of the group that received monoolein gel showed small quantities of new bone and a great amount of fibrous connective tissue, indicating slow bone healing in the groups without rhBMP-2 (H&E stain). (**e**) rhBMP-2/CS gel treated MTS micrograph showed organized and mineralized bone in the occlusal region. (**f**) Representative photomicrograph of CS gel treated group revealed a small amount of new bone growth between the fibrous connective tissue (MTS). (Reproduced from Ref. [192] with permission)

exhibited a burst release, followed by a steady release of PDGF-BB for 28 days. This prolonged release of GF from the scaffolds increased bone formation, and this study was carried out in rats [201]. A controlled DDS composed of PLLA and CS was manifested in a study involving PDGF-BB as the payload. The porous PLLA-CS scaffolds were developed, and the GF was loaded onto them. CS in the scaffolds had led to a controlled release of the drug and accelerated bone regeneration in rats with calvarial defects [202].

Bone regeneration is a sequential process that requires the presence of a multitude of GFs at temporal precision. Hence, DDS capable of delivering multiple mitogens at once might have a significant effect on promoting bone formation [15]. A novel injectable CS nanoparticles-integrated into gellan xanthan hydrogel was fabricated, and the delivery of two GFs, BMP-7 and bFGF, was demonstrated. The study concluded that dual GFs delivery significantly increased osteoblast differentiation. The entrapment of CS nanoparticles in the gels resulted in a slow drug release rate, thus amplifying the effect of GFs [203]. CS-based 3D-fiber mesh scaffolds were used as a delivery system for the sequential delivery of two GFs, BMP-2 and BMP-7. PLGA and poly 3-hydroxybutyrate-co-3-hydroxyvalerate nanocapsules were loaded with BMP-2 and BMP-7, respectively, and these drug carriers were trapped into the CS mesh. The in vitro studies using rat bone marrow MSCs showed that the sequential delivery of GFs improved osteogenesis, and the CS fibers had prolonged release of GFs from the nanocapsules [204]. A comparative study was done using porous CS scaffolds as a delivery medium for the osteogenic GFs such as BMP-2 and IGF-1. The GFs were impregnated into individual scaffolds and were tested on in vivo rabbit models with tibial bone defects. The results revealed that the CS scaffolds' utilization had helped in a controlled release of the GFs, thus enhancing bone regeneration and healing [205]. The delivery of these two GFs has been demonstrated with CS gel/Gel microspheres biocomposites in a similar study. The study evidenced that incorporating GFs-loaded Gel microspheres in CS gel resulted in a controlled release of the drug. The sequential action of BMP-2 and IGF-1 showed increased osteoblast differentiation [206].

Demineralized bone matrix power (DMP), primarily composed of BMPs and collagen, is a potent osteoinductive used for bone regeneration. However, handling difficulties, stability, and site-specific drug release are the primary concern while using DMP. Hence, numerous attempts have been made to utilize this osteogenic compound effectively. In a recent study, the delivery of DMP was demonstrated with a thermo-sensitive CS-based hydrogel. The incorporation of DMP with the hydrogel increased its stability and prevented unwanted migration of the compound. There were also enhanced mechanical properties, degradability, and elevated bone formation in rats [207]. All these therapeutic strategies involving CS as the delivery system had made a considerable impact on regenerative medicine as they advise novel routes towards the efficient repair of bone defects.

6.2 *Nucleic Acid Delivery*

Gene-based therapeutics has expanded the scope to alter the cell cytology at its genetic and epigenetic levels. An effective gene delivery ensured the delivery of DNA into the cells and converted into a functional protein via transcription and translation [208]. The gene delivery can be accomplished by either a viral or non-viral vector. Usually, viral vectors cause immunogenicity and oncogenic reactions. Non-viral vectors overshadow these drawbacks. In non-viral vectors, usually, cationic polymers such as CS are preferred. This polymer electrostatically interacts with the negatively charged nucleic acids, forms a polyelectrolyte complex with DNA, and protects it from degradation by the cell nucleases [209]. Several factors, such as the molecular weight of CS, the concentration of plasmid DNA, external pH, and CS/DNA ratio, may influence the transfection efficiency of CS. The molecular weight of CS is high, and hence, it is hard to degrade the polymer and release the DNA. The interaction between CS and DNA is due to the electrostatic attraction between the amino group in CS and the phosphate group in DNA [210]. An acidic pH environment can alter this interaction by enhancing the charge density of CS and eventually can lead to the formation of a stable complex with DNA [210, 211]. During endocytosis of the CS/DNA complex, CS is protonated due to acidification in the lysosome and leads to the repulsion of the amine moieties and causes expansion of the polymer, which finally leads to rupture of the endosome. After the disruption of the endosome, the complex is released in the nucleus [211]. During these steps, CS delivers the gene safely without degradation into the nucleus. This non-viral CS-based gene delivery has been successful in the control of tissue-specific gene expression. In the context of skeletal tissue regeneration, this method has been employed for successful gene transfection that could potentially influence the induction of osteogenesis and improvise bone repair [212].

CS is being used in various forms that can efficiently act as a gene carrier for osteogenesis. A nano complex (NC) was formulated using CS and pDNA that encodes for BMP-2. This pDNA containing CS nanocomplex was immobilized in an MMP (matrix metalloproteinase)-sensitive peptide-functionalized PCL scaffold. During the culture of MSCs on NC containing scaffold, MMP2 secreted by MSCs cleaves the MMP-sensitive peptide and dispenses the NC. The released NC is accessible to the MSCs, and subsequently, NC was internalized. The expression level of BMP-2 and its targets was significantly increased, indicating the cell's successful gene transfection. There was a significant increase in Runx2, a bone transcription factor, and its downstream genes such as ALP, COL1, OCN, and OPN [213]. This study proved that CS nanocomplexes are efficient in gene transfection and would lead to the cell's osteogenic fate. In another study, the BMP-2 gene-containing plasmid was incorporated into CS nanoparticles, and Col/HAp scaffolds supported this complex. These composites containing the BMP-2 gene-containing plasmid promoted ALP activity and calcium deposition in mMSCs [214]. Similarly, the nano-polyplexes complex was formed using water-soluble modified CS and polyethylenimine (PEI). This composite was endowed with BMP-2 plasmid and transfected into human osteoblastic cells (MG-63). Confocal fluorescent microscopy analysis revealed the presence of polyplexes in the cytoplasm, denoting the cellular uptake, and encouragingly there was a significant increase in BMP-2 gene expression in successfully transfected MG-63 cells [215]. The BMP-2/pcDNA3.1 plasmid was incorporated into CS films to improve gene transfection and osteoinduction. The MC3T3-E1 cells were treated with the films containing BMP-2/pcDNA3.1 plasmid, and the immunofluorescent staining results showed an increase in Runx2 protein in the nucleus. The histological analysis revealed a new thick bone formation twice the size of the control [216]. Some previous studies proved CS as a successful dual gene delivery system. CS nanoparticles were combined with both BMP-2 and VEGF plasmids, and these nanocarriers were exposed to MSCs. The cells showed improved ALP activity. In vivo results demonstrated the rapid bone formation in rat models with the calvarial defect. Since bone is a vascularized tissue, osteogenesis happens simultaneously with angiogenesis, and therefore transfection of VEGF and BMP-2 in MSCs showed enhanced osteogenesis [217].

Several microRNAs (miRNAs) are known to regulate osteogenesis effectively. miRNAs are non-coding RNAs that contain 21–25 nucleotides, and they regulate gene expression at the post-transcriptional level. miRNAs based delivery was considered an improvised strategy to enhance osteogenesis by inhibiting its antagonists [218]. The miR-199a-5p-loaded CS nanoparticles effectively facilitated osteogenesis by inhibiting HIF1a and Twist1 late-stage osteogenic inhibitors. HIF1a and twist1 are known to regulate the expression of Runx2 negatively. The hMSCs treated with miR-199a-5p-loaded CS nanoparticles showed increased osteogenic expression such as Runx2, ALP, SP7, and OCN. In vivo studies using the above miRNA-loaded CS nanoparticles revealed new bone formation at the tibial defect upon treatment [219]. In another study, mMSCs grown in the presence of miR-590-5p and CS/n-HAp/n-zirconium dioxide biocomposite scaffolds promoted osteoblast differentiation. This osteogenic effect was due to miR-590-5p-mediated inhibition of Smad7, an antagonist of TGF-β1 signaling [220].

Some endogenous miRNAs can negatively regulate osteogenesis, and there are studies on inhibiting these miRNAs by the delivery of specific antagonists. Pertinent to the information, CS nanoparticles were used as a carrier for antimiRNA-138 to impede the effect of endogenous miRNA-138, and osteogenesis was promoted (Fig. 4). The miRNA-138 negatively regulated Runx2 by arresting the ERK 1/2 pathway. The MSCs transfected with antimiRNA-138 showed a significant increase in the expression of osteogenic genes [221]. In a similar demonstration, antimiRNA-138 was incorporated in CS nanoparticles and SDF-1 α (Stromal derived factor-1alpha)-loaded hydrogel (Fig. 4). MSCs cultured on these composites increased osteogenic gene expression, and in vivo analysis reported new bone formation in the rat model system. Immunohistochemistry analysis showed positive staining of COL-1, OPN, and OCN on the tissue formed at the periphery of the implanted composites. These remarkable outcomes are due to rapid MSC migration to the defective site and subsequent commitment to osteogenic lineage after exposure to composites [222]. The above studies supported that CS is an efficient and successful vector in the nucleic acid delivery system.

6.3 *Phytocompound Delivery*

Traditional medicine around the globe practiced by different ethnicities has a characteristic reliance on indigenous medicinal plants. These plants' therapeutic

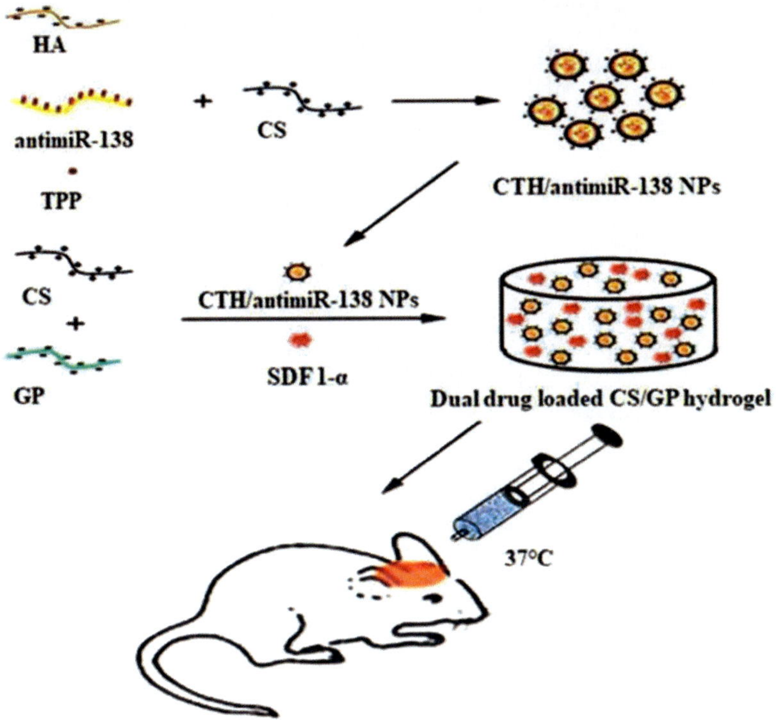

Fig. 4 Graphical representation of SDF-1α and CS/antimiRNA-138 incorporated NPs in CS/GP hydrogel biocomposite showed enhanced bone regeneration in rat critical-sized calvarial defect. (Reproduced from Ref. [222] with permission)

potential, regarded as common knowledge for generations, is now being investigated to identify the specific biological factors that exhibit the required medicinal property. One such prevalent chemical constituent of plant extracts happens to be phytocompounds. Phytocompounds are plant-educed metabolites currently being investigated for their pharmacological activity to treat complex diseases including cancer, autoimmune disorders, diabetes, etc. These plant derivatives exhibit functional redundancy, making them useful in treating a wide variety of ailments. As a naturally occurring product, phytocompounds exhibit minimal toxicity, biocompatibility and are endorsed by their availability and ease of preparation [223]. The phytocompounds have been generally classified into phenolics, carotenoids, terpenes, alkaloids, organosulfur compounds, and nitrogen-containing compounds [224]. This class of plant metabolites acting as "Phyto-drugs" can desirably regulate stem cell differentiation, enticing their tissue regeneration applications.

CS is a versatile biopolymer popular in regenerative medicine. CS-based scaffolds exhibit excellent porosity, especially when lyophilized, accompanied by an anisotropic architecture and an extensive network of interconnected pores. This porous structure is instrumental in promoting cell attachment and other subsequent

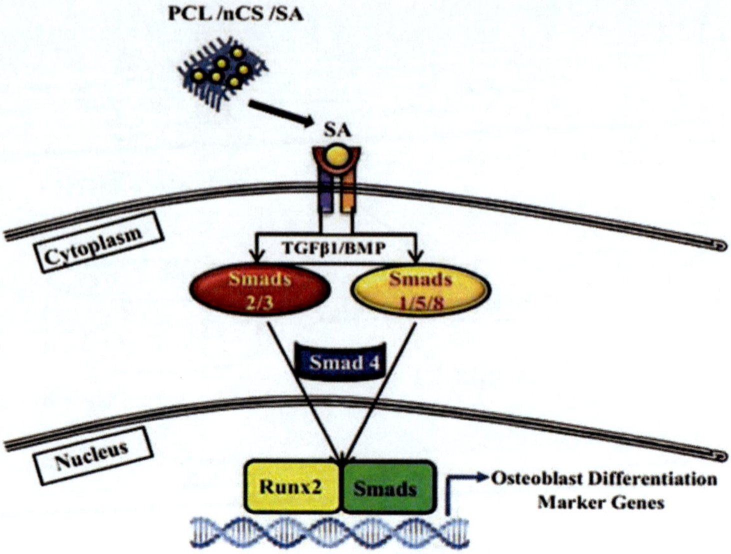

Fig. 5 Diagrammatic representation of osteoblast differentiation by sinapic acid. SA release from PCL/nCS electrospun scaffold activates the TGFβ 1/BMP signalling pathways and their components mMSCs. (Reproduced from Ref. [25] with permission)

cellular functions. Drug modulated tissue repair is a novel strategy in BTE that invokes drug delivery models to enhance the incorporated scaffolds' bio-functionality. The incorporation of nanoparticles can improve the osteoconductive and mechanical properties of such biopolymeric scaffolds. Since CS is bestowed with unique features like acidic pH tolerance, mucoadhesive property, and chemical flexibility, it can be considered an ideal material for constructing sophisticated DDS, including nanofibers and microspheres nanoparticles, NGs, and so on [225]. The recently discovered bioactive phytocompounds capable of acting as stem cell inducers have been delivered via the CS-based systems to facilitate tissue regeneration.

Veratric acid, an osteoinductive phenolic compound, was delivered in a sustained manner via the CS nanoparticles incorporated in coaxially electrospun PCL/PVP fibers. The results showed enhanced bone formation as evidenced by the up-regulation of Runx2 and its downstream osteogenic marker genes along with subsequent mineralization [26]. Hydroxycinnamic acid is known to have a significant impact on bone metabolism. SA, a derivative of hydroxycinnamic acid, promoted osteogenic differentiation of mMSCs by the up-regulation of Runx2. This effect was mediated by the TGFβ1/BMP/Smad signaling pathway (Fig. 5). When encapsulated in CS nanoparticles and steadily released by the surface erosion of the embedded PCL fibers, this osteogenic compound significantly increased its bioavailability, thus improving bone repair as evidenced by in vitro and in vivo bone formation [25].

Chrysin, a flavonoid found in mushrooms known to reduce osteoclastogenesis and induce bone formation, was incorporated in a biocomposite blend of CS, CMC, and nano-HAp. A sustained release of chrysin promoted osteogenic differentiation of mMSCs endorsing its potential in BTE applications [30]. Silibinin is a phenolic phytocompound capable of inducing the differentiation of mMSCs towards osteoblast lineage. Its reduced bioavailability owing to its hydrophobic nature severely diminishes its overall therapeutic potential. Hence, the effective delivery mechanisms become a prerequisite for the efficient exploitation of this phyto-drug. Due to its mucoadhesive and cell-permeable nature, CS has been used in encapsulating silibinin, facilitating its intracellular delivery. Alg and Gel have naturally derived biopolymers with inherent osteoconductive and biomimetic properties proving useful in the reconstruction of complex tissues. The silibinin-loaded CS nanoparticles were fabricated into an Alg/Gel biocomposite, and the increased bioavailability of the silibinin was hugely useful in triggering osteogenesis in vitro [226].

Icarin, a glycosylated flavonoid, is similar to chrysin in regulating the process of bone modeling with simultaneous inhibition of bone resorption. This potent osteoinducer was delivered in a stable manner protracting its osteogenic effect with a CS/HAp hybrid delivery module. The inclusion of HAp in this venture stabilized the degradation rate of CS, facilitating uniform surface erosion, achieving a sustained release of icariin, which enabled new bone formation by hBMSCs. This osteoregenerative scaffold could help in treating destructive bone disorders [227]. Berberine is a naturally occurring plant alkaloid that functions as an antipyretic and detoxifying agent in traditional Chinese herbs. This molecule is also known for its antioxidant, anti-inflammatory, and antibacterial properties. Recently, its ability to induce the osteogenic differentiation of stem cells has been discovered [228]. Delivering with polymeric nanoparticles can significantly improve its intracellular presence to facilitate osteoblast differentiation. Researchers have utilized berberine in constructing an osteoinductive/bacteriostatic gel infused with negatively charged CMC/CS copolymeric nanoparticles along with positively charged CS nanoparticles. This construction provided nano and micro-pores for the loaded cells to flourish and enabled cellular networking. The researchers witnessed an enhancement in berberine-mediated osteogenic differentiation of BMSCs correlated with the parallel delivery of BMP-2, a well-known bone-forming GF. Berberine also demonstrated its bacteriostatic activity in this scaffold, thus paving the way for bone healing in a sterile microenvironment [229].

Coumarins are well known for their anti-osteoporotic activities. Osthole (OST) is one such coumarin derivative clinically involved in preventing active bone resorption by suppressing RANKL-mediated osteoclastogenesis. Recent studies suggested that it also induces osteogenesis by up-regulating Osx via the cAMP/CREB signaling pathway [230]. Its poor solubility is overcome by encapsulating in suitable DDS. Polymeric micelles are a class of colloidal drug carriers useful in the cellular trafficking of hydrophilic and hydrophobic molecules. An OST encapsulated N-octyl-O-sulfonyl CS micelle DDS was synthesized for the effective delivery of OST to improve its therapeutic efficiency in osteoporosis. In vitro studies using bone marrow macrophages and in vivo studies involving ovariectomized SD rats suggested that the anti-osteoporotic activity was greatly enhanced when OST was delivered with CS micelles, establishing CS as a highly lucrative polymer in bone repair strategies [231]. Naringin is a flavonoid known for its efficient inhibition of bone resorption. It is also known to stimulate bone formation, as evidenced by the up-regulation of ALP, OCN, and other osteoblast-specific marker genes in M3C3T3-E1 cells. It is also known to induce osteoblast differentiation of BMSCs by downregulating peroxisome proliferator-activated receptor ϒ, an adipogenic transcription factor [232]. Parthenolide is another plant metabolite that up-regulated osteogenesis via the Wnt signaling and suppressed osteoclastogenesis by downregulating NF-kB [233]. A dual DDS was developed, wherein naringin was incorporated in CS microspheres and parthenolide in the PLLA matrix. This composite showed significant bone formation in the SD rat periodontal fenestration defect model due to the sustained release of both osseous factors, endorsing its potential in guided bone regeneration [234].

Salvianolic acid B is a major biochemical constituent in the Chinese herb *Salvia miltiorrhiza*, useful in treating cardiovascular diseases. It is known to exert a certain level of neuroprotection and also therapeutic effect towards liver fibrosis. The osteogenic potential of salvianolic acid B has been recently reported, along with its angiogenic properties [235]. A salvianolic acid B loaded CS/HAp scaffold showed promising in vitro and in vivo bone formation along with vascularization, where the sustained release of organic acid was facilitated by the uniform degradation of CS [236]. The studies mentioned above demonstrated that CS has effects on increasing the bioavailability and the subsequent therapeutic activity of pharmacologically effective phyto-drugs. The CS-based delivery system for other miscellaneous drugs has been mentioned in Table 1.

7 Summary

Bone is a vital organ that plays a crucial role in providing structural integrity, protection to fundamental organs, and calcium homeostasis. Bone defects that severely cripple the skeletal system are hard to treat with conventional grafts. BTE is a revolutionary discipline that has contributed to the recent advancements in treating clinically challenging bone defects. It utilizes various strategies invoking adept biomaterials and bioactive molecules to stimulate the regeneration of incapacitated bone. Bioactive molecules are distinct biochemical entities capable of regulating several cellular and molecular pathways consistently. Recruiting these biomolecules can significantly enhance bone repair, and for this, a potent delivery system for targeted therapy is required. Drug delivery is a developing field that aims to treat various diseases by efficient administration of established therapeutics. Different processes can be formulated depending on specific applications such as wound healing, bone and cartilage regeneration, etc. CS is one such polymer that is flexible enough to be utilized to formulate several effective drug carriers. As a natural polymer, CS features excellent biological properties like increased

Table 1 CS delivery system for miscellaneous drugs in bone tissue engineering

S. No.	Drugs	Original function	CS formulation	Biocomposite(s)	In vitro studies	In vivo studies	Effect on bone	Reference
1.	Alendronate	Osteoporosis	Microspheres	nHAp/PLLA/CS microspheres	hADSC	Radial defect in Rabbit model	• Increased ALP expression • Increased calcium deposition • Bone repaired in 8 weeks	[237]
2.	Atrovastatin	Hyperlipidemia	Hydrogels	TIPGS nanoemulsions-loaded CS hydrogels	Human osteoblasts	Rat calvarial defect model	• Increased BSP2 expression • Increased neo-bone formation	[238]
3.	Dexamethasone	Inflammatory diseases	2D-film	CS/DEX film	ADSCs	–	• Increased ALP expression	[239]
4.	Heparin	Thrombosis	Freeze-dried scaffolds	Heparin immobilized CS scaffolds	MC3T3-E1	–	• Increased ALP and OCN expression	[240]
5	Lovastatin	Hypercholestrolemia	Nanoparticles	Lovastatin-loaded CS nanoparticles	–	Femoral defects in female Wistar rats	• Increased bone density	[241]
6	Raloxifene	Osteoporosis	Scaffold constituent	Raloxifene microspheres-embedded Col/β-TCP/CS	MC3T3-E1	–	• Increased ALP activity • Increased mineral deposition	[242]

(continued)

Table 1 (continued)

S. No.	Drugs	Original function	CS formulation	Biocomposite(s)	In vitro studies	In vivo studies	Effect on bone	Reference
7	Risedronate	Osteoporosis	Intra-pocket dental film	CS/Risedronate/zinc-HAp	–	Alveolar bone defect model in Wistar rats	• Increased ALP and OCN expression • Remarkable inhibition of alveolar bone loss	[243]
8	Rosuvastatin	Hypercholestrolemia	CS/chondroitin sulfate nanoparticles (CS/CTS)	CS/CTS nanoparticles-loaded Pluronic F127/ hyaluronic acid hydrogel	MG-63	–	• Stimulated osteoblast proliferation • Increased mineralization density	[244]
9	Simavastatin	Hyper cholestrolemia	CS nanoparticles	–	–	Intra-bony defect in albino rabbits	• Increased bone formation • Promoted angiogenesis	[245]
10	Zoledronate	Hypercalcemia	Scaffold	CS/nHAp/zoledronic acid	hBMSCs	–	• Increased ALP activity	[246]

biocompatibility, minimal toxicity, and mucoadhesion. Due to the presence of multiple free amino and OH groups in the CS, it could serve as a repository for various biomolecules. Hence, CS-based delivery systems such as nanoparticles, nanofibers, hydrogels, microneedles, and microspheres have significantly been considered for developing medications for tissue engineering and regenerative medicine. These drug transport systems have been successfully used to bolster the therapeutic efficiency of bioactive factors such as GFs, nucleic acids, and phytochemicals by achieving precise delivery. Thus, utilizing CS-based DDS has made an enormous impact on the field of BTE. Further investigations on this unique polymer as a delivery vehicle will acknowledge its potential prospects in future biomedical applications.

References

1. Florencio-Silva R, Sasso GRDS, Sasso-Cerri E, Simões MJ, Cerri PS (2015) Biology of bone tissue: structure, function, and factors that influence bone cells. Biomed Res Int 2015:421746
2. Osterhoff G, Morgan EF, Shefelbine SJ, Karim L, McNamara LM, Augat P (2016) Bone mechanical properties and changes with osteoporosis. Injury 47:S11–S20
3. Gilbert SF (2017) Developmental biology, the stem cell of biological disciplines. PLoS Biol 15(12):e2003691
4. Wiese A, Pape HC (2010) Bone defects caused by high-energy injuries, bone loss, infected nonunions, and nonunions. Orthop Clin N Am 41:1
5. Amini AR, Laurencin CT, Nukavarapu SP (2012) Bone tissue engineering: recent advances and challenges. Crit Rev Biomed Eng 40
6. Winkler T, Sass FA, Duda GN, Schmidt-Bleek K (2018) A review of biomaterials in bone defect healing, remaining shortcomings and future opportunities for bone tissue engineering: the unsolved challenge. Bone Joint Res 7:232–243
7. Qu H, Fu H, Han Z, Sun Y (2019) Biomaterials for bone tissue engineering scaffolds: a review. RSC Adv 9:26252–26262
8. Chocholata P, Kulda V, Babuska V (2019) Fabrication of scaffolds for bone-tissue regeneration. Materials 12:568
9. Sughanthy AP, Ansari MNMNM, Siva APS, Ansari MNMNM (2015) A review on bone scaffold fabrication methods. Int Res J Eng Technol 2:1232–1238
10. Koons GL, Diba M, Mikos AG (2020) Materials design for bone-tissue engineering. Nat Rev Mater:1–20
11. Guaadaoui A, Benaicha S, Elmajdoub N, Bellaoui M, Hamal A (2014) What is a bioactive compound? A combined definition for a preliminary consensus. Int J Food Sci Nutr 3:174–179
12. Biesalski HK, Dragsted LO, Elmadfa I, Grossklaus R, Müller M, Schrenk D, Weber P (2009) Bioactive compounds: definition and assessment of activity. Nutr J 25:1202–1205
13. Mitchell AC, Briquez PS, Hubbell JA, Cochran JR (2016) Engineering growth factors for regenerative medicine applications. Acta Biomater 30:1–12
14. De Witte TM, Fratila-Apachitei LE, Zadpoor AA, Peppas NA (2018) Bone tissue engineering via growth factor delivery: from scaffolds to complex matrices. Regen Biomater 5:197–211
15. Farokhi M, Mottaghitalab F, Shokrgozar MA, Ou KL, Mao C, Hosseinkhani H (2016) Importance of dual delivery systems for bone tissue engineering. J Control Release 225:152–169

16. Sahoo S, Ang LT, Goh JCH, Toh SL (2010) Growth factor delivery through electrospun nanofibers in scaffolds for tissue engineering applications. J Biomed Mater Res A 93:1539–1550
17. Kempen DH, Lu L, Heijink A, Hefferan TE, Creemers LB, Maran A, Dhert WJ (2009) Effect of local sequential VEGF and BMP-2 delivery on ectopic and orthotopic bone regeneration. Biomaterials 30:2816–2825
18. Lienemann PS, Lutolf MP, Ehrbar M (2012) Biomimetic hydrogels for controlled biomolecule delivery to augment bone regeneration. Adv Drug Deliv Rev 64:1078–1089
19. Potts JT (2005) Parathyroid hormone: past and present. J Endocrinol 187:311–325
20. Wojda SJ, Donahue SW (2018) Parathyroid hormone for bone regeneration. J Orthop 36:2586–2594
21. Arrighi I, Mark S, Alvisi M, von Rechenberg B, Hubbell JA, Schense JC (2009) Bone healing induced by local delivery of an engineered parathyroid hormone prodrug. Biomaterials 30:1763–1771
22. Arisawa EAL, Brandão AAH, Almeida JD, da Rocha RF (2008) Calcitonin in bone-guided regeneration of mandibles in ovariectomized rats: densitometric, histologic and histomorphometric analysis. Int J Oral Maxillofac Surg 37:47–53
23. Nascimento SB, Cardoso CA, Ribeiro TP, Almeida JD, Albertini R, Munin E, Arisawa EAL (2010) Effect of low-level laser therapy and calcitonin on bone repair in castrated rats: a densitometric study. Photomed Laser Surg 28:45–49
24. Yoon SJ, Park KS, Kim MS, Rhee JM, Khang G, Lee HB (2007) Repair of diaphyseal bone defects with calcitriol-loaded PLGA scaffolds and marrow stromal cells. Tissue Eng 13:1125–1133
25. Balagangadharan K, Trivedi R, Vairamani M, Selvamurugan N (2019) Sinapic acid-loaded chitosan nanoparticles in polycaprolactone electrospun fibers for bone regeneration in vitro and in vivo. Carbohydr Polym 216:1–16
26. Sruthi R, Balagangadharan K, Selvamurugan N (2020) Polycaprolactone/polyvinylpyrrolidone coaxial electrospun fibers containing veratric acid-loaded chitosan nanoparticles for bone regeneration. Colloids Surf B Biointerfaces 193:111110
27. Srinaath N, Balagangadharan K, Pooja V, Paarkavi U, Trishla A, Selvamurugan N (2019) Osteogenic potential of zingerone, a phenolic compound in mouse mesenchymal stem cells. Biofactors 45:575–582
28. Shadamarshan RP, Balaji H, Rao HS, Balagangadharan K, Chandran SV, Selvamurugan N (2018) Fabrication of PCL/PVP electrospun fibers loaded with trans-anethole for bone regeneration in vitro. Colloids Surf B Biointerfaces 171:698–706
29. Chandran SV, Vairamani M, Selvamurugan N (2019) Osteostimulatory effect of biocomposite scaffold containing phytomolecule diosmin by Integrin/FAK/ERK signaling pathway in mouse mesenchymal stem cells. Sci Rep 9:1–13
30. Menon AH, Soundarya SP, Sanjay V, Chandran SV, Balagangadharan K, Selvamurugan N (2018) Sustained release of chrysin from chitosan-based scaffolds promotes mesenchymal stem cell proliferation and osteoblast differentiation. Carbohydr Polym 195:356–367
31. Akshaya N, Prasith P, Abinaya B, Ashwin B, Chandran SV, Selvamurugan N (2021) Valproic acid, a potential inducer of osteogenesis in mouse mesenchymal stem cells. Curr Mol Pharmacol 14(1):27–35
32. Ganesh S, Sidharthan DS, Pranavkrishna S, Pranavadithya S, Abhinandan R, Akshaya RL, Balagangadharan K, Siddabathuni N, Srinivasan S, Selvamurugan N (2020) An osteoinductive effect of phytol on mouse mesenchymal stem cells (C3H10T1/2) towards osteoblasts. Bioorg Med Chem Lett:127137
33. Ketonis C, Barr S, Shapiro IM, Parvizi J, Adams CS, Hickok NJ (2011) Antibacterial activity of bone allografts: comparison of a new vancomycin-tethered allograft with allograft loaded with adsorbed vancomycin. Bone 48:631–638
34. Zhang Y, Ma W, Zhan Y, Mao C, Shao X, Xie X, Lin Y (2018) Nucleic acids and analogs for bone regeneration. Bone Res 6:1–9

35. Wei G, Ma PX (2004) Structure and properties of nano-hydroxyapatite/polymer composite scaffolds for bone tissue engineering. Biomaterials 25:4749–4757
36. Sonia TA, Sharma CP (2011) Chitosan and its derivatives for drug delivery perspective. In: Chitosan for biomaterials I. Springer, Berlin, pp 23–53
37. Dutta PK, Dutta J, Tripathi VS (2004) Chitin and chitosan: chemistry, properties and applications. J Sci Ind Res 63:20–31
38. Ruiz GAM, Corrales HFZ (2017) Chitosan, chitosan derivatives and their biomedical applications. Biological activities and application of marine polysaccharides. IntechOpen, 87
39. Muxika A, Etxabide A, Uranga J, Guerrero P, De La Caba K (2017) Chitosan as a bioactive polymer: processing, properties and applications. Int J Biol Macromol 105:1358–1368
40. Balagangadharan K, Rao H, Shadamarshan P, Balaji H, Selvamurugan N (2019) Composites containing marine biomaterials for bone tissue repair. In: Marine-derived biomaterials for tissue engineering applications. Springer, Singapore, pp 357–382
41. Khaleda Firdous Swati Chakraborty (2017) A review: naturally available sources of chitosan and analysis of chitosan derivatives for its antimicrobial activity. Int J Recent Sci Res 8:15773–15776
42. Al Sagheer FA, Al-Sughayer MA, Muslim S, Elsabee MZ (2009) Extraction and characterization of chitin and chitosan from marine sources in Arabian gulf. Carbohydr Polym 77:410–419
43. Venkatesan J, Kim SK (2010) Chitosan composites for bone tissue engineering an overview. Mar Drugs 8:2252–2266
44. Muley AB, Chaudhari SA, Mulchandani KH, Singhal RS (2018) Extraction and characterization of chitosan from prawn shell waste and its conjugation with cutinase for enhanced thermo-stability. Int J Biol Macromol 111:1047–1058
45. Paul S, Jayan A, Sasikumar CS, Cherian SM (2014) Extraction and purification of chitosan from chitin isolated from sea prawn Fenneropenaeus indicus. Extraction 7
46. Acosta N, Jiménez C, Borau V, Heras A (1993) Extraction and characterization of chitin from crustaceans. Biomass Bioenergy 5:145–153
47. De Queiroz Antonino RSCM, Lia Fook BRP, De Oliveira Lima VA, De Farias Rached RÍ, Lima EPN, Da Silva Lima RJ, Lia Fook MV (2017) Preparation and characterization of chitosan obtained from shells of shrimp (Litopenaeus vannamei Boone). Mar Drugs 15:141
48. Zhu KY, Merzendorfer H, Zhang W, Zhang J, Muthukrishnan S (2016) Biosynthesis, turnover, and functions of chitin in insects. Annu Rev Entomol 61:177–196
49. Kaya M, Baran T (2015) Description of a new surface morphology for chitin extracted from wings of cockroach (Periplaneta americana). Int J Biol Macromol 75:7–12
50. Kaya M, Baran T, Asan-Ozusaglam M, Cakmak YS, Tozak KO, Mol A, Sezen G (2015) Extraction and characterization of chitin and chitosan with antimicrobial and antioxidant activities from cosmopolitan Orthoptera species (Insecta). Biotechnol Bioproc E 20:168–179
51. Hou L, Shi Y, Zhai P, Le G (2007) Antibacterial activity and in vitro anti-tumor activity of the extract of the larvae of the housefly (Musca domestica). J Ethnopharmacol 111:227–231
52. Ai H, Wang F, Yang Q, Zhu F, Lei C (2008) Preparation and biological activities of chitosan from the larvae of housefly, Musca domestica. Carbohydr Polym 72:419–423
53. Rane KD, Hoover DG (1993) Production of chitosan by fungi. Food Biotechnol 7:11–33
54. Ghormade V, Pathan EK, Deshpande MV (2017) Can fungi compete with marine sources for chitosan production? Int J Biol Macromol 104:1415–1421
55. Synowiecki J, Al-Khateeb NAAQ (1997) Mycelia of Mucor rouxii as a source of chitin and chitosan. Food Chem 60:605–610
56. White SA, Farina PR, Fulton I (1979) Production and isolation of chitosan from Mucor rouxii. Appl Environ Microbiol 38:323–328
57. Dhillon GS, Kaur S, Brar SK, Verma M (2013) Green synthesis approach: extraction of chitosan from fungus mycelia. Crit Rev Biochem Mol Biol 33:379–403

58. Amorim RVDS, Pedrosa RP, Fukushima K, Martínez CR, Ledingham WM, Campos-Takaki-D, Maria G (2006) Alternative carbon sources from sugar cane process for submerged cultivation of Cunninghamella bertholletiae to produce chitosan. Food Technol Biotechnol 44
59. Cardoso A, Lins CIM, Dos Santos ER, Silva MCF, Campos-Takaki GM (2012) Microbial enhance of chitosan production by Rhizopus arrhizus using agroindustrial substrates. Molecules 17:4904–4914
60. Rinaudo M, Domard A (1989) Chitin and chitosan. In: Solution properties of chitosan. Elsevier, London, pp 71–86
61. El Knidri H, Belaabed R, Addaou A, Laajeb A, Lahsini A (2018) Extraction, chemical modification and characterization of chitin and chitosan. Int J Biol Macromol 120:1181–1189
62. Hamed I, Özogul F, Regenstein JM (2016) Industrial applications of crustacean by-products (chitin, chitosan, and chitooligosaccharides): a review. Trends Food Sci Technol 48:40–50
63. Yuan Y, Chesnutt BM, Haggard WO, Bumgardner JD (2011) Deacetylation of chitosan: material characterization and in vitro evaluation via albumin adsorption and pre-osteoblastic cell cultures. Materials 4:1399–1416
64. Arbia W, Arbia L, Adour L, Amrane A (2013) Chitin extraction from crustacean shells using biological methods – a review. Food Technol Biotechnol 51:12–25
65. Rinaudo M (2006) Chitin and chitosan: properties and applications. Prog Polym Sci 31:603–632
66. Croisier F, Jerome C (2013) Chitosan-based biomaterials for tissue engineering. Eur Polym J 49:780–792
67. Ibrahim HM, El-Zairy EMR (2015) Chitosan as a biomaterial – structure, properties, and electrospun nanofibers. Concepts, compounds and the alternatives of antibacterials. IntechOpen, pp 81–101
68. Power KA, Fitzgerald KT, Gallagher WM (2010) Examination of cell–host–biomaterial interactions via high-throughput technologies: a re-appraisal. Biomaterials 31:6667–6674
69. Rodrigues S, Dionísio M, López CR, Grenha A (2012) Biocompatibility of chitosan carriers with application in drug delivery. J Funct Biomater 3:615–641
70. Funkhouser JD, Aronson NN (2007) Chitinase family GH18: evolutionary insights from the genomic history of a diverse protein family. BMC Evol Biol 7:96
71. Punitha S, Girish Y (2010) Polymers in mucoadhesive buccal drug delivery system – a review. Int J Pharm Sci Res 1
72. Boddupalli BM, Mohammed ZN, Nath RA, Banji D (2010) Mucoadhesive drug delivery system: an overview. J Adv Pharm Technol Res 1:381
73. Bertoni FA, González JC, García SI, Sala LF, Bellú SE (2018) Application of chitosan in removal of molybdate ions from contaminated water and groundwater. Carbohydr Polym 180:55–62
74. Sarode S, Upadhyay P, Khosa MA, Mak T, Shakir A, Song S, Ullah A (2019) Overview of wastewater treatment methods with special focus on biopolymer chitin-chitosan. Int J Biol Macromol 121:1086–1100
75. Vakili M, Rafatullah M, Salamatinia B, Abdullah AZ, Ibrahim MH, Tan KB, Amouzgar P (2014) Application of chitosan and its derivatives as adsorbents for dye removal from water and wastewater: a review. Carbohydr Polym 113:115–130
76. Park JU, Song EH, Jeong SH, Song J, Kim HE, Kim S (2018) Chitosan-based dressing materials for problematic wound management. In: Novel biomaterials for regenerative medicine. Springer, Singapore, pp 527–537
77. Singh R, Shitiz K, Singh A (2017) Chitin and chitosan: biopolymers for wound management. Int Wound J 14:1276–1289
78. Balagangadharan K, Dhivya S, Selvamurugan N (2017) Chitosan based nanofibers in bone tissue engineering. Int J Biol Macromol 104:1372–1382
79. Anitha A, Rani VD, Krishna R, Sreeja V, Selvamurugan N, Nair SV, Jayakumar R (2009) Synthesis, characterization, cytotoxicity and antibacterial studies of chitosan,

O-carboxymethyl and N, O-carboxymethyl chitosan nanoparticles. Carbohydr Polym 78:672–677
80. Soundarya SP, Menon AH, Chandran SV, Selvamurugan N (2018) Bone tissue engineering: scaffold preparation using chitosan and other biomaterials with different design and fabrication techniques. Int J Biol Macromol 119:1228–1239
81. Banerjee R (2018) Nanotechnology in drug delivery: present status and a glimpse into the future. Ther Deliv 9(4):231–232
82. Deshpande AA, Rhodes CT, Shah NH, Malick AW (1996) Controlled-release drug delivery systems for prolonged gastric residence: an overview. Drug Dev Ind Pharm 22:531–539
83. Tibbitt MW, Dahlman JE, Langer R (2016) Emerging frontiers in drug delivery. J Am Chem Soc 138:704–717
84. Moritera T, Ogura Y, Honda Y, Wada R, Hyon SH, Ikada Y (1991) Microspheres of biodegradable polymers as a drug-delivery system in the vitreous. Invest Ophthalmol Vis Sci 32:1785–1790
85. Napper DH (1983) Polymeric stabilization of colloidal dispersions, vol 3. Academic Press, Cambridge
86. Patil YP, Jadhav S (2014) Novel methods for liposome preparation. Chem Phys Lipids 177:8–18
87. Akbarzadeh A, Rezaei Sadabady R, Davaran S, Joo SW, Zarghami N, Hanifehpour Y, Nejati Koshki K (2013) Liposome: classification, preparation, and applications. Nanoscale Res Lett 8:102
88. Lee Y, Thompson DH (2017) Stimuli-responsive liposomes for drug delivery. Wiley Interdiscip Rev Nanomed Nanobiotechnol 9:e1450
89. Li SP, Kowarski CR, Feld KM, Grim WM (1988) Recent advances in microencapsulation technology and equipment. Drug Dev Ind Pharm 14:353–376
90. Dai C, Wang B, Zhao H (2005) Microencapsulation peptide and protein drugs delivery system. Colloids Surf B Biointerfaces 41:117–120
91. Burke PA, Klumb LA, Herberger JD, Nguyen XC, Harrell RA, Zordich M (2004) Poly (lactide-co-glycolide) microsphere formulations of darbepoetin alfa: spray drying is an alternative to encapsulation by spray-freeze drying. Pharm Res 21:500–506
92. Okochi H, Nakano M (2000) Preparation and evaluation of w/o/w type emulsions containing vancomycin. Adv Drug Deliv Rev 45:5–26
93. Sinha VR, Trehan A (2003) Biodegradable microspheres for protein delivery. J Control Release 90:261–280
94. Rathore KS, Nema RK (2009) An insight into ophthalmic drug delivery system. Int J Pharm Sci Drug Res 1:1–5
95. Amreddy N, Babu A, Muralidharan R, Panneerselvam J, Srivastava A, Ahmed R, Ramesh R (2018) Recent advances in nanoparticle-based cancer drug and gene delivery. Advances cancer research, vol 137. Academic Press, Cambridge, pp 115–170
96. Wang J, Hu X, Xiang D (2018) Nanoparticle drug delivery systems: an excellent carrier for tumor peptide vaccines. Drug Deliv 25:1319–1327
97. Ashfaq UA, Riaz M, Yasmeen E, Yousaf MZ (2017) Recent advances in nanoparticle-based targeted drug-delivery systems against cancer and role of tumor microenvironment. Crit Rev Ther Drug 34
98. Nagpal K, Singh SK, Mishra DN (2010) Chitosan nanoparticles: a promising system in novel drug delivery. Chem Pharm Bull 58:1423–1430
99. Severino P, da Silva CF, Andrade LN, de Lima Oliveira D, Campos J, Souto EB (2019) Alginate nanoparticles for drug delivery and targeting. Curr Pharm Des 25:1312–1334
100. Danafar H (2016) Applications of copolymeric nanoparticles in drug delivery systems. Drug Dev Res 66:506–519
101. Ulbrich K, Hola K, Subr V, Bakandritsos A, Tucek J, Zboril R (2016) Targeted drug delivery with polymers and magnetic nanoparticles: covalent and noncovalent approaches, release control, and clinical studies. Chem Rev 116:5338–5431

102. Manjunath K, Reddy JS, Venkateswarlu V (2005) Solid lipid nanoparticles as drug delivery systems. Methods Find Exp Clin Pharmacol 27:127–144
103. Sripriyalakshmi S, Jose P, Ravindran A, Anjali CH (2014) Recent trends in drug delivery system using protein nanoparticles. Cell Biochem Biophys 70:17–26
104. Aliabadi HM, Lavasanifar A (2006) Polymeric micelles for drug delivery. Expert Opin Drug Deliv 3:139–162
105. Tanbour R, Martins MA, Pitt GW, Husseini AG (2016) Drug delivery systems based on polymeric micelles and ultrasound: a review. Curr Pharm Des 22:2796–2807
106. Jones MC, Leroux JC (1999) Polymeric micelles–a new generation of colloidal drug carriers. Eur J Pharm Biopharm 48:101–111
107. Biswas S, Kumari P, Lakhani PM, Ghosh B (2016) Recent advances in polymeric micelles for anti-cancer drug delivery. Eur J Pharm Sci 83:184–202
108. Zhou Q, Zhang L, Yang T, Wu H (2018) Stimuli-responsive polymeric micelles for drug delivery and cancer therapy. Int J Nanomedicine 13:2921
109. Gupta KC, Haider A, Choi YR, Kang IK (2014) Nanofibrous scaffolds in biomedical applications. Biomater Res 18:1–11
110. Miguel SP, Figueira DR, Simões D, Ribeiro MP, Coutinho P, Ferreira P, Correia IJ (2018) Electrospun polymeric nanofibres as wound dressings: a review. Colloids Surf B Biointerfaces 169:60–71
111. Zhang Z, Hu J, Ma PX (2012) Nanofiber-based delivery of bioactive agents and stem cells to bone sites. Adv Drug Deliv Rev 64:1129–1141
112. Ranganathan S, Balagangadharan K, Selvamurugan N (2019) Chitosan and gelatin-based electrospun fibers for bone tissue engineering. Int J Biol Macromol 133:354–364
113. Ye K, Kuang H, You Z, Morsi Y, Mo X (2019) Electrospun nanofibers for tissue engineering with drug loading and release. Pharmaceutics 11:182
114. Goonoo N, Bhaw-Luximon A, Jhurry D (2014) Drug loading and release from electrospun biodegradable nanofibers. J Biomed Nanotechnol 10:2173–2199
115. Fréchet JM, Tomalia DA (eds) (2001) Dendrimers and other dendritic polymers. Wiley, New York, p 647
116. Sherje AP, Jadhav M, Dravyakar BR, Kadam D (2018) Dendrimers: a versatile nanocarrier for drug delivery and targeting. Int J Pharm 548:707–720
117. De Brabander-Van Den Berg EM, Meijer EW (1993) Poly (propylenimin)-dendrimere: Synthese in größerem Maßstab durch heterogen katalysierte Hydrierungen. Angew Chem Int Ed 105:1370–1372
118. Lothian-Tomalia MK, Hedstrand DM, Tomalia DA, Padias AB, Hall Jr HK (1997) A contemporary survey of covalent connectivity and complexity. The divergent synthesis of poly (thioether) dendrimers. Amplified, genealogically directed synthesis leading to the de Gennes dense packed state. Tetrahedron 53:15495–15513
119. Seebach D, Herrmann GF, Lengweiler UD, Bachmann BM, Amrein W (1996) Synthesis and enzymatic degradation of dendrimers from (R)-3-hydroxybutanoic acid and trimesic acid. Angew Chem Int Ed Engl 35:2795–2797
120. Lim J, Kostiainen M, Maly J, da Costa VC, Annunziata O, Pavan GM, Simanek EE (2013) Synthesis of large dendrimers with the dimensions of small viruses. J Am Chem Soc 135:4660–4663
121. Kambhampati SP, Kannan RM (2013) Dendrimer nanoparticles for ocular drug delivery. J Ocul Pharmacol Ther 29:151–165
122. Akbarzadeh A, Khalilov R, Mostafavi E, Annabi N, Abasi E, Kafshdooz T, Davaran S (2018) Role of dendrimers in advanced drug delivery and biomedical applications: a review. Exp Oncol 40:178–183
123. Bunggulawa EJ, Wang W, Yin T, Wang N, Durkan C, Wang Y, Wang G (2018) Recent advancements in the use of exosomes as drug delivery systems. J Nanobiotechnol 16:1–13

124. Ha D, Yang N, Nadithe V (2016) Exosomes as therapeutic drug carriers and delivery vehicles across biological membranes: current perspectives and future challenges. Acta Pharm Sin B 6:287–296
125. Hessvik NP, Llorente A (2018) Current knowledge on exosome biogenesis and release. Cell Mol Life Sci 75:193–208
126. Rani S, Ryan AE, Griffin MD, Ritter T (2015) Mesenchymal stem cell-derived extracellular vesicles: toward cell-free therapeutic applications. Mol Ther 23:812–823
127. Batrakova EV, Kim MS (2015) Using exosomes, naturally-equipped nanocarriers, for drug delivery. J Control Release 219:396–405
128. Rosen Y, Elman NM (2009) Carbon nanotubes in drug delivery: focus on infectious diseases. Expert Opin Drug Deliv 6:517–530
129. Wong BS, Yoong SL, Jagusiak A, Panczyk T, Ho HK, Ang WH, Pastorin G (2013) Carbon nanotubes for delivery of small molecule drugs. Adv Drug Deliv Rev 65(15):1964–2015
130. Singh RP, Sharma G, Singh S, Patne SC, Pandey BL, Koch B, Muthu MS (2016) Effects of transferrin conjugated multi-walled carbon nanotubes in lung cancer delivery. Mater Sci Eng C 67:313–325
131. Ali-Boucetta H, Al-Jamal KT, McCarthy D, Prato M, Bianco A, Kostarelos K (2008) Multiwalled carbon nanotube–doxorubicin supramolecular complexes for cancer therapeutics. Chem Commun (Camp) 4:459–461
132. McDevitt MR, Chattopadhyay D, Kappel BJ, Jaggi JS, Schiffman SR, Antczak C, Scheinberg DA (2007) Tumor targeting with antibody-functionalized, radiolabeled carbon nanotubes. J Nucl Med 48:1180–1189
133. Adithya SP, Sidharthan DS, Abhinandan R, Balagangadharan K, Selvamurugan N (2020) Nanosheets-incorporated bio-composites containing natural and synthetic polymers/ceramics for bone tissue engineering. Int J Biol Macromol 164:1960–1972
134. Zhang F, Peng F, Qin L, Yang D, Li R, Jiang S, Zhang P (2019) pH/near infrared dual-triggered drug delivery system based black phosphorus nanosheets for targeted cancer chemo-photothermal therapy. Colloids Surf B Biointerfaces 180:353–361
135. Hatanaka T, Saito T, Fukushima T, Todo H, Sugibayashi K, Umehara S, Okamura Y (2019) Potential of biocompatible polymeric ultra-thin films, nanosheets, as topical and transdermal drug delivery devices. Int J Pharm 565:41–49
136. Peng L, Mei X, He J, Xu J, Zhang W, Liang R, Duan X (2018) Monolayer nanosheets with an extremely high drug loading toward controlled delivery and cancer theranostics. Adv Mater 30:1707389
137. Suhail M, Rosenholm JM, Minhas MU, Badshah SF, Naeem A, Khan KU, Fahad M (2019) Nanogels as drug-delivery systems: a comprehensive overview. Ther Deliv 10:697–717
138. Hamidi M, Rafiei P, Azadi A, Mohammadi-Samani S (2011) Encapsulation of valproate-loaded hydrogel nanoparticles in intact human erythrocytes: a novel nano-cell composite for drug delivery. J Pharm Sci 100:1702–1711
139. Sultana F, Manirujjaman M, Imran-Ul-Haque MA, Sharmin S (2013) An overview of nanogel drug delivery system. J Appl Pharm Sci 3:95–105
140. Kazakov S, Levon K (2006) Liposome nanogel structures for future pharmaceutical applications. Curr Pharm Des 12:4713–4728
141. Nguyen MH, Tran TT, Hadinoto K (2016) Controlling the burst release of amorphous drug–polysaccharide nanoparticle complex via crosslinking of the polysaccharide chains. Eur J Pharm Biopharm 104:156–163
142. Kaban K, Salva E, Akbuga J (2017) *In vitro* dose studies on chitosan nanoplexes for microRNA delivery in breast cancer cells. Nucleic Acids Ther 27:45–55
143. Kalhapure RS, Suleman N, Mocktar C, Seedat N, Govender T (2015) Nanoengineered drug delivery systems for enhancing antibiotic therapy. J Pharm Sci 104:872–905
144. Nakamura M, Tahara Y, Ikehara Y, Murakami T, Tsuchida K, Iijima S, Yudasaka M (2011) Single-walled carbon nanohorns as drug carriers: adsorption of prednisolone and anti-inflammatory effects on arthritis. Nanotechnology 22:465102

145. Zhu S, Xu G (2010) Single-walled carbon nanohorns and their applications. Nanoscale 2:2538–2549
146. Murakami T, Ajima K, Miyawaki J, Yudasaka M, Iijima S, Shiba K (2004) Drug-loaded carbon nanohorns: adsorption and release of dexamethasone in vitro. Mol Pharm 1:399–405
147. Felt O, Buri P, Gurny R (1998) Chitosan: a unique polysaccharide for drug delivery. Drug Dev Ind Pharm 24:979–993
148. Bernkop-Schnürch A, Dünnhaupt S (2012) Chitosan-based drug delivery systems. Eur J Pharm Biopharm 81:463–469
149. Salomon C, Goycoolea FM, Moerschbacher B (2017) Recent trends in the development of chitosan-based drug delivery systems. AAPS PharmSciTech 18(4):933–935
150. Prabaharan M, Mano JF (2004) Chitosan-based particles as controlled drug delivery systems. Drug Deliv 12:41–57
151. Li J, Cai C, Li J, Li J, Li J, Sun T, Yu G (2018) Chitosan-based nanomaterials for drug delivery. Molecules 23:2661
152. Kajdič S, Planinšek O, Gašperlin M, Kocbek P (2019) Electrospun nanofibers for customized drug-delivery systems. J Drug Deliv Sci Technol 51:672–681
153. Kosaraju SL, D'ath L, Lawrence A (2006) Preparation and characterisation of chitosan microspheres for antioxidant delivery. Carbohydr Polym 64:163–116
154. Zhao LM, Shi LE, Zhang ZL, Chen JM, Shi DD, Yang J, Tang ZX (2011) Preparation and application of chitosan nanoparticles and nanofibers. Abbreviated as Braz. J Chem Eng 28:353–362
155. Chandra Hembram K, Prabha S, Chandra R, Ahmed B, Nimesh S (2016) Advances in preparation and characterization of chitosan nanoparticles for therapeutics. Artif Cells Nanomed Biotechnol 44:305–314
156. Gomathi T, Sudha PN, Florence JAK, Venkatesan J, Anil S (2017) Fabrication of letrozole formulation using chitosan nanoparticles through ionic gelation method. Int J Biol Macromol 104:1820–1832
157. Divya K, Jisha MS (2018) Chitosan nanoparticles preparation and applications. Environ Chem Lett 16:101–112
158. Shukla SK, Mishra AK, Arotiba OA, Mamba BB (2013) Chitosan-based nanomaterials: a state-of-the-art review. Int J Biol Macromol 59:46–58
159. Lang X, Wang T, Sun M, Chen X, Liu Y (2020) Advances and applications of chitosan-based nanomaterials as oral delivery carriers: a review. Int J Biol Macromol 154:433–445
160. Kalantari K, Afifi AM, Jahangirian H, Webster TJ (2019) Biomedical applications of chitosan electrospun nanofibers as a green polymer – review. Carbohydr Polym 207:588–600
161. Moreno MA, Gómez-Mascaraque LG, Arias M, Zampini IC, Sayago JE, Ramos LLP, Isla MI (2018) Electrosprayed chitosan microcapsules as delivery vehicles for vaginal phytoformulations. Carbohydr Polym 201:425–437
162. Ifuku S (2014) Chitin and chitosan nanofibers: preparation and chemical modifications. Molecules 19:18367–18380
163. Sedghi R, Shaabani A, Mohammadi Z, Samadi FY, Isaei E (2017) Biocompatible electrospinning chitosan nanofibers: a novel delivery system with superior local cancer therapy. Carbohydr Polym 159:1–10
164. Hamedi H, Moradi S, Hudson SM, Tonelli AE (2018) Chitosan based hydrogels and their applications for drug delivery in wound dressings: a review. Carbohydr Polym 199:445–460
165. Bhattarai N, Gunn J, Zhang M (2010) Chitosan-based hydrogels for controlled, localized drug delivery. Adv Drug Deliv Rev 62:83–99
166. Martínez-Martínez M, Rodríguez-Berna G, Bermejo M, Gonzalez-Alvarez I, Gonzalez-Alvarez M, Merino V (2019) Covalently crosslinked organophosphorous derivatives-chitosan hydrogel as a drug delivery system for oral administration of camptothecin. Eur J Pharm Biopharm 136:174–183

167. Jeddi MK, Mahkam M (2019) Magnetic nano carboxymethyl cellulose-alginate/chitosan hydrogel beads as biodegradable devices for controlled drug delivery. Int J Biol Macromol 135:829–838
168. Hanna DH, Lotfy VF, Basta AH, Saad GR (2020) Comparative evaluation for controlling release of niacin from protein-and cellulose-chitosan based hydrogels. Int J Biol Macromol 150:228–237
169. George D, Maheswari PU, Begum KMS (2020) Cysteine conjugated chitosan based green nanohybrid hydrogel embedded with zinc oxide nanoparticles towards enhanced therapeutic potential of naringenin. React Funct Polym 148:104480
170. Dehghan-Baniani D, Chen Y, Wang D, Bagheri R, Solouk A, Wu H (2020) Injectable in situ forming kartogenin-loaded chitosan hydrogel with tunable rheological properties for cartilage tissue engineering. Colloids Surf B:111059
171. Eicher AC, Dobler D, Kiselmann C, Schmidts T, Runkel F (2019) Dermal delivery of therapeutic DNAzymes via chitosan hydrogels. Int J Pharmaceut 563:208–216
172. Iftime MM, Tartau LM, Marin L (2020) New formulations based on salicyl-imine-chitosan hydrogels for prolonged drug release. Int J Biol Macromol 160:398–408
173. Bao Z, Jiang C, Wang Z, Ji Q, Sun G, Bi S, Liu Y, Chen X (2017) The influence of solvent formulations on thermosensitive hydroxybutyl chitosan hydrogel as a potential delivery matrix for cell therapy. Carbohydr Polym 170:80–88
174. Sami AJ, Khalid M, Jamil T, Aftab S, Mangat SA, Shakoori AR, Iqbal S (2018) Formulation of novel chitosan guargum based hydrogels for sustained drug release of paracetamol. Int J Biol Macromol 108:324–332
175. Sadeqi A, Nejad HR, Kiaee G, Sonkusale S (2018) Cost-effective fabrication of chitosan microneedles for transdermal drug delivery. In 2018 40th annual international conference of the IEEE engineering in medicine and biology society (EMBC), pp 5737–5740
176. AbuAmmouneh L, Kaddoura Z, AbuHantash F, Alkhalidi BA, Al-Halhouli A (2019) Fabrication of dissolvable microneedle patches using an innovative laser-cut mould design to shortlist potentially transungual delivery systems: In vitro evaluation. AAPS Pharmscitech 20:215
177. Indermun S, Luttge R, Choonara YE, Kumar P, Du Toit LC, Modi G, Pillay V (2014) Current advances in the fabrication of microneedles for transdermal delivery. J Control Release 185:130–138
178. Chen MC, Ling MH, Lai KY, Pramudityo E (2012) Chitosan microneedle patches for sustained transdermal delivery of macromolecules. Biomacromolecules 13:4022–4031
179. Serrano-Castañeda P, Escobar-Chávez JJ, Rodríguez-Cruz IM, Melgoza LM, Martinez-Hernandez J (2018) Microneedles as enhancer of drug absorption through the skin and applications in medicine and cosmetology. J Pharm Pharm Sci 21:73–93
180. Chen BZ, Ashfaq M, Zhang XP, Zhang JN, Guo XD (2018) In vitro and in vivo assessment of polymer microneedles for controlled transdermal drug delivery. J Drug Target 26:720–729
181. Al-najjar BY, Hussain SA (2017) Chitosan microspheres for the delivery of chemotherapeutic agents: paclitaxel as a model. Asian J Pharm Clin Res 10:1–5
182. Rajawat GS, Shinde UA, Nair HA (2016) Chitosan-N-acetyl cysteine microspheres for ocular delivery of acyclovir: synthesis and in vitro/in vivo evaluation. J Drug Deliv Sci Technol 35:333–342
183. Kas HS (1997) Chitosan: properties, preparations and application to microparticulate systems. J Microencapsul 14:689–711
184. Shanmuganathan S, Shanumugasundaram N, Adhirajan N, Lakshmi TR, Babu M (2008) Preparation and characterization of chitosan microspheres for doxycycline delivery. Carbohydr Polym 73:201–211
185. He T, Wang W, Chen B, Wang J, Liang Q, Chen B (2020) 5-Fluorouracil monodispersed chitosan microspheres: microfluidic chip fabrication with crosslinking, characterization, drug release and anticancer activity. Carbohydr Polym:116094
186. Tayalia P, Mooney DJ (2009) Controlled growth factor delivery for tissue engineering. Adv Mater 21(32–33):3269–3285

187. Lee SH, Shin H (2007) Matrices and scaffolds for delivery of bioactive molecules in bone and cartilage tissue engineering. Adv Drug Deliv Rev 59:339–359
188. Dang M, Saunders L, Niu X, Fan Y, Ma PX (2018) Biomimetic delivery of signals for bone tissue engineering. Bone Res 6:1–12
189. Kao HJ, Lin HR, Lo YL, Yu SP (2006) Characterization of pilocarpine loaded chitosan/Carbopol nanoparticles. J Pharm Pharmacol 58:179–186
190. Schmidt-Bleek K, Willie BM, Schwabe P, Seemann P, Duda GN (2016) BMPs in bone regeneration: less is more effective, a paradigm-shift. Cytokine Growth F R 27:141–148
191. Arosarena OA, Collins WL (2005) Bone regeneration in the rat mandible with bone morphogenetic protein-2: a comparison of two carriers. Otolaryngol Head Neck Surg 132(4):592–597
192. Issa JPM, do Nascimento C, Bentley MVLB, Del Bel EA, Iyomasa MM, Sebald W, de Albuquerque Jr RF (2008) Bone repair in rat mandible by rhBMP-2 associated with two carriers. Micron 39:373–379
193. Ferrand A, Eap S, Richert L, Lemoine S, Kalaskar D, Demoustier-Champagne S, Kuhn L (2014) Osteogenetic properties of electrospun nanofibrous PCL scaffolds equipped with chitosan-B ased nanoreservoirs of growth factors. Macromol Biosci 14:45–55
194. Park YJ, Kim KH, Lee JY, Ku Y, Lee SJ, Min BM, Chung CP (2006) Immobilization of bone morphogenetic protein-2 on a nanofibrous chitosan membrane for enhanced guided bone regeneration. Biotechnol Appl Biochem 43:17–24
195. Lee EJ, Shin DS, Kim HE, Kim HW, Koh YH, Jang JH (2009) Membrane of hybrid chitosan–silica xerogel for guided bone regeneration. Biomaterials 30:743–750
196. Lee EJ, Kim HE (2016) Accelerated bony defect healing by chitosan/silica hybrid membrane with localized bone morphogenetic protein-2 delivery. Mater Sci Eng C Mater 59:339–345
197. Bae IH, Jeong BC, Kook MS, Kim SH, Koh JT (2013) Evaluation of a thiolated chitosan scaffold for local delivery of BMP-2 for osteogenic differentiation and ectopic bone formation. Biomed Res Int 2013:878930
198. Shah P, Keppler L, Rutkowski J (2014) A review of platelet derived growth factor playing pivotal role in bone regeneration. J Oral Implantol 40:330–340
199. Lee YM, Park YJ, Lee SJ, Ku Y, Han SB, Klokkevold PR, Chung CP (2000) The bone regenerative effect of platelet-derived growth factor-BB delivered with a chitosan/tricalcium phosphate sponge carrier. J Periodontol 71:418–424
200. Park YJ, Lee YM, Park SN, Sheen SY, Chung CP, Lee SJ (2000) Platelet derived growth factor releasing chitosan sponge for periodontal bone regeneration. Biomaterials 21:153–159
201. Im SY, Cho SH, Hwang JH, Lee SJ (2003) Growth factor releasing porous poly (ε--caprolactone)-chitosan matrices for enhanced bone regenerative therapy. Arch Pharm Res 26:76–82
202. Lee JY, Nam SH, Im SY, Park YJ, Lee YM, Seol YJ, Lee SJ (2002) Enhanced bone formation by controlled growth factor delivery from chitosan-based biomaterials. J Control Release 78:187–197
203. Dyondi D, Webster TJ, Banerjee R (2013) A nanoparticulate injectable hydrogel as a tissue engineering scaffold for multiple growth factor delivery for bone regeneration. Int J Nanomedicine 8:47
204. Yilgor P, Tuzlakoglu K, Reis RL, Hasirci N, Hasirci V (2009) Incorporation of a sequential BMP-2/BMP-7 delivery system into chitosan-based scaffolds for bone tissue engineering. Biomaterials 21:3551–3559
205. Nandi SK, Kundu B, Basu D (2013) Protein growth factors loaded highly porous chitosan scaffold: a comparison of bone healing properties. Mater Sci Eng C Mater 33:1267–1275
206. Kim S, Kang Y, Krueger CA, Sen M, Holcomb JB, Chen D, YangY (2012) Sequential delivery of BMP-2 and IGF-1 using a chitosan gel with gelatin microspheres enhances early osteoblastic differentiation. Acta Biomater 8:1768–1777
207. Tian M, Yang Z, Kuwahara K, Nimni ME, Wan C, Han B (2012) Delivery of demineralized bone matrix powder using a thermogelling chitosan carrier. Acta Biomater 8:753–762
208. dos Santos Rodrigues B, Lakkadwala S, Sharma D, Singh J (2019) Chitosan for gene, DNA vaccines, and drug delivery. In: Materials for biomedical engineering. Elsevier, Amsterdam, pp 515–550

209. Buschmann MD, Merzouki A, Lavertu M, Thibault M, Jean M, Darras V (2013) Chitosans for delivery of nucleic acids. Adv Drug Deliv Rev 65:1234–1270
210. Raftery R, O'brien FJ, Cryan SA (2013) Chitosan for gene delivery and orthopedic tissue engineering applications. Molecules 18:5611–5647
211. Lu CH, Chang YH, Lin SY, Li KC, Hu YC (2013) Recent progresses in gene delivery-based bone tissue engineering. Biotechnol Adv 31:1695–1706
212. Bez M, Pelled G, Gazit D (2020) BMP gene delivery for skeletal tissue regeneration. Bone:115449
213. Malek-Khatabi A, Javar HA, Dashtimoghadam E, Ansari S, Hasani-Sadrabadi MM, Moshaverinia A (2020) In situ bone tissue engineering using gene delivery nanocomplexes. Acta Biomater 108:326–336
214. Raftery RM, Mencía-Castaño I, Sperger S, Chen G, Cavanagh B, Feichtinger GA, Redl H, Hacobian A, O'Brien FJ (2018) Delivery of the improved BMP-2-advanced plasmid DNA within a gene-activated scaffold accelerates mesenchymal stem cell osteogenesis and critical size defect repair. J Control Release 283:20–31
215. Saengkrit N, Sajomsang W, Tencomnao T (2011) Nano-polyplex as a non-viral gene carrier for the expression of bone morphogenetic protein in osteoblastic cells. Carbohydr Polym 86:587–593
216. Li J, Lin J, Yu W, Song X, Hu Q, Xu JH, Wang H, Mehl C (2017) BMP-2 plasmid DNA-loaded chitosan films – a new strategy for bone engineering. J Cranio Maxill Surg 45:2084–2091
217. Raftery RM, Castaño IM, Chen G, Cavanagh B, Quinn B, Curtin CM, Cryan SA, O'Brien FJ (2017) Translating the role of osteogenic-angiogenic coupling in bone formation: highly efficient chitosan-pDNA activated scaffolds can accelerate bone regeneration in critical-sized bone defects. Biomaterials 149:116–127
218. Sriram M, Sainitya R, Kalyanaraman V, Dhivya S, Selvamurugan N (2015) Biomaterials mediated microRNA delivery for bone tissue engineering. Int J Biol Macromol 74:404–412
219. Chen X, Gu S, Chen BF, Shen WL, Yin Z, Xu GW, Hu JJ, Zhu T, Li G, Wan C, Ouyang HW (2015) Nanoparticle delivery of stable miR-199a-5p agomir improves the osteogenesis of human mesenchymal stem cells via the HIF1a pathway. Biomaterials 53:239–250
220. Balagangadharan K, Chandran SV, Arumugam B, Saravanan S, Venkatasubbu GD, Selvamurugan N (2018) Chitosan/nano-hydroxyapatite/nano-zirconium dioxide scaffolds with miR-590-5p for bone regeneration. Int J Biol Macromol 111:953–958
221. Wu G, Feng C, Hui G, Wang Z, Tan J, Luo L, Xue P, Wang Q, Chen X (2016) Improving the osteogenesis of rat mesenchymal stem cells by chitosan-based-microRNA nanoparticles. Carbohydr Polym 138:49–58
222. Wu G, Feng C, Quan J, Wang Z, Wei W, Zang S, Kang S, Hui G, Chen X, Wang Q (2018) In situ controlled release of stromal cell-derived factor-1α and antimiR-138 for on-demand cranial bone regeneration. Carbohydr Polym 182:215–224
223. Dadashpour M, Pilehvar-Soltanahmadi Y, Zarghami N, Firouzi-Amandi A, Pourhassan-Moghaddam M, Nouri M (2017) Emerging importance of phytochemicals in regulation of stem cells fate via signaling pathways. Phytother Res 31:1651–1668
224. Saxena M, Saxena J, Nema R, Singh D, Gupta A (2013) Phytochemistry of medicinal plants. J Pharmacogn Phytochem 1
225. Ali A, Ahmed S (2018) A review on chitosan and its nanocomposites in drug delivery. Int J Biol Macromol 109:273–286
226. Leena RS, Vairamani M, Selvamurugan N (2017) Alginate/gelatin scaffolds incorporated with Silibinin-loaded chitosan nanoparticles for bone formation in vitro. Colloids Surf B 158:308–318
227. Li Y, Liu T, Zheng J, Xu X (2013) Glutaraldehyde-crosslinked chitosan/hydroxyapatite bone repair scaffold and its application as drug carrier for icariin. J Appl Polym Sci 130:1539–1547
228. Xin BC, Wu QS, Jin S, Luo AH, Sun DG, Wang F (2019) Berberine promotes osteogenic differentiation of human dental pulp stem cells through activating EGFR-MAPK-Runx2 pathways. Pathol Oncol Res:1–9

229. Cai B, Zou Q, Zuo Y, Mei Q, Ma J, Lin L, Li Y (2018) Injectable gel constructs with regenerative and anti-infective dual effects based on assembled chitosan microspheres. ACS Appl Mater Interfaces 10:25099–25112
230. Zhang ZR, Leung WN, Li G, Kong SK, Lu X, Wong YM, Chan CW (2017) Osthole enhances osteogenesis in osteoblasts by elevating transcription factor osterix via cAMP/CREB signaling in vitro and in vivo. Nutrients 9:588
231. Wang L, Zheng S, Huang G, Sun J, Pan Y, Si Y, Guo Y (2020) Osthole-loaded N-octyl-O-sulfonyl chitosan micelles (NSC-OST) inhibits RANKL-induced osteoclastogenesis and prevents ovariectomy-induced bone loss in rats. J Cell Mol Med 24:4105–4117
232. Fan J, Li J, Fan Q (2015) Naringin promotes differentiation of bone marrow stem cells into osteoblasts by upregulating the expression levels of microRNA-20a and downregulating the expression levels of PPARγ. Mol Med Rep 12:4759–4765
233. Zhang X, Chen Q, Liu J, Fan C, Wei Q, Chen Z, Mao X (2017) Parthenolide promotes differentiation of osteoblasts through the Wnt/β-catenin signaling pathway in inflammatory environments. J Interferon Cytokine Res 37:406–414
234. Guo Z, Bo D, He P, Li H, Wu G, Li Z, Li Q (2017) Sequential controlled-released dual-drug loaded scaffold for guided bone regeneration in a rat fenestration defect model. J Mater Chem B 5:7701–7710
235. Xu D, Xu L, Zhou C, Lee WY, Wu T, Cui L, Li G (2014) Salvianolic acid B promotes osteogenesis of human mesenchymal stem cells through activating ERK signaling pathway. Int J Biochem Cell B 51:1–9
236. Ji C, Bi L, Li J, Fan J (2019) Salvianolic acid B-loaded chitosan/hydroxyapatite scaffolds promotes the repair of segmental bone defect by angiogenesis and osteogenesis. Int J Nanomedicine 14:8271
237. Wu H, Lei P, Liu G, Zhang YS, Yang J, Zhang L, Hu Y (2017) Reconstruction of large-scale defects with a novel hybrid scaffold made from poly (L-lactic acid)/nanohydroxyapatite/alendronate-loaded chitosan microsphere: in vitro and in vivo studies. Sci Rep 7:1–14
238. Petit C, Batool F, Stutz C, Anton N, Klymchenko A, Vandamme T, Huck O (2020) Development of a thermosensitive statin loaded chitosan-based hydrogel promoting bone healing. Int J Pharmaceut:119534
239. Chiang ZC, Yu SH, Chao AC, Dong GC (2012) Preparation and characterization of dexamethasone-immobilized chitosan scaffold. J Biosci Bioeng 113:654–660
240. Gümüşderelioglu M, Aday S (2011) Heparin-functionalized chitosan scaffolds for bone tissue engineering. Carbohydr Res 346:606–613
241. Zhu P, Huang G, Zhang B, Zhang W, Dang M, Huang Z (2019) Assessment of fracture healing properties of lovastatin loaded nanoparticles: preclinical study in rat model. Acta Biochim Pol 66:71–76
242. Zhang ML, Cheng J, Xiao YC, Yin RF, Feng X (2017) Raloxifene microsphere-embedded collagen/chitosan/β-tricalcium phosphate scaffold for effective bone tissue engineering. Int J Pharm 518:80–85
243. Khajuria DK, Zahra SF, Razdan R (2018) Effect of locally administered novel biodegradable chitosan based risedronate/zinc-hydroxyapatite intra-pocket dental film on alveolar bone density in rat model of periodontitis. J Biomater Sci Polym E 29:74–91
244. Rezazadeh M, Parandeh M, Akbari V, Ebrahimi Z, Taheri A (2019) Incorporation of rosuvastatin-loaded chitosan/chondroitin sulfate nanoparticles into a thermosensitive hydrogel for bone tissue engineering: preparation, characterization, and cellular behavior. Pharm Dev Technol 24:357–367
245. Delan WK, Zakaria M, Elsaadany B, ElMeshad AN, Mamdouh W, Fares AR (2020) Formulation of simvastatin chitosan nanoparticles for controlled delivery in bone regeneration: optimization using Box-Behnken design, stability and in vivo study. Int J Pharmaceut:119038
246. Lu Y, Li M, Li L, Wei S, Hu X, Wang X, Yin Q (2018) High-activity chitosan/nano hydroxyapatite/zoledronic acid scaffolds for simultaneous tumor inhibition, bone repair and infection eradication. Mater Sci Eng C Mater 82:225–233

Adv Polym Sci (2021) 288: 163–190
https://doi.org/10.1007/12_2021_92

Published online: 14 May 2021

Chitosan Based Biomaterials for Periodontal Therapy

Arun Kumar Rajendran and R. Jayakumar

Contents

Abstract Periodontal diseases remain one of the major causes for tooth loss. Untreated periodontal disease could be debilitating for an individual and it could also lead to various other systemic health issues such as diabetes, stroke, cardiovascular diseases, Alzheimer's disease, and obesity. This would greatly reduce the quality of life of an individual. The loss of tooth structures and an unhealthy periodontium will lead to loss or reduced function of mastication and aesthetics. Treatment of periodontitis is a multistage process which requires intervention by debridement and prevention of microbial load by antimicrobials, delivery of bio-active growth enhancers, providing tissue growth supporting scaffolds, and also tissue growth guiding membranes and/or constructs to properly control the spatio-temporal regeneration of periodontal complex. Chitosan is one such a potential candidate to be used in the periodontal therapy at various stages. The versatility of chitosan to be synthesized into different forms, made into different formulations

A. K. Rajendran and R. Jayakumar (✉)
Centre for Nanosciences and Molecular Medicine, Amrita Vishwa Vidyapeetham, Kochi, India
e-mail: rjayakumar@aims.amrita.edu

for easy handling and application by the clinicians, the ability to incorporate multitude of drugs, growth factors, biodegradability, and so forth make it an interesting biomaterial. Some of the chitosan based materials for periodontal therapy include nano formulations, gels, hydrogels, scaffolds, membranes, fibers, and so forth. This chapter reviews recently developed chitosan based materials with different strategies used for enhancing the therapeutic outcomes of periodontal treatment.

Keywords Chitosan · Fibers · Hydrogels · Osteogenesis · Periodontal · Scaffolds · Tissue engineering

1 Introduction

> "Behind every smile there's teeth."
> -Confucius

A great smile is a window to oneself and it can create a long-lasting impression when a new person is met. A great smile requires a healthy set of teeth and for a set of teeth to be healthy, the periodontal health is of prime importance. A healthy periodontium and thereby healthy set of teeth are also prime importance for functions such as speech, mastication [1]. The periodontium or the tissues surrounding the teeth in the oral cavity include the gums, the jaw bones, the cementum, and the element which links the alveolar bone and the cementum of the tooth, the periodontal ligament fibers [1, 2]. The combined effect of these two hard tissues and two soft tissues results in proper fixation of the tooth in its position without any mobility, helps to distribute the masticatory load in an ideal way, helps in proper positioning of the jaw during physiological functions, and also provides a barrier around the neck of the tooth, so that the apical regions of the tooth are protected from microbial activity [3, 4]. However, every healthy system is prone to disease due to improper care, neglect of very early signs and symptoms, other underlying diseases, trauma, and aging and periodontium is not an exception to this [5–8]. Once this healthy periodontium is diseased, it causes a major decrease in quality of life of an individual, due to conditions such as bad breath, painful gums and teeth, shaking teeth, and even loss of teeth [9, 10]. This results in loss of physiological function and aesthetics of an individual. Once there is a loss of periodontal tissues, various strategies have been practiced to regain the healthy periodontal tissues as much as possible [11]. As the science progresses, we understand better how the periodontal tissues are being lost and how this issues can be addressed with newer biomaterials [12–16]. Chitosan, a biopolymer, is one such biomaterial which has been explored a lot as a choice of biomaterial in the field of tissue engineering and regenerative medicine [17–21]. It has been extensively used in commercial preparations of hemostatic products, dressings, bandages, creams for burn wounds, scar repair creams, and in some anti-ageing creams [22–29]. Due to its biocompatibility [30, 31], beneficial biodegradability [32], and easy processability, it has been explored in various applications in periodontal therapy. This chapter discusses the advances that have been made in the last few years, using chitosan as a biomaterial for use in periodontal therapies.

2 Periodontal Problems

Periodontal disease is one of the most commonly found problems worldwide. The prevalence ranges from 20 to 50% worldwide [33]. The periodontal problems mostly arise due to the negligence of the oral health such as improper brushing or improper oral hygiene practices [2]. However, there could be other etiologies such as trauma [8], underlying disease conditions [6], radiation [34], smoking [35], or developmental anomalies [36]. Periodontal problems have also been known to arise due to diabetes and poor periodontal health could also aggravate the underlying diabetes [37]. Similarly, studies have shown that periodontitis has been linked to aggravate conditions such as heart and lung diseases, stroke, respiratory problems, and obesity [38]. Therefore, maintaining a healthy periodontium is of prime importance.

Most often the periodontal problems arising due to negligent oral care start as a mild accumulation of debris and plaque attached to the dentinogingival junction [5, 9, 10]. This debris or the plaque then gets accumulated into the gingival sulcus and causes slight inflammation of the gingival tissues [39]. This is characterized by the classical inflammatory signs such as rubor, tumor, calor, and dolor. The accumulated debris and plaque also attract different species of bacteria and aid as a colonizing ground, thus enabling bacteria to establish themselves and release toxins. The plaque usually consists of higher percentage of gram-negative bacteria and can also include motile anaerobic rods, filamentous and cocci bacteria [40]. Due to these, the tissue environment around gingival margins and gingival sulcus becomes acidic in pH. If proper oral cleaning techniques to remove the debris or plaque accumulation are not followed, then this would lead to swelling of the gums around the teeth, which is visualized as loss of scalloped gingival margins and increased redness in the gingival margins along the facial and lingual surfaces of the teeth. Bleeding from the gingival tissues around the teeth while brushing and also on probing the gingival sulcus could be seen upon clinical examination. This initial inflammation of the gingival tissues is known as gingivitis. In the advanced stages of the gingivitis, the gums start to recede and the probing depth increases in the gingival sulcus known as the deepened pocket (Fig. 1) [9, 10].

Gingivitis, when left untreated, progresses to a next stage known as periodontitis. The accumulated plaque and debris hardens and gets strongly attached to the tooth surface, especially at the dento-enamel junction and the gingival margins. The gingival sulcus starts to deepen down more, leading to more accumulation of debris. The debris, plaque, and the bacteria combined together lead to hardened structure known as calculus. Once the calculus is established, the by-products from the bacterial colonies start to destroy the anchoring structures around the tooth such as the periodontal ligaments. Thus the gums start to recede exposing more amount of the tooth root. This results in a clear interdental space, i.e. space between the teeth

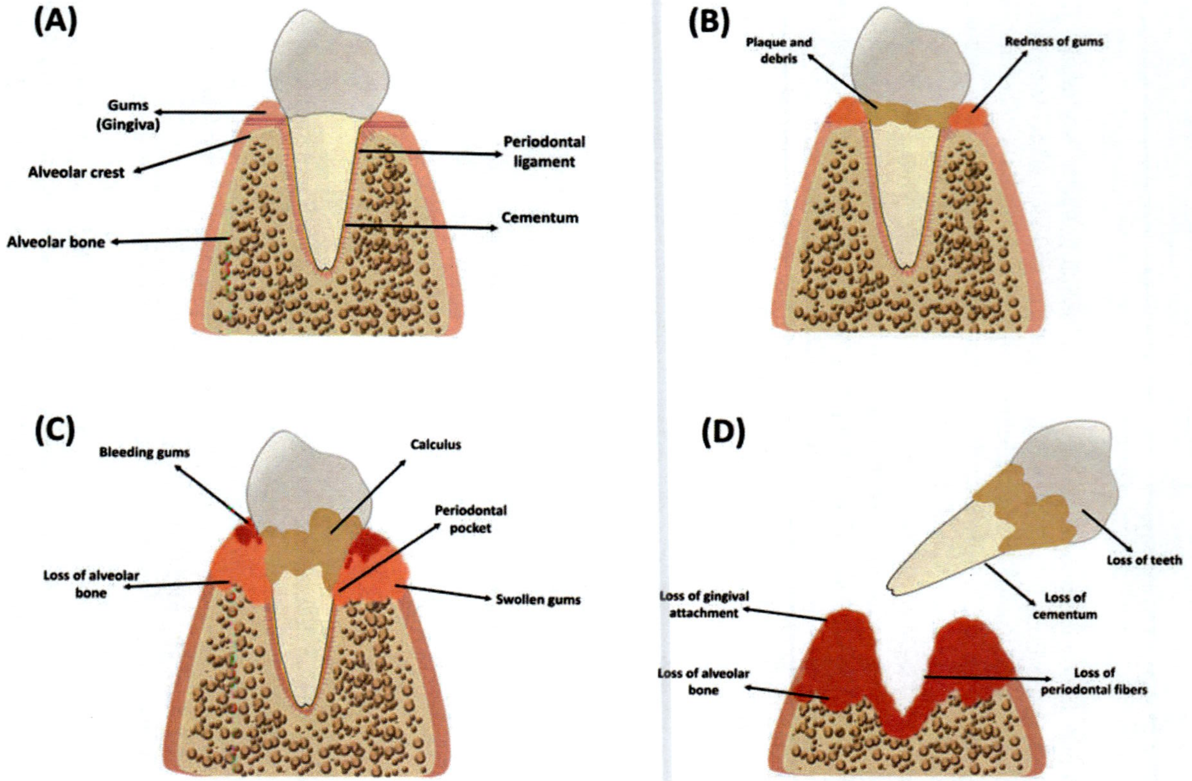

Fig. 1 Periodontium and stages of periodontal disease. (**a**) Diagrammatic representation of healthy periodontal complex; (**b**) Gingivitis, the initial stages of periodontal disease; (**c**) Periodontitis, characterized by accumulation of hardened calculus, swollen gums, bleeding, loss of alveolar crestal bone and periodontal pocket; (**d**) Advanced periodontitis, characterized by loss of teeth, severe loss of periodontal elements

and forms a periodontal pocket [41]. As a domino effect, the alveolar bone starts to resorb in the dento-alveolar ridge regions. This leads to loss of attachment and results in different grades of mobility of the teeth depending on the level of loss of periodontal structures. When still left untreated, the disease progresses to advanced periodontitis. This stage of the disease exhibits extreme mobility of teeth due to almost complete loss of attachment, severe resorption of gingival tissue, exposure of tooth roots, severe alveolar bone loss, thereby leading to loss of teeth [8].

2.1 *Current Treatment Modalities and Their Shortcomings*

As the saying goes, prevention is better than cure is much true in the case of periodontal diseases. Regular dental checkup visits and following proper oral hygiene practices will definitely help to mitigate the inflammation to the periodontal tissues. However due to unavoidable circumstances such as underlying health conditions, improper oral care routines, and missing the regular visits to dentists, thereby missing the early diagnosis can lead to various stages of periodontitis. Furthermore, proper oral hygiene practices such as correct brushing techniques, cleaning the oral cavity after consumption of food, and use of plaque reducing agents can effectively reduce the incidence of gingivitis [42]. The treatment modalities usually vary from patient to patient, depending on the extent of the gingivitis or periodontitis when the patients present themselves to the dentist.

The treatment of initial stages of gingivitis usually requires cleaning of the tooth root surface and the removal of plaque and debris, followed by proper oral hygiene instructions. This would lead to reversal of the inflammation of the gingival tissues and the swollen gums, bleeding, and bad breath returns to healthy condition [43]. In advanced cases, periodontal surgery is done, followed by thorough debridement and curettage of the periodontal pocket, application of bone grafts if needed, use of gingival flaps and antibiotic dressings (Fig. 2).

2.2 *Use of Biomaterials in Periodontal Therapy*

An effective periodontal therapy involves uneventful healing and regeneration of the gingiva, alveolar bone, periodontal ligaments, and the cementum. The very common treatment of periodontitis involves the debridement of the calculus around the roots, removal of very high mobile tooth, removal of the diseased gingival tissues, alveolar bone followed by application of antimicrobials and closure using different gingival flap designs to enhance the healing and restore the gum line as much as possible [11, 42]. One of the ideal expected outcomes is the regeneration of the lost alveolar bone, which in turn can restore the gingival height and thereby hold the tooth firmly. This would result in the most aesthetic and functional outcome. However, stages of periodontal disease such as periodontitis and advanced periodontitis would require

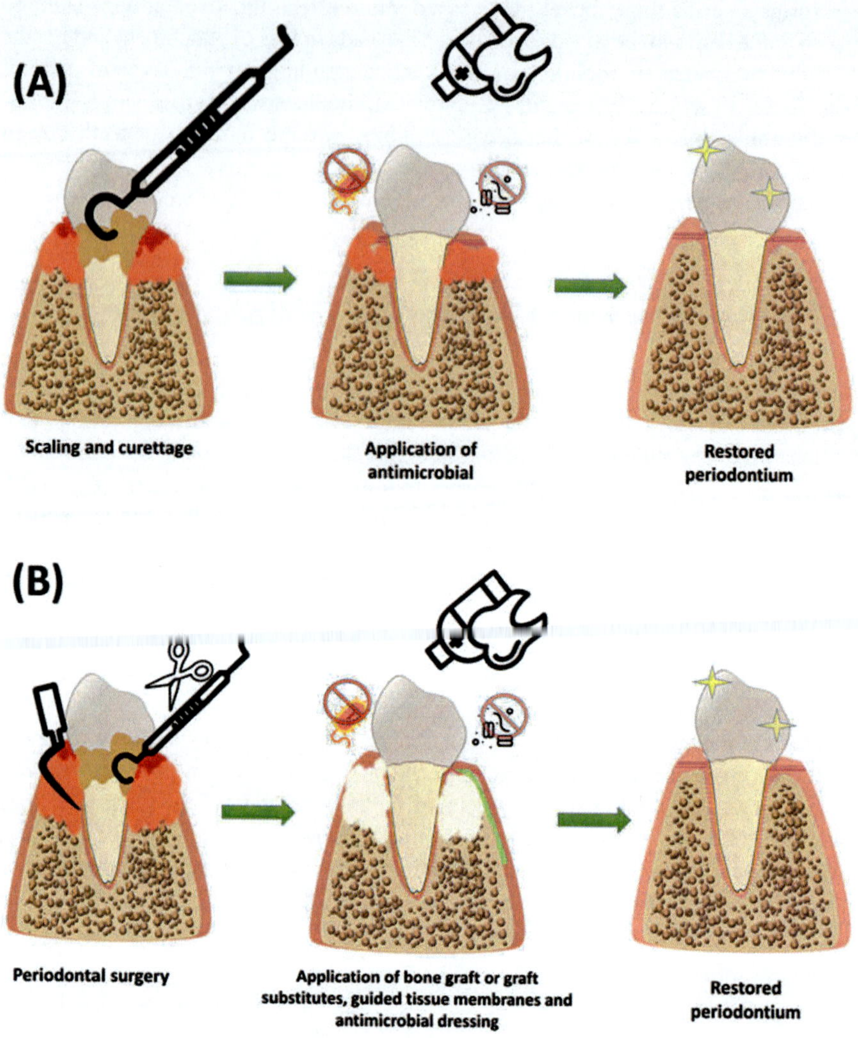

Fig. 2 Treatment modalities for periodontal disease. **(a)** In case of early periodontitis, scaling is done, then application of antimicrobial mouthwashes is prescribed. Depending on the severity, antimicrobial dressings are also given. **(b)** when there is severe periodontitis, associated with bone loss, periodontal surgery is done, necrotic debridement is done, bone graft or graft substitutes are placed, if necessary GTR/GBR membranes are placed, and antimicrobial dressings are given

augmentative procedures such as autograft, which is still the gold standard for periodontal regeneration, which would greatly improve the treatment outcomes. Autografts excel in tissue regeneration due to the well-known fact that they could serve as osteoconductive, osteoinductive as well as osteogenic scaffolds and also greatly reduce the graft rejections due to immune compatibility [44]. Although autografts are accepted as gold standards for periodontal regeneration, due to the lack of enough tissues from the same patient and also considering the donor site morbidity, clinicians have always looked for alternatives such as allografts or xenografts [45]. However still there are shortcomings in using these, such as graft compatibility, availability of donors, and immune rejections. To overcome these issues, researchers have always been looking out for newer materials which could be used as viable substitutes for autografts which could enhance the alveolar bone regeneration in periodontal treatment. This has led to experimentation of various categories of materials such as minerals, bioceramics, polymers, and composites thereof as biological substitutes [46, 47]. Some of the biomaterials used for this purpose include, but not limited to, ceramics such as calcium sulfate, calcium phosphate, calcium silicate, polymers such as poly(ethylene glycol) (PEG), poly (vinyl alcohol) (PVA), PLLA, PCL, PLGA [48]. Researchers have also explored the possibilities of using biopolymers such as alginate, agar, agarose, chitin, chitosan, collagen, and lots more, as they provide comparatively better biocompatibility and could easily mimic the building blocks of extracellular matrix [49]. These biomaterials have been formulated into various forms such as powders, sponges, or injectable gels to aid in application to the periodontal disease site to enhance the healing [17]. Additionally, with the advance of research and technologies, nanoformulations of the above listed materials and inclusion of various growth factors to promote alveolar bone regeneration have been shown [16, 50].

While efforts have been made to improve the alveolar bone regeneration, to augment the periodontal regeneration, efforts have been made to enhance the soft tissue healing using various soft tissue mucosal flaps or grafts [51]. Another strategy employed to improve the outcome of the treatment is to utilize guided tissue regeneration (GTR) or guided bone regeneration (GBR) [52]. The GBR or GTR membrane helps to contain the ingrowth of the fibrous tissues before the regeneration of alveolar bone tissue, thereby aiding in the proper spatial regeneration of bone tissue. Various synthetic polymers such as PEG, PVA, PLLA, PLGA, PTFE, and bio polymers such as collagen, gelatin, silk, alginate, chitosan, chitin, keratin, fibroin, cellulose, and so forth have been experimented to design a potential GTR or GBR membranes [53, 54]. By adjusting the various parameters of the polymer compositions, researchers have managed to control the degradation of the membrane in a controlled manner, which will aid in the spatio-temporal control of the regenerating tissues, and will help in proper compartmentalization of the new growing tissues, thereby leading to new tissues which closely resemble the native tissue/organ architecture. Furthermore, specific drugs such as minocycline, doxycycline, and similar antibiotics could be incorporated into these preparations which could slowly release the drug aiding in better microbial control in the periodontal pockets, which would significantly improve the outcome of the treatment [55–59]. Further, when combined with the growth factor delivery systems, this could greatly enhance the periodontal regeneration (Fig. 3).

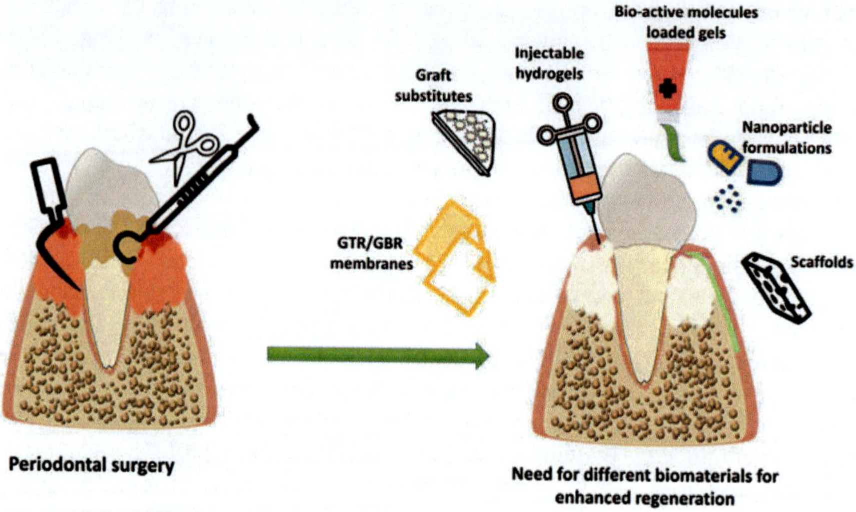

Fig. 3 Schematic showing the various types of biomaterials that could be applied, after periodontal surgery, to enhance the periodontal regeneration

3 Chitosan as a Biomaterial

As the world turns away from non-renewable sources for obtaining polymers, extracting biopolymers from waste materials and utilizing them for biomedical applications is gaining great interest [60–64]. One such biopolymer which could be obtained from large amounts of seafood waste is chitin and its derivative chitosan, which has a very promising future in the field of biomaterials and tissue regeneration [17, 65]. Chitin forms the exoskeleton of crustaceans such as crabs, shrimp, squid and is also found in exoskeleton of insects [59]. Chitin is made up of 2-(acetylamino)-2-deoxy-D-glucose, which is a derivative of glucose, molecular structure resembling cellulose and its function could be compared to that of keratin. Chitin and chitosan have been widely used as a food additive as a stabilizer, thickener as well as emulsifier [66]. As a biomaterial it has been greatly explored due to its extensive availability as well as its versatility in processing into different forms such as powders [67], solutions [68], gels [69], hydrogels [70], sponges [71], bandages [29], membranes [72], fibers [61], sutures [73], beads [74], nanoparticles [75], drug carriers [76], and so forth (Fig. 4). Further, chitin and chitosan also exhibit excellent processing compatibility with other polymers such a PCL [77], PLLA [78], PHBV [79], PLGA [50]; with ceramics such as calcium sulfate [80], calcium phosphates [77], calcium silicates [81], and as such, to synthesize composite biomaterials for various tissue engineering applications. Researchers have also explored and have shown successful incorporation of drugs [29], small molecules [82], growth factors [83], DNA [84], and peptides [85] for controlled delivery of these molecules in tissues [86].

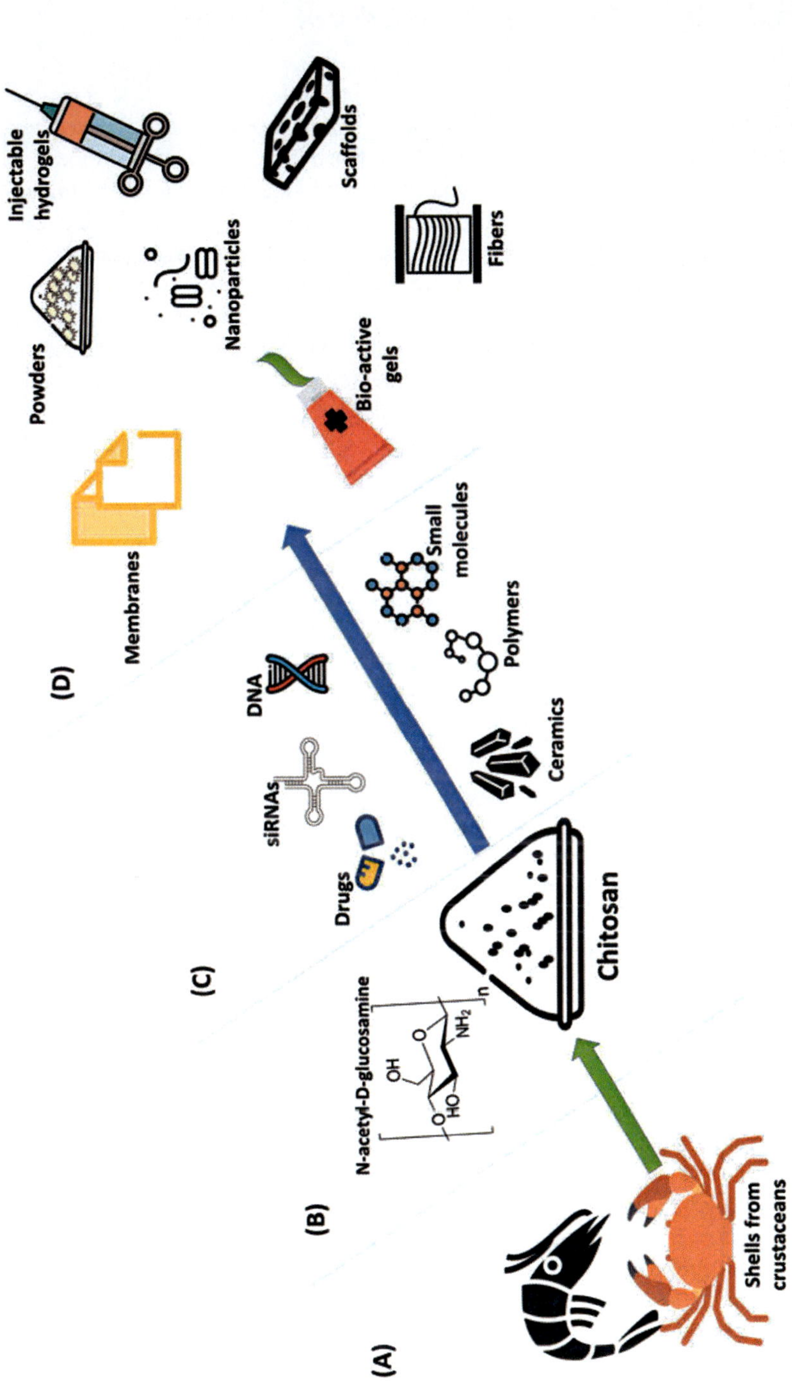

Fig. 4 Utilization of chitosan in the field of biomaterials. (**a**) Denotes the source of extraction; (**b**) Chitosan in its unprocessed powder form; (**c**) the additives that could be incorporated into chitosan while processing; (**d**) various forms of biomaterials that could be obtained depending on the medical application

3.1 Advantages of Chitosan in Periodontal Therapy

As seen in the previous sections, chitosan has a very high versatility to be combined with different polymers, ceramics, made into different forms and can be used as a carrier for drugs. This is very advantageous when utilized in the treatment of periodontal therapy for the following reasons. Periodontal disease involves inflammation and degeneration of two soft tissues and two hard tissues which has very different architectures [1]. Thus, for an effective treatment and regeneration of all these structures, different approaches and formulations will be required. The regeneration of periodontal structures also involves strategic spatio-temporal release of multiple growth factors to initiate and sustain the regeneration of gingival tissues, alveolar bone, periodontal ligament, and cementum to result in complete healing [87]. It has been shown that addition of bioceramics to the chitosan membranes alters the mechanical properties, stiffness and extensibility and also enhances the osteoconductivity [88–90]. Similarly, growth factors such as platelet derived growth factor (PDGF) which greatly enhances bone regeneration have been loaded into chitosan sponges which showed a promising alveolar bone regeneration [91]. The porous nature of the chitosan sponges proves advantageous for the loading of small molecules and peptides into these pores and slowly released at a predetermined time interval [92]. Furthermore, the presence of pores and the ability to tune these pore sizes also facilitate the cells from the adjacent healthy tissues to move into these spaces, utilize the growth factors, and can form bone tissues, thus regenerating the tissue. Since the breakdown units of chitosan are D-glucosamine and N-acetyl-D-glucosamine, which are one of the basic building blocks of native extra cellular matrix, the endogenous cells can utilize and process these molecules to build the tissues [93]. Thus chitosan can serve a great purpose even after degradation in vivo.

Furthermore, to mitigate the presence of huge amounts of microbial colonies in the affected periodontal pockets, it would be also a great choice to choose a material, such as chitosan, into which antimicrobial drugs can be loaded and slowly released over a course of healing period [10]. Interestingly, it has been found that chitosan, in its non-crosslinked form, itself can exhibit antibacterial activity against *P. gingivalis* and *S. mutans*, by rapid adsorption and clustering of bacteria on the chitosan surface due to its positive charge [94]. It has been shown that antimicrobials such as chlorhexidine gluconate can be effectively incorporated into bioadhesive chitosan gels, which are then injected into the periodontal pockets for efficient treatment against *Porphyromonas gingivalis* [95]. Likewise, a chitosan/collagen composite membrane has been developed which was incorporated with minocycline loaded nanoparticle serving as an antibacterial GBR membrane [96]. This membrane was proved to enhance angiogenesis as well as bone regeneration. Similarly, lipophilic drugs such as ipriflavone have been incorporated into chitosan-PLGA film to give a sustained release for at least 20 days, indicating that such systems could be of great potential as GBR membrane which prevents bone resorption [97]. Thus we could see that using chitosan as a biomaterial or part of a biomaterial has great advantages for utilizing in periodontal therapies. Some of the research strategies utilizing these properties of chitosan into an advantageous situation for use in periodontal therapy will be discussed in the upcoming sections.

4 Application of Chitosan in Periodontal Therapy

4.1 Scaffold

Almost all the complex structures which are being built require a temporary structure which can help in supporting the building crew, help in transport and storage of building materials, etc. Such temporary structures are termed as scaffoldings. Similarly, when a tissue or an organ is being regenerated in a void space, scaffolds provide a temporary structure in which the cells, which are the building blocks and crew, can attach themselves, carry out their functions, secrete the specific extracellular matrix, and thus accomplish building a healthy tissue [98]. Chitosan with its great versatile properties for being a biomaterial, as discussed previously, could act as a great scaffold. This is due to one of the facts that chitosan is greatly biodegradable, i.e. it stays temporarily in the defect site, gets degraded by the cells and its degradation products get absorbed and serve as building blocks for new tissue [99]. Furthermore, chitosan can be made into porous sponges in which the parameters such as the mechanical strength, pore size, pore alignment, number of pores, pore continuity, and so forth could be easily tuned according to the needs. Since the periodontium consists of gingiva, alveolar bone, periodontal ligament, and cementum, all of which have different tissue organization and architecture from each other, using chitosan would be a great choice to tailor design scaffolds, according to the need of each periodontal tissues, which would greatly improve the outcome [100]. Also, it has been known that each periodontal tissue requires different growth factor stimulations in order to regenerate them efficiently. Thus it would be a great idea to incorporate different bio-active molecules such as growth factor or tissue specific growth inducers into the specifically designed chitosan scaffolds to aid in complete regeneration. In this section let us see how various researchers have utilized chitosan scaffolds either without or with bio-active molecules to accomplish the never-ending goal of synthesizing a near ideal scaffolds for periodontal regeneration.

4.1.1 Without Bio-active Molecules

Chitosan by itself can be a great biomaterial due to the reasons discussed above. Arancibia et al. demonstrated that chitosan particles alone can have good antibacterial as well as anti-inflammatory properties in relation to the periodontal therapy [101]. Their studies showed that chitosan particles synthesized through ionic gelation were effectively inhibiting the most common periodontal pathogens such as *Porphyromonas gingivalis* and *Aggregatibacter actinomycetemcomitans*. They have

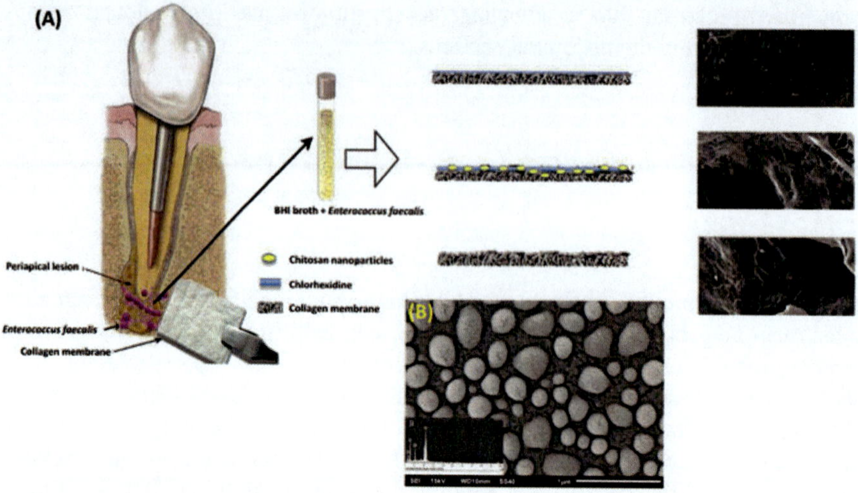

Fig. 5 A antimicrobial collagen membrane. (**a**) Graphical representation of collagen membrane loaded with chlorhexidine-chitosan nanoparticles. (**b**) SEM image showing the size and shape of the chitosan nanoparticle [75]. Reproduced with kind permission from Elsevier

gone ahead and showed that the same chitosan formulation was able to function as anti-inflammatory agents by modulating the JNK pathway. Similarly, Costa et al. studied the effect of high and low molecular weight chitosan solutions against different periodontal pathogens such as *Porphyromonas gingivalis, Prevotella intermedia, Prevotella buccae, Tannerella forsythensis,* and *Aggregatibacter actinomycetemcomitans* [102]. Their results indicated that chitosan was effective in reduction of multispecies biofilm formation, which is the initial step in periodontal disease, and this activity was seen even after 168 h. They suggested that this effect could be due to the inhibition of quorum sensing reporter system by chitosan. Chitosan nanoparticles containing chlorhexidine were synthesized and tested against *Enterococcus faecalis* (Fig. 5) [75]. It was found that when chlorhexidine was encapsulated into chitosan nanoparticles a synergistic antibacterial effect could be seen, thus proving the enhancement of an already existing drug by chitosan nanoparticle. It was suggested that this system could be used to coat the GBR membranes prior to their implantation, so that the microbial load will be reduced in the periodontal pocket. Lee et al. synthesized a dual drug loaded sequential delivery system for the treatment of chronic periodontitis [103]. This system had PLGA-lovastatin-chitosan-tetracycline nanoparticles which measured about 111.5 nm and had good antibacterial effect against *A. actinomycetemcomitans* and *P. nigrescens.* The prepared nanoparticles were loaded into gelatin and placed in the canine periodontal defect model and the bone formation was found to be higher when this system was used. Gjoseva et al. developed a mucoadhesive drug delivery system using chitosan microparticles loaded with doxycycline, a potent drug used in the treatment of periodontal disease [104]. These chitosan microparticles were coated with ethyl cellulose and thus the drug release rate was sustained for at least 20 days. Hurt et al. developed a potential GTR membrane made out of chitosan and silver exchanged tobermorite mineral by solvent casting [105]. This membrane was able to form support the formation of hydroxyapatite on its surface and had significant antibacterial activity against *S. aureus, P. aeruginosa*, and *E. coli* indicating its potential to be used as a GTR membrane during periodontal treatment.

Zang et al. showed that it is possible to synthesize a thermosensitive chitosan based hydrogel by autoclaving either the chitosan powder or the chitosan solution and then mixing it with β-glycerophosphate [106]. This hydrogel exhibited an interesting sol-gel transition at physiological temperature, i.e. at 37°C. Freeze drying of this hydrogel yielded them as porous scaffold with pore size ranging from 140 to 230 μm. They had implanted this scaffold into a Class III periodontal furcation defect in canine model and showed that it promoted periodontal regeneration. Miranda et al. synthesized a hybrid scaffold consisting of succinyl modified chitosan and oxidized hyaluronic acid having a regular porous structure [107]. They also indicated that the protein adsorption on these scaffold was enhanced due to the positive charges imparted by the chitosan's amine groups. Although they had carried out their experiments only in vitro, their results suggested that designing such a hybrid scaffold would result in higher fluid absorption by the scaffold thereby improving the chances of cell migration into the scaffold and cell survivability, thus having a great potential for periodontal regeneration. Qasim et al. demonstrated freeze gelation technique to produce a functionally graded membrane [108]. These functionally graded scaffolds were loaded with doxycycline and crosslinked using glutaraldehyde and showed a sustained drug release for periodontal regeneration. Interestingly Varoni et al. have shown that it is possible to fabricate a 3-layered scaffold using chitosan for periodontal regeneration [109]. The authors have utilized medium molecular weight and low molecular weight chitosan, crosslinked with genipin and freeze dried them to obtain two different porous compartments capable of aiding regeneration of bone and gingiva. They went a step further and utilized electrochemical deposition of chitosan to fabricate a third compartment consisting of highly oriented microchannels, to guide the periodontal fiber growth. They pre-seeded these compartments with tissue specific cells and successfully showed it is possible to regenerate multiple periodontal tissue architecture at the same time in vivo. However, they couldn't evidently show the PDL fiber orientation which is much needed for proper integration of alveolar bone to the cementum of the tooth structure. Liao et al. have tried to include β-tricalcium phosphate in chitosan and synthesized a three-dimensional scaffolds using freeze-drying technique. The scaffolds exhibited 91% porosity (pore size around 120 μm) and showed the ability to support the differentiation of human periodontal ligament cells [110]. Gümüşderelioğlu et al. have recently developed an interesting double faced chitosan membrane for GTR applications [111]. In this work a chitosan scaffold was developed which was coated on one side with bone mimicking hydroxyapatite and on the other it was coated with PCL fibers. This barrier membrane could support osteogenic potential on bone facing side and can act as a barrier for epithelial cells.

Although the above-mentioned examples clearly indicate the possibilities of using chitosan as a scaffold in periodontal tissue engineering, these systems mostly served as a passive platform for the endogenous cells to home in and then differentiate into specific tissue and build the specific extracellular matrices. It was envisioned that by incorporating bio-active molecules to chitosan formulation, a better and active scaffold systems could be made.

4.1.2 With Bio-active Molecules

The essential functions of cells such as adhesion to a substrate, survivability, growth, migration, cell division, differentiation, and much more require the help of bio-active molecules. Bio-active molecules function by either attaching themselves to the cell surface receptors or by getting processed inside the cell after uptake. The bio-active molecules include small molecules, peptides, growth factors, cytokines, enzymes, hormones, DNA, siRNA, and also immunomodulation molecules. Delivering the appropriate molecules at the defect site and tuning their spatio-temporal release can greatly affect the way the cells respond. For example, human platelet lysate, which contains a cocktail of growth factors, was mixed with calcium phosphate cement and chitosan scaffold. This system showed that it can enhance the proliferation, osteogenic differentiation and apatite deposition by human periodontal ligament stem cells, indicating a great potential of this system for use in periodontal therapy (Fig. 6a) [112]. Duruel et al. developed a hybrid scaffold consisting of chitosan, alginate, and PLGA which exhibited sequentially delivery of two growth factors, namely IGF-1 and BMP-6. They found that this system supported the cell attachment, tissue formation and also enhanced the cementoblast differentiation [113]. It was shown by Amir et al. that by using a RGD peptide modified chitosan scaffold, and by including cells sheets made out of periodontal ligament cells, formation of periodontal tissue can be greatly enhanced (Fig. 6b) [114]. This RGD modified chitosan scaffold was implanted in periodontal defects of *Macaque nemestrina* monkeys and resulted in increased alveolar bone density and reduced the distance between the apical crest of alveolar bone to the cementoenamel junction, indicative of periodontal attachment gain and restored periodontium. Inclusion of recombinant human amelogenin, one of the components of enamel matrix protein, into a mesoporous hydroxyapatite and chitosan scaffold was shown by Liao et al. [115]. This scaffold showed sustained release of amelogenin enhancing the cementogenic differentiation and also showed antibacterial effect against *Fusobacterium nucleatum* and *Porphyromonas gingivalis*. Sowmya et al. had developed a tissue specific tri-layered chitin-PLGA scaffold in which each layers was tailored for cementum, periodontal ligament, and alveolar bone regeneration [83]. It also delivered specific growth factors such as cementum protein 1, fibroblast growth factor 2, and platelet rich plasma for enhancing the regeneration of cementum, periodontal ligament, and alveolar bone. Implantation of these scaffolds in periodontal defect model of rabbits indicated the concurrent formation of cementum, fibrous periodontal ligament, and alveolar bone. Zang et al. developed chitosan and anorganic bovine bone based scaffold [116]. They seeded this scaffold with human bone marrow derived mesenchymal stem cells and implanted into infrabony

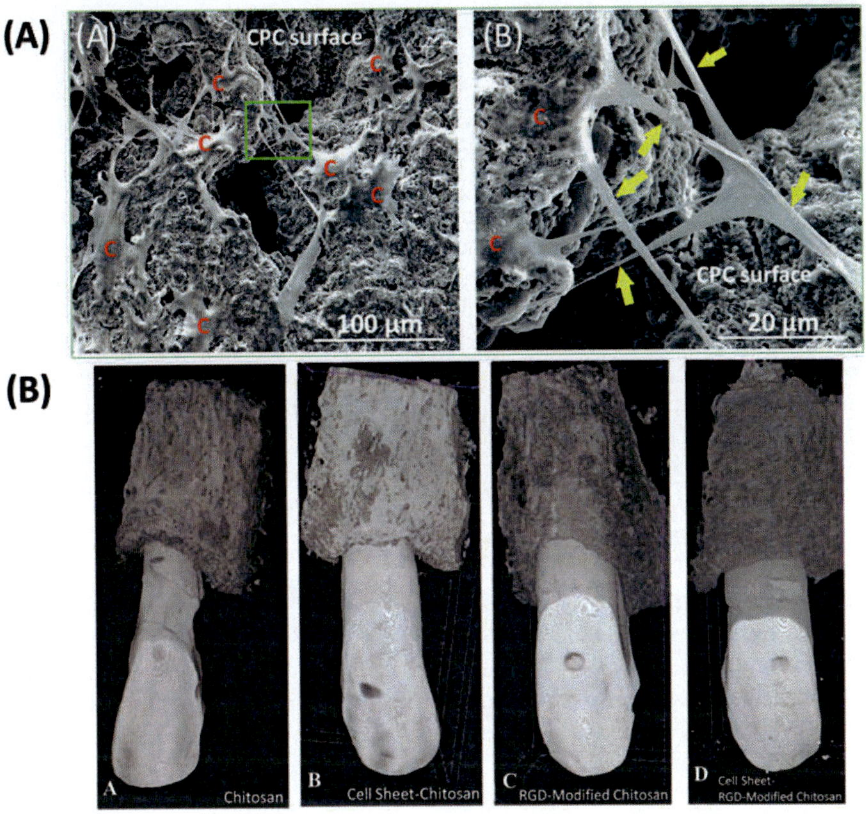

Fig. 6 Representative images of bio-active molecules loaded chitosan scaffolds developed for periodontal therapy. (**a**) SEM images showing the human periodontal ligament stem cells on CPC-chitosan scaffold. Right panel in top row is an enlarged view of the inset box. Yellow arrows indicate cytoplasmic extensions [112]. (**b**) Micro CT images showing the healing of horizontal periodontal defect when periodontal ligament cell sheet added RGD modified chitosan was used [114]. Reproduced with kind permission from RSC and Thieme

periodontal defects in beagle dogs. Their study concluded that this cell seeded scaffold system stimulated periodontal regeneration. It has been shown by Li et al. that it is possible to encapsulate epigenetic modifying molecules such as trichostatin into temperature dependent chitosan scaffolds [117]. This system showed that it is possible to successfully deliver epigenetic modifiers in rat periodontal defects and promote periodontal repair.

4.2 *Hydrogels*

Hydrogels are networked structures of one or more polymers, which can absorb water and retain it between their polymeric networks [118]. By carefully designing and customizing the polymeric networks, it is possible to control the amount of water absorbed, amount of water retained, and the amount of swelling from a dry state. These hydrated networks act as excellent microenvironments for cells to survive and very good reservoirs for various bio-active molecules, drugs, and tissue fluids. As discussed earlier, chitosan can be easily made into hydrogel, e.g., by neutralizing an acidic solution of chitosan with sodium hydroxide [61] or by allowing chitosan solution to react with pentasodium tripolyphosphate solutions [119]. Based on the form and factor of application, chitosan hydrogels can be divided into non-injectable and injectable forms. This section brings out examples of these categories of chitosan hydrogel used in periodontal therapy.

4.2.1 Non-injectable Hydrogels

Chitosan hydrogels have been synthesized through enzymatical solidification method, by using urease and urea solution [120]. Loading of periodontal ligament cells was possible during the synthesis process of the hydrogel and this was transplanted into intrabony periodontal defect in rats. Interestingly they found that this even without cells was enough for the formation of periodontal ligament, indicating its potential. Echazú et al. synthesized thymol loaded chitosan hydrogel exhibiting antimicrobial and anti-oxidant property for use in periodontal treatment [121]. Ma et al. have shown that thermosensitive chitosan hydrogel can be effectively used a siRNA reservoir for treating periodontitis [122]. It was found that the chitosan/β-glycerophosphate gel was able to maintain the functional RANK siRNA and exhibited a release more than 15 days in vitro and up to 15 days in vivo thus proving its potential to be used in periodontal tissue engineering. Jayash et al. formulated a chitosan hydrogel with human osteoprotegerin protein which showed sustained protein release in vitro, good cytocompatibility for human periodontal ligament fibroblasts, and induced osteogenesis [69].

4.2.2 Injectable Hydrogels

Peng et al. developed a hydroxypropyl methylcellulose-chitosan composite injectable hydrogel which had mucoadhesive property [123]. Further they showed that this hydrogel augments the killing of bacterial biofilms by photodynamic inactivation. A chitosan/gelatin/β-glycerophosphate injectable hydrogel was synthesized along with the incorporation of a potent antibacterial agents vancomycin and metronidazole [124]. Shen et al. incorporated exosomes derived from the dental pulpal stem cells into chitosan/β-glycerophosphate hydrogel and injected into mice periodontitis model [125]. It was found that this system could alter the macrophage polarization, reduced the periodontal inflammation, and promoted alveolar bone formation. Utilizing this chitosan/β-glycerophosphate injectable hydrogel system, Zang et al. added BMP-7 and ornidazole into this and showed it was possible for sustained release of the peptide and the molecule [126]. The injectable chitosan hydrogel with these molecules had antibacterial activity and also enhanced periodontal regeneration in Class III periodontal defects in beagle dogs. A similar chitosan/β-glycerophosphate/gelatin was showcased by Xu et al. [127]. They showed that this injectable hydrogel can promote the regeneration of periodontium by incorporating aspirin and erythropoietin into the hydrogel for a controlled release. An in situ photocrosslinkable hydrogel was synthesized by Chichiricco et al., using carboxymethyl chitosan, silanized hydroxypropyl methylcellulose, and riboflavin phosphate [128]. This hydrogel system could be injected into periodontal pockets and then can be crosslinked by visible dental light. After photocrosslinking, the hydrogel showed cell barrier property thus a potential candidate for GTR membrane (Fig. 7).

4.3 Fibers

Chitosan, either alone or combined with other polymers, can be drawn into micro or nanofibers using the electrospinning technique. Advances in electrospinning techniques have greatly helped to synthesize fiber and fiber mats with different configurations such as uniform size distributed fibers, multi size-distribution fibers, multifilament fibers, random fibers, aligned fibers, micro and nanofibers in same system, and so on. The main advantage of chitosan fibers is that they can form mesh like structures wherein the mesh size can be easily controlled, very high surface area to volume ratios, mimicking the fiber bundles of native tissue architecture and the ability to be further processed into threads, sutures, gauze, sheet, and membranes. Apart from electrospinning technique, wet spinning method can also be employed to obtain fibers. In this method, chitosan solution is mixed with other polymers such as collagen and is sprayed through a spinneret into alkaline solutions such as ammonia to obtain the fibers [129].

Shalumon et al. developed a chitosan-PCL based nanofiber system, which consists of incorporation of nanobioglass and hydroxyapatite, both of which are potent osteoconductive materials [130]. Their studies showed that this chitosan-bioceramic based nanofiber system was able to enhance the osteoblastic activity of human periodontal ligament fibroblastic cells, thus suggesting for use in periodontal treatment. Similarly, an electrospun PLLA/chitosan fiber membrane was synthesized by Chen et al. [131]. This membrane was able to act as a barrier for fibroblasts while

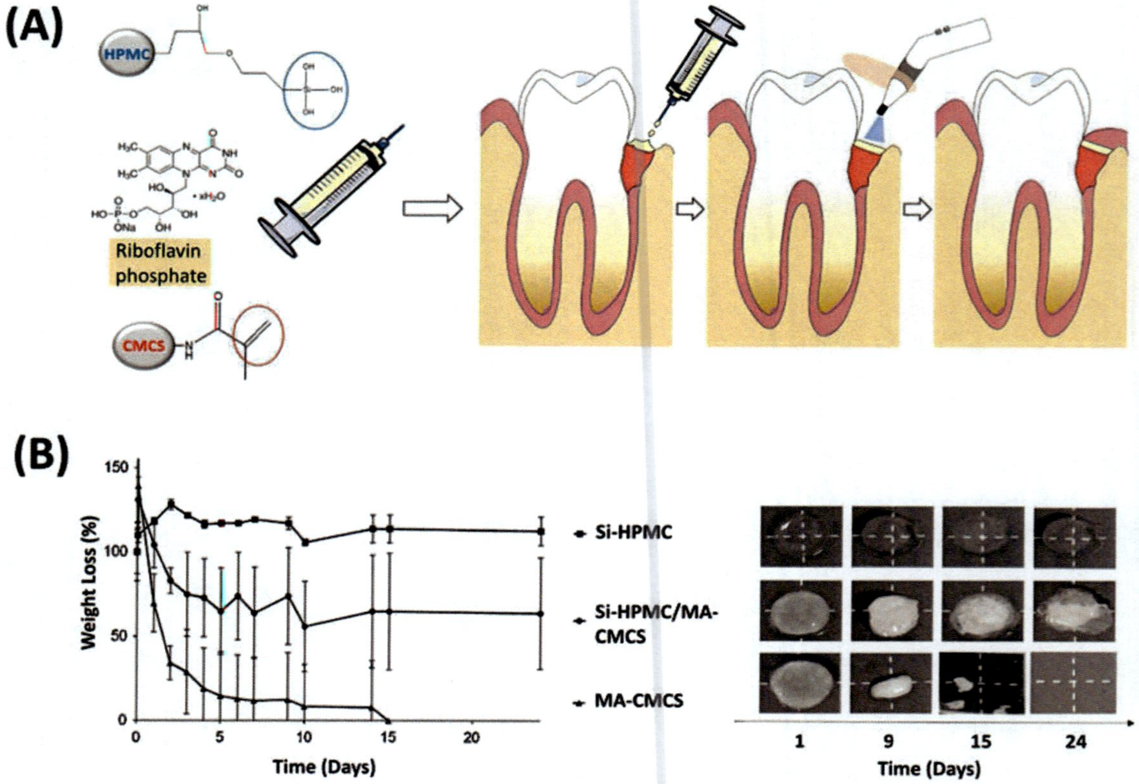

Fig. 7 A photocrosslinkable injectable barrier membrane. (**a**) The schematics explaining the chemistry of the photocrosslinkable chitosan based injectable gel; (**b**) silanized hydroxypropyl methylcellulose methacrylated carboxymethyl chitosan gel showed stability for 24 days and thereafter slow degradation, indicating the tenability of degradation [128]. Reproduced with kind permission from Elsevier

supporting the newly formed tissues, thus a potential candidate for GTR membrane for use during periodontal therapy. Sundaram et al. have explored the possibility of recreating the complex architecture of periodontal tissues by using a PCL electrospun fiber membrane along with a second layer of chitosan and $CaSO_4$ [132]. Their bilayer membrane construct was able to support the osteoblastic differentiation of human dental follicle stem cells thus proving useful for simultaneous regenerations in periodontal tissues. Qasim et al. showed that a graded electrospun membrane coating made out of chitosan and polyethylene oxide supported the viability of cells and also the mineral deposition, having a potential for periodontal regeneration [133]. Emulsion electrospinning has been employed by Shen et al. to produce nanofibers of chitosan/PLA [134]. They have utilized chitosan nanoparticles to reinforce the PLA nanofibers. This fiber system enhanced the osteogenic differentiation of bone marrow stem cells. Recently Niu et al. synthesized polyamide-6/chitosan electrospun nanofibers and used this as a reinforcement for nano-hydroxyapatite/polyamide-6 membranes for periodontal GTR application [135]. This bilayer electrospun chitosan scaffold showed great bioactivity and promoted osteoconductivity (Fig. 8). These are some of the recent studies showing the great potential of chitosan fibers for use in periodontal regeneration.

5 Summary and Future Directions

From the above overview, it is very clear that chitosan serves a great biomaterial for periodontal regeneration. It could be used in various stages and in different applications of periodontal treatment. It could be used as an antibacterial gel or solution to prevent the progress of the periodontal disease. As the disease progresses, chitosan constructs in the form of scaffolds, gels, or fibers could be used to promote the regeneration of the specific tissue of periodontal complex. In order to further enhance the tissue repair, periodontal tissue specific growth factors or molecules could be incorporated and finely tuned for the proper spatio-temporal release. With the advances in polymer chemistry day by day, chitosan based shape memory fibers [136], suture filaments improved by addition of chitosan [73], flavanols loaded chitosan gels [137], core-shell chitosan nanospheres [138], sulfonated chitosan oligosaccharides [139], core sheathed nanostructured chitosan non-woven sheets [140], and much more would provide a wide array of choices for the dental clinicians to effectively treat and repair the periodontal diseases.

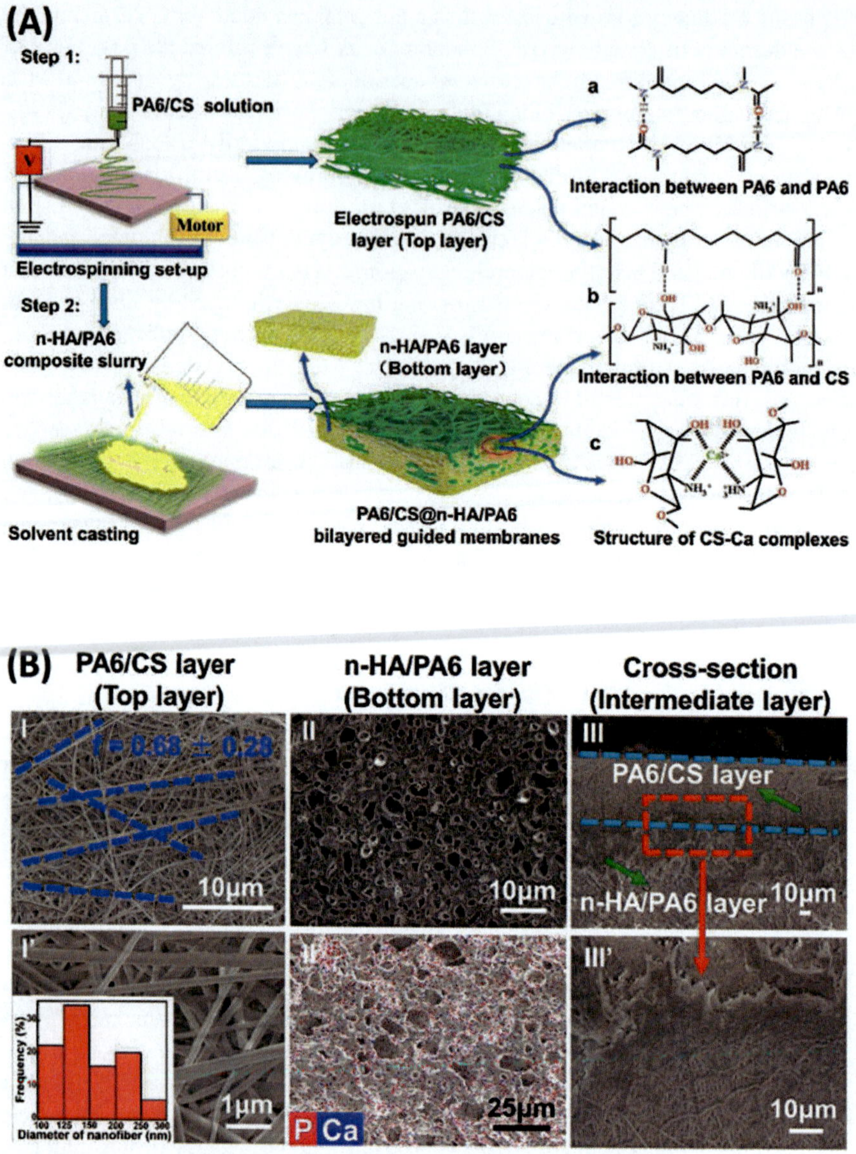

Fig. 8 Bilayer fibrous membrane for guided bone regeneration. (**a**) Schematic explaining the compositing of an electrospun fibrous layer with nano-hydroxyapatite solvent casting layer, yielding a bilayered membrane. (**b**) SEM images of the different layers and the interface in the membrane [135]. Reproduced with kind permission from Elsevier

References

1. Cope G, Cope A (2011) The periodontium: an anatomical guide. Dental Nursing 7:376–378
2. Hughes FJ (2015) Periodontium and periodontal disease. Stem cell biology and tissue engineering in dental sciences. Academic Press, Boston, pp 433–444
3. Nanci A, Bosshardt DD (2006) Structure of periodontal tissues in health and disease. Periodontol 2000 40:11–28
4. Muhlemann HR (1967) Tooth mobility – a review of clinical aspects and research findings. J Periodontol 38:686–713
5. Baelum V, Lopez R (2003) Defining and classifying periodontitis: need for a paradigm shift? Eur J Oral Sci 111:2–6
6. Genco RJ, Borgnakke WS (2020) Diabetes as a potential risk for periodontitis: association studies. Periodontol 2000 83:40–45
7. Romandini M, Shin HS, Romandini P, Lafori A, Cordaro M (2020) Hormone-related events and periodontitis in women. J Clin Periodontol 47:429–441
8. Lindhe J, Svanberg G (1974) Influence of trauma from occlusion on progression of experimental periodontitis in the beagle dog. J Clin Periodontol 1:3–14
9. Flemmig TF (1999) Periodontitis. Ann Periodontol 4:32–37
10. Kinane DF (2001) Causation and pathogenesis of periodontal disease. Periodontol 2000 2000 (25):8–20
11. Sweeting LA, Davis K, Cobb CM (2008) Periodontal treatment protocol (PTP) for the general dental practice. J Dent Hyg 82(Suppl 3):16–26
12. Chen X, Wu GF, Feng ZH, Dong Y, Zhou W, Li B, Bai SZ, Zhao YM (2016) Advanced biomaterials and their potential applications in the treatment of periodontal disease. Crit Rev Biotechnol 36:760–775
13. Ausenda F, Rasperini G, Acunzo R, Gorbunkova A, Pagni G (2019) New perspectives in the use of biomaterials for periodontal regeneration. Materials 12:2197
14. Sun HH, Qu TJ, Zhang XH, Yu Q, Chen FM (2012) Designing biomaterials for in situ periodontal tissue regeneration. Biotechnol Prog 28:3–20
15. Liang YX, Luan XH, Liu XH (2020) Recent advances in periodontal regeneration: a biomaterial perspective. Bioact Mater 5:297–308
16. Sowmya S, Bumgardener JD, Chennazhi KP, Nair SV, Jayakumar R (2013) Role of nanostructured biopolymers and bioceramics in enamel, dentin and periodontal tissue regeneration. Prog Polym Sci 38:1748–1772
17. Jayakumar R, Prabaharan M, Muzzarelli RA (2011) Chitosan for biomaterials I. Springer, Berlin
18. Nivedhitha Sundaram M, Deepthi S, Mony U, Shalumon KT, Chen JP, Jayakumar R (2019) Chitosan hydrogel scaffold reinforced with twisted poly(L lactic acid) aligned microfibrous bundle to mimic tendon extracellular matrix. Int J Biol Macromol 122:37–44
19. Mohandas A, Sun W, Nimal TR, Shankarappa SA, Hwang NS, Jayakumar R (2018) Injectable chitosan-fibrin/nanocurcumin composite hydrogel for the enhancement of angiogenesis. Res Chem Intermediat 44:4873–4887
20. Mohandas A, Rangasamy J (2021) Nanocurcumin and arginine entrapped injectable chitosan hydrogel for restoration of hypoxia induced endothelial dysfunction. Int J Biol Macromol 166:471–482
21. Vignesh S, Sivashanmugam A, Mohandas A, Janarthanan R, Iyer S, Nair SV, Jayakumar R (2018) Injectable deferoxamine nanoparticles loaded chitosan-hyaluronic acid coacervate hydrogel for therapeutic angiogenesis. Colloid Surface B 161:129–138
22. Sundaram MN, Amirthalingam S, Mony U, Varma PK, Jayakumar R (2019) Injectable chitosan-nano bioglass composite hemostatic hydrogel for effective bleeding control. Int J Biol Macromol 129:936–943
23. Sanandam M, Salunkhe A, Shejale K, Patil D (2013) Chitosan bandage for faster blood clotting and wound healing. Int J Adv Biotechnol Res 4:47–50

24. Matica MA, Aachmann FL, Tøndervik A, Sletta H, Ostafe V (2019) Chitosan as a wound dressing starting material: antimicrobial properties and mode of action. Int J Mol Sci 20:5889
25. Rong NN, Yang RY, Lu YZ (2020) Chitosan-based composite nanosilver gel application in nursing care of intensive care unit patients with severe burns. Nanosci Nanotech Let 12:939–945
26. Khan NR, Shah KU, Nawaz A, Wong TW (2020) Chitosan based composites and their applications in tissue engineering. Encyclopedia Marine Biotechnol 2:979–1006
27. Zhang NH, Gao T, Wang Y, Liu J, Zhang JW, Yao RJ, Wu F (2020) Modulating cationicity of chitosan hydrogel to prevent hypertrophic scar formation during wound healing. Int J Biol Macromol 154:835–843
28. Moeini A, Pedram P, Makvandi P, Malinconico M, d'Ayala GG (2020) Wound healing and antimicrobial effect of active secondary metabolites in chitosan-based wound dressings: a review. Carbohydr Polym 233:115839
29. Raveendran NT, Mohandas A, Menon RR, Menon AS, Biswas R, Jayakumar R (2019) Ciprofloxacin and fluconazole containing fibrin nanoparticle incorporated chitosan bandages for the treatment of polymicrobial wound infections. ACS Appl Bio Mater 2:243–254
30. Peniche C, Arguelles-Monal W, Peniche H, Acosta N (2003) Chitosan: an attractive biocompatible polymer for microencapsulation. Macromol Biosci 3:511–520
31. Landriscina A, Rosen J, Friedman AJ (2015) Biodegradable chitosan nanoparticles in drug delivery for infectious disease. Nanomedicine 10:1609–1619
32. Tigli RS, Karakecili A, Gumusderelioglu M (2007) In vitro characterization of chitosan scaffolds: influence of composition and deacetylation degree. J Mater Sci Mater M 18:1665–1674
33. Nazir M, Al-Ansari A, Al-Khalifa K, Alhareky M, Gaffar B, Almas K (2020) Global prevalence of periodontal disease and lack of its surveillance. Sci World J 2020:2146160
34. Khaw A, Logan R, Keefe D, Bartold M (2014) Radiation-induced oral mucositis and periodontitis – proposal for an inter-relationship. Oral Dis 20:e7–e18
35. Leite FRM, Nascimento GG, Scheutz F, Lopez R (2018) Effect of smoking on periodontitis: a systematic review and meta-regression. Am J Prev Med 54:831–841
36. Hou GL, Tsai CC, Huang JS (1997) Relationship between molar root fusion and localized periodontitis. J Periodontol 68:313–319
37. Preshaw PM, Bissett SM (2019) Periodontitis and diabetes. Br Dent J 227:577–584
38. Seymour G, Ford P, Cullinan M, Leishman S, Yamazaki K (2007) Relationship between periodontal infections and systemic disease. Clin Microbiol Infect 13:3–10
39. Wolff L, Dahlén G, Aeppli D (1994) Bacteria as risk markers for periodontitis. J Periodontol 65:498–510
40. Tanner ACR, Haffer C, Bratthall G, Visconti R, Socransky S (1979) A study of the bacteria associated with advancing periodontitis in man. J Clin Periodontol 6:278–307
41. Niemiec BA (2008) Periodontal disease. Top Companion Anim M 23:72–80
42. Loesche WJ, Grossman NS (2001) Periodontal disease as a specific, albeit chronic, infection: diagnosis and treatment. Clin Microbiol Rev 14:727–752
43. Drisko CH (2001) Nonsurgical periodontal therapy. Periodontol 2000 2000(25):77–88
44. Rosenberg E, Rose LF (1998) Biologic and clinical considerations for autografts and allografts in periodontal regeneration therapy. Dent Clin N Am 42:467–490
45. Sukumar S, Drizhal I (2008) Bone grafts in periodontal therapy. Acta Medica (Hradec Kralove) 51:203–207
46. Reynolds MA, Aichelmann-Reidy ME, Branch-Mays GL (2010) Regeneration of periodontal tissue: bone replacement grafts. Dent Clin N Am 54:55–71
47. Sheikh Z, Hamdan N, Ikeda Y, Grynpas M, Ganss B, Glogauer M (2017) Natural graft tissues and synthetic biomaterials for periodontal and alveolar bone reconstructive applications: a review. Biomater Res 21:1–20
48. Shue L, Yufeng Z, Mony U (2012) Biomaterials for periodontal regeneration: a review of ceramics and polymers. Biomatter 2:271–277

49. Park SB, Lih E, Park KS, Joung YK, Han DK (2017) Biopolymer-based functional composites for medical applications. Prog Polym Sci 68:77–105
50. Sivashanmugam A, Charoenlarp P, Deepthi S, Rajendran A, Nair SV, Iseki S, Jayakumar R (2017) Injectable shear-thinning $CaSO_4$/FGF-18-incorporated chitin PLGA hydrogel enhances bone regeneration in mice cranial bone defect model. ACS Appl Mater Interfaces 9:42639–42652
51. Zucchelli G, Mounssif I (2015) Periodontal plastic surgery. Periodontol 2000 68:333–368
52. Bottino MC, Thomas V, Schmidt G, Vohra YK, Chu TMG, Kowolik MJ, Janowski GM (2012) Recent advances in the development of gtr/gbr membranes for periodontal regeneration-a materials perspective. Dent Mater 28:703–721
53. Florjanski W, Orzeszek S, Olchowy A, Grychowska N, Wieckiewicz W, Malysa A, Smardz J, Wieckiewicz M (2019) Modifications of polymeric membranes used in guided tissue and bone regeneration. Polymers 11:782
54. Jiang YL, Deng Y, Tu Y, Ay B, Sun XD, Li YB, Wang XH, Chen XC, Zhang L (2019) Chitosan-based asymmetric topological membranes with cell-like features for healthcare applications. J Mater Chem B 7:2634–2642
55. Nithya S, Nimal TR, Baranwal G, Suresh MK, Anju CP, Kumar VA, Mohan CG, Jayakumar R, Biswas R (2018) Preparation, characterization and efficacy of lysostaphin-chitosan gel against *staphylococcus aureus*. Int J Biol Macromol 110:157–166
56. Anjana J, Mohandas A, Seethalakshmy S, Suresh MK, Menon R, Biswas R, Jayakumar R (2018) Bi-layered nanocomposite bandages for controlling microbial infections and overproduction of matrix metalloproteinase activity. Int J Biol Macromol 110:124–132
57. Elbi S, Nimal TR, Rajan VK, Baranwal G, Biswas R, Jayakumar R, Sathianarayanan S (2017) Fucoidan coated ciprofloxacin loaded chitosan nanoparticles for the treatment of intracellular and biofilm infections of *Salmonella*. Colloid Surface B 160:40–47
58. Nimal TR, Baranwal G, Bavya MC, Biswas R, Jayakumar R (2016) Anti-*Staphylococcal* activity of injectable nano tigecycline/chitosan-prp composite hydrogel using *Drosophila melanogaster* model for infectious wounds. ACS Appl Mater Interfaces 8:22074–22083
59. Mohandas A, Deepthi S, Biswas R, Jayakumar R (2018) Chitosan based metallic nanocomposite scaffolds as antimicrobial wound dressings. Bioact Mater 3:267–277
60. Isola C, Sieverding HL, Raghunathan R, Sibi MP, Webster DC, Sivaguru J, Stone JJ (2017) Life cycle assessment of photodegradable polymeric material derived from renewable bioresources. J Clean Prod 142:2935–2944
61. Deepthi S, Jeevitha K, Sundaram MN, Chennazhi KP, Jayakumar R (2015) Chitosan-hyaluronic acid hydrogel coated poly(caprolactone) multiscale bilayer scaffold for ligament regeneration. Chem Eng J 260:478–485
62. Deepthi S, Gafoor AAA, Sivashanmugam A, Nair SV, Jayakumar R (2016) Nanostrontium ranelate incorporated injectable hydrogel enhanced matrix production supporting chondrogenesis in vitro. J Mater Chem B 4:4092–4103
63. Anisha BS, Sankar D, Mohandas A, Chennazhi KP, Nair SV, Jayakumar R (2013) Chitosan-hyaluronan/nano chondroitin sulfate ternary composite sponges for medical use. Carbohydr Polym 92:1470–1476
64. Mohandas A, Anisha BS, Chennazhi KP, Jayakumar R (2015) Chitosan-hyaluronic acid/VEGF loaded fibrin nanoparticles composite sponges for enhancing angiogenesis in wounds. Colloid Surface B 127:105–113
65. Mohan K, Ganesan AR, Muralisankar T, Jayakumar R, Sathishkumar P, Uthayakumar V, Chandirasekar R, Revathi N (2020) Recent insights into the extraction, characterization, and bioactivities of chitin and chitosan from insects. Trends Food Sci Technol 105:17–42
66. No HK, Meyers SP, Prinyawiwatkul W, Xu Z (2007) Applications of chitosan for improvement of quality and shelf life of foods: a review. J Food Sci 72:R87–R100
67. Rawal T, Parmar R, Tyagi RK, Butani S (2017) Rifampicin loaded chitosan nanoparticle dry powder presents an improved therapeutic approach for alveolar tuberculosis. Colloid Surf B 154:321–330

68. Li BX, Wang J, Gui Q, Yang H (2020) Drug-loaded chitosan film prepared via facile solution casting and air-drying of plain water-based chitosan solution for ocular drug delivery. Bioact Mater 5:577–583
69. Jayash SN, Hashim NM, Misran M, Baharuddin NA (2017) Formulation and in vitro and in vivo evaluation of a new osteoprotegerin-chitosan gel for bone tissue regeneration. J Biomed Mater Res A 105:398–407
70. Pillai NSM, Eswar K, Amirthalingam S, Mony U, Varma PK, Jayakumar R (2019) Injectable nano whitlockite incorporated chitosan hydrogel for effective hemostasis. ACS Appl Bio Mater 2:865–873
71. Sankar PCK, Rajmohan G, Rosemary MJ (2017) Physico-chemical characterisation and biological evaluation of freeze dried chitosan sponge for wound care. Mater Lett 208:130–132
72. Mao HQ, Wei C, Gong YY, Wang SQ, Ding WW (2019) Mechanical and water-resistant properties of eco-friendly chitosan membrane reinforced with cellulose nanocrystals. Polymers 11(1):166
73. Mohammadi H, Alihosseini F, Hosseini SA (2020) Improving physical and biological properties of nylon monofilament as suture by chitosan/hyaluronic acid. Int J Biol Macromol 164:3394–3402
74. Qu B, Luo YC (2020) Chitosan-based hydrogel beads: preparations, modifications and applications in food and agriculture sectors – a review. Int J Biol Macromol 152:437–448
75. Barreras US, Mendez FT, Martinez REM, Valencia CS, Rodriguez PRM, Rodriguez JPL (2016) Chitosan nanoparticles enhance the antibacterial activity of chlorhexidine in collagen membranes used for periapical guided tissue regeneration. Mat Sci Eng C Mater 58:1182–1187
76. Skorik YA, Golyshev AA, Kritchenkov AS, Gasilova ER, Poshina DN, Sivaram AJ, Jayakumar R (2017) Development of drug delivery systems for taxanes using ionic gelation of carboxyacyl derivatives of chitosan. Carbohydr Polym 162:49–55
77. Arun Kumar R, Sivashanmugam A, Deepthi S, Iseki S, Chennazhi K, Nair SV, Jayakumar R (2015) Injectable chitin-poly (ε-caprolactone)/nanohydroxyapatite composite microgels prepared by simple regeneration technique for bone tissue engineering. ACS Appl Mater Interfaces 7:9399–9409
78. Deepthi S, Sundaram MN, Kadavan JD, Jayakumar R (2016) Layered chitosan-collagen hydrogel/aligned PLLA nanofiber construct for flexor tendon regeneration. Carbohydr Polym 153:492–500
79. Abd El-Fattah A, Mansour A (2018) Viscoelasticity, mechanical properties, and in vitro biodegradation of injectable chitosan-poly(3-hydroxybutyrate-co-3-hydroxyvalerate)/nanohydroxyapatite composite hydrogel. B Mater Sci 41:1–10
80. Kumar A, Sivashanmugam A, Deepthi S, Bumgardner JD, Nair SV, Jayakumar R (2016) Nano-fibrin stabilized $CaSO_4$ crystals incorporated injectable chitin composite hydrogel for enhanced angiogenesis & osteogenesis. Carbohydr Polym 140:144–153
81. Boschetto F, Doan HN, Vo PP, Zanocco M, Zhu WL, Sakai W, Adachi T, Ohgitani E, Tsutsumi N, Mazda O, Kinashi K, Marin E, Pezzotti G (2020) Antibacterial and osteoconductive effects of chitosan/polyethylene oxide (PEO)/bioactive glass nanofibers for orthopedic applications. Appl Sci 10:2360
82. Saranya TS, Rajan VK, Biswas R, Jayakumar R, Sathianarayanan S (2018) Synthesis, characterisation and biomedical applications of curcumin conjugated chitosan microspheres. Int J Biol Macromol 110:227–233
83. Sowmya S, Mony U, Jayachandran P, Reshma S, Kumar RA, Arzate H, Nair SV, Jayakumar R (2017) Tri-layered nanocomposite hydrogel scaffold for the concurrent regeneration of cementum, periodontal ligament, and alveolar bone. Adv Healthc Mater 6:1601251
84. Pakornpadungsit P, Prasopdee T, Swainson NM, Chworos A, Smitthipong W (2020) DNA: chitosan complex, known as a drug delivery system, can create a porous scaffold. Polym Test 83:106333
85. Wong CY, Al-Salami H, Dass CR (2018) The role of chitosan on oral delivery of peptide-loaded nanoparticle formulation. J Drug Target 26:551–562

86. Anjana J, Rajan VK, Biswas R, Jayakumar R (2017) Controlled delivery of bioactive molecules for the treatment of chronic wounds. Curr Pharm Design 23:3529–3537
87. Vaquette C, Saifzadeh S, Farag A, Hutmacher DW, Ivanovski S (2019) Periodontal tissue engineering with a multiphasic construct and cell sheets. J Dent Res 98:673–681
88. Mota J, Yu N, Caridade SG, Luz GM, Gomes ME, Reis RL, Jansen JA, Walboomers XF, Mano JF (2012) Chitosan/bioactive glass nanoparticle composite membranes for periodontal regeneration. Acta Biomater 8:4173–4180
89. Tang S, Jiang LY, Ma BL, Tang CY, Wen Y, Zhang N, Zhang Y, Su SP (2020) Preparation and characterization of bamboo fiber/chitosan/nano-hydroxyapatite composite membrane by ionic crosslinking. Cellul 27:5089–5100
90. Peter M, Binulal NS, Soumya S, Nair SV, Furuike T, Tamura H, Jayakumar R (2010) Nanocomposite scaffolds of bioactive glass ceramic nanoparticles disseminated chitosan matrix for tissue engineering applications. Carbohydr Polym 79:284–289
91. Park YJ, Lee YM, Park SN, Sheen SY, Chung CP, Lee SJ (2000) Platelet derived growth factor releasing chitosan sponge for periodontal bone regeneration. Biomaterials 21:153–159
92. Anisha BS, Biswas R, Chennazhi KP, Jayakumar R (2013) Chitosan-hyaluronic acid/nano silver composite sponges for drug resistant bacteria infected diabetic wounds. Int J Biol Macromol 62:310–320
93. Sakaidani Y, Nomura T, Matsuura A, Ito M, Suzuki E, Murakami K, Nadano D, Matsuda T, Furukawa K, Okajima T (2011) *O*-linked-*n*-acetylglucosamine on extracellular protein domains mediates epithelial cell-matrix interactions. Nat Commun 2:1–9
94. Li YH, Chi YQ, Yu CH, Xie Y, Xia MY, Zhang CL, Han XL, Peng Q (2020) Drug-free and non-crosslinked chitosan scaffolds with efficient antibacterial activity against both gram-negative and gram-positive bacteria. Carbohydr Polym 241:116386
95. Ikinci G, Senel S, Akincibay H, Kas S, Ercis S, Wilson CG, Hincal AA (2002) Effect of chitosan on a periodontal pathogen *Porphyromonas gingivalis*. Int J Pharm 235:121–127
96. Ma SQ, Adayi A, Liu ZH, Li M, Wu MY, Xiao LH, Sun YC, Cai Q, Yang XP, Zhang X, Gao P (2016) Asymmetric collagen/chitosan membrane containing minocycline-loaded chitosan nanoparticles for guided bone regeneration. Sci Rep 6:1–10
97. Perugini P, Genta I, Conti B, Modena T, Pavanetto F (2003) Periodontal delivery of ipriflavone: new chitosan/PLGA film delivery system for a lipophilic drug. Int J Pharm 252:1–9
98. Hollister SJ (2009) Scaffold design and manufacturing: from concept to clinic. Adv Mater 21:3330–3342
99. da Silva MC, da Silva HN, Cruz RDAL, Amoah SKS, Silva SMD, Fook MVL (2019) *N*-acetyl-*d*-glucosamine-loaded chitosan filaments biodegradable and biocompatible for use as absorbable surgical suture materials. Materials 12:1807
100. Park CH, Kim KH, Lee YM, Seol YJ (2016) Advanced engineering strategies for periodontal complex regeneration. Materials 9:57
101. Arancibia R, Maturana C, Silva D, Tobar N, Tapia C, Salazar JC, Martinez J, Smith PC (2013) Effects of chitosan particles in periodontal pathogens and gingival fibroblasts. J Dent Res 92:740–745
102. Costa EM, Silva S, Pina C, Tavaria FK, Pintado M (2014) Antimicrobial effect of chitosan against periodontal pathogens biofilms. SOJ Microbiol Infect Dis 2:1–6
103. Lee BS, Lee CC, Wang YP, Chen HJ, Lai CH, Hsieh WL, Chen YW (2016) Controlled-release of tetracycline and lovastatin by poly(D,L-lactide-co-glycolide acid)-chitosan nanoparticles enhances periodontal regeneration in dogs. Int J Nanomedicine 11:285–297
104. Gjoseva S, Geskovski N, Sazdovska SD, Popeski-Dimovski R, Petrusevski G, Mladenovska K, Goracinova K (2018) Design and biological response of doxycycline loaded chitosan microparticles for periodontal disease treatment. Carbohydr Polym 186:260–272
105. Hurt AP, Kotha AK, Trivedi V, Coleman NJ (2015) Bioactivity, biocompatibility and antimicrobial properties of a chitosan-mineral composite for periodontal tissue regeneration. Polimeros 25:311–316

106. Zang SQ, Dong GY, Peng B, Xu J, Ma ZW, Wang XW, Liu LX, Wang QT (2014) A comparison of physicochemical properties of sterilized chitosan hydrogel and its applicability in a canine model of periodontal regeneration. Carbohydr Polym 113:240–248
107. Miranda DG, Malmonge SM, Campos DM, Attik NG, Grosgogeat B, Gritsch K (2016) A chitosan-hyaluronic acid hydrogel scaffold for periodontal tissue engineering. J Biomed Mater Res B 104:1691–1702
108. Qasim SSB, Nogueria LP, Fawzy AS, Daood U (2020) The effect of cross-linking efficiency of drug-loaded novel freeze gelated chitosan templates for periodontal tissue regeneration. AAPS PharmSciTech 21:1–9
109. Varoni EM, Vijayakumar S, Canciani E, Cochis A, De Nardo L, Lodi G, Rimondini L, Cerruti M (2018) Chitosan-based trilayer scaffold for multitissue periodontal regeneration. J Dent Res 97:303–311
110. Liao F, Chen Y, Li Z, Wang Y, Shi B, Gong Z, Cheng X (2010) A novel bioactive three-dimensional β-tricalcium phosphate/chitosan scaffold for periodontal tissue engineering. J Mater Sci Mater Med 21:489–496
111. Gumusderelioglu M, Sunal E, Demirtas TT, Kiremitci AS (2020) Chitosan-based double-faced barrier membrane coated with functional nanostructures and loaded with bmp-6. J Mater Sci Mater M 31:1–14
112. Zhao ZQ, Liu J, Weir MD, Zhang N, Zhang L, Xie XJ, Zhang C, Zhang K, Bai YX, Xu HHK (2019) Human periodontal ligament stem cells on calcium phosphate scaffold delivering platelet lysate to enhance bone regeneration. RSC Adv 9:41161–41172
113. Duruel T, Cakmak AS, Akman A, Nohutcu RM, Gumusderelioglu M (2017) Sequential IGF-1 and BMP-6 releasing chitosan/alginate/PLGA hybrid scaffolds for periodontal regeneration. Int J Biol Macromol 104:232–241
114. Amir LR, Soeroso Y, Fatma D, Sunarto H, Sulijaya B, Idrus E, Rahdewati H, Tjokroyonco AM, Izumi K, Abbas B, Latief FDE (2020) Periodontal ligament cell sheets and RGD-modified chitosan improved regeneration in the horizontal periodontal defect model. Eur J Dent 14:306–314
115. Liao Y, Li HX, Shu R, Chen HW, Zhao LP, Song ZC, Zhou W (2020) Mesoporous hydroxyapatite/chitosan loaded with recombinant-human amelogenin could enhance antibacterial effect and promote periodontal regeneration. Front Cell Infect Microbiol 10:180
116. Zang S, Jin L, Kang S, Hu X, Wang M, Wang J, Chen B, Peng B, Wang Q (2016) Periodontal wound healing by transplantation of jaw bone marrow-derived mesenchymal stem cells in chitosan/anorganic bovine bone carrier into one-wall infrabony defects in beagles. J Periodontol 87:971–981
117. Li Q, Liu F, Dang R, Feng C, Xiao R, Hua Y, Wang W, Jia Z, Liu D (2020) Epigenetic modifier trichostatin a enhanced osteogenic differentiation of mesenchymal stem cells by inhibiting NF-κB (p65) DNA binding and promoted periodontal repair in rats. J Cell Physiol 235:9691–9701
118. Sivashanmugam A, Kumar RA, Priya MV, Nair SV, Jayakumar R (2015) An overview of injectable polymeric hydrogels for tissue engineering. Eur Polym J 72:543–565
119. Sacco P, Borgogna M, Travan A, Marsich E, Paoletti S, Asaro F, Grassi M, Donati I (2014) Polysaccharide-based networks from homogeneous chitosan-tripolyphosphate hydrogels: synthesis and characterization. Biomacromolecules 15:3396–3405
120. Yan XZ, van den Beucken JJJP, Cai XJ, Yu N, Jansen JA, Yang F (2015) Periodontal tissue regeneration using enzymatically solidified chitosan hydrogels with or without cell loading. Tissue Eng Pt A 21:1066–1076
121. Echazu MIA, Olivetti CE, Anesini C, Perez CJ, Alvarez GS, Desimone MF (2017) Development and evaluation of thymol-chitosan hydrogels with antimicrobial-antioxidant activity for oral local delivery. Mat Sci Eng C Mater 81:588–596
122. Ma ZW, Yang CX, Song W, Wang QT, Kjems J, Gao S (2014) Chitosan hydrogel as siRNA vector for prolonged gene silencing. J Nanobiotechnol 12:1–9
123. Peng PC, Hsieh CM, Chen CP, Tsai T, Chen CT (2016) Assessment of photodynamic inactivation against periodontal bacteria mediated by a chitosan hydrogel in a 3D gingival model. Int J Mol Sci 17:1821

124. Pakzad Y, Ganji F (2016) Thermosensitive hydrogel for periodontal application: in vitro drug release, antibacterial activity and toxicity evaluation. J Biomater Appl 30:919–929
125. Shen ZS, Kuang SH, Zhang Y, Yang MM, Qin W, Shi XT, Lin ZM (2020) Chitosan hydrogel incorporated with dental pulp stem cell-derived exosomes alleviates periodontitis in mice via a macrophage-dependent mechanism. Bioact Mater 5:1113–1126
126. Zang S, Mu R, Chen F, Wei X, Zhu L, Han B, Yu H, Bi B, Chen B, Wang Q, Jin L (2019) Injectable chitosan/beta-glycerophosphate hydrogels with sustained release of BMP-7 and ornidazole in periodontal wound healing of class iii furcation defects. Korean J Couns Psychother 99:919–928
127. Xu XW, Gu ZY, Chen X, Shi C, Liu CW, Liu M, Wang L, Sun ML, Zhang K, Liu QL, Shen YQ, Lin CT, Yang B, Sun HC (2019) An injectable and thermosensitive hydrogel: promoting periodontal regeneration by controlled-release of aspirin and erythropoietin. Acta Biomater 86:235–246
128. Chichiricco PM, Riva R, Thomassin JM, Lesoeur J, Struillou X, Le Visage C, Jerome C, Weiss P (2018) In situ photochemical crosslinking of hydrogel membrane for guided tissue regeneration. Dent Mater 34:1769–1782
129. Hirano S, Zhang M, Nakagawa M, Miyata T (2000) Wet spun chitosan-collagen fibers, their chemical n-modifications, and blood compatibility. Biomaterials 21:997–1003
130. Shalumon KT, Sowmya S, Sathish D, Chennazhi KP, Nair SV, Jayakumar R (2013) Effect of incorporation of nanoscale bioactive glass and hydroxyapatite in PCL/chitosan nanofibers for bone and periodontal tissue engineering. J Biomed Nanotechnol 9:430–440
131. Chen S, Hao YT, Cui WG, Chang J, Zhou Y (2013) Biodegradable electrospun PLLA/chitosan membrane as guided tissue regeneration membrane for treating periodontitis. J Mater Sci 48:6567–6577
132. Sundaram MN, Sowmya S, Deepthi S, Bumgardener JD, Jayakumar R (2016) Bilayered construct for simultaneous regeneration of alveolar bone and periodontal ligament. J Biomed Mater Res B 104:761–770
133. Qasim SB, Najeeb S, Delaine-Smith RM, Rawlinson A, Rehman IU (2017) Potential of electrospun chitosan fibers as a surface layer in functionally graded GTR membrane for periodontal regeneration. Dent Mater 33:71–83
134. Shen RZ, Xu WH, Xue YX, Chen LY, Ye HC, Zhong EY, Ye ZC, Gao J, Yan YR (2018) The use of chitosan/PLA nano-fibers by emulsion eletrospinning for periodontal tissue engineering. Artif Cell Nanomed B 46:419–430
135. Niu X, Wang L, Xu M, Qin M, Zhao L, Wei Y, Hu Y, Lian X, Liang Z, Chen S, Chen W, Huang D (2021) Electrospun polyamide-6/chitosan nanofibers reinforced nano-hydroxyapatite/polyamide-6 composite bilayered membranes for guided bone regeneration. Carbohydr Polym 260:117769
136. Zhu KK, Wang Y, Lu A, Fu Q, Hu JL, Zhang LN (2019) Cellulose/chitosan composite multifilament fibers with two-switch shape memory performance. ACS Sustain Chem Eng 7:6981–6990
137. Arpornmaeklong P, Sareethammanuwat M, Apinyauppatham K, Boonyuen S (2021) Characteristics and biologic effects of thermosensitive quercetin-chitosan/collagen hydrogel on human periodontal ligament stem cells. J Biomed Mater Res B. https://doi.org/10.1002/jbm.b.34823
138. Chang PC, Tai WC, Luo HT, Lai CH, Lin HH, Lin ZJ, Chang YC, Lee BS (2020) Core-shell poly-(D,l-lactide-co-glycolide)-chitosan nanospheres with simvastatin-doxycycline for periodontal and osseous repair. Int J Biol Macromol 158:627–635
139. Li YF, Yu FL, Liu Y, Liang Q, Huang YD, Xiang Q, Zhang QH, Su ZJ, Yang Y, Zhao YP (2020) Sulfonated chitosan oligosaccharide alleviates the inhibitory effect of basic fibroblast growth factor on osteogenic differentiation of human periodontal ligament stem cells. J Periodontol 91:975–985
140. dos Santos DM, Chagas PAM, Leite IS, Inada NM, de Annunzio SR, Fontana CR, Campana SP, Correa DS (2020) Core-sheath nanostructured chitosan-based nonwovens as a potential drug delivery system for periodontitis treatment. Int J Biol Macromol 142:521–534

Adv Polym Sci (2021) 288: 191–212
https://doi.org/10.1007/12_2021_97

Published online: 28 May 2021

Generations of Chitosan: The Progress in Drug Delivery

Eva Sanchez Armengol and Flavia Laffleur

Contents

Abstract The discovery of sulfhydryl-modified polymers named thiomers in the late twentieth century resulted in a reframing conceptualization of polymers and their pharmaceutical use. As it is well known, the mucosal barrier in the body has a high influence in drug delivery, leading to loss of drug or low bioavailability. Chitosan is a wide known polymer commonly used in the pharmaceutical and medical field due to its great properties. However, thiolated chitosan, as first generation, presents

E. Sanchez Armengol and F. Laffleur (✉)
Department of Pharmaceutical Technology, University of Innsbruck, Innsbruck, Austria
e-mail: eva.sanchez-armengol@uibk.ac.at; Flavia.Laffleur@uibk.ac.at

improvement in its mucoadhesive, as well as other properties compared to the native chitosan. S-protected or second generation of chitosan overcomes stability and oxidation problems. However, recent development of a third generation, less-reactive, chitosan changes the recent drug delivery field completely.

Hence, second and third generation thiomers present enhanced mucoadhesive, efflux pump inhibitory and controlled drug release properties, without (or with a very small, almost non-notable) cell toxicity. In this article, the syntheses of thiolated, S-protected and less reactive chitosan are elucidated. Furthermore, its properties, more in detail, the cell viability and toxicity, permeation enhancement, efflux pump inhibition, mucoadhesive properties and controlled drug release, are stated. Finally, an overview of in vivo studies for buccal, nasal, ocular, oral and vaginal drug delivery is given within this work as a proof of the concept.

Keywords Chitosan · Drug delivery · Mucoadhesion · Polymer · Thiol · Thiomers

1 Introduction

Chitosan, a well-known positively charged polysaccharide obtained from the alkaline deacetylation of chitin, is easily found in nature in the skeleton of crustaceous or in some fungi, for instance. The composition of chitosan is based on a random distribution of β(1–4)-linked D-glucosamine and *N*-acetyl-D-glucosamine units [1]. This fact equips chitosan with some chemical properties such as its high charge density at pH < 6.5, its adherence to negatively charged surfaces, its chelating capacity to certain transition metals, or the reactivity of its amino and hydroxyl groups. Furthermore, this compound is biocompatible, biodegradable, safe and non-toxic, as well as versatile and presents antimicrobial activity [2–4].

The impressive research interest in chitosan is due to its great properties and pharmaceutical potential. Thus, in the last decades, chitosan has played a very important role in the pharmaceutical field for different purposes. It is widely used as therapeutic agent carrier [5], as an absorption and permeation enhancer, as well as a great mucoadhesive compound [6–8]. Although the presented features boost chitosan to a great candidate for pharmaceutical and biomedical applications, a first generation of thiolated chitosan, also known as thiomers, was developed in the early 2000s to further develop its permeation, cellular uptake and mucoadhesive enhancing properties [9, 10]. The method described by Bernkop-Schnürch et al. is based on the immobilization of sulfhydryl groups in the primary amino group of the glucosamine subunits of chitosan by a covalent attachment. Hence, amide or amidine bonds are formed, depending on the synthesis pathway. The advantages of thiolated chitosan in comparison to the unmodified compound are the following: thiomers present enhanced mucoadhesive properties (I), thiolated excipients serve as permeation and drug absorption enhancers (II), as well as inhibitors of the efflux pump (III). Moreover, these compounds have the ability to form covalent linkages with the mucus. Nevertheless, thiolated chitosan is unstable towards oxidation and in aqueous solutions, a fact that was overcome with the development of a second generation of thiomers, named preactivated thiomers [11, 12].

Preactivated thiomers were firstly synthesized in 2012. These compounds presented the possibility to protect the bound thiol groups towards oxidation. Moreover, this resulted into upgraded properties in terms of mucoadhesiveness compared to unmodified ones. However, thiol group protecting polymers or S-protected thiomers are still not advantageous due to their high reactivity. Hence, a less reactive S-protected thiomers named third generation is currently being studied and developed, which should not only solve the reactiveness problem, but also improve the mucoadhesive properties of the preactivated thiomers [13].

Research on chemical functionalization on chitosan's backbone is paving the mucoadhesive street for drug delivery research areas.

2 Synthesis of Thiolated Chitosan

2.1 *First Generation: Thiolated Chitosan*

The thiolation process of chitosan can be classified taking into account the radicals affected by the substitution as it follows methods that affect the hydroxyl groups, methods that affect the amino moieties, or the methods affecting both the hydroxyl and amino radicals of chitosan. Furthermore, a deeper grouping can be done as per the way how the sulfhydryl groups are introduced, either by direct substitution or by the attachment of a thiol bearing ligand [14, 15]. However, the mostly used procedure for the thiomerization of this compound is based on the previous description to covalently attach thiol bearing ligands, such as L-cysteine or thioglycolic acid, to the polymeric backbone [9, 16].

Briefly, the polymer is hydrated and the thiol bearing ligand is added. It is of great importance that there is a huge variety of ligands being utilized for the thiolation procedure (Table 1). Afterwards, the addition of *N*-(3-dimethylaminopropyl)-*N'*-ethylcarbodiimide hydrochloride (EDAC) is necessary in order to activate the carboxylic moieties and the mixture at pH 4.5 is left under stirring at room temperature for an incubation period. The obtained conjugate is later dialysed and lyophilized to obtain a white powder [1, 17].

2.2 *Second Generation: Preactivated/S-Protected Chitosan*

The second generation of thiolated polymers was launched to the pharmaceutical platform in 2012. These are also known as preactivated, or S-protected, thiomers, and present the capacity of protecting the thiol groups towards oxidation. In contrast to the first generation of thiomers, these S-protected thiolated compounds have

Table 1 Summary of the different thiolated chitosan depending on the thiol bearing ligand used for the chemical reaction to take place

Chitosan (R_2)	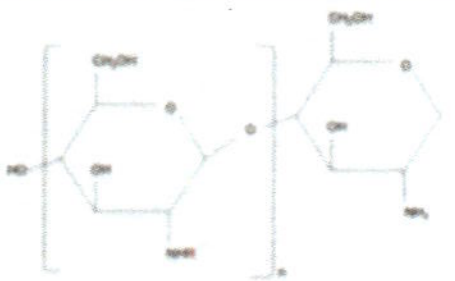						
Derivate depending on the R	Cysteine (CYS)	Thioglycolic acid (TGA)	2-Thiomethyl amidine (TEA)	4-Thiobutyl-amidine (TBA)	N-acetylcysteine (NAC)	6-Mercaptonicotinic acid (MNA)	4-Mercaptobenzoic acid (MBA)
Chemical structure of the thiolated chitosan	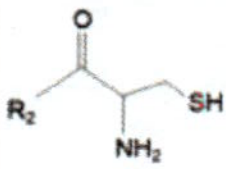	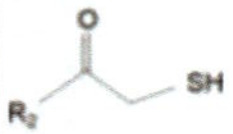	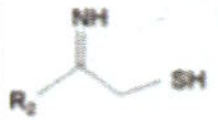	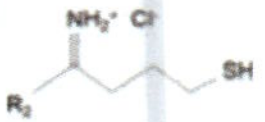	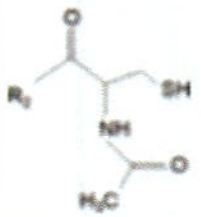	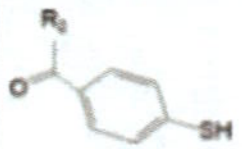	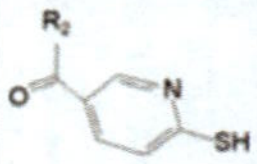
SH (μmol/g of polymer)	100	855 ± 7	224.82 ± 12.89	292	325.5 ± 41.8	452.78	98.49
References	[1]	[18]	[19]	[20]	[21]	[22]	[23]

Table 2 Comparison of the two of the methods used in order to obtain preactivated thiolated chitosan [24–26]

	Method 1: classic	Method 2: EDTA
Materials	Chitosan, thioglycolic acid, EDAC, acetic acid, NaOH, HCl, NaCl, MNA, hydrogen peroxide	Chitosan, EDTA, acetic acid, L-cysteine, hydrochloric acid, sodium hydroxide, hydrogen peroxide, 2MNA
Synthesis	Step 1: Synthesis of thiolated chitosan (Ch-TGA) Step 2: Synthesis of the dimeric ligand Step 3: Synthesis of preactivated thiolated chitosan (Ch-TGA-2MNA)	Step 1: Synthesis of Ch-EDTA conjugates Step 2: Synthesis of thiolated Ch-EDTA conjugated (Ch-EDTA-cys) Step 3: Synthesis of preactivated Ch-EDTA-cys conjugated (Ch-EDTA-cys-2MNA)
Purification	Dialysis (MWCO 12 kDa) Step 1 and step 2 follow the same procedure: One time against 5 mM HCl, two times against 5 mM HCl (1% NaCl), two times against 1 mM HCl.	Dialysis (MWCO 10–20 kDa) Step 1: One time against demineralized water, two times against 1 mM NaOH with 1% NaCl, two times against 1 mM NaOH, one time against demineralized water Step 2: Two times against 1 mM HCl, two times against 1 mM HCl with 1% NaCl and one time against 0.2 mM HCl. Step 3: Five times against demineralized water
SH (μmol/g of polymer)	Ch-TGA: 334 ± 24 Ch-TGA-MNA: 40.05 ± 6	Ch-EDTA-cys: 1,541 ± 252.5 Ch-EDTA-cys-2MNA: 1,560.5 ± 291.8

Ch Chitosan, *TGA* Thioglycolic acid, *cys* Cysteine, *EDAC* N-(3-Dimethylaminopropyl)-N′-ethyl-carbodiimide, *EDTA* Ethylenediaminetetraacetic acid, *NAOH* Sodium hydroxide, *HCl* Hydrochloric acid, *NaCl* Sodium chloride, *MNA* Mercaptonicotinic acid, *MWCO* Molecular weight cut-off

improved mucoadhesive properties, among others that are also upgraded. However, an important drawback of these thiomers is the high reactivity presented [13].

As shown in Table 2, the classic method to obtain an S-protected thiolated polymer encompasses the use of mercaptonicotinic acid. This compound is the most used to protect the thiol groups of the substances. Indeed, in the second method shown, the procedure is almost the same, but EDTA is used to activate the carboxylic moieties of the chitosan. The main difference between method one and method two relies on the use of EDTA, indeed.

2.2.1 Method 1: Chitosan-TGA Conjugates

The classic method to preactivate thiolated chitosan goes as following: the previously thiolated chitosan with thioglycolic acid is dissolved in water in a 1% (w/v) final concentration. Then, the dimeric ligand, previously prepared by agitating a solution of mercaptonicotinic acid in water with hydrogen peroxide (26 mM) during

1 h at pH 9.0 and afterwards a lyophilisation process, is added dropwise in a final concentration of 2.7 mM. This mixture is then left under constant stirring at room temperature to ensure that the reaction takes place, followed by a dialysis and posterior lyophilisation process [24].

2.2.2 Method 2: Chitosan-EDTA-cys Conjugates

Ethylenediaminetetraacetic acid (EDTA) is well known as a chelating agent. Among all the advantageous properties of this chemical compound, it is to mention the capacity of complexation for bivalent ion metals as well as its high adhesive properties when combining it with therapeutic polymers [27].

In order to obtain the chitosan-EDTA (Ch-EDTA) conjugate, a covalent attachment via amide bond formation is the most commonly used method. Thus, a first step in the synthesis reaction has to be performed. This is based on the dissolution of chitosan in hydrochloric acid, where EDTA is added in order to activate the carboxylic moieties. This mixture has to be left under stirring for 5 h at room temperature. Afterwards, a purification process based on dialysis has to be carried out. The second step in the synthesis process is the thiolation of these Ch-EDTA conjugates. For this purpose, the previously published thiolation process using L-cysteine was executed, where Ch-EDTA is firstly hydrated and the carboxylic moieties are activated, step followed by the addition of L-cysteine. The mixture is left under stirring at room temperature for 3 h followed by purification via dialysis. The third step is performed in order to obtain the preactivated Ch-EDTA-cys conjugates (Ch-EDTA-cys-2MNA) [25]. In brief, the Ch-EDTA-cys is mixed with a prepared 2MNA dimer, and the reaction mixture is mixed overnight at room temperature. So as to obtain the product, a purification process via dialysis has to be carried out [26].

2.3 *Third Generation: Less Reactive S-Protected Chitosan*

Recently, Netsomboon et al. have developed an approach to a third generation of thiolated chitosan in order to solve the reactivity problem that the second generation of these compounds presents. The hypothesis supports the fact that the reactivity of preactivated thiolated polymers is high with the thiol groups presented in the superficial area of the mucus gel layer. Hence, the penetration of these compounds is hindered, problem that a less reactive S-protected thiomers could solve.

The study performed by the research group involves two methods for the synthesis of S-protected chitosan (Table 3). The most prominent difference is the kind of variety in the leaving group, which is corresponding to NAC-6-MNA and NAC-NAC dimer for the first and second method, respectively. However, the results reported no significant difference regarding the amount of immobilized thiol groups and disulphide bond formation, as shown in Table 3 [13].

Table 3 Comparison of the two methods used in order to obtain less reactive S-protected thiolated polymer [13]

	Method 1: CS-NAC-MNA	Method 2: CS-NAC-NAC
Materials	Chitosan, 6-chloronicotinamide, thiourea, NAC-6-MNA, S-(5-carbamyl-2-pyridyl) thiouranium chloride, 6-MNA, 6,6′-dithionicotinamide	Chitosan, NAC-NAC dimer, deionized water, EDAC
Synthesis	Synthesis of CS-NAC-MNA	Step 1: NAC-NAC dimer preparation Step 2: Synthesis of CS-NAC-NAC
Purification	Dialysis (MWCO: 10–20 kDa), freeze and lyophilize. Storage at room temperature	Dialysis, freeze and lyophilize. Storage at room temperature.
Yield (%)	85.4	64.7
SH (μmol/g of polymer)	Not detectable	Not detectable
S-S (μmol/g of polymer)	566.7 ± 32.2	610.0 ± 91.3

Ch Chitosan, *TGA* Thioglycolic acid, *NAC* N-acetyl cysteine, *EDAC* N-(3-Dimethylaminopropyl)-N′-ethyl-carbodiimide, *MNA* Mercaptonicotinic acid, *MWCO* Molecular weight cut-off

3 Properties

3.1 *Toxicity and Cell Viability*

Cytotoxicity assays are commonly performed as a first-step indicator to evaluate the effectiveness of the new synthesized polymer in terms of potential harm. To evaluate cell viability, MTT (3-[4,5-dimethylthiazol-2-yl]-2,5 diphenyltetrazolium bromide) and resazurin assays are usually carried out (Fig. 1).

Perrone et al. [28] studied the cytotoxic effect of preactivated thiolated glycol chitosan using the MTT assay protocol. Results showed an increase in cell viability with lower tested concentration of polymer. Indeed, the viability was over 85% for samples with 0.02 mg of modified polymer/mL of medium and lowered to about 80% for those samples with a concentration of 2 mg/mL, resulting in not being harmful for Caco-2 cells [28]. In contrast, Netsomboon et al. [26] carried out the cell viability test by the use of the resazurin test in Caco-2 cells. Results exhibited a lower cell viability for cells treated with Ch-EDTA in comparison to cells treated with the thiolated and preactivated conjugate of this compound. Indeed, for a concentration of 0.5% (w/v), cell viability was observed to be 89%, 99% and 95% after 4 h for Ch-EDTA, Ch-EDTA-cys and Ch-EDTA-cys-2MNA, respectively [26].

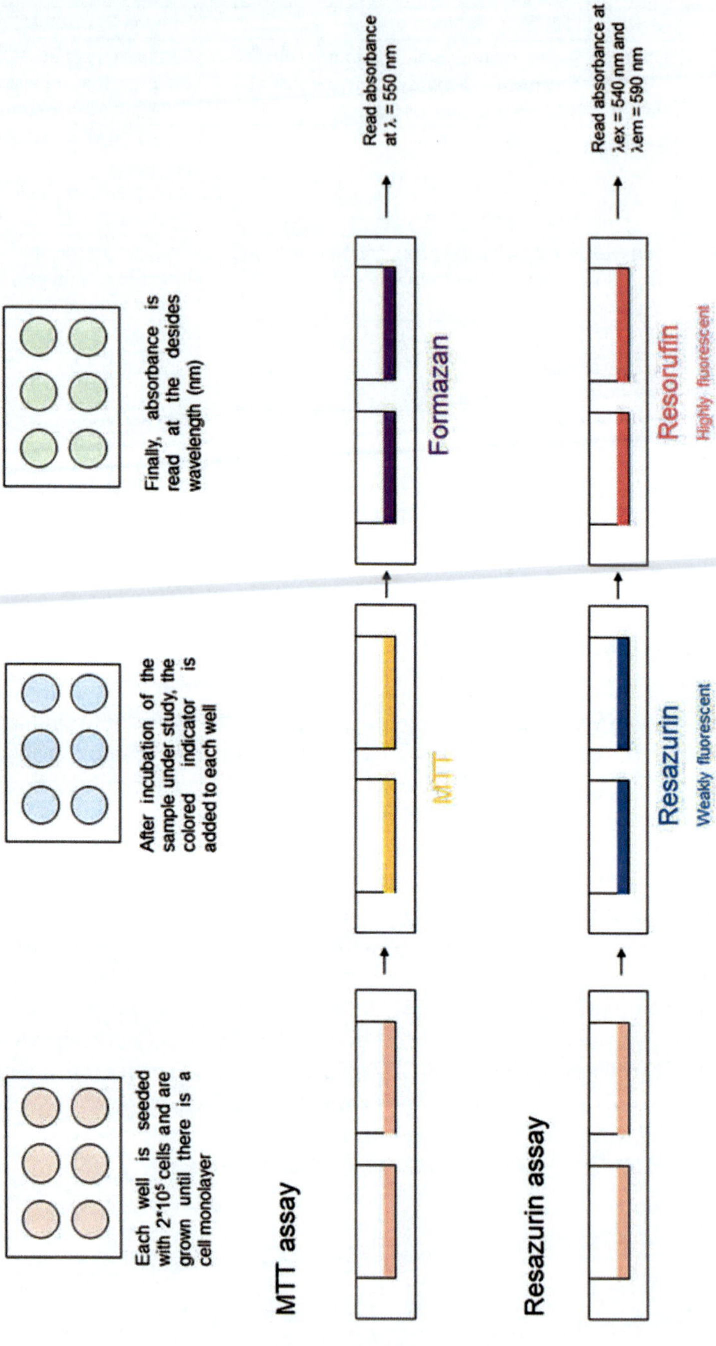

Fig. 1 Comparison of the process for performing a cell viability assay using the MTT and the resazurin protocol

3.2 *Mucoadhesive Properties*

In vitro mucoadhesive studies commonly include the rotating cylinder method and tensile studies. When performing these assays, the mucosa of interest can be used taking into account the desired application of the compound. On the one hand, the rotating cylinder method is used to evaluate the capacity of the polymer under study to adhere to mucosal surfaces. The method is a good option to simulate in vivo conditions, such as gastrointestinal motion or the application of a shear force, which can be provoked by fluids. On the other hand, tensile studies are used to determine two very important properties: the maximum detachment force (MDF), in mN, and the total work of adhesion (TWA), in μJ. The first one is the necessary force that has to be applied so that the polymer is detached from the mucosal surface, while the second one corresponds to the sum of the necessary work to detach the polymer, calculated as force multiplied by distance.

For instance, Perrone et al. [28] obtained a 1.4-fold increase in the residence time of S-preactivated thiolated glycol chitosan (GC-NAC-MNA) in comparison to the thiolated polymer (GC-NAC) and a 4.6-fold enhanced residence time for the S-preactivated thiolated chitosan modified with glutathione (GC-GSH-MNA) in comparison to the thiolated precursor (GC-GSH). These results were confirmed with tensile studies, exhibiting a 2.4-fold and twofold enhanced MDF for the S-preactivated polymer GC-NAC-MNA and GC-GSH-MNA in comparison to the correspondent thiolated compound, respectively. Furthermore, TWA showed a 4.88-fold increase for GC-NAC-MNA in comparison to GC-NAC and a onefold enhancement for GC-GSH-MNA in comparison to GC-GSH (Fig. 2) [28].

Mucoadhesive studies carried out by Netsomboon et al. [26] showed an increase in the residence time of Ch-EDTA-cys-2MNA, compared to Ch-EDTA and Ch-EDTA-cys, of 5.6-fold and 3.6-fold higher, respectively. Tensile studies also exhibited an enhancement of MDF and TWA of 1.4- and 1.3-fold of Ch-EDTA-cys in comparison to Ch-EDTA, respectively. Moreover, Ch-EDTA-cys-2MNA compared to Ch-EDTA exhibited a 1.7- and 2.6-fold increased MDF and TWA, respectively [26].

3.3 *Rheological Studies*

Rheology assays are performed in order to study the interaction between mucus and polymer of interest due to the change in the flow behaviour of the mixture [29]. By this method, viscoelasticity of the polymer under study can be determined under certain controlled environmental conditions and other properties such as the dynamic viscosity or the elastic modulus can be measured [30].

As it is a worldwide used technique, most research groups perform rheological studies when characterizing a compound. Indeed, Hintzen et al. determined a 1,087-fold increased dynamic viscosity of thiolated chitosan (Chi-TGA) after 1 h of testing

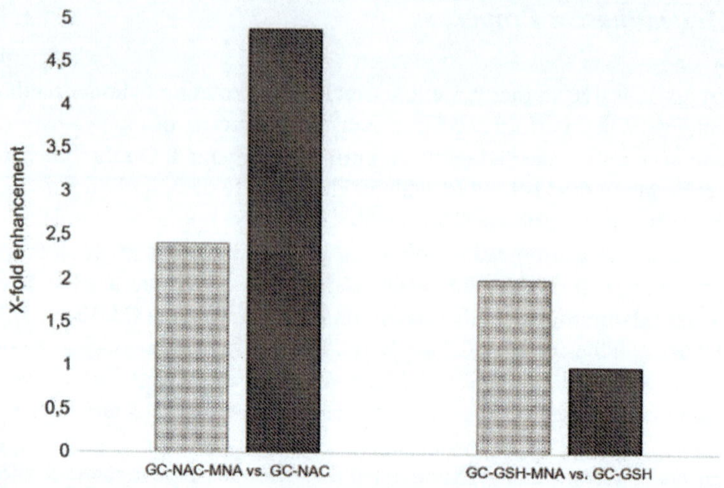

Fig. 2 X-fold enhancement in the residence time of S-preactivated thiolated chitosan in rat intestine mucosa. GC-NAC-MNA and GC-GSH-MNA correspond to second generation, S-protected, thiolated chitosan carried out by two different methods, while GC-NAC and GC-GSH correspond to the first generation, thiolated chitosan [28]

[17]. Friedl et al. were able to determine that the mucoadhesiveness of preactivated chitosan (Chi-TGA-MNA) is 1.5-fold improved in comparison to commercial formulations [30].

3.4 *Permeation Enhancement Effect*

Chitosan presents permeation-enhancing properties due to its positive charges, which are capable of interacting with the cell membrane resulting in a structural reorganization of the proteins associated to tight junctions. Moreover, molecular mass and degree of deacetylation of the compound play an important role in the permeation enhancement capacity, as well as the toxicity of the molecule [31]. In Fig. 3 the improvement in the oral bioavailability of hydrophilic drugs is shown when the administration is performed with only chitosan, chitosan with a co-administrator such as dodecylsulfate, or thiolated chitosan [7].

Illum et al. firstly reported the permeation enhancing properties of chitosan in 1994. This ability is of great interest for the increase in the paracellular absorption route for hydrophilic drugs [32]. However, the good properties of chitosan can be improved by performing a thiolation of this polymer. Indeed, an improvement ratio of up to 5 when testing in Caco-2 monolayers in comparison to unmodified chitosan was achieved by the immobilization of thioglycolic acid on chitosan with a modification degree (μM/g) of 27 [34]. Another study showed a permeation enhancing

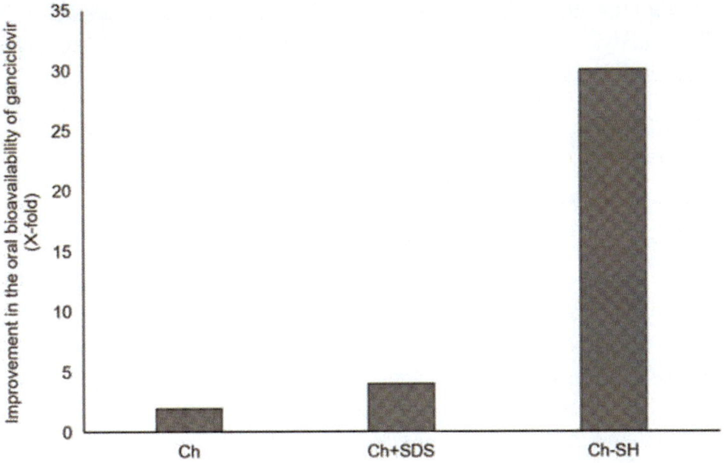

Fig. 3 X-fold improvement in the oral bioavailability of ganciclovir due to the permeation enhancement effect of chitosan. Ch: unmodified chitosan; Ch + SDS: chitosan administered with sodium dodecylsulfate; Ch-SH: first generation, thiolated chitosan [32–34]

effect of chitosan-4-thio-butyl-amidine conjugates (Chitosan-TBA) with an improvement ratio in comparison to unmodified chitosan of over 140 with a degree of modification of 95 μM of TBA per gram of unmodified polymer [35].

3.5 Efflux Pump Inhibition

The efflux pump is composed of the P-glycoprotein (P-gp), encoded by a multidrug resistance gene 1 (MDR1). Anionic thiomers have started being attractive P-gp inhibitors. In recent studies performed by Bernkop-Schnürchs' research group, the P-gp inhibitory properties of synthesized chitosan-thioglycolic acid (Chi-TGA) and its S-protected derivative (Chi-TGA-6MNA) were studied. For this, intracellular accumulation of *Rhodamine-123*, and *Rho-123* transportation across a Caco-2 cell monolayer and excised rat intestines were measured. On the one hand, results for the experiments performed in Caco-2 cell monolayers exhibited a noteworthy enhancement in the intracellular *Rho-123* and more transportation was observed from the absorptive direction and less in the secretory. On the other hand, results observed in freshly excised rat intestine transportation exhibited a concentration-dependent transportation of *Rho-123*. Four features were taken into account to determine the P-gp inhibitory mechanism of thiolated and S-protected chitosan. The first one was the level of expression of P-gp. The second, the ATPase activity of the efflux pump. Thirdly, the intracellular levels of ATP were considered and lastly, the plasma membrane fluidity. Outcomes of the experiments did not exhibit any change in the

P-gp expression or intracellular ATP, while plasma membrane fluidization and the P-gp ATPase activity decreased, leading to the thought that these factors affect the efflux pump inhibitory effect of chitosan [36, 37].

3.6 Controlled Drug Release Properties

Release studies are carried out in order to evaluate the controlled drug liberation properties of the system. Thus, dissolution experiments are usually performed mimicking the biological conditions of the target site. The fact that chitosan has great mucoadhesive properties makes this a polymer of interest in terms of controlled drug release. Requirements for a controlled release comprise cohesion and stability of the system. Thiolated chitosan present these claims due to the capability to form disulphide bonds between the mucosal glycoproteins and the thiol groups of their polymeric backbone.

Results obtained by Perrone et al. [28] showed that the preactivation of thiolated polymers conduce to a more intimate contact with the epithelial barrier, increasing the time of residence of the drug and controlling its release. In Fig. 4 the results of the dissolution efficiency (DE) and the percentage of dissolved drug after 30 min of incubation are shown [28].

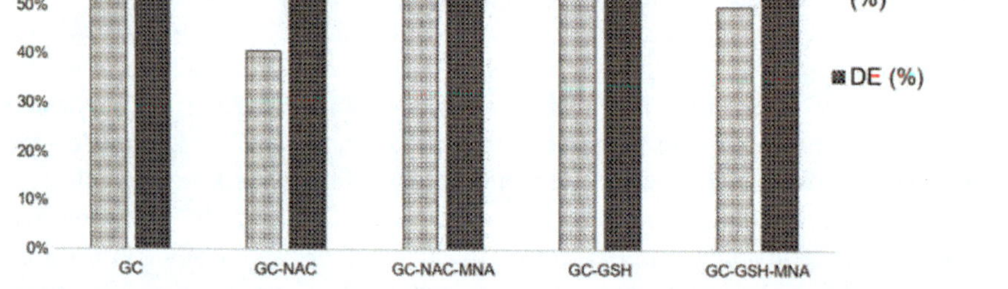

Fig. 4 Dissolution efficiency (DE) and drug dissolved in 30 min, all in terms of percentage [28]. *GC* Glycol chitosan, *GC-NAC* Thiolated glycol chitosan, *GC-NAC-MNA* Preactivated thiolated glycol chitosan, *GC-GSH* Thiolated glycol chitosan, *GC-GSH-MNA* Preactivated thiolated glycol chitosan

4 In Vivo Studies: Proof of Concept

To evaluate the safety of the systems, as well as other properties such as mucoadhesiveness or bioavailability, in vivo studies have to be performed. This is important because in some cases, in vitro studies can be promising, but drug efficacy or safety can be a failure in the in vivo experiments due to the multiple metabolic processes that are continuously taking place in the living organism. As there are many different administration routes (Table 4), the in vivo studies will be different for each one of them.

Table 4 Comparison of advantages and disadvantages of some delivery routes

	Advantages	Disadvantages
Buccal drug delivery	• Rich vascularization • Better access for administration and removal of the dosage form • High patient acceptability • Avoids hydrolysis in the GI tract • Avoids "first-pass" effect • Rapid cell recovery	• Low permeability • Smaller surface area • Saliva secretion (0.5–2 L/day) results in a dilution of the drug, involuntary removal of the dosage form, or loss of the drug due to swallowing
Nasal drug delivery	• Large absorptive surface area • High vascularity of nasal mucosa • Direct transport of drug into the systemic circulation • Avoids "first-pass" effect • Avoids degradation in the GI tract	• The normal physiology of the nasal cavity presents several barriers to peptide and protein drug absorption • Mucociliary clearance mechanism • Enzymatic degradation • Low permeability of the nasal epithelium
Ocular drug delivery	• Local effect	• High tear turnover and pre-corneal tear turnover lead to loss of drug
Oral drug delivery	• No specialist needed • Good targeted effect • Very studied and researched	• GI tract enzymatic degradation • Stomach acidic hydrolysis • First-pass metabolism
Vaginal drug delivery	• Local effect • Easy drug application • High contact surface area • High vascularity • Good permeability to several substances • Avoids first-pass effect • Avoids GI tract side effects	• Infection with Candida albicans • Hormonal changes • Menstrual cycle • Histology and physiology of the vagina varies with age • Endometrial fluids

4.1 Buccal Drug Delivery

The buccal administration route is a great alternative to the oral drug delivery route in order to overcome the first-pass metabolism, as well as the degradation of the compound in the stomach and small intestine due to the acidic conditions. One of the most problematic drawbacks presented in this administration route is the short residence time of the drug in the oral cavity. Hence, the use of polymeric compounds is of great interest when the mucoadhesive properties of the polymer are adequate. Moreover, the absorption through this administration route can occur either in the sublingual or local regions. The main applications of this drug delivery route enclose from toothache to bacterial infections [7, 38].

In order to understand the properties of buccal drug delivery systems, such as the permeability or mucoadhesion, it is of great interest to have knowledge of the anatomy of the buccal mucosa. It is composed of three main layers: the epithelium, which is divided into two areas: the keratinized and the non-keratinized, the basement membrane, which serves as mechanical support for the first layer, and the connective tissues, which include the blood vessels and present a rich arterial blood supply derived from the carotid artery. The whole oral cavity is covered by mucus, which is composed of water-insoluble glycoproteins and serves as a protective barrier for the cells [39, 40].

So as to release the drug through this administration route, there are different types of systems that can be used: buccal tablets, patches, films, gels and ointments, and excipients, such as nanoparticles [38, 41].

In 2014, Mortazavian et al. designed, developed and characterized nanoparticles composed of thiolated chitosan to delivery insulin through the buccal route in order to avoid injections and enhance the patient compliance for insulin administration. To characterize, several studies were carried out by in vitro and ex vivo methods. The general results show that when using thiolated chitosan in the nanoparticles formulation, the drug release is up to 97.18% after 8 h, and the burst release appears to be 14.39%, which is notably higher in comparison to unmodified chitosan, demonstrating sustained release [42]. Another type of tested drug delivery system are buccal films. Koland et al. evaluated in vitro and in vivo chitosan buccal films of ondansetron hydrochloride. Results show that formulations with polyvinylpyrrolidone (PVP K30) present enhanced drug release properties. The formulation containing chitosan (0.8 g), PVP (0.09 g) and glycerine (0.4 mL) shows the best release in vitro, with a percentage of up to 98.5 after five h. Moreover, when performing in vivo studies, this film formulation showed a slower, but greater extent of release in the same period of time and enriched mucoadhesive properties in comparison to other formulations used. By these studies, the pharmacokinetic parameters of ondansetron hydrochloride were tested in rabbits. When comparing the buccal film to an oral solution with the same drug, a significant increase in the C_{max}, T_{max} and $AUC_{0\text{-}\infty}$ of 8.67, 1.75 and 118.349, respectively, was observed [43].

4.2 Nasal Drug Delivery

Among other problems presented by most administration routes, the degradation of drugs in the gastrointestinal sites and low bioavailability are the most common. A convenient and reliable administration route to overcome this might be the intranasal drug delivery. In this case, the drug can be directly absorbed into the blood vessels present in the nasal cavity, which also presents a very high surface area and vascularity of mucosa. Thus, drugs can be directly transported into the systematic circulation and the first-pass metabolism can be avoided. However, the nasal cavity has also peptide and protein absorption barrier systems in its physiology [7, 44].

Studies performed by Wu et al. to investigate a new nasal drug delivery system based on a, at room temperature, sprayable thermosensitive hydrogel composed by chitosan, which, in contact with the nasal mucosa and the elevated temperature (37°C) turned into a viscous gel-like compound showed that this new formulation is of great interest for nasal drug delivery. Moreover, it apparently presents no toxicity and has a high patient compliance. Furthermore, the controlled drug release experiments exhibited that the most promising formulation of hydrogel was when using PEG 6000, showing an up to 60% of drug release in within 40 days [44].

Krauland et al. [45] researched and developed a microparticulate delivery system based on thiolated chitosan. The chitosan-TBA conjugate presented 304.89 ± 63.45 μmol of thiol groups per gram of unmodified polymer. Although the results of studies with insulin as model drug showed a bioavailability of $7.24 \pm 0.76\%$ in living rats, these are low compared to previous studies performed by other research groups [45]. For instance, Callens and Remon reached 14.4% of bioavailability in rabbits [46]. A chitosan-powder based formulation exhibited to be up to 17.0% bioavailable in sheep [47].

4.3 Ocular Drug Delivery

In order to treat superficial eye diseases or provide an intra-ocular treatment through the cornea, the topical ocular drug delivery route is the preferred one. However, this presents disadvantages such as the pre-corneal loss of drug due to drainage and high tear turnover. Moreover, chitosan presents permeation-enhancing properties and no toxicity, rendering in a great candidate for ocular drug delivery in forms of liquid or nanoparticulate formulations as well as hydrogels [7, 48].

Studies carried out by De Campos et al. [49] showed that the interaction between chitosan and the ocular surface is higher and stronger when the polymer is presented in a nanoparticulate formulation. In vivo experiments were performed in rabbit cornea and confocal fluorescence images indicated a paracellular mechanism of transportation for chitosan. Moreover, these nanoparticles composed of chitosan showed an ability of associating to the mucosal surface and a longer residence time, enhancing and controlling the drug release in the cornea [49].

Further experimental procedures using chitosan as main polymer for the ocular delivery of pharmaceutical compounds led to the statement, that, when using nanoparticles as carrier systems, the size is a very important variable. Indeed, this parameter has a close relationship with the ability of the systems to interact with mucosal surfaces in general, but in particular with the ocular mucosa. Evaluations of these phenomena were performed by Diebold et al. [50], resulting in a maintained stability of the systems and higher size, up to 755.3 ± 30.0, 491.2 ± 1.9 and 407.8 ± 9.6 nm when developing liposome-chitosan nanoparticulate systems composed of distearoylphosphatidylcholine (DSPC), dipalmitoylphosphatidylserine (DPPS), dipalmitoylphosphatidylcholine (DPPC) and cholesterol (CHOL) in a proportion of 6:0:1:4, 6:0:0:4 and 0:0:6:4, respectively. Nanoparticles made of chitosan were obtained by the mixture of this compound with pentasodium tripolyphosphate in a 3:1 ratio [50].

Although chitosan exhibits mucoadhesive properties, when thiolation occurred those features become even more enhanced. As thiolated chitosan presents several advantageous features in comparison to unmodified chitosan, nanoparticles thereof are convincing as drug delivery systems with improved features. Thus, nanoparticles composed of thiolated chitosan were developed and studied by Zhu et al. [51]. Results obtained after experimenting with the modified nanoparticles in rat corneas exhibited that these are capable of delivering more drug into the cornea in comparison to unmodified nanoparticles at the same time of operation [51].

4.4 *Oral Drug Delivery*

As already stated the oral administration route is the most popular and used one among all the drug delivery pathways due to the ease of administration and the patience compliance. Nevertheless, the oral administration route presents a wide variety of drawbacks, such as the poor absorption of macromolecular drugs, like peptides, or the limited controlled drug delivery [5, 7]. As most of the gastrointestinal tract is covered by a mucus layer, it is of great interest to use thiolated vehicles for the transport of the drug to the target site and, due to its great mucoadhesive properties, enable a controlled and prolonged drug release. Thus, the oral is the most investigated mucosal drug delivery administration route. Drug carriers can go from nanoparticles to micelles or liposomes [52].

Krauland et al. carried out in vivo evaluation studies of a delivery system based on thiolated chitosan in 2004. Results exhibited a threefold enhancement, which corresponds to a relative pharmacological efficacy of 1.69 ± 0.42%, in rats. Moreover, these systems presented excellent mucoadhesive properties [33].

Millotti et al. [53] evaluated in vivo different properties and parameters of interest. Results showed an increase of the AUC of 4.78-fold and 21.02-fold for thiolated chitosan of 20 kDa and thiolated chitosan of 400 kDa, respectively, in comparison to unmodified chitosan. Moreover, an improvement in the mucoadhesive properties and an influence of the molecular mass of the polymer on the mucoadhesion extent was observed. Other properties tested were the C_{max} and the bioavailability. On the one hand, results for the 20 kDa thiolated chitosan in comparison to unmodified chitosan showed a 19.48-fold enhancement in the C_{max} and a 4.86-fold increase in bioavailability. On the other hand, results for the 400 kDa modified chitosan in comparison to non-thiolated chitosan presented a 14.34-fold and 24.5-fold increase for C_{max} and bioavailability, respectively [53]. Bioavailability studies using Wistar rats performed by Fabiano et al. showed that nanoparticles composed of second generation, S-protected, thiolated chitosan presented a significantly higher bioavailability in comparison to nanoparticles composed of unmodified chitosan. In fact, maximal concentration (C_{max}) and maximal residence time (T_{max}) increased in 0.24 μg/mL and 1 h, respectively [52].

More studies carried out with preactivated thiolated chitosan exhibited a 157.35 μg/mL and 1 h increase in the C_{max} and T_{max} compared to native chitosan in rabbits. Moreover, the elimination half-life was enhanced in a fourfold improvement. This research group also studied the $AUC_{0\text{-}t}$, $AUC_{0\text{-}\infty}$ and $AUMC_{0\text{-}\infty}$, obtaining results of 1,532.83, 2,124.5 and 36,794.5 μg/mL, respectively, which, compared to the unmodified drug solution, exhibit an increase of 1,347.18, 1,905.75 and 35,947.18 μg/mL, respectively [24].

4.5 *Vaginal Drug Delivery*

An alternative for the treatment of diseases, with local or systemic effect, is the vaginal delivery. This local drug release method is commonly used in pharmaceutical applications such as the administration of antibacterials, antifungals, antivirals or steroids due to its advantages, being the ease of drug absorption, high blood supply, high contact surface area and the avoidance of the first-pass metabolism. However, the drawbacks of this alternative administration route are of high importance at the moment of performing in vivo studies. Such disadvantages embed the histological and physiological variation of the vagina with age, menstrual cycle and hormonal changes. Moreover, there is also present the secretion of endometrial fluids [54].

Safety studies performed in vivo by Meng et al. [55] showed that microspheres composed of thiomers had an eightfold increased drug effective dose. Optical microscopy examination of the vaginal epithelium 1 and 7 days after the incorporation of the drug was carried out showing an increase in lymphocyte infiltration, which indicated inflammatory response to the system. Although results are not the optimal, thiolated microspheres are considered as a good candidate as drug delivery systems of small-molecule and water-soluble therapeutic molecules, as well as a promising vaginal delivery system to prevent HIV transmission [55].

5 Conclusions

Within this work, three main topics have been elucidated, namely synthesis of three generations of chitosan, their obtained properties as well as their proof of concept.

Regarding the synthesis of thiolated chitosan, different techniques are available depending on the conjugate of interest. However, Chi-TGA is the most promising one, having obtained up to 855 $\pm$ 7 μmol of thiol group per gram of polymer. Second generations are also known as preactivated thiolated chitosans. There are also divergent synthesis pathways, although the most common ones are the classic method and the EDTA-based method. Less reactive S-protected chitosans, or the third generation of chitosan thiomers, are the most newly studied conjugates. Two methods have been developed recently and both are based on the non-detectability of thiol groups and a disulphide bond detected between the range of 550 and 610 μmol of S-S per gram of polymer.

Concerning the obtained features, it has been proved that cell viability of thiolated chitosan and preactivated thiolated chitosan is up to 99 and 95% when using a concentration of 0.5% (w/v) after 4 h of incubation. In regard to the mucoadhesive properties, the S-protected chitosan conjugate Ch-EDTA-cys-2MNA presents a 5.6-fold and 3.6-fold increased residence time in comparison to Ch-EDTA and Ch-EDTA-cys, respectively. Furthermore, rheological studies exhibited a 1.5-fold higher mucoadhesiveness of Ch-TGA-MNA in comparison to already commercially available formulations. The permeation enhancing properties of chitosan were already stated. However, thiolated chitosan shows ability of approximately 30-fold increase compared to native chitosan. In addition, studies lead to the thoughts that chitosan has P-gp inhibitory effects. Another interesting characteristic of these compounds is the controlled drug release capacity. It has been under study that preactivated thiolated chitosan is capable of having a more intimate contact with the epithelial barrier, which results in a higher residence time and thus, a controlled drug release.

With respect to the in vivo proof of concept studies, several conclusions can be established. The buccal drug delivery is a great alternative to the oral administrative route, presenting an 8.67, 1.75 and 118.349-fold increase of the C_{max}, T_{max} and $AUC_{0\text{-}\infty}$, respectively, when tested in rabbits. About the nasal delivery route, bioavailability in rats has been found to be 7.24%. However, studies in rabbits and sheep have also been performed. Considering the ocular drug delivery, enhanced release has been found out to be when administrating the drug with nanoparticles composed of thiolated chitosan and preactivated thiolated chitosan. The oral administration route is the preferred drug delivery route. Studies determined an enhanced C_{max} and T_{max} of thiolated and preactivated chitosan in comparison to the unmodified compound in rabbits. Finally, vaginal drug delivery presents to be a good option when administrating microspheres composed of thiolated chitosan for the delivery of small and water-soluble molecules.

In conclusion, although thiolated chitosan provides enhanced properties regarding its mucoadhesiveness, efflux pump inhibition and controlled drug release, preactivated or second generation of thiolated chitosan present greater and enhanced properties. Moreover, to overcome some limitations of these, a third generation is being studied and first reports exhibit an increase in most properties. Research on chemical functionalization on chitosan's backbone is paving the mucoadhesive delivery avenue in the not too far future.

References

1. Sarti F, Bernkop-Schnürch A (2011) Chitosan for Biomaterials I
2. Rhoades J, Roller S (2000) Antimicrobial actions of degraded and native chitosan against spoilage organisms in laboratory media and foods. Appl Environ Microbiol 66:80–86. https://doi.org/10.1128/AEM.66.1.80-86.2000
3. Hejazi R, Amiji M (2003) Chitosan-based gastrointestinal delivery systems. J Control Release 89:151–165. https://doi.org/10.1016/S0168-3659(03)00126-3
4. Pellá MCG, de Lima HHC, Rinaldi AW, Fajardo AR, Tenório-Neto ET, Guilherme MR, Rubira AF, Lima-Tenório MK (2019) Chitosan-based hydrogels for drug delivery. https://doi.org/10.1007/978-981-15-0263-7_6
5. Felt O, Buri P, Gurny R (1998) Chitosan: a unique polysaccharide for drug delivery. Drug Dev Ind Pharm 24:979–993. https://doi.org/10.3109/03639049809089942
6. Thanou M, Verhoef JC, Junginger HE (2001) Oral drug absorption enhancement by chitosan and its derivatives. Adv Drug Deliv Rev 52:117–126. https://doi.org/10.1016/S0169-409X(01)00231-9
7. Bernkop-Schnürch A, Dünnhaupt S (2012) Chitosan-based drug delivery systems. Eur J Pharm Biopharm 81:463–469. https://doi.org/10.1016/j.ejpb.2012.04.007
8. Cheung RCF, Ng TB, Wong JH, Chan WY (2015) Chitosan: an update on potential biomedical and pharmaceutical applications. https://doi.org/10.3390/md13085156
9. Bernkop-Schnürch A (2005) Thiomers: a new generation of mucoadhesive polymers. Adv Drug Deliv Rev 57:1569–1582. https://doi.org/10.1016/j.addr.2005.07.002
10. Zhang J, Xia W, Liu P, Cheng Q, Tahirou T, Gu W, Li B (2010) Chitosan modification and pharmaceutical/biomedical applications. Mar Drugs 8:1962–1987. https://doi.org/10.3390/md8071962
11. Iqbal J, Shahnaz G, Dünnhaupt S, Müller C, Hintzen F, Bernkop-Schnürch A (2012) Preactivated thiomers as mucoadhesive polymers for drug delivery. Biomaterials 33:1528–1535. https://doi.org/10.1016/j.biomaterials.2011.10.021
12. Laffleur F, Fischer A, Schmutzler M, Hintzen F, Bernkop-Schnürch A (2015) Evaluation of functional characteristics of preactivated thiolated chitosan as potential therapeutic agent for dry mouth syndrome. ACTA Biomater. https://doi.org/10.1016/j.actbio.2015.04.016
13. Netsomboon K, Jalil A, Laffleur F, Hupfauf A, Gust R, Bernkop-Schnürch A (2020) Thiolated chitosans: are Cys-Cys ligands key to the next generation? Carbohydr Polym 242. https://doi.org/10.1016/j.carbpol.2020.116395
14. Dünnhaupt S, Barthelmes J, Thurner CC, Waldner C, Sakloetsakun D, Bernkop-Schnürch A (2012) S-protected thiolated chitosan: synthesis and in vitro characterization. Carbohydr Polym 90:765–772. https://doi.org/10.1016/j.carbpol.2012.05.028
15. Federer C, Kurpiers M, Bernkop-Schnürch A (2020) Thiolated chitosans – a multi-talented class of polymers for various applications. Biomacromolecules. https://doi.org/10.1021/acs.biomac.0c00663
16. Hornof MD, Kast CE, Bernkop-Schnürch A (2003) In vitro evaluation of the viscoelastic properties of chitosan-thioglycolic acid conjugates. Eur J Pharm Biopharm 55:185–190. https://doi.org/10.1016/S0939-6411(02)00162-5

17. Hintzen F, Laffleur F, Sarti F, Shahnaz G, Bernkop-Schnürch A (2012) Thiomers: influence of molar mass on in situ gelling properties. Int J Pharm 436:120–126. https://doi.org/10.1016/j.ijpharm.2012.05.073
18. Bernkop-Schnürch A, Hopf TE (2001) Synthesis and in vitro evaluation of chitosan-thioglycolic acid conjugates. Sci Pharm 69:109–118. https://doi.org/10.3797/scipharm.aut-01-12
19. Kafedjiiski K, Hoffer M, Werle M, Bernkop-Schnürch A (2006) Improved synthesis and in vitro characterization of chitosan- thioethylamidine conjugate. Biomaterials 27:127–135. https://doi.org/10.1016/j.biomaterials.2005.05.075
20. Guggi D, Langoth N, Hoffer MH, Wirth M, Bernkop-Schnürch A (2004) Comparative evaluation of cytotoxicity of a glucosamine-TBA conjugate and a chitosan-TBA conjugate. Int J Pharm 278:353–360. https://doi.org/10.1016/j.ijpharm.2004.03.016
21. Schmitz T, Grabovac V, Palmberger TF, Hoffer MH, Bernkop-Schnürch A (2008) Synthesis and characterization of a chitosan-N-acetyl cysteine conjugate. Int J Pharm 347:79–85. https://doi.org/10.1016/j.ijpharm.2007.06.040
22. Millotti G, Perera G, Vigl C, Pickl K, Sinner FM, Bernkop-Schnürch A (2011) The use of chitosan-6-mercaptonicotinic acid nanoparticles for oral peptide drug delivery. Drug Deliv 18:190–197. https://doi.org/10.3109/10717544.2010.522611
23. Millotti G, Samberger C, Fröhlich E, Sakloetsakun D, Bernkop-Schnürch A (2010) Chitosan-4-mercaptobenzoic acid: synthesis and characterization of a novel thiolated chitosan. J Mater Chem 20:2432–2440. https://doi.org/10.1039/b916528b
24. Maria S, Sarwar HS, Sohail MF, Imran M, Salman Qureshi O, Raza A, Ahmad NM, Iqbal A, Shahnaz G (2020) Synthesis and characterization of pre-activated thiolated chitosan nanoparticles for oral delivery of octreotide. J Drug Deliv Sci Technol:58. https://doi.org/10.1016/j.jddst.2020.101807
25. Netsomboon K, Laffleur F, Netsomboon K, Laffleur F (2015) P-glycoprotein inhibitors: synthesis and in vitro evaluation of a preactivated thiomer P-glycoprotein inhibitors: synthesis and in vitro evaluation of a preactivated thiomer, 9045. https://doi.org/10.3109/03639045.2015.1075025
26. Netsomboon K, Suchaoin W, Laffleur F, Prüfert F, Bernkop-Schnürch A (2017) Multifunctional adhesive polymers: preactivated thiolated chitosan-EDTA conjugates. Eur J Pharm Biopharm 111:26–32. https://doi.org/10.1016/j.ejpb.2016.10.029
27. Bernkop-schnürch A, Scerbe-Saiko A (1998) Synthesis and in vitro evaluation of chitosan-EDTA-protease-inhibitor conjugates which might be useful in oral delivery of peptides and proteins. Pharm Res 15
28. Perrone M, Lopalco A, Lopedota A, Cutrignelli A, Laquintana V, Franco M, Bernkop-Schnürch A, Denora N (2018) S-preactivated thiolated glycol chitosan useful to combine mucoadhesion and drug delivery. Eur J Pharm Biopharm 132:103–111. https://doi.org/10.1016/j.ejpb.2018.09.015
29. Mackie AR, Goycoolea FM, Menchicchi B, Caramella CM, Saporito F, Lee S, Stephansen K, Chronakis IS, Hiorth M, Adamczak M, Waldner M, Nielsen HM, Marcelloni L (2017) Innovative methods and applications in Mucoadhesion research. Macromol Biosci 17:1–32. https://doi.org/10.1002/mabi.201600534
30. Friedl HE, Dünnhaupt S, Waldner C, Bernkop-Schnürch A (2013) Preactivated thiomers for vaginal drug delivery vehicles. Biomaterials 34:7811–7818. https://doi.org/10.1016/j.biomaterials.2013.06.021
31. Schipper NGM, Olsson S, Hoogstraate JA, DeBoer AG, Varum KM, Artursson P (1997) Chitosans as absorption enhancers for poorly absorbable drugs 2: mechanism of absorption enhancement. Pharm Res 14
32. Illum L, Farraj NF, Davis SS (1994) Chitosan as a novel nasal delivery system for peptide drugs. Pharm Res 11

33. Krauland AH, Guggi D, Bernkop-Schnürch A (2004) Oral insulin delivery: the potential of thiolated chitosan-insulin tablets on non-diabetic rats. J Control Release 95:547–555. https://doi.org/10.1016/j.jconrel.2003.12.017
34. Kast CE, Bernkop-Schnürch A (2001) Thiolated polymers – thiomers: development and in vitro evaluation of chitosan-thioglycolic acid conjugates. Biomaterials 22:2345–2352. https://doi.org/10.1016/S0142-9612(00)00421-X
35. Bernkop-Schnürch A, Hornof M, Zoidl T (2003) Thiolated polymers – thiomers: synthesis and in vitro evaluation of chitosan-2-iminothiolane conjugates. Int J Pharm 260:229–237. https://doi.org/10.1016/S0378-5173(03)00271-0
36. Chen X, Zhang Y, Yuan L, Zhang H, Dai W, He B, Wang X, Zhang Q (2015) The P-glycoprotein inhibitory effect and related mechanisms of thiolated chitosan and its S-protected derivative. RSC Adv 5:104228–104238. https://doi.org/10.1039/c5ra19418k
37. Sakloetsakun D, Iqbal J, Millotti G, Vetter A, Bernkop-Schnürch A (2011) Thiolated chitosans: influence of various sulfhydryl ligands on permeation-enhancing and P-gp inhibitory properties. Drug Dev Ind Pharm 37:648–655. https://doi.org/10.3109/03639045.2010.534484
38. Salamat-Miller N, Chittchang M, Johnston TP (2005) The use of mucoadhesive polymers in buccal drug delivery. Adv Drug Deliv Rev 57:1666–1691. https://doi.org/10.1016/j.addr.2005.07.003
39. Squier CA, Nanny D (1985) Measurement of blood flow in the oral mucosa and skin of the rhesus monkey using radiolabelled microspheres. Arch Oral Biol 30:313–318. https://doi.org/10.1016/0003-9969(85)90003-2
40. Haas J, Lehr CM (2002) Developments in the area of bioadhesive drug delivery systems. Expert Opin Biol Ther 2:287–298. https://doi.org/10.1517/14712598.2.3.287
41. Laffleur F (2014) Mucoadhesive polymers for buccal drug delivery. Drug Dev Ind Pharm 40:591–598. https://doi.org/10.3109/03639045.2014.892959
42. Mortazavian E, Dorkoosh FA, Rafiee-Tehrani M (2014) Design, characterization and ex vivo evaluation of chitosan film integrating of insulin nanoparticles composed of thiolated chitosan derivative for buccal delivery of insulin. Drug Dev Ind Pharm 40:691–698. https://doi.org/10.3109/03639045.2014.886590
43. Koland M, Vijayanarayana K, Charyulu R, Prabhu P (2011) In vitro and in vivo evaluation of chitosan buccal films of ondansetron hydrochloride. Int J Pharm Investig 1:164. https://doi.org/10.4103/2230-973x.85967
44. Wu J, Wei W, Wang LY, Su ZG, Ma GH (2007) A thermosensitive hydrogel based on quaternized chitosan and poly(ethylene glycol) for nasal drug delivery system. Biomaterials. 28:2220–2232. https://doi.org/10.1016/j.biomaterials.2006.12.024
45. Krauland AH, Guggi D, Bernkop-Schnürch A (2006) Thiolated chitosan microparticles: a vehicle for nasal peptide drug delivery. Int J Pharm 307:270–277. https://doi.org/10.1016/j.ijpharm.2005.10.016
46. Callens C, Remon JP (2000) Evaluation of starch-maltodextrin-Carbopol(®) 974 P mixtures for the nasal delivery of insulin in rabbits. J Control Release 66:215–220. https://doi.org/10.1016/S0168-3659(99)00271-0
47. Dyer AM, Hinchcliffe M, Watts P, Castile J, Jabbal-Gill I, Nankervis R, Smith A, Illum L (2002) Nasal delivery of insulin using novel chitosan based formulations: a comparative study in two animal models between simple chitosan formulations and chitosan nanoparticles. Pharm Res 19:998–1008. https://doi.org/10.1023/A:1016418523014
48. Alonso MJ, Sánchez A (2003) The potential of chitosan in ocular drug delivery. J Pharm Pharmacol 55:1451–1463. https://doi.org/10.1211/0022357022476
49. De Campos AM, Diebold Y, Carvalho ELS, Sánchez A, Alonso MJ (2004) Chitosan nanoparticles as new ocular drug delivery systems: in vitro stability, in vivo fate, and cellular toxicity. Pharm Res 21:803–810. https://doi.org/10.1023/B:PHAM.0000026432.75781.cb
50. Diebold Y, Jarrín M, Sáez V, Carvalho ELS, Orea M, Calonge M, Seijo B, Alonso MJ (2007) Ocular drug delivery by liposome-chitosan nanoparticle complexes (LCS-NP). Biomaterials 28:1553–1564. https://doi.org/10.1016/j.biomaterials.2006.11.028

51. Zhu X, Su M, Tang S, Wang L, Liang X, Meng F, Hong Y, Xu Z (2012) Synthesis of thiolated chitosan and preparation nanoparticles with sodium alginate for ocular drug delivery. Mol Vis 18:1973–1982
52. Fabiano A, Maria A, Uccello-barretta G, Balzano F, Cesari A, Testai L, Citi V, Zambito Y (2018) Impact of mucoadhesive polymeric nanoparticulate systems on oral bioavailability of a macromolecular model drug. Eur J Pharm Biopharm 130:281–289. https://doi.org/10.1016/j.ejpb.2018.07.010
53. Millotti G, Laffleur F, Perera G, Vigl C, Pickl K, Sinner F, Urch AB (2014) In vivo evaluation of thiolated chitosan tablets for oral insulin delivery. pp 3165–3170. https://doi.org/10.1002/jps.24102
54. Frank LA, Sandri G, D'Autilia F, Contri RV, Bonferoni MC, Caramella C, Frank AG, Pohlmann AR, Guterres SS (2014) Chitosan gel containing polymeric nanocapsules: a new formulation for vaginal drug delivery. Int J Nanomed 9:3151–3161. https://doi.org/10.2147/IJN.S62599
55. Meng J, Agrahari V, Ezoulin MJ, Zhang C, Purohit SS, Molteni A, Dim D, Oyler NA, Youan BBC (2016) Tenofovir containing thiolated chitosan core/shell nanofibers: in vitro and in vivo evaluations. Mol Pharm 13:4129–4140. https://doi.org/10.1021/acs.molpharmaceut.6b00739

Adv Polym Sci (2021) 288: 213–250
https://doi.org/10.1007/12_2021_106

Published online: 22 July 2021

Recent Progress in Biomedical Applications of Chitosan Derivatives as Gene Carrier

Pu-Song Zhao, Yi Wang, Wenshuang Sun, Lian-Yu Qi, Li-Fan Hu, Tian-Jiao Zhou, Lei Xing, Ki-Hyun Cho, Chengjun Li, Chong-Su Cho, and Hu-Lin Jiang

Contents

Pu-Song Zhao, Yi Wang, and Wenshuang Sun contributed equally to this work.

P.-S. Zhao, L.-Y. Qi, L.-F. Hu, and T.-J. Zhou
State Key Laboratory of Natural Medicines, Department of Pharmaceutics, China Pharmaceutical University, Nanjing, China

Y. Wang, L. Xing, and H.-L. Jiang (✉)
State Key Laboratory of Natural Medicines, Department of Pharmaceutics, China Pharmaceutical University, Nanjing, China

Jiangsu Key Laboratory of Druggability of Biopharmaceuticals, China Pharmaceutical University, Nanjing, China

W. Sun and C. Li (✉)
Department of Orthopedics, Jinling Hospital, School of Medicine, Nanjing University, Nanjing, China

K.-H. Cho
Department of Plastic Surgery, Institute of Dermatology and Plastic Surgery, Cleveland Clinic, Cleveland, OH, USA

C.-S. Cho (✉)
Department of Agricultural Biotechnology, Research Institute of Agriculture and Life Sciences, Seoul National University, Seoul, South Korea
e-mail: chocs@snu.ac.kr

Abstract Appropriate gene carriers are vital components of gene therapy for treating numerous intractable diseases. Compared with viral carriers, non-viral carriers exhibit lower immunogenicity and higher biosafety, which have received increasing attention. In particular, among non-viral vectors, gene vectors based on chitosan derivatives are considered a promising gene therapy tool and have been extensively studied due to the biocompatibility, biodegradability, and modifiability of chitosan. In this review, we first discussed the influence of chitosan parameters on the efficiency of gene therapy. More importantly, we summarized the recent research progress in various diseases of chitosan derivatives to deliver different gene therapy substances including plasmid (pDNA), short interfering RNA (siRNA), short hairpin RNA (shRNA), microRNA (miRNA), and clustered regularly interspaced short palindromic repeats-associated endonuclease 9 (CRISPER-Cas9) system. Finally, the current challenges and future directions of chitosan derivatives as gene carriers are also proposed.

Keywords Chitosan derivatives · Gene carriers · Gene editing · MicroRNA · Plasmid DNA · Short hairpin RNA · Short interfering RNA

1 Introduction

Gene therapy is a prospective medical method that has huge therapeutic potential to cure a variety of inherited and acquired diseases at the genetic level [1–4]. One of the critical steps for gene therapy is to deliver a sufficient amount of nucleic acids [plasmid DNA (pDNA), short interfering RNA (siRNA), short hairpin RNA (shRNA), microRNA (miRNA), etc.] into targeted sites. However, nucleic acids are unstable and prone to be degraded in vivo. Besides, since nucleic acids are anionic biological macromolecules, naked therapeutic genes are burdensome to pass through cell membranes and easily degraded by nucleases in the cytoplasm. Thus, gene delivery vehicles are very crucial in gene therapy [5, 6].

Currently, gene vectors are mainly divided into two categories: viral carriers and non-viral ones. Viral vectors can realize long-term expression and relatively efficient transfection efficiency [7]. However, complex production processes, immunogenicity, low loading capacity, and lurking infectious probability have become the main obstacles for the widespread use of viral vectors [8]. Compared with viral vectors, non-viral vectors are expected to solve various limitations of viral vectors due to their simple production, low immunogenicity, and better biosafety, which have been extensively studied [9–13]. Among non-viral systems, cationic polymers are one of the most concerned categories in gene therapy. Polymers such as polyethylenimine (PEI), chitosan (CS), polyamide (amidoamine) (PAMAM), and poly (L-lysine) (PLL) possess several free amine groups and therefore are positively charged [14, 15]. Cationic polymers based on high positive charge densities are efficient for several important steps of transfection including gene loading, cellular uptake, and endosomal escape, but have rather high toxicity, resulting in cell death [11, 16].

It's worth noting that chitosan is a special case of cationic polymers because it is biodegradable, biocompatible, and almost non-toxic in animals as well as humans [14, 17–19]. Chitosan is a kind of natural cationic polysaccharide materials consisting of repeating D-glucosamine and *N*-acetyl-D-glucosamine units connected by (1–4) glycosidic bonds and is obtained by deacetylation of chitin which can be isolated from the exoskeleton of crustaceans (crabs, krill, shrimp, etc.), insects, and from cell walls of fungi [14, 20]. In addition to its biodegradability and non-toxicity, chitosan can effectively condense negatively charged nucleic acid through its abundant amine groups and prevent nucleic acid from being degraded by nucleases or in serum [21]. Mumper et al. have attempted to use chitosan as a gene therapy vector for the first time [22]. Since then, chitosan as one of the gene carriers has gained increasing interest attributed to the aforementioned proper qualities of chitosan for gene delivery [23]. However, there are some limitations in the use of chitosan as a gene delivery vector. The pKa of the amine group is in the range of 6.0–6.5, which makes chitosan to have low solubility under physiological conditions. Only in acidic media, chitosan can be effectively protonated and soluble. Thus, the application of unmodified chitosan as a gene carrier is restricted [24]. Fortunately, chitosan is rich in modifiable amino groups and hydroxyl groups. The extensive research on chitosan derivatives has effectively made up for the defects of chitosan materials and gives it new functions such as targeting and enhanced cell uptake ability and endosomal escape ability, thus to improve transfection efficiency [17, 20]. Currently, chitosan-based delivery systems are widely used in varied aspects of gene therapy including the delivery of siRNA, shRNA, miRNA, and clustered regularly interspaced short palindromic repeats-associated endonuclease 9 (CRISPER-Cas9) system, not just limited to pDNA delivery [20, 22].

In this review, we mainly focus on the recent progress of chitosan and chitosan derivatives as gene therapy carriers in biomedical applications. Besides, we summarize the role of the parameters of chitosan for the enhancement of transfection efficiency. Importantly, the research applications of various gene delivery systems constructed by chitosan derivatives to deliver different gene therapy substances are also discussed here. Finally, we propose the current challenges and possible future directions of chitosan derivatives as gene carriers. Scheme 1 summarizes the various modification types of chitosan derivatives in current gene therapy and their enhanced functions.

2 Factors Affecting the Transfection Efficiency of Chitosan

For efficient in vitro and in vivo gene transfection, polycation-based gene vectors like chitosan are obligated to assist gene to targeted sites and then complete the process of effective cell uptake, endosomal escape, and gene release. Many factors are affecting the transfection efficiency of chitosan, and early studies have shown

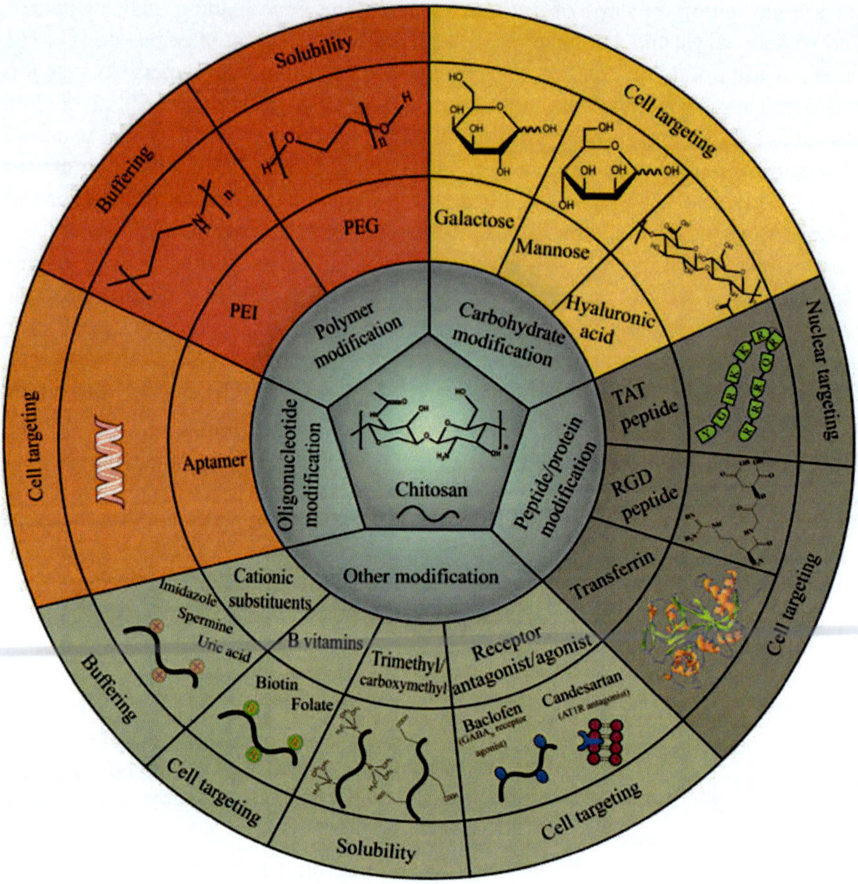

Scheme 1 Diagram showing various types of chitosan derivatives in current gene therapy and their enhanced functions

that adjusting the tunable parameters of chitosan [molecular weight (MW), the degree of deacetylation (DDA), the molar ratio between chitosan amine and plasmid phosphate (N/P ratio)], and some external factors (including pH, serum content) can effectively change the physicochemical properties (particle size, surface ζ-potential, the binding strength of the nanoparticles and nucleic acids) and biological functions (serum resistance, cell uptake, nucleic acid release) of chitosan, and then influencing the transfection efficiency of chitosan nanoparticles [25, 26].

MW is an important factor influencing the transfection efficiency of chitosan although it has a blurred effect on transfection efficiency. The increase in MW of chitosan is often considered to be more effective in protecting nucleic acids for higher cellular uptake, and ultimately increasing the transfection efficiency [27, 28]. Research by Huang et al. have proved that with the increase of MW, the cellular uptake of chitosan/DNA complexes and transfection efficiency have been improved. In their researches, chitosan/DNA nanoparticles with high MW (MW 213 kDa) possessed the highest transfection efficiency (12.1%), while the transfection efficiency of chitosan with 98 kDa was 8.3% [29]. However, in some studies, low MW chitosan showed a higher transfection efficiency than high MW chitosan [30, 31]. Yang et al. reported that 17 kDa chitosan was more efficient than 173 and 425 kDa chitosan in HEK293 cells [30]. Although low MW chitosan has a poor ability to compress and protect DNA, it can more easily achieve the intracellular release of DNA, while high MW chitosan is just the opposite. At present, there is no consensus on which MW of chitosan is more suitable for transfection. Anyway, it is undeniable that a balance between nucleic acid protection and intracellular release is more likely to achieve the best transfection effect.

DDA is another vital element affecting the transfection efficiency of chitosan since it can significantly affect the charge density of the chitosan chain. A higher DDA allows chitosan to have a higher charge density, thus making it more effectively bind to nucleic acids and interact with cell membranes to promote cell uptake [29, 32]. Some researchers showed that it is easier to form stable chitosan/DNA complexes with above 65% DDA [33].

It is not comprehensive to evaluate the transfection efficiency of chitosan based on MW and DDA alone, because the transfection efficiency is also affected by some other factors. N/P is defined as the molar ratio between chitosan amine and phosphates in nucleic acids which potentially adjusts the charge density of the complexes [14, 34]. In addition, pH value and external salt concentration directly control the protonation of chitosan and the stability of chitosan/nucleic acid complexes. Since the protonation of amine groups in chitosan needs an acidic pH range (6.0–6.5), a low acid pH environment improves the DNA binding efficiency of chitosan, thereby improving the transfection efficiency [35, 36]. Some studies have confirmed that adding NaCl to the transfection medium would give better results in gene transfection because NaCl shielded the surface charge of chitosan and reduced the repulsive force between the same charge making the complexes more stable [37].

The anionic plasma proteins are considered to compete with the nucleic acid in chitosan binding, leading to instability of the complexes and declining the transfection efficiency. Interestingly, Sato et al. found that the transfection efficiency of chitosan in the presence of 20% serum was 2–3 times higher than that of serum-free, while 50% serum content reduced transfection. They discussed that albumin and globulin in serum had a promoting effect on transfection and believed that serum might enhance cell function and thus improve transfection efficiency [25].

Therefore, the transfection efficiency of chitosan is affected by many factors mentioned above, including MW, DDA, N/P, pH value, salt concentration, and serum content. A single element is not enough to commendably control the transfection efficiency of chitosan, and the same chitosan formulation also has differences of transfection according to the different cells [25]. Adjusting various parameters to achieve the balance of nucleic acid protection and intracellular release has been proven to be more conducive to achieve the best DNA transfection effect. Therefore, higher DDA (>80%), N/P (>1), and appropriate pH (about 6.5) are required to maintain the positive charge of the complexes in the optimal formulation. Also, to obtain higher transfection efficiency, we need to comprehensively consider the influence of chitosan preparation parameters on transfection. When delivering various types of nucleic acids, these formulation parameters also have different performances on transfection efficiency [38].

3 Chitosan Derivatives as Gene Carriers

Efficient gene therapy often requires several key links, including cellular uptake, endosomal escape, and intracellular gene release. The transfection efficiency of the chitosan vector for gene therapy can be improved by adjusting the chitosan formulation parameters although there still exists a gap between chitosan and satisfactory gene therapy, due to the poor water solubility, low cell specificity, and low protonation ability under physiological conditions of pristine chitosan. Fortunately, the free amine groups and hydroxyl groups on the chitosan backbone are highly reactive, which can be easily chemically modified by different functional chemical groups and ligands [39]. Currently, the interests of researchers have been shifted toward chitosan derivatives-based vector systems for gene therapy [17].

Different modification methods to improve the function of chitosan have been extensively studied and summarized. For example, the modification of chitosan by transferrin, folic acid, hyaluronic acid, galactose, mannose, and other specific ligands has been applied to enhance the targeting ability; modification of chitosan by poly (ethylene glycol) (PEG), carboxymethyl, and *N, N, N*-trimethyl was favorable for increasing the water solubility of the chitosan; modification of chitosan by PEI, imidazole, spermine, urocanic acid improved the buffering function of chitosan itself [20, 40]. In the following sections, we will further discuss recent researches on chitosan derivative-based vectors in gene therapy for several kinds of diseases.

3.1 Chitosan Derivatives for DNA Delivery

3.1.1 Chitosan Derivatives for DNA Delivery in Cancer

The potential of chitosan derivatives for DNA delivery has been widely reported. The purpose of DNA delivery is to introduce exogenous genes into target cells, resulting in the transcription and translation of proteins with special therapeutic effects, and ultimately achieving the expected gene therapy. Recently, DNA delivery mediated by chitosan derivatives has a wide range of applications, including cancer treatment, tissue regeneration, and immune vaccines (summarized in Table 1).

Gene therapy has exhibited great potential in the treatment of cancer, which is a kind of heterogeneous disease caused by unregulated gene damage and mutations. Seven of the 16 gene therapy drugs launched in the market are for cancer treatment [52]. Different chemical modification methods have been harnessed to modify

Table 1 Summary of chitosan derivatives as DNA carriers

Application	Used gene and drug	Carrier	Modification group	Enhanced function	Results	Ref.
Pancreatic cancer	p53	Chitosan-graft-polyethylenimine-candesartan (CPC)	Candesartan; PEI	Cell targeting; buffering	CPC/wt-p53 complexes exhibited enhanced targeting and buffering capacity, effectively inhibiting angiogenesis and exerting anti-tumor effects	[41]
Cervical cancer	p53; Doxorubicin (DOX)	Grafted poly-(*N*-3-carbobenzyloxy-lysine) with TAT-chitosan (TAT-CPCL)	Poly-(*N*-3-carbobenzyloxy-lysine); TAT	Nuclear targeting; buffering	In vitro, TAT-CPCL co-delivery of p53 gene and DOX had achieved efficient gene transfection and promoted HeLa cell apoptosis	[42]
Liver cancer	p53; DOX	Chitosan grafted poly-(N-3-carbobenzyloxy-lysine)-amidized TAT (aTAT-CPCL)	Poly-(N-3-carbobenzyloxy-lysine); TAT	Nuclear targeting; buffering; acid responsiveness; prolonged blood circulation	In vivo, the anti-cancer effect of aTAT-CPCL co-delivery of p53 gene and DOX system was enhanced compared with TAT-CPCL	[43]
Glioblastoma cancer	pING4 (inhibitor of growth 4)	*N*-octyl-*N*-quaternary chitosan-PEI-K12 (OTMCS-PEI-K12)	*N*-octyl-*N*-quaternary; PEI	Amphiphilic; buffering; nuclear targeting; cell targeting	In vivo experiments showed that OTMCS-PEI-K12/pING4 complexes effectively shrunk the tumor volume in mice	[44]
Breast cancer	p53	FA-PEG-CCTS/PEI/NLS/pDNA (FPCPNDs)	Folate; PEG	Solubility; prolonged blood circulation; acid responsiveness; buffering; nuclear targeting	In vivo anti-tumor experiments showed that FPCPNDs group had the highest anti-tumor ability, with a tumor inhibition rate of 59.6%, which was significantly higher than other groups	[45]

(continued)

Table 1 (continued)

Application	Used gene and drug	Carrier	Modification group	Enhanced function	Results	Ref.
Skin regeneration	Ang-1(pANG); Trp-rich peptides (PSI)	Poly (ethylene glycol) diacrylate (PEGDA)/ thiolated chitosan (TCS) hydrogels	Thiol groups	Bioadhesion; sustained release	The hydrogel PEGDA/TCS showed sustained release function and tissue adhesion. When loaded with pANG, it could promote microangiogenesis and accelerated rats' skin tissue healing	[46]
Neural regeneration	pCMV-EGFP-Ntf3	RGD/TAT-chitosan-graft-PEI-PEG (CPPP)	RGD; TAT; PEI; PEG	Solubility; cellular uptake; DNA compression; buffering	Transfection of CPPP/ pCMV-EGFP-Ntf3 complexes into 293T cells accumulated neurotrophin-3 and promoted the differentiation of neural stem cells	[47]
Neural regeneration	Brain-derived nerve growth factor (BDNF)	Thiolated trimethyl chitosan (TMCSH)-non-toxic carboxylic fragment of the tetanus neurotoxin (HC) (TMCSH-HC)	Thiolated trimethyl; HC	Solubility; cell targeting	Intramuscular injection of TMCSH-HC/BDNF nanoparticles was a safe and effective method to prevent long-term neurodegeneration and accelerate nerve regeneration	[48]
Aeromonas hydrophila DNA vaccine	Aerolysin gene (aerA)	Oleoyl-carboxymethyl-chitosan (OCMCS)/ hyaluronic acid (HA) (OHA)	Oleoyl; carboxymethyl; hyaluronic acid	Sustained release; cell targeting	The OHA NPs enhanced the penetration of DNA and effectively induced antibody production in carp through oral administration	[49]

Tuberculosis DNA vaccine	pPES (with 5T-cell-epitopes from H37Rv)	Mannosylated chitosan	Mannose	Cell targeting	MCS-DNA induced a powerful and enhanced lung SIgA response and pulmonary multifunctional CD4 +/CD8 + T response compared to CS-DNA and subcutaneous bacillus Calmette-Guerin (BCG) vaccination	[50]
Newcastle disease virus DNA vaccine	Newcastle disease virus (NDV) F plasmid; C3d6 molecular adjuvant	*N*-2-hydroxypropyl trimethylammoniu-m chloride chitosan (*N*-2-HACC)-*N*,*O*-carboxymethyl chitosan (CMC)	*N*-2-hydroxypro-pyl trimethylammonium chloride; carboxymethyl	Solubility; enhanced encapsulation	Intranasal immunization of chickens with DNA vaccines produced higher anti-NDV antibodies than other groups and induced higher levels of interleukin (IL)-2, IL-4, and interferon γ (IFN-γ)	[51]

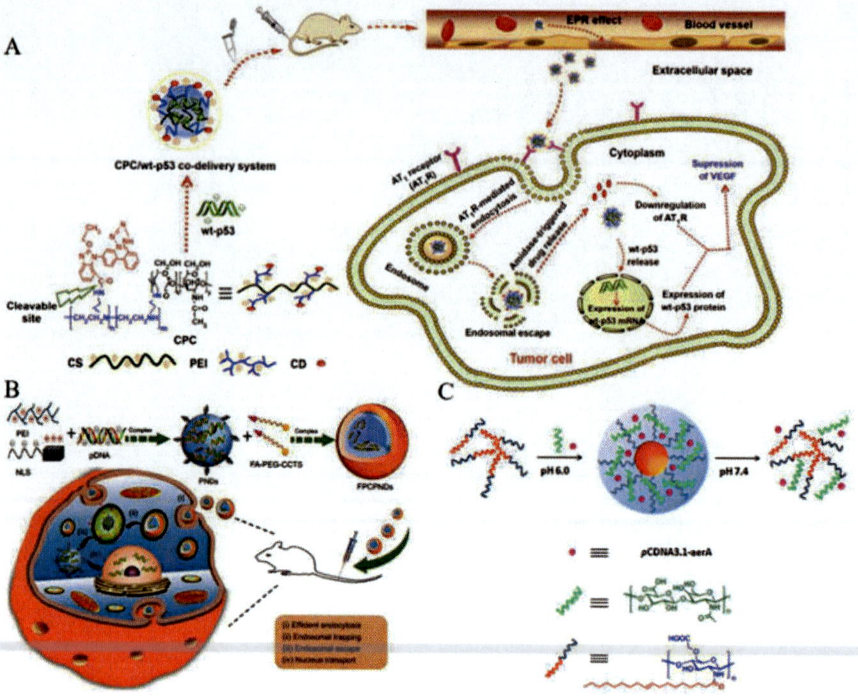

Fig. 1 Chitosan derivatives for DNA delivery. (**a**) Illustration of the multifunctional delivery system CPC/wt-p53 complexes and its self-assembly, tumor tissue accumulation, intracellular transport, and cooperative anti-angiogenesis mechanism [41]; (**b**) Illustration of the constituent of FPCPNDs nanoparticles and the intracellular trafficking for the treatment of tumor [45]; (**c**) Illustration of the formation of OCMCS/hyaluronic acid (HA) as DNA vaccines carriers and releasing DNA in PBS under different pH conditions in vitro [49]. Permissions From [41, 45, 49]

chitosan for overcoming various barriers in tumor treatment. Lin et al. grafted PEG onto galactosylated chitosan to target liver cancer cells for liver cancer treatment. PEG was expected to shield nanoparticles from being taken up by the reticuloendothelium system (RES) to increase circulation time and the accumulation of nanoparticles in the tumor site, while galactose was used to identify the overexpressed asialoglycoprotein receptors (ASGPR) on the surface of the liver cancer cells. The results exhibited cell uptake of nanoparticles in liver cancer cells had increased [53]. Also, Bao and colleagues prepared chitosan with strong buffering capacity and targeting ability via grafting candesartan (CD)-conjugated PEI arms on the chitosan backbone named as CPC (Fig. 1a). In this study, candesartan was an angiotensin II type 1 receptor (AT1R) antagonist, which could specifically bind to tumor-overexpressed AT1R, promoting the cellular uptake of nanoparticles and eventually reducing angiogenesis. And the acid-base titration experiments proved that CPC had a stronger buffering capacity than chitosan and PEI alone, which was beneficial to endosomal escape. The improvement of buffering capacity could be explained as the increase in the number of protonatable amino groups in the nanoparticles due to the introduction of the CD. In vivo studies on tumor-bearing PANC-1 xenograft nude mice confirmed that the CPC/wt-p53 complexes had the high tumor-targeted ability and strong anti-tumor activity [41]. The above-mentioned enhanced capabilities of chitosan derivatives, including prolonged circulation time, cell-targeted ability, and endosomal escape, are all conducive to overcome various barriers in gene therapy of cancer such as RES interception, insufficient cellular uptake, and poor intracellular release.

However, for DNA delivery, there is an additional barrier to overcome, which is nuclear import. DNA needs to enter into the nucleus for subsequent transcription [54], but the nuclear pores on the nuclear membrane have a much smaller size than overwhelming gene-loading complexes. Hence, the nuclear membrane is an important barrier to transfection. Some transfection reagents without nuclear uptake function performed certain transfection effects in vitro because the cell nucleus loses its integrity during mitosis. Meanwhile, in some cases, the gene will be liberated from complexes and be random diffusion through the nuclear pore into the nucleus [44]. However, the nuclear uptake of these methods is limited, hence, adding nuclear uptake function into DNA delivery carriers is expected to achieve effective nuclear import, thereby improving the effect of tumor treatment [55]. For example, Wang et al. designed a transactivator of transcription (TAT)-modified chitosan-grafted poly-(*N*-3-carbobenzyloxy-lysine) (CPCL) cationic micelle (CPCL-TAT) to encapsulate the p53 gene and doxorubicin (DOX). Herein, TAT was a member of the nuclear target localization sequence and could mediate efficient nuclear import through the nuclear pore complex. The images of confocal laser scanning microscopy (CLSM) indicated that the nuclear-targeting ability of CPCL-TAT was better than CPCL alone [42]. However, the primary amine of the lysine residue on the TAT peptide triggered a non-specific reaction and led to the unsatisfactory anti-tumor effect of CPCL-TAT in vivo. To solve this problem, they further used the succinyl chloride linked by the hydrazone bond to modify TAT for covering the primary amine in the TAT peptide, and the hydrazone bond could be degraded in the acidic microenvironment in the tumor. And the new nanocomposite aTAT-CCL circulated widely in the blood and had high anti-tumor activity in HepG2 tumor-bearing nude mice [43]. In addition to TAT peptides, other nuclear targeting sequences have also been applied for chitosan derivatives to deliver DNA for tumor therapy. The bifunctional peptide tLyP-1-NLS (K12), which combined tumor-targeting peptide (tLyP) and nuclear localization sequence (NLS), was prepared to modify the PEI-grafted chitosan amphiphilic derivative (*N*-octyl-*N*-quaternary chitosan) obtaining (OTMCS)-PEI-K12. As the authors expected, OTMCS-PEI-K12/pING4 had superior tumor suppressor activity in vivo compared to OTMCS-PEI/pING4 [44]. Besides, some researchers constructed a pH-sensitive conjugated folate-polyethylene glycol-carboxylated chitosan (FA-PEG-CCTS) to wrap the inner layer of PEI/NLS/pDNA (PNDs) for facilitating nuclear-targeting ability (Fig. 1b). The FPCPNDs nanoparticles maintained electrical neutrality as a whole, and folic acid targeted the folate receptor overexpressed on the tumor surface to promote cell uptake. In endosomes, the outer layer of FA-PEG-CCTS was separated from PNDs and then realized endosomal escape assisted by PEI. Finally, the NLS in PNDs promoted DNA to enter into the nucleus. Both in vitro and in vivo experiments have proved the anti-tumor activity of the delivery system [45].

To sum up, there is no doubt that the anti-tumor effect in vivo has stricter requirements for carrier function compared to in vitro. However, the biocompatibility and modifiability of chitosan enable the design of a multifunctional, safe, and efficient DNA delivery platform, which is in favor of achieving the high-efficiency anti-tumor in vitro and in vivo.

3.1.2 Chitosan Derivatives for DNA Delivery in Tissue Engineering

Gene therapy allows cells to continuously express and secrete growth factors that promote tissue regeneration or stem cell differentiation, which is an important strategy for tissue engineering. Owing to the unique biocompatibility, biodegradability, and antibacterial ability, chitosan plays a vital role in tissue regeneration. Numerous studies have wielded a resorbable and gene-active chitosan scaffold to carry the bone morphogenetic protein (BMP-2) gene to achieve bone regeneration [56, 57]. At the same time, chitosan scaffolds were also used in skin regeneration due to their reported function to promote granulation tissue formation, collagen expression, inflammatory cell infiltration, and hemostasis [58, 59]. However, most of these studies used chitosan rather than chitosan derivatives. Chitosan derivatives also possess special functions in tissue regeneration. For example, Huang and his companions developed a dual-network hydrogel with polyethylene glycol diacrylate (PEGDA) and thiolated chitosan (TCS). By carrying Trp-rich peptides (PSI) with broad-spectrum antibacterial effect and pANG (expressing angiopoietin 1) plasmid coated by *N*-trimethyl chitosan chloride, the hydrogel could achieve excellent skin tissue regeneration. The introduction of thiol groups made PEGDA/TCS own stronger tissue adhesion and sustained drug release than PEGDA/CS. In in vivo wound healing experiments, the PEGDA/TCS-PSI-pANG group had a better effect of accelerating skin regeneration than other groups in a full-thickness skin defect model. Subsequent histological analysis demonstrated that the hydrogel PEGDA/TCS-PSI-pANG could promote skin healing by reducing inflammation and increasing microvascular regeneration [46]. Also, there were researchers developing chitosan/PEI patch to load epidermal growth factor (EGF) and epidermal growth factor receptor (EGFR) genes (EErP-CPs) for tympanic membrane regeneration substituting for surgical treatment of tympanic membrane perforation. The addition of lower cytotoxicity 1.3 kDa PEI significantly increased the cell adhesion ability of EerP-CPs. After treatment, EGF could be released quickly to elevate the survival rate of tympanic membrane cells. EGFR gene played a role after 3 days of treatment and further increased the viability of tympanic membrane cells. The synergistic effect of EGF and EGFR enhanced cell viability, cell migration, and cell adhesion [60].

For the application of chitosan derivatives in nerve regeneration, the main strategy is to deliver and successfully express neurotrophins, which are a family of growth factors that have an important impact on neurogenesis after injury. The group of Wu constructed a multifunctional gene nanocarrier (CPPP) composed of chitosan-graft-PEI-PEG modified with RGD/TAT. In brief, 293T cells transfected by the CPPP/pCMV-EGFP-Ntf3 complexes could continuously produce neurotrophin-3 (NT-3), which accelerated the differentiation of neural stem cells and the growth of nerve protrusions [47]. To achieve nerve repair in vivo, targeting neurons is helpful for precise nerve regeneration treatment. Pereira Gomes et al. prepared thiolated trimethyl chitosan nanoparticles grafted with non-toxic carboxyl fragments of tetanus neurotoxin (HC) to deliver brain-derived nerve growth factor (BDNF) DNA plasmids for the treatment of peripheral nerve damage. In this study, HC has been reported to enable nerve targeting [61]. And the TMCSH-HC/BDNF nanoparticles were used to treat peripheral nerve injury in BALB/c mice by intramuscular injection. After 21 days of treatment, compared with the non-targeted nanoparticles and the blank control group, the expression of BDNF, growth-associated protein GAP-43, and neurofilament protein were augmented, and the mice of the TMCSH-HC/BDNF group showed functional recovery, indicating the potential of chitosan-based gene therapy in nerve regeneration [48].

3.1.3 Chitosan Derivatives for DNA Vaccine Delivery

DNA can enter the nucleus to drive protein synthesis and induce an immune response against pathogens, which is always considered as DNA vaccine [22]. DNA vaccines have received increasing attention because of their simple production, stability at room temperature, and the ability to induce humoral and cellular immunity, without the risk of virulence reversal similar to that of live attenuated vaccines [51, 62]. Importantly, chitosan derivatives can be applied in the field of DNA vaccine stemmed from its superior safety and cationic characteristics [63]. Most DNA vaccines delivered by chitosan derivatives are administered by a mucosal route such as oral [49] and nasal [50, 51, 64] due to the unique mucoadhesive property of chitosan [65]. The mucosal administration method based on chitosan derivatives has unique advantages such as being more suitable for mass immunization with increased patient compliance than needle immunization [66, 67]. For oral DNA vaccine delivery, polymeric nanoparticles (NPs) composed of oleoylcarboxymethyl chitosan (OCMCS)/hyaluronic acid (HA) have been elaborated by Liu et al. to immunize fish against Aeromonas hydrophila infection by wrapping aerolysin gene (aerA). The introduction of negatively charged HA weakens the interaction between nanoparticles and DNA for promoting the release of DNA in a pH 7.4 environment. Interestingly, the polymeric NPs/DNA complexes remained stable under the acidic condition (pH 6.0), and released DNA at pH 7.4 (Fig. 1c). The antigen-specific antibodies in the serum of carp after oral administration of OCMCS-HA/aerA NPs were higher than those immunized with OCMCS/aerA NPs and DNA alone [49]. Also, the mannose-modified chitosan nanoparticles could form a multi-T-epitope DNA vaccine, which was administered intranasally to combat *Mycobacterium tuberculosis*. The delivery system effectively mediated humoral and cellular immunity, which was superior to subcutaneous injection of bacillus Calmette-Guerin (BCG). It is worth mentioning that mannose was introduced to target antigen-presenting cells (APC) in this study since the surface of APC cells is rich in mannose receptors and APC ones preferentially internalize pathogens or particles containing mannose residues [50]. Similar APC targeting strategies have also been used in previous studies of chitosan derivatives for delivering a hepatitis B DNA vaccine [68].

It is well known that DNA vaccines are often combined with adjuvants to pursue better immune effects [62]. Zhao et al. prepared water-soluble chitosan nanoparticles to co-deliver Newcastle disease virus (NDV) F gene and C3d6 molecular adjuvant through nasal administration and achieved effective immunity of Newcastle disease virus, which confirmed molecular adjuvants improved the immune response of DNA vaccines [51]. On the other hand, chitosan has been exploited in combination with poly (lactic-co-glycolic acid) (PLGA) or adjuvant of [Quil-A (extracted from Quillaja Saponaria)] to deliver DNA vaccines [69, 70]. Notably, chitosan itself is also an adjuvant to induce dendritic cell maturation and enhance cell-mediated immunity, which is regulated by intracellular DNA sensors as the enzyme cyclic-di-GMP-AMP synthase (cGAS) and stimulator of type I interferon genes (STING) [71]. However, the mechanism of chitosan as an adjuvant is not yet complete, and few studies have been conducted on the synergistic effect between chitosan itself and DNA vaccines [62]. At present, chitosan derivatives as adjuvants or DNA vaccine carriers are still in their infancy. With the further elucidation of the chitosan adjuvant mechanism, chitosan derivatives will have better application prospects as DNA vaccine delivery vehicles.

3.2 Chitosan Derivatives for RNA Delivery

RNA interference (RNAi) has become a popular gene therapy method, which can be used for knocking out special genes in a sequence-specific manner to suppress gene expression [99, 100]. Just like DNA delivery, the delivery of RNA molecule also requires suitable carriers to resist degradation by nucleases in vivo and ultimately be effectively internalized by target cells and organelles. Currently, the safety and effective gene silencing of chitosan derivative vectors have got a great degree of recognition. And chitosan derivatives have been introduced into numerous studies to treat cancer and inflammatory diseases by delivering different RNA interference molecules including siRNA, shRNA, miRNA. (summarized in Table 2).

3.2.1 Chitosan Derivatives in siRNA Delivery System

siRNA, as a double-stranded RNA obtained by chemical synthesis containing 21–23 base pairs, is considered to be the most effective gene silencing substance in RNAi molecules [101]. Once siRNA enters cells, it can bind with Dicer and then be cut in 21–23 nt short siRNA which integrates into the RNA-induced silencing complex (RISC) and then is unwound to become single-stranded siRNA. The remained guide

Table 2 Summary of chitosan derivatives to deliver RNA for biomedicine applications

Application	Used gene and drug	Carrier	Modification group	Enhanced function	Results	Ref.
Liver cancer	Vascular endothelial growth factor (VEGF)-siRNA	Urocanic acid-galactosylated trimethyl chitosan (UA-GT)/poly (allylamine hydrochloride)-citraconic anhydride (PAH-Cit)	Urocanic acid; galactose; PAH-Cit	Buffering; cell targeting; siRNA compression; solubility; charge reversal	The imidazole residue of urocanic acid enhanced the endosomal escape process of the complexes, while PAH-Cit realized charge reversal and could quickly release siRNA into the cytoplasm	[72]
Breast cancer	VEGF-siRNA and placental growth factor (PIGF)-siRNA	PEG-mannose modified trimethyl chitosan/poly (allylamine hydrochloride)-citraconic anhydride (PAH-Cit, PC)	PEG; mannose; PAH-Cit	Cell targeting; solubility; charge reversal	Nanoparticles were taken up by tumor-associated macrophages and breast cancer cells through active targeting involving mannose and passive targeting, respectively, to achieve tumor microenvironment reversal	[73]
Non-small cell lung cancer	Survivin-siRNA	Baclofen-trimethyl chitosan (Bac-TMC)/tripolyphosphate (TPP)	Baclofen; trimethyl	Cell targeting; solubility	Bac-TMC/TPP nanoparticles could target GABAB receptor on the surface of non-small cell lung cancer through baclofen, by effectively delivering survivin siRNA to promote cancer cell apoptosis	[74]

(continued)

Table 2 (continued)

Application	Used gene and drug	Carrier	Modification group	Enhanced function	Results	Ref.
Non-small cell lung cancer	B-cell lymphoma/Leukemia 2 (BCL2)-siRNA	Hyaluronic acid-modified chitosan	Hyaluronic acid	Cell targeting	Nanoparticles could enter cells through internalization mediated by the CD44 receptor on the tumor surface, via effectively reducing the BCL2 gene in A549 cells	[75]
Melanoma cancer	Interleukin (IL)-6 siRNA and signal transducer and activator of transcription (STAT) 3-siRNA	Hyaluronic acid-trimethyl chitosan	Hyaluronic acid; trimethyl	Cell targeting; solubility	Synergistic silencing of IL-6 and STAT3 effectively inhibited the proliferation of cancer cells, which was better than silencing a single gene	[76]
Liver cancer	VEGF-siRNA; Doxorubicin (DOX)	Galactose-modified trimethyl chitosan-cysteine (GTC)/PAH-Cit/TAT peptide-mesoporous silica (TAT-MSN)	Galactose; trimethyl; cysteine; PAH-Cit; TAT-MSN	Cell targeting; solubility; charge reversal; nuclear targeting; hydrophobic drug encapsulation	Multilayer nanoparticles could achieve rapid intracellular release of siRNA and DOX nuclear import which was mediated by TAT-MSN	[77]
Cervical cancer	BCL-2-siRNA; Lonidamine (LND)	Triphenylphosphine-chitosan-graft-PEI-lonidamine (TPP-CP-LND)	Triphenylphosphine; PEI; lonidamine	Mitochondrial targeting; buffering; anti-tumor activity	After treatment, the Bcl-2 protein in tumor cells was significantly down-regulated, which mediated cell apoptosis and relieved the inhibition of LND induced by Bcl-2 protein	[78]

Liver cancer	P-glycoprotein (P-GP)-siRNA; DOX	*N*-succinyl chitosan-poly-L-lysine-palmitic acid	Succinyl; poly-L-lysine; palmitic acid	Solubility; siRNA compression; hydrophobic drug encapsulation	By silencing the P-GP gene, the collaborative therapy effectively reversed the multidrug resistance of HepG2/ADM cells and exhibited high anti-tumor activity	[79]
Melanoma cancer; Colorectal cancer	Hypoxia-inducible factor (HIF)-1α-siRNA; Dinaciclib	Carboxylated graphene oxide (CGO)-trimethyl chitosan (TMC)-hyaluronate (HA)	Hyaluronate; carboxylated graphene oxide; trimethyl	Cell targeting; solubility; hydrophobic drug encapsulation	The CGO element provided an effective load for the hydrophobic drug CDK inhibitor dinaciclib, and the new type of synergistic therapy exerted a highly effective anti-cancer ability	[80]
Inflammation induced acute liver injury	Tumor necrosis factor-α (TNF)-α-siRNA	Mannose-modified trimethyl chitosan-cysteine (MTC)	Mannose; Trimethyl; cysteine	Stability; solubility; mucoadhesion	The cysteine residues on the nanoparticles could realize the formation of disulfide bonds with the mucin glycoprotein in the intestine to promote the adhesion between the nanoparticles and the mucosa and enhance the intestinal penetration	[81, 82]
Ulcerative colitis	TNF-α-siRNA	Galactosylated trimethyl chitosan-cysteine (GTC)	Galactose; trimethyl; cysteine	Stability; solubility; mucoadhesion	GTC nanoparticles with a particle size of 450 nm could exert the best therapeutic effect on ulcerative colitis because of a suitable siRNA affinity	[83]

(continued)

Table 2 (continued)

Application	Used gene and drug	Carrier	Modification group	Enhanced function	Results	Ref.
Anti-inflammatory	TNF-α-siRNA	*N*-(2-hydroxy) propyl-3-trimethyl ammonium chitosan chlorides (HTCC)/TPP	*N*-(2-hydroxy) propyl; trimethyl; TPP	Solubility; siRNA compression; cell uptake; buffering	In vitro experiments have shown that TPP could significantly enhance the cellular uptake of HTCC nanoparticles and increase the silencing efficiency on macrophages (about 85% to 90%, much higher than lipofectamine/siRNA)	[84]
Anti-inflammatory	CD98-siRNA	Urocanic acid-modified chitosan (UAC)/TPP	Urocanic acid; TPP	siRNA compression; cell uptake; buffering	In vitro experiments had shown that UAC/siCD98 NPs with a mass ratio of 60:1 exhibited the best anti-inflammatory ability, can effectively silence the expression of CD98, and down-regulate TNF-α	[85]
Rheumatoid arthritis	TNF-α-siRNA	PEG-folic acid labeled diethylaminoethyl chitosan (FA-PEG-DEAE-CH)	PEG; folic acid; diethylamino-ethyl	Solubility; cell targeting; buffering	In vivo and in vitro experiments had confirmed that FA-PEG-DEAE-CH/siRNA nanoparticles could effectively silence TNF-α gene and reduce inflammation with sufficient safety	[86–88]

Breast cancer	Baculoviral IAP repeat-containing 5 (Birc5)-shRNA plasmid	Disulfide bond-hydrazone bond-dual stimulus-response chitosan-PEG-PEI (SCS-*g*-PEI-*g*-PEG)	Disulfide bond; hydrazone bond; PEG; PEI	Acid responsiveness; glutathione responsiveness; solubility; buffering	The dual stimulus responsiveness of the nanoparticles realized the rapid release of shRNA in cells and mediated the long-term silencing of Birc5 protein in vivo	[89]
Nasopharyngeal carcinoma	Matrix metallopeptidase 9 (MMP-9)-shRNA plasmid	Transferrin-chitosan-graft-poly (L-lysine) (CS-PLLD-Tf)	Transferrin; poly (L-lysine)	Cell targeting; stability	CS-PLLD-Tf nanoparticles could achieve a transfection rate of more than 40% on HNE-1 cells through transferrin-mediated cell targeting	[90]
Liver cancer	Survivin-shRNA plasmid; DOX	α-Tocopherol succinate (α-TOS)-grafted *N*-trimethyl chitosan chloride	α-TOS; trimethyl	Anti-tumor activity; solubility	Blank nanoparticles had the function of promoting tumor apoptosis and inhibiting tumor growth, while the delivery system played a further synergistic anti-tumor effect after loading drug and shRNA	[91]
Breast cancer	P-glycoprotein-shRNA plasmid; paclitaxel	Folic acid-mediated chitosan-PEG-PEI	Folic acid; PEG; PEI	Cell targeting; solubility; buffering	The co-delivery of paclitaxel and P-shRNA effectively reversed the tolerance of drug-resistant cancer cells to paclitaxel and enhanced anti-cancer activity	[92]

(continued)

Table 2 (continued)

Application	Used gene and drug	Carrier	Modification group	Enhanced function	Results	Ref.
Non-small cell lung cancer	Survivin-shRNA plasmid; erlotinib	Heptamethine cyanine dyes (Cy7)-CQ/6-*N*, *N*, *N*-trimethyltriazole chitosan (CQ)	6-*N*, *N*, *N*-trimethyltriazole; Cy7	Solubility; buffering; photothermal therapy; imaging	This nano-platform combined chemotherapy, gene therapy, photodynamic therapy, and imaging with significant anti-cancer advantages	[93]
Gastrointestinal stromal tumors	miRNA-218	O-carboxymethyl chitosan (OCMC)-tocopherol	Carboxymeth-yl; tocopherol	Solubility; anti-tumor activity	Nanoparticles could achieve effective intracellular delivery of miR-218, exhibiting obvious tumor suppression ability and resistance to tumor invasion	[94]
Melanoma cancer	miRNA33a/ miRNA199a	Graphene oxide-chitosan-cellular penetrating peptide (MPG) (GO-CS-MPG)	Graphene oxide; MPG	Stability; cell uptake	GO-CS-MPG-miR33a/ miR199a drug-loaded nanoparticles had the potential to treat melanoma and reduce the recurrence of malignant melanoma	[95]
Breast cancer	miRNA-145	AS1411 aptamer-chitosan-thiolate-d dextran	AS1411 aptamer; disulfide bond; thiolated dextran	Cell targeting; stability; glutathione responsiveness	Nanoparticles achieved the rapid release of miR-145 in the cytoplasm through AS1411 aptamer-mediated tumor cell targeting, and effectively inhibited tumor progression	[96]

Liver cancer	miRNA-122	Galactosylated-chitosan-5-fluorouracil	Galactose; 5-fluorouracil	Cell targeting; anti-tumor activity	The nano-platform co-delivery of miRNA-122 and 5-fluorouracil had played a significant synergistic anti-cancer effect with satisfactory safety and stability	[97]
Ulcerative colitis	miRNA-146b mimic	Mannose-modified trimethyl chitosan	Mannose; trimethyl	Cell targeting; solubility	Oral administration of nanoparticles could effectively mediate the delivery of miRNA-146b mimic in macrophages and reduce the expression of pro-inflammatory factors of TNF-α, IL-6, and IL-1β	[98]

strand can guide the RISC to cleave specific mRNA or prevent its subsequent translation [102]. Except that it does not require to enter the nucleus, the delivery of siRNA is the same as DNA delivery. Both require the carrier to possess targeting ability, achieving the successful endosomal escape after being internalized into the cells, and finally releasing the gene into the cytoplasm for serving functions.

Chitosan derivatives with different targeting ligands were usually applied in cancer by delivering siRNA that could silence anti-apoptosis, pro-angiogenesis, or immunosuppressive related molecules [72–76]. As a $GABA_B$ receptor agonist, baclofen was introduced to modify trimethyl chitosan for survivin (a kind of anti-apoptotic gene) siRNA delivery by Ni et al. eventually treating non-small cell lung cancer. Baclofen could target $GABA_B$ receptor overexpressed on the surface of non-small cell lung cancer for enabling the gene silencing system to effectively inhibit tumor growth [74]. Han et al. have designed galactose-modified multifunctional nanocarrier, including urocanic acid-grafted galactose trimethyl chitosan derivative (UA-GT), charge reversal crosslinker poly (allylamine hydrochloride)-citraconic anhydride (PAH-Cit), and vascular endothelial growth factor (VEGF) siRNA. In their research, the introduction of galactose could recognize liver cancer cells for effective cellular uptake, urocanic acid accelerated nanocarriers endosomal escape, and PAH-Cit could trigger the charge reversal by the acidic condition of endosomes. All of these promoted siRNA delivery to the cytoplasm. Their nanocarriers have proven effective gene silencing and anti-tumor activity in vivo without obvious systemic toxicity [72]. Hyaluronic acid-modified chitosan usually can target the tumor cells overexpressing CD44 to not only improve targeting ability but also increase the stability of nanoparticles and reduce serum protein adsorption [75, 76]. It is worth mentioning that in addition to hyaluronic acid, mannose-modified trimethyl chitosan has also been synthesized to improve internalization of nanoparticles in tumor-associated macrophages, thus relieving tumor immunosuppression and hindering tumor progression by inhibiting the expression of vascular endothelial growth factor (VEGF) and placental growth of factor (PIGF) [73].

Recently, many studies have got a better result by co-delivering chemotherapeutics and siRNA than single treatment via chitosan derivatives [77, 79, 80, 103]. The co-delivery of chemotherapeutics and siRNA can simultaneously regulate multiple signal pathways and exert a synergistic anti-tumor effect [104]. Since the physicochemical properties of small molecule drug and siRNA are diverse, it is a challenge to deliver sufficient siRNA and related hydrophobic chemotherapeutics at the same time. A hyaluronic acid-grafted carboxylated graphene oxide-modified trimethyl chitosan carrier (CGO-TMC-HA) was constructed to encapsulate hypoxia-inducible factor (HIF)-1α siRNA and dinaciclib as a kind of CDK inhibitor (Fig. 2a). The overexpression of HIF promotes the formation of a tumor hypoxic microenvironment, which can cooperate with cyclin-dependent kinase (CDK) 1, 2, and 5 to accelerate tumor cell division and proliferation [80]. In this study, graphene oxide, as a hydrophobic material with a large surface area and good stability, loaded the hydrophobic drug dinaciclib through a π-π stacking method. The combination with trimethyl chitosan can simultaneously be delivered siRNA. Finally, this delivery

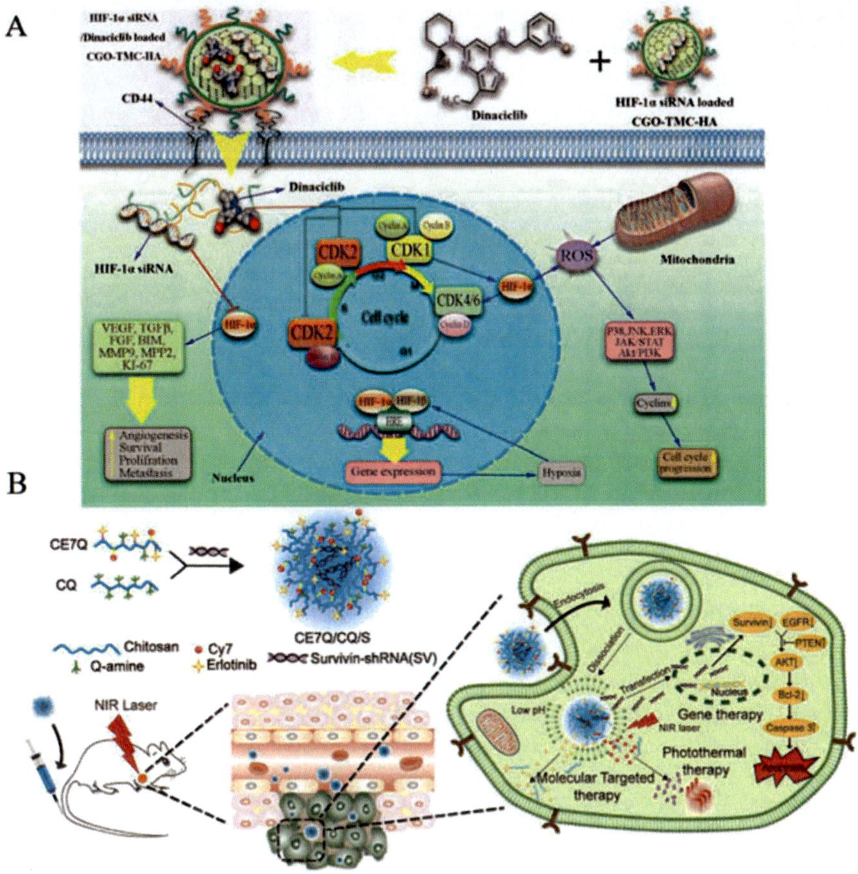

Fig. 2 Chitosan derivatives for RNA delivery. (**a**) Illustration of the anti-tumor mechanism and intracellular transport of CGO-TMC-HA NPs co-delivering HIF-1α-siRNA and the CDK inhibitor Dinaciclib [80]; (**b**) Illustration of the formation of CE7Q/CQ/S and its simultaneous chemo/gene/photothermal tri-therapies functions [93]. Permissions From [80, 93]

system was proven to be able to effectively inhibit the invasion and proliferation of murine melanoma, colorectal cancer, and breast cancer cells.

What's more, there are some strategies through the co-delivery of chemical drugs and siRNA reversing the multidrug resistance (MDR) that is a predominant barrier to limit the efficacy of chemotherapeutic drugs in the clinic [105]. Zhang et al. designed a kind of polymeric micelles consisted of poly-L-lysine-palmitic acid-modified *N*-succinyl chitosan derivative (NSC-PLL-PA) to deliver DOX and P-glycoprotein (P-GP, a multidrug resistance-related efflux transporter) siRNA. The structure of *N*-succinyl chitosan as a hydrophilic shell facilitated the NPs transportation in the blood and palmitic acid as the core could carry DOX. The IC50 values of free DOX, DOX-micelle, and siRNA-DOX-micelle were 5.23, 4.21, and 1.28 μg/mL, respectively, after 48 h of incubation with HepG2/ADM cells (overexpression of P-GP), which demonstrated the synergistic anti-tumor effect of the co-delivery of P-GP siRNA and DOX [79]. However, DOX and siRNA play a role in different regions. Just like DOX needs to enter the nucleus, and siRNA generally works in the cytoplasm. To solve this problem, multilayer dual-targeting chitosan nanoparticles were prepared by Han et al. The outer galactose-modified trimethyl chitosan could target the tumor cells through galactose, the middle layer of poly (allylamine hydrochloride)-citraconic anhydride (PAH-Cit) could promote endosomal escape and release siRNA into the cytoplasm, and the inner TAT-modified mesoporous silicon could carry DOX into the nucleus. It was found that after intravenous injection of low-dose nanocomplexes, there was a powerful anti-tumor effect with no obvious toxicity. Therefore, these nanocomplexes can be used as an effective and safe carrier to maximize the synergy between chemotherapeutics and siRNA [77]. In addition, our group elaborated a mitochondrial targeting copolymer TPP-CP-LND (TCPL) to deliver siRNA into the cytoplasm and mitochondria-damaging chemotherapeutics to the mitochondrial matrix. The copolymer TCPL consisted of triphenylphosphine (TPP, mitochondrial targeting ligand) and lonidamine (LND, mitochondrial-damage chemotherapeutics) conjugated to the polyethylenimine in chitosan-graft-PEI (CP). After coating the tumor-targeting copolymer poly (acrylic acid)-polyethylene glycol-folic acid (PPF), the delivery system could synergistically activate the mitochondrial apoptosis pathway by LND and effectively mediate tumor cell apoptosis via Bcl-2-siRNA in the cytoplasm [78]. In general, multifunctional carriers and co-delivery strategies of siRNA and chemical drugs can often play a more significant anti-tumor effect than single-functional carriers and single siRNA delivery. In multilayer delivery systems, targeting and hydrophilic modified chitosan derivatives are often used as the outermost layer of nanostructures [72, 77, 79, 80]. Chitosan is also used with materials that possess special functions but with toxic side effects, such as graphene oxide and poly-L-lysine-palmitic [79, 80]. Due to the excellent biocompatibility of chitosan, the performance of the delivery system can be effectively improved while reducing the toxicity of these materials. Interestingly, in these studies, trimethyl chitosan, a quaternized chitosan derivative, has become the most popular modification method, which may derive from its superiority in terms of positive charge and water solubility [72–74, 76, 77, 80, 103].

Tumor necrosis factor-α (TNF-α) is a pro-inflammatory cytokine mainly secreted by macrophages. Inducing the recruitment of inflammatory cells plays a key role in the pathogenesis of various inflammatory diseases, such as acute liver injury, enteritis, and rheumatoid arthritis [106]. And chitosan derivatives have been proven to effectively treat a variety of inflammatory diseases by inhibiting the expression of TNF-α in macrophages via effective delivery of TNF-α siRNA [81–84, 86–88]. He et al. prepared mannose-grafted trimethyl chitosan-cysteine (MTC) oral nanoparticles encapsulating TNF-α siRNA. And in vivo, their treatment effectively protected mice from acute liver injury and death induced by inflammation under the condition of low-dose siRNA (3.75 nM/kg) [81]. No doubt targeting ligand plays an important role in the delivery of MTC nanoparticles. Subsequently, they synthesized MTC materials with different mannose densities, namely MTC-4, MTC-13, and MTC-21 containing mannose density of 4%, 13%, and 21% respectively.

Interestingly, MTC-4 NPs exhibited the best in vivo TNF-α knockout efficiency because the internalization of MTC-4 in peritoneal exudate cell macrophages (PECs) was superior to other groups in vivo. This phenomenon was explained that the mannose-mediated PEC uptake of MTC-4 NPs was saturated under the condition of limited chitosan receptors on PEC, and too many mannose residues shielded the positive charge of NPs and led to steric hindrance, which was not conducive to the cellular uptake. Hence, the ligand density is the key factor in gene silencing, and the conclusion in this study may be a handbook for targeted chitosan derivatives-mediated nanocarriers in the anti-inflammatory treatment [82].

Recently, more and more cationic ligands were carried out to modify chitosan derivatives, such as *N*-(2-hydroxy)propyl-3-trimethyl ammonium chitosan chlorides (HTCCs) [84], urocanic acid-modified chitosan [85], and diethyl aminoethyl chitosan (DEAE-CH) [86], to ameliorate the gene silencing efficiency of siRNA in macrophages, curing inflammation. These cationic modifications were designed to enhance the buffer capacity of chitosan and strengthen the interaction between chitosan and siRNA to form more stable nanoparticles. Among them, DEAE-CH with different molecular weights was made by Shi, which possessed stronger siRNA encapsulation ability than pure chitosan [86]. By further modifying folic acid and PEG, the final folate-PEG-CH-DEAE/TNFα siRNA effectively inhibited the expression of TNFα and significantly reduced inflammation in a mouse collagen antibody-induced arthritis model [87]. Besides, the safety studies have confirmed that the nanoparticles met the threshold standards of the existing International Standard Organization (ISO) and American Society for Testing Material (ASTM) guidelines [88], which represents a highly biocompatible material based on chitosan derivatives possessing good clinical application prospects.

3.2.2 Chitosan Derivatives in shRNA Delivery System

Usually, the expression of shRNA is realized by delivery of plasmid DNA or through viral vectors, and the generated pre-shRNA is departed from the nucleus through Exportin 5, which is then treated by Dicer and imported into the RISC. As the degradation of the sense strand, the antisense strand guides RISC to find mRNA with a complementary sequence. After perfect complementarity, the targeted mRNA is cleaved by RISC. Unlike direct delivery of siRNA, shRNA delivery can achieve continuous silencing of targeted mRNA, while siRNA requires multiple administrations [102, 107].

At present, the application of chitosan derivative-based vectors to deliver shRNA is still relatively scarce and mainly focuses on tumors [89–93, 108, 109]. By cross-linking hydrazone bonds and disulfide bonds with chitosan, and then immobilizing PEI and PEG on them, Chen et al. proposed a dual stimulus-response (acidic pH condition and reducing environment) carrier SCS-*g*-PEI-*g*-PEG, which was harnessed to load Birc5-shRNA for promoting tumor cell apoptosis [89]. In vitro transfection results manifested that the silencing efficiency of SCS-*g*-PEI-*g*-PEG/Birc5-shRNA on MCF-7 cells reached 40% which was higher than CS-*g*-PEI-*g*-PEG and PEI 25K. Thanks to the rapid expression of Birc5-shRNA and the continuous inhibition of the Birc5 gene, SCS-*g*-PEI-*g*-PEG/Birc5-shRNA also achieved the best therapeutic effect in MCF-7 cell tumor-bearing nude mice with a single administration in vivo compared with the blank control. However, the carrier in this research lacks the deficiency of targeting group, while it has obvious uptake in the liver and spleen. Liu et al. prepared transferrin-functionalized chitosan-graft-poly (L-lysine)/ MMP-9 shRNA plasmid complexes to treat nasopharyngeal carcinoma. By targeting the transferrin receptors on the surface of HNE-1 cells, the CS-PLLD-Tf/MMP-9 shRNA complexes could be highly effective to silence the MMP-9 protein (more than 40%) and inhibit tumor development with satisfactory stability and safety [90].

Similar to siRNA, chitosan derivatives can be used to co-deliver shRNA and chemotherapeutics to achieve synergistic anti-tumor effects. Jia et al. designed a folate-modified chitosan-disulfide bond-PEI combined with silica to load paclitaxel and P-GP shRNA (P-shRNA). Nanoparticles significantly down-regulated the expression of P-GP (expression rate decreased by 63%) and successfully reversed the resistance of MCF-7 cells to paclitaxel (resistance reversal index 50.59) [92]. Also, Song's group delivered DOX and survivin shRNA by grafting α-tocopherol (α-TOS) succinate onto trimethyl chitosan (TAS). α-TOS has been reported that it could induce the apoptosis of a variety of tumor cells and yet be safe for normal cells. In this study, TAS vectors selectively killed tumor cells, and TAS/DOX/survivin shRNA NPs had strong anti-tumor activity through multiple synergistic treatments. No doubt modifying substances with the anti-cancer ability on the carrier opens up a promising idea for shRNA delivery and combination therapy [91]. What's more, other researchers exploited a water-soluble chitosan quaternary amine derivative with high transfection efficiency, 6-*N*, *N*, *N*-trimethyl triazole chitosan (CQ). Then, they used the click chemistry to connect near-infrared fluorescence heptamethine cyanine dyes (Cy7s) on CQ and prepared a multifunctional carrier platform CE7Q/CQ/S to deliver survivin shRNA and erlotinib, which included chemo/gene/photothermal tri-therapies functions (Fig. 2b). Both in vitro and in vivo results indicated that the suppression of survivin expression could play a synergistic role with erlotinib to induce powerful anti-cancer effects in erlotinib-resistant NSCLC cells [93].

3.2.3 Chitosan Derivatives in MiRNA Delivery System

Unlike siRNA, miRNA is an endogenous non-coding single-stranded RNA, as post-transcriptional regulation of cellular gene expression. Normally, miRNA is produced from longer precursor molecules (pre-miRNA) and mainly binds to the 3-′-untranscribed region of target mRNA, while siRNA can interact with any part of mRNA [102]. By pairing with mRNA antisense complementarity, miRNAs have been confirmed to regulate cell metabolism including cell migration, proliferation, autophagy, and stem cell self-maintenance [110]. However, owing to the instability, metabolic toxicity, and non-specific distribution of miRNA, a suitable carrier is necessary for the delivery of miRNA in the treatment of many diseases. Some successful cases have been suggested that chitosan derivatives are promising as powerful carriers for miRNA delivery.

Recently, miRNA-based clinical cancer treatment has attracted more and more attention [111, 112], and chitosan derivatives have achieved exciting results for delivering miRNAs to treat cancer [94–97]. Liu et al. designed a hybrid carrier (GO-CS-MPG) containing graphene oxide, chitosan, and cell-penetrating peptide (MPG) to deliver miR33a/miR199a. GO-CS-MPG by in situ injections had a significant inhibitory effect on melanoma [95]. But due to in situ administration in the tumor site, the delivery system was not modified with solubility groups for better retention effect, which also limited the further application of this delivery system in the clinic. Tekie et al. designed chitosan-thiolated dextran nanoparticles. And miR-145 was conjugated on thiolated dextran via a reduction-responsive disulfide bond. Finally, the targeted aptamer AS1411 was modified on the surface of the nanoparticles. Unlike electrostatic adsorption of nucleic acid, the method of conjugated modification makes the nucleic acid more stable and reduces the risk of leakage [96]. Also, a synergistic treatment strategy was adopted by Liu et al. via modifying 5-fluorouracil to galactosylated chitosan and co-delivering 5-fluorouracil and miRNA-122 (GC-FU/miR-122) to treat hepatocellular carcinoma. The co-delivery system had a superior ability to inhibit the growth of liver cancer cells and promote cancer cell apoptosis compared to the single treatments [97]. What's more, Tu et al. used a similar strategy to deliver miRNA-218 through tocopherol-modified carboxymethyl chitosan, in which tocopherol could cooperate with miRNA-218 to promote cancer apoptosis [94]. And the flexible modification methods of chitosan derivatives could provide innovative ideas for the delivery of miRNA and might produce better results for cancer treatment.

Apart from cancer, ulcerative colitis is also a field of application for miRNA [81]. Deng et al. used mannose-trimethyl chitosan to load miR-146b mimic (MTC-miR146b). After oral administration, miR146b could be effectively delivered to the colon and reduce the secretion of pro-inflammatory factors in M1 macrophages including TNF-α, IL-6, and IL-1β via inhibiting the Toll-like receptor 4 pathway. Finally, MTC-miR146b was confirmed for effective treatment of colitis [98]. At present, there are still few studies on chitosan derivatives to deliver miRNA in areas other than cancer and ulcerative colitis. It is undeniable that miRNA-based gene therapy technology will develop rapidly with the further study of the mechanism of miRNA by researchers. Besides, based on the biocompatibility and flexibility of chitosan, chitosan derivatives have great potential to become ideal carriers for miRNA delivery.

3.3 Chitosan Derivatives for Gene Editing

Gene editing is a breakthrough technology that can alter cell fates through precise manipulation of specific gene sequences in the organism's genome [113]. With the rapid development of delivery vectors, gene editing has become a beneficial weapon for genetic disease treatment. CRISPR/Cas9 was first proposed as an emerging genome editing tool in mammalian cells in 2013, its application has been continuously expanded. CRISPR/Cas9 can not only modify the genome sequence of cells and organisms but also achieve epigenetically and transcription modification [114]. Like other gene therapy vectors, the delivery vectors of the CRISPR-Cas9 system are also divided into two types: viral vectors and non-viral vectors. Viral vectors like adeno-associated viral (AAV), adenoviral (AD), and lentiviral vectors face critical tests of safety issues and limited capacity. Non-viral vectors such as PEIs, Lipofectamine, etc. are relatively safe and have a high capacity, but with unsatisfactory biocompatibility and low delivery efficiency. Chitosan is a biocompatible gene delivery vector with great potential, but there are a few studies on the delivery of CRISPR/Cas9 gene-editing system by chitosan vector [115–119]. In most of these studies, CRISPR/Cas9 plasmids were delivered and then transfected to express cas9 protein in cells [115, 117, 118]. However, Qiao et al. considered that the unexpected immune response may be triggered by the sustaining expression of Cas9 enzyme in the cells, leading to the risk of other negative side effects such as carcinogenesis. So, they prepared chitosan-coated red fluorescent protein (RFP@CS) to co-deliver CRISPR/Cas9 ribonucleoproteins (RNPs) and single-stranded DNA (ssDNA) donors. The introduction of negatively charged RFP in chitosan facilitates the intracellular release of CRISPR/Cas9 ribonucleoproteins and may protect its activity. The safety of this delivery system has been sufficiently verified, and the gene-editing capability of RFP@CS was similar to the commercial reagent Lipofectamine CRISPRMAX, in a variety of cells (HEK293T, RAW264.7, HeLa, U2OS and A549 cells) [116].

The occurrence of cancer involves a variety of gene mutations and oncogene activation. CRISPR-Cas9, as a simple and highly specific multifunctional genome editing platform, has proven to be a powerful tool in cancer therapy [114]. Cheng et al. successfully published three studies on chitosan-based CRISPR/Cas9 gene therapy vectors for tumor therapy. The carboxymethyl chitosan was modified by the AS1411 aptamer to target nucleolin overexpressed in the tumor cell membrane and nucleus [115, 118, 119]. Based on AS1411 aptamer-modified carboxymethyl chitosan, they studied the effects of multifunctional drug delivery systems modified with different cell-penetrating peptides and nuclear targeting peptides (Table 3). And it appeared that successful cellular uptake and nuclear localization could effectively promote CRISPR-Cas9 plasmid-mediated gene editing.

However, the instability and off-target effects of CRISPR-Cas9 limit its application [120]. The delivery of RNPs by natural chitosan may hinder its release and cause its inactivation. Therefore, it is necessary to modify effective molecules on chitosan to ensure the activity of CRISPR/Cas9 ribonucleoprotein and the successful intracellular release, which will become a promising research interest in the future.

Table 3 Summary of chitosan derivatives to deliver CRISPR-Cas9 plasmid for biomedicine applications

Application	Used gene	Carrier	Modification groups	Enhanced function	Results	Ref.
Breast cancer	CRISPR-Cas9 plasmid (for CDK11 (cyclin-dependent kinase 11) knockout)	BCMC/ACMC/PS/$CaCO_3$/CaP	BCMC (biotinylated carboxymethyl chitosan); ACMC (AS1411 aptamer incorporated carboxymethyl chitosan); PS/$CaCO_3$/CaP (protamine sulfate/$CaCO_3$/CaP)	Biotin targets diverse tumor cells. AS1411 aptamer targets nucleolin overexpressed in tumor cell membrane and nucleus. Protamine sulfate, rich in arginine sequence, promotes membrane penetration and nucleic acid encapsulation. $CaCO_3$/CaP, inorganic components, form nano-sized co-precipitates with CRISPR-Cas9 plasmid	The dual-targeting CRISPR-Cas9 plasmid delivery system successfully delivered the CRISPR-Cas9 plasmid into MCF-7 tumor cells and nuclei, achieving efficient gene editing and reducing the expression of CDK11 protein by 90%	[118]
Breast cancer	CRISPR-Cas9 plasmid (for CDK11 knockout)	ACMC/KALA / PS	ACMC (AS1411 aptamer incorporated carboxymethyl chitosan); KALA (endosmotic and fusogenic peptide); PS (protamine sulfate)	AS1411 aptamer targets nucleolin overexpressed in tumor cell membrane and nucleus. KALA facilitates cellular uptake and endosomal escape. Protamine sulfate promotes membrane penetration and nucleic acid encapsulation	In vitro transfection experiments show that CDK11 knockout is related to the plasmid concentration. When the plasmid concentration is equal to or higher than 1 μg mL-1, CDK11 expression is significantly down-regulated	[119]
Cervical cancer	CRISPR-Cas9 plasmid (for CTNNB1 gene (encoding β-catenin) knockout)	ACMC/TCMC/PS/$CaCO_3$	ACMC (AS1411 aptamer incorporated carboxymethyl chitosan); TCMC (cell-penetrating peptide (TAT)-functionalized carboxymethyl chitosan); PS/$CaCO_3$ (protamine sulfate/$CaCO_3$)	AS1411 aptamer targets nucleolin overexpressed in tumor cell membrane and nucleus. TAT facilitates cellular uptake and endosomal escape. Protamine sulfate promotes membrane penetration and nucleic acid encapsulation. $CaCO_3$ provides plasmid protection and endosomal escape	The CRISPR-Cas9 delivery system can effectively knock out CTNNB1 gene in HeLa cells, thereby reducing the expression of β-catenin. What's more, contributed to AS1411-mediated tumor cell/nuclear targeting and TAT-induced enhanced endocytosis, cell uptake and nuclear transport increased significantly	[115]

4 Challenges and Future Directions

Chitosan, a natural cationic polysaccharide material, has been widely used as a gene therapy vector due to its excellent biocompatibility and biodegradability. It has been proved that the water solubility, charge strength, and even transfection efficiency of chitosan were settled by the parameters of chitosan. Adjusting the parameters to achieve a balance between nucleic acid protection and release is regarded as an effective way to optimize chitosan. However, due to the size difference between nucleic acid substances (such as pDNA and siRNA), the parameters of chitosan have various rules for the delivery of different nucleic acids. At present, there is still a lack of a unified and systematic evaluation of the best parameters for chitosan to deliver different nucleic acids.

More importantly, it cannot be ignored that the low solubility and limited transfection efficiency of pristine chitosan under physiological conditions have impeded its further application in the clinic. Luckily, a variety of chitosan derivatives make up for these shortcomings and endow chitosan with more functions, such as cell targeting, nuclear targeting, and hydrophobic drug encapsulation. The preponderances of chitosan derivatives are based on the modifiability, biocompatibility, and biodegradability of chitosan, which can be combined with some materials possessing special functions to ameliorate the properties of chitosan itself. For example, the introduction of PEI increases the transfection efficiency, and the combined use of graphene enhances the encapsulation ability of hydrophobic drugs to achieve combined therapy while reducing the toxic side effects of PEI and graphene materials themselves. What's more, multifunctional chitosan derivatives modified by targeting ligands such as mannose and galactose, water-soluble groups such as quaternization and PEG, and buffering enhancement groups such as urocanic acid and poly-(*N*-3-carbobenzyloxy-lysine) have generally shown better effects in biomedical applications as gene carriers.

Currently, chitosan derivatives have been widely used in many fields to deliver gene drugs, including tumor treatment, inflammation, tissue regeneration, and DNA vaccines. But it is undeniable that there are still many challenges that we need to solve to facilitate the clinical transformation of chitosan derivatives as gene carriers. The degree of grafting of chitosan derivatives targeting groups still needs systematic evaluation. Because of membrane receptor saturation, too many modified targeting ligands may cause steric hindrance to adversely affect the membrane uptake. However, due to a large number of applicable ligands in the complex system of chitosan derivatives, only rare studies have evaluated this aspect. Also, more methods are needed to optimize the design details of chitosan derivatives to achieve better gene therapy such as how to deliver siRNA or miRNA into the cytoplasm and chemotherapy drugs into the special organelles (such as nucleus or mitochondria, etc.), respectively, in the synergy treatment of tumors. Because some chemotherapeutics (such as DOX and lonidamine) often need to enter the nucleus or mitochondria to play a role, while siRNA or miRNA takes effect in the cytoplasm. Hence, the co-delivery of genes and drugs with different target endpoints in one carrier is still a challenge, and the targeting and high transfection ability of the delivery system are also the key points in the various disorder treatment. Importantly, chitosan's flexible modifiability has shown great potentials for the construction of such multifunctional gene delivery carriers. Besides, chitosan has been proposed to have an adjuvant effect that can induce the maturation of dendritic cells and trigger an immune response. Currently, its immune mechanism has not been fully understood, however, it is foreseeable that chitosan derivatives have huge prospects as nucleic acid vaccine carriers stemming from their potential adjuvant capabilities.

Undeniably, the research of chitosan derivatives for miRNA, shRNA, and CRISPR/Cas9 delivery is still far less than that for DNA and siRNA. However, with the further elucidation of the molecular mechanism in miRNA and shRNA and the development of the CRISPR/Cas9 gene-editing system, chitosan derivatives have sufficient potential to deliver different gene drugs for varying disorder treatment, including but not limited to tumor treatment, tissue regeneration, immune vaccine, and inflammation disease.

References

1. Gong JH, Wang Y, Xing L, Cui PF, Qiao JB, He YJ, Jiang HL (2018) Biocompatible fluorinated poly (β-amino ester)s for safe and efficient gene therapy. Int J Pharm 535:180–193
2. Luo CQ, Jang Y, Xing L, Cui PF, Qiao JB, Lee AY, Kim HJ, Cho MH, Jiang HL (2016) Aerosol delivery of folate-decorated hyperbranched polyspermine complexes to suppress lung tumorigenesis via Akt signaling pathway. Int J Pharm 513:591–601
3. Kim YD, Park TE, Singh B, Cho KS, Sangshetti JN, Choi YJ, Arote RB, Cho CS (2016) Efficient gene transfection to liver cells via the cellular regulation of a multifunctional polylactitol-based gene transporter. J Mater Chem B 4:2208–2218
4. Islam MA, Kim S, Firdous J, Lee AY, Hong SH, Seo MK, Park TE, Yun CH, Choi YJ, Chae C, Cho CS, Cho MH (2016) A high affinity kidney targeting by chitobionic acid-conjugated polysorbitol gene transporter alleviates unilateral ureteral obstruction in rats. Biomaterials 102:43–57
5. Liang H, Chen X, Jin R, Ke B, Barz M, Ai H, Nie Y (2020) Integration of indocyanine green analogs as near-infrared fluorescent carrier for precise imaging-guided gene delivery. Small 16:e1906538
6. Rajendrakumar SK, Uthaman S, Cho CS, Park IK (2017) Trigger-responsive gene transporters for anticancer therapy. Nanomaterials (Basel) 7:120
7. Dunbar CE, High KA, Joung JK, Kohn DB, Ozawa K, Sadelain M (2018) Gene therapy comes of age. Science 359:eaan4672
8. Yang S, Ou C, Wang L, Liu X, Yang J, Wang X, Wang M, Shen M, Wu Q, Gong C (2020) Virus-esque nucleus-targeting nanoparticles deliver trojan plasmid for release of anti-tumor shuttle protein. J Control Release 320:253–264
9. Bez M, Foiret J, Shapiro G, Pelled G, Ferrara KW, Gazit D (2019) Nonviral ultrasound-mediated gene delivery in small and large animal models. Nat Protoc 14:1015–1026
10. Jiang HL, Islam MA, Xing L, Firdous J, Cao W, He YJ, Zhu Y, Cho KH, Li HS, Cho CS (2017) Degradable polyethylenimine-based gene carriers for cancer therapy. Top Curr Chem (Cham) 375:34
11. Cui PF, Qi LY, Wang Y, Yu RY, He YJ, Xing L, Jiang HL (2019) Dex-Aco coating simultaneously increase the biocompatibility and transfection efficiency of cationic polymeric gene vectors. J Control Release 303:253–262

12. Zhou Z, Li H, Wang K, Guo Q, Li C, Jiang H, Hu Y, Oupicky D, Sun M (2017) Bioreducible cross-linked hyaluronic acid/calcium phosphate hybrid nanoparticles for specific delivery of siRNA in melanoma tumor therapy. ACS Appl Mater Interfaces 9:14576–14589
13. Park TE, Singh B, Li H, Lee JY, Kang SK, Choi YJ, Cho CS (2015) Enhanced BBB permeability of osmotically active poly (mannitol-co-PEI) modified with rabies virus glycoprotein via selective stimulation of caveolar endocytosis for RNAi therapeutics in Alzheimer's disease. Biomaterials 38:61–71
14. Babu A, Ramesh R (2017) Multifaceted applications of chitosan in cancer drug delivery and therapy. Mar Drugs 15:96
15. Xie RL, Jang YJ, Xing L, Zhang BF, Wang FZ, Cui PF, Cho MH, Jiang HL (2015) A novel potential biocompatible hyperbranched polyspermine for efficient lung cancer gene therapy. Int J Pharm 478:19–30
16. Zhang M, Kim YK, Cui P, Zhang J, Qiao J, He Y, Lyu J, Luo C, Xing L, Jiang H (2016) Folate-conjugated polyspermine for lung cancer-targeted gene therapy. Acta Pharm Sin B 6:336–343
17. Qindeel M, Ahmed N, Khan GM, Rehman AU (2019) Ligand decorated chitosan as an advanced nanocarrier for targeted delivery: a critical review. Nanomedicine (Lond) 14:1623–1642
18. Pereira LA, da Silva RL, Batista FA, Mendes AN, Osajima JA, Silva-Filho EC (2019) Biological properties of chitosan derivatives associated with the ceftazidime drug. Carbohydr Polym 222:115002
19. Jiang HL, Cui PF, Xie RL, Cho CS (2014) Chemical modification of chitosan for efficient gene therapy. Adv Food Nutr Res 73:83–101
20. Chuan D, Jin T, Fan R, Zhou L, Guo G (2019) Chitosan for gene delivery: methods for improvement and applications. Adv Colloid Interface Sci 268:25–38
21. Nguyen MA, Wyatt H, Susser L, Geoffrion M, Rasheed A, Duchez AC, Cottee ML, Afolayan E, Farah E, Kahiel Z, Côté M, Gadde S, Rayner KJ (2019) Delivery of microRNAs by chitosan nanoparticles to functionally alter macrophage cholesterol efflux in vitro and in vivo. ACS Nano 13:6491–6505
22. Santos-Carballal B, Fernández Fernández E, Goycoolea FM (2018) Chitosan in non-viral gene delivery: role of structure, characterization methods, and insights in cancer and rare diseases therapies. Polymers (Basel) 10:444
23. Moran HBT, Turley JL, Andersson M, Lavelle EC (2018) Immunomodulatory properties of chitosan polymers. Biomaterials 184:1–9
24. Wang W, Meng Q, Li Q, Liu J, Zhou M, Jin Z, Zhao K (2020) Chitosan derivatives and their application in biomedicine. Int J Mol Sci 21:487
25. Sato T, Ishii T, Okahata Y (2001) In vitro gene delivery mediated by chitosan. Effect of pH, serum, and molecular mass of chitosan on the transfection efficiency. Biomaterials 22:2075–2080
26. Strand SP, Lelu S, Reitan NK, de Lange DC, Artursson P, Vårum KM (2010) Molecular design of chitosan gene delivery systems with an optimized balance between polyplex stability and polyplex unpacking. Biomaterials 31:975–987
27. Kean T, Thanou M (2010) Biodegradation, biodistribution and toxicity of chitosan. Adv Drug Deliv Rev 62:3–11
28. Kritchenkov AS, Andranovit S, Skorik YA (2017) Chitosan and its derivatives: vectors in gene therapy. Russ Chem Rev 86:231–239
29. Huang M, Fong CW, Khor E, Lim LY (2005) Transfection efficiency of chitosan vectors: effect of polymer molecular weight and degree of deacetylation. J Control Release 106:391–406
30. Yang X, Yuan X, Cai D, Wang S, Zong L (2009) Low molecular weight chitosan in DNA vaccine delivery via mucosa. Int J Pharm 375:123–132

31. Weecharangsan W, Opanasopit P, Ngawhirunpat T, Apirakaramwong A, Rojanarata T, Ruktanonchai U, Lee RJ (2008) Evaluation of chitosan salts as non-viral gene vectors in CHO-K1 cells. Int J Pharm 348:161–168
32. Prasitsilp M, Jenwithisuk R, Kongsuwan K, Damrongchai N, Watts P (2000) Cellular responses to chitosan in vitro: the importance of deacetylation. J Mater Sci Mater Med 11:773–778
33. Köping-Höggård M, Tubulekas I, Guan H, Edwards K, Nilsson M, Vårum KM, Artursson P (2001) Chitosan as a nonviral gene delivery system. Structure-property relationships and characteristics compared with polyethylenimine in vitro and after lung administration in vivo. Gene Ther 8:1108–1121
34. Alameh M, Dejesus D, Jean M, Darras V, Thibault M, Lavertu M, Buschmann MD, Merzouki A (2012) Low molecular weight chitosan nanoparticulate system at low N:P ratio for nontoxic polynucleotide delivery. Int J Nanomedicine 7:1399–1414
35. Nimesh S, Thibault MM, Lavertu M, Buschmann MD (2010) Enhanced gene delivery mediated by low molecular weight chitosan/DNA complexes: effect of pH and serum. Mol Biotechnol 46:182–196
36. Bravo-Anaya LM, Fernández-Solís KG, Rosselgong J, Nano-Rodríguez JLE, Carvajal F, Rinaudo M (2019) Chitosan-DNA polyelectrolyte complex: influence of chitosan characteristics and mechanism of complex formation. Int J Biol Macromol 126:1037–1049
37. Kolonko AK, Bangel-Ruland N, Goycoolea FM, Weber WM (2020) Chitosan nanocomplexes for the delivery of ENaC antisense oligonucleotides to airway epithelial cells. Biomol Ther 10:553
38. Alameh M, Lavertu M, Tran-Khanh N, Chang CY, Lesage F, Bail M, Darras V, Chevrier A, Buschmann MD (2018) siRNA delivery with chitosan: influence of chitosan molecular weight, degree of deacetylation, and amine to phosphate ratio on in vitro silencing efficiency, hemocompatibility, biodistribution, and in vivo efficacy. Biomacromolecules 19:112–131
39. Yazdian-Robati R, Bayat P, Oroojalian F, Zargari M, Ramezani M, Taghdisi SM, Abnous K (2020) Therapeutic applications of AS1411 aptamer, an update review. Int J Biol Macromol 155:1420–1431
40. Xing L, Fan YT, Shen LJ, Yang CX, Liu XY, Ma YN, Qi LY, Cho KH, Cho CS, Jiang HL (2019) pH-sensitive and specific ligand-conjugated chitosan nanogels for efficient drug delivery. Int J Biol Macromol 141:85–97
41. Bao X, Wang W, Wang C, Wang Y, Zhou J, Ding Y, Wang X, Jin Y (2014) A chitosan-graft-PEI-candesartan conjugate for targeted co-delivery of drug and gene in anti-angiogenesis cancer therapy. Biomaterials 35:8450–8466
42. Wang GH, Cai YY, Du JK, Li L, Li Q, Yang HK, Lin JT (2018) TAT-conjugated chitosan cationic micelle for nuclear-targeted drug and gene co-delivery. Colloids Surf B Biointerfaces 162:326–334
43. Lin JT, Chen H, Wang D, Xiong L, Li JZ, Chen GH, Chen GB (2019) Nuclear-targeted p53 and DOX co-delivery of chitosan derivatives for cancer therapy in vitro and in vivo. Colloids Surf B Biointerfaces 183:110440
44. Zhang M, Hu J, Zou Y, Wu J, Yao Y, Fan H, Liu K, Wang J, Gao S (2018) Modification of degradable nonviral delivery vehicle with a novel bifunctional peptide to enhance transfection in vivo. Nanomedicine (Lond) 13:9–24
45. Zhao M, Li J, Ji H, Chen D, Hu H (2019) A versatile endosome acidity-induced sheddable gene delivery system: increased tumor targeting and enhanced transfection efficiency. Int J Nanomedicine 14:6519–6538
46. Huang L, Zhu Z, Wu D, Gan W, Zhu S, Li W, Tian J, Li L, Zhou C, Lu L (2019) Antibacterial poly (ethylene glycol) diacrylate/chitosan hydrogels enhance mechanical adhesiveness and promote skin regeneration. Carbohydr Polym 225:115110
47. Wu D, Zhang Y, Xu X, Guo T, Xie D, Zhu R, Chen S, Ramakrishna S, He L (2018) RGD/TAT-functionalized chitosan-graft-PEI-PEG gene nanovector for sustained delivery of NT-3 for potential application in neural regeneration. Acta Biomater 72:266–277

48. Lopes CDF, Gonçalves NP, Gomes CP, Saraiva MJ, Pêgo AP (2017) BDNF gene delivery mediated by neuron-targeted nanoparticles is neuroprotective in peripheral nerve injury. Biomaterials 121:83–96
49. Liu Y, Wang FQ, Shah Z, Cheng XJ, Kong M, Feng C, Chen XG (2016) Nano-polyplex based on oleoyl-carboxymethy-chitosan (OCMCS) and hyaluronic acid for oral gene vaccine delivery. Colloids Surf B Biointerfaces 145:492–501
50. Wu M, Zhao H, Li M, Yue Y, Xiong S, Xu W (2017) Intranasal vaccination with mannosylated chitosan formulated DNA vaccine enables robust IgA and cellular response induction in the lungs of mice and improves protection against pulmonary mycobacterial challenge. Front Cell Infect Microbiol 7:445
51. Zhao K, Han J, Zhang Y, Wei L, Yu S, Wang X, Jin Z, Wang Y (2018) Enhancing mucosal immune response of newcastle disease virus DNA vaccine using N-2-hydroxypropyl trimethylammonium chloride chitosan and N,O-carboxymethyl chitosan nanoparticles as delivery carrier. Mol Pharm 15:226–237
52. Picanço-Castro V, Pereira CG, Covas DT, Porto GS (2020) Emerging patent landscape for non-viral vectors used for gene therapy. Nat Biotechnol 38:151–157
53. Lin WJ, Hsu WY (2015) Pegylation effect of chitosan based polyplex on DNA transfection. Carbohydr Polym 120:7–14
54. Bai H, Lester GMS, Petishnok LC, Dean DA (2017) Cytoplasmic transport and nuclear import of plasmid DNA. Biosci Rep 37:BSR20160616
55. Gong N, Teng X, Li J, Liang XJ (2019) Antisense oligonucleotide-conjugated nanostructure-targeting incRNA MALAT1 inhibits cancer metastasis. ACS Appl Mater Interfaces 11:37–42
56. Raftery RM, Mencía Castaño I, Chen G, Cavanagh B, Quinn B, Curtin CM, Cryan SA, O'Brien FJ (2017) Translating the role of osteogenic-angiogenic coupling in bone formation: highly efficient chitosan-pDNA activated scaffolds can accelerate bone regeneration in critical-sized bone defects. Biomaterials 149:116–127
57. Li H, Ji Q, Chen X, Sun Y, Xu Q, Deng P, Hu F, Yang J (2017) Accelerated bony defect healing based on chitosan thermosensitive hydrogel scaffolds embedded with chitosan nanoparticles for the delivery of BMP2 plasmid DNA. J Biomed Mater Res A 105:265–273
58. Lord MS, Ellis AL, Farrugia BL, Whitelock JM, Grenett H, Li C, O'Grady RL, DeCarlo AA (2017) Perlecan and vascular endothelial growth factor-encoding DNA-loaded chitosan scaffolds promote angiogenesis and wound healing. J Control Release 250:48–61
59. Lou D, Luo Y, Pang Q, Tan WQ, Ma L (2020) Gene-activated dermal equivalents to accelerate healing of diabetic chronic wounds by regulating inflammation and promoting angiogenesis. Bioact Mater 5:667–679
60. Lee MC, Seonwoo H, Garg P, Jang KJ, Pandey S, Park SB, Kim HB, Lim J, Choung YH, Chung JH (2018) Chitosan/PEI patch releasing EGF and the EGFR gene for the regeneration of the tympanic membrane after perforation. Biomater Sci 6:364–371
61. Pereira Gomes C, Leiro V, Ferreira Lopes CD, Spencer AP, Pêgo AP (2018) Fine tuning neuronal targeting of nanoparticles by adjusting the ligand grafting density and combining PEG spacers of different length. Acta Biomater 78:247–259
62. Poecheim J, Barnier-Quer C, Collin N, Borchard G (2016) Ag85A DNA vaccine delivery by nanoparticles: influence of the formulation characteristics on immune responses. Vaccines (Basel) 4:32
63. Xing L, Fan YT, Zhou TJ, Gong JH, Cui LH, Cho KH, Choi YJ, Jiang HL, Cho CS (2018) Chemical modification of chitosan for efficient vaccine delivery. Molecules 23:229
64. Chandrasekar SS, Kingstad-Bakke B, Wu CW, Suresh M, Talaat AM (2020) A novel mucosal adjuvant system for immunization against avian coronavirus causing infectious bronchitis. J Virol 94:e01016–e01020
65. Singh B, Maharjan S, Cho KH, Cui L, Park IK, Choi YJ, Cho CS (2018) Chitosan-based particulate systems for the delivery of mucosal vaccines against infectious diseases. Int J Biol Macromol 110:54–64

66. Singh B, Maharjan S, Sindurakar P, Cho KH, Choi YJ, Cho CS (2018) Needle-free immunization with chitosan-based systems. Int J Mol Sci 19:3639
67. Islam MA, Park TE, Reesor E, Cherukula K, Hasan A, Firdous J, Singh B, Kang SK, Choi YJ, Park IK, Cho CS (2015) Mucoadhesive chitosan derivatives as novel drug carriers. Curr Pharm Des 21:4285–4309
68. Layek B, Lipp L, Singh J (2015) APC targeted micelle for enhanced intradermal delivery of hepatitis B DNA vaccine. J Control Release 207:143–153
69. Jesus S, Soares E, Borchard G, Borges O (2017) Poly-ϵ-caprolactone/chitosan nanoparticles provide strong adjuvant effect for hepatitis B antigen. Nanomedicine (Lond) 12:2335–2348
70. Leya T, Ahmad I, Sharma R, Tripathi G, Kurcheti PP, Rajendran KV, Bedekar MK (2020) Bicistronic DNA vaccine macromolecule complexed with poly lactic-co-glycolic acid-chitosan nanoparticles enhanced the mucosal immunity of Labeo rohita against Edwardsiella tarda infection. Int J Biol Macromol 156:928–937
71. Carroll EC, Jin L, Mori A, Muñoz-Wolf N, Oleszycka E, Moran HBT, Mansouri S, McEntee CP, Lambe E, Agger EM, Andersen P, Cunningham C, Hertzog P, Fitzgerald KA, Bowie AG, Lavelle EC (2016) The vaccine adjuvant chitosan promotes cellular immunity via DNA sensor cGAS-STING-dependent induction of type I interferons. Immunity 44:597–608
72. Han L, Tang C, Yin C (2015) Enhanced antitumor efficacies of multifunctional nanocomplexes through knocking down the barriers for siRNA delivery. Biomaterials 44:111–121
73. Song Y, Tang C, Yin C (2018) Combination antitumor immunotherapy with VEGF and PIGF siRNA via systemic delivery of multi-functionalized nanoparticles to tumor-associated macrophages and breast cancer cells. Biomaterials 185:117–132
74. Ni S, Liu Y, Tang Y, Chen J, Li S, Pu J, Han L (2018) GABA(B) receptor ligand-directed trimethyl chitosan/tripolyphosphate nanoparticles and their pMDI formulation for survivin siRNA pulmonary delivery. Carbohydr Polym 179:135–144
75. Zhang W, Xu W, Lan Y, He X, Liu K, Liang Y (2019) Antitumor effect of hyaluronic-acid-modified chitosan nanoparticles loaded with siRNA for targeted therapy for non-small cell lung cancer. Int J Nanomedicine 14:5287–5301
76. Masjedi A, Ahmadi A, Atyabi F, Farhadi S, Irandoust M, Khazaei-Poul Y, Ghasemi Chaleshtari M, Edalati Fathabad M, Baghaei M, Haghnavaz N, Baradaran B, Hojjat-Farsangi-M, Ghalamfarsa G, Sabz G, Hasanzadeh S, Jadidi-Niaragh F (2020) Silencing of IL-6 and STAT3 by siRNA loaded hyaluronate-N,N,N-trimethyl chitosan nanoparticles potently reduces cancer cell progression. Int J Biol Macromol 149:487–500
77. Han L, Tang C, Yin C (2015) Dual-targeting and pH/redox-responsive multi-layered nanocomplexes for smart co-delivery of doxorubicin and siRNA. Biomaterials 60:42–52
78. Zhang BF, Xing L, Cui PF, Wang FZ, Xie RL, Zhang JL, Zhang M, He YJ, Lyu JY, Qiao JB, Chen BA, Jiang HL (2015) Mitochondria apoptosis pathway synergistically activated by hierarchical targeted nanoparticles co-delivering siRNA and lonidamine. Biomaterials 61:178–189
79. Zhang CG, Zhu WJ, Liu Y, Yuan ZQ, Yang SD, Chen WL, Li JZ, Zhou XF, Liu C, Zhang XN (2016) Novel polymer micelle mediated co-delivery of doxorubicin and P-glycoprotein siRNA for reversal of multidrug resistance and synergistic tumor therapy. Sci Rep 6:23859
80. Izadi S, Moslehi A, Kheiry H, Karoon Kiani F, Ahmadi A, Masjedi A, Ghani S, Rafiee B, Karpisheh V, Hajizadeh F, Atyabi F, Assali A, Mirzazadeh Tekie FS, Namdar A, Ghalamfarsa G, Sojoodi M, Jadidi-Niaragh F (2020) Codelivery of HIF-1α siRNA and dinaciclib by carboxylated graphene oxide-trimethyl chitosan-hyaluronate nanoparticles significantly suppresses cancer cell progression. Pharm Res 37:196
81. He C, Yin L, Tang C, Yin C (2013) Multifunctional polymeric nanoparticles for oral delivery of TNF-α siRNA to macrophages. Biomaterials 34:2843–2854
82. Chu S, Tang C, Yin C (2015) Effects of mannose density on in vitro and in vivo cellular uptake and RNAi efficiency of polymeric nanoparticles. Biomaterials 52:229–239

83. Cheng W, Tang C, Yin C (2015) Effects of particle size and binding affinity for small interfering RNA on the cellular processing, intestinal permeation and anti-inflammatory efficacy of polymeric nanoparticles. J Gene Med 17:244–256
84. Xiao B, Ma P, Ma L, Chen Q, Si X, Walter L, Merlin D (2017) Effects of tripolyphosphate on cellular uptake and RNA interference efficiency of chitosan-based nanoparticles in raw 264.7 macrophages. J Colloid Interface Sci 490:520–528
85. Xiao B, Ma P, Viennois E, Merlin D (2016) Urocanic acid-modified chitosan nanoparticles can confer anti-inflammatory effect by delivering CD98 siRNA to macrophages. Colloids Surf B Biointerfaces 143:186–193
86. de Souza R, Picola IPD, Shi Q, Petrônio MS, Benderdour M, Fernandes JC, Lima AMF, Martins GO, Martinez Junior AM, de Oliveira Tiera VA, Tiera MJ (2018) Diethylaminoethyl-chitosan as an efficient carrier for siRNA delivery: improving the condensation process and the nanoparticles properties. Int J Biol Macromol 119:186–197
87. Shi Q, Rondon-Cavanzo EP, Dalla Picola IP, Tiera MJ, Zhang X, Dai K, Benabdoune HA, Benderdour M, Fernandes JC (2018) In vivo therapeutic efficacy of TNFα silencing by folate-PEG-chitosan-DEAE/siRNA nanoparticles in arthritic mice. Int J Nanomedicine 13:387–402
88. Rondon EP, Benabdoun HA, Vallières F, Segalla Petrônio M, Tiera MJ, Benderdour M, Fernandes JC (2020) Evidence supporting the safety of pegylated diethylaminoethyl-chitosan polymer as a nanovector for gene therapy applications. Int J Nanomedicine 15:6183–6200
89. Chen H, Ma Y, Lan H, Zhao Y, Zhi D, Cui S, Du J, Zhang Z, Zhen Y, Zhang S (2018) Dual stimuli-responsive saccharide core based nanocarrier for efficient Birc5-shRNA delivery. J Mater Chem B 6:7530–7542
90. Liu T, Chen S, Zhang S, Wu X, Wu P, Miao B, Cai X (2018) Transferrin-functionalized chitosan-graft-poly (l-lysine) dendrons as a high-efficiency gene delivery carrier for nasopharyngeal carcinoma therapy. J Mater Chem B 6:4314–4325
91. Song Y, Tang C, Yin C (2018) Enhanced antitumor efficacy of arginine modified amphiphilic nanoparticles co-delivering doxorubicin and iSur-pDNA via the multiple synergistic effect. Biomaterials 150:1–13
92. Jia L, Li Z, Zheng D, Li Z, Zhao Z (2021) A targeted and redox/pH-responsive chitosan oligosaccharide derivatives based nanohybrids for overcoming multidrug resistance of breast cancer cells. Carbohydr Polym 251:117008
93. Li Z, Zhu L, Liu W, Zheng Y, Li X, Ye J, Li B, Chen H, Gao Y (2020) Near-infrared/pH dual-responsive nanocomplexes for targeted imaging and chemo/gene/photothermal tri-therapies of non-small cell lung cancer. Acta Biomater 107:242–259
94. Tu L, Wang M, Zhao WY, Zhang ZZ, Tang DF, Zhang YQ, Cao H, Zhang ZG (2017) miRNA-218-loaded carboxymethyl chitosan-tocopherol nanoparticle to suppress the proliferation of gastrointestinal stromal tumor growth. Korean J Couns Psychother 72:177–184
95. Liu C, Xie H, Yu J, Chen X, Tang S, Sun L, Chen X, Peng D, Zhang X, Zhou J (2018) A targeted therapy for melanoma by graphene oxide composite with microRNA carrier. Drug Des Devel Ther 12:3095–3106
96. Tekie FSM, Soleimani M, Zakerian A, Dinarvand M, Amini M, Dinarvand R, Arefian E, Atyabi F (2018) Glutathione responsive chitosan-thiolated dextran conjugated miR-145 nanoparticles targeted with AS1411 aptamer for cancer treatment. Carbohydr Polym 201:131–140
97. Ning Q, Liu YF, Ye PJ, Gao P, Li ZP, Tang SY, He DX, Tang SS, Wei H, Yu CY (2019) Delivery of liver-specific miRNA-122 using a targeted macromolecular prodrug toward synergistic therapy for hepatocellular carcinoma. ACS Appl Mater Interfaces 11:10578–10588
98. Deng F, He S, Cui S, Shi Y, Tan Y, Li Z, Huang C, Liu D, Zhi F, Peng L (2019) A molecular targeted immunotherapeutic strategy for ulcerative colitis via dual-targeting nanoparticles delivering miR-146b to intestinal macrophages. J Crohns Colitis 13:482–494
99. Edson JA, Ingato D, Wu S, Lee B, Kwon YJ (2018) Aqueous-soluble, acid-transforming chitosan for efficient and stimuli-responsive gene silencing. Biomacromolecules 19:1508–1516

100. Qiao JB, Fan QQ, Xing L, Cui PF, He YJ, Zhu JC, Wang L, Pang T, Oh YK, Zhang C, Jiang HL (2018) Vitamin A-decorated biocompatible micelles for chemogene therapy of liver fibrosis. J Control Release 283:113–125
101. Li L, Hu X, Zhang M, Ma S, Yu F, Zhao S, Liu N, Wang Z, Wang Y, Guan H, Pan X, Gao Y, Zhang Y, Liu Y, Yang Y, Tang X, Li M, Liu C, Li Z, Mei X (2017) Dual tumor-targeting nanocarrier system for siRNA delivery based on pRNA and modified chitosan. Mol Ther Nucleic Acids 8:169–183
102. Xin Y, Huang M, Guo WW, Huang Q, Zhang LZ, Jiang G (2017) Nano-based delivery of RNAi in cancer therapy. Mol Cancer 16:134
103. Nikkhoo A, Rostami N, Farhadi S, Esmaily M, Moghadaszadeh Ardebili S, Atyabi F, Baghaei M, Haghnavaz N, Yousefi M, Aliparasti MR, Ghalamfarsa G, Mohammadi H, Sojoodi M, Jadidi-Niaragh F (2020) Codelivery of STAT3 siRNA and BV6 by carboxymethyl dextran trimethyl chitosan nanoparticles suppresses cancer cell progression. Int J Pharm 581:119236
104. Cui PF, Xing L, Qiao JB, Zhang JL, He YJ, Zhang M, Lyu JY, Luo CQ, Jin L, Jiang HL (2016) Polyamine metabolism-based dual functional gene delivery system to synergistically inhibit the proliferation of cancer. Int J Pharm 506:79–86
105. Ross NT, Lohmann F, Carbonneau S, Fazal A, Weihofen WA, Gleim S, Salcius M, Sigoillot F, Henault M, Carl SH, Rodríguez-Molina JB, Miller HR, Brittain SM, Murphy J, Zambrowski M, Boynton G, Wang Y, Chen A, Molind GJ, Wilbertz JH, Artus-Revel CG, Jia M, Akinjiyan FA, Turner J, Knehr J, Carbone W, Schuierer S, Reece-Hoyes JS, Xie K, Saran C, Williams ET, Roma G, Spencer M, Jenkins J, George EL, Thomas JR, Michaud G, Schirle M, Tallarico J, Passmore LA, Chao JA, Beckwith REJ (2020) CPSF3-dependent pre-mRNA processing as a druggable node in AML and Ewing's sarcoma. Nat Chem Biol 16:50–59
106. Waljee AK, Higgins PDR, Jensen CB, Villumsen M, Cohen-Mekelburg SA, Wallace BI, Berinstein JA, Allin KH, Jess T (2020) Anti-tumour necrosis factor-α therapy and recurrent or new primary cancers in patients with inflammatory bowel disease, rheumatoid arthritis, or psoriasis and previous cancer in Denmark: a nationwide, population-based cohort study. Lancet Gastroenterol Hepatol 5:276–284
107. Rao DD, Vorhies JS, Senzer N, Nemunaitis J (2009) siRNA vs. shRNA: similarities and differences. Adv Drug Deliv Rev 61:746–759
108. Yu B, Tang C, Yin C (2014) Enhanced antitumor efficacy of folate modified amphiphilic nanoparticles through co-delivery of chemotherapeutic drugs and genes. Biomaterials 35:6369–6378
109. Javan B, Atyabi F, Shahbazi M (2018) Hypoxia-inducible bidirectional shRNA expression vector delivery using PEI/chitosan-TBA copolymers for colorectal Cancer gene therapy. Life Sci 202:140–151
110. Leite-Moreira AM, Lourenço AP, Falcão-Pires I, Leite-Moreira AF (2013) Pivotal role of microRNAs in cardiac physiology and heart failure. Drug Discov Today 18:1243–1249
111. Song Q, Pang H, Qi L, Liang C, Wang T, Wang W, Li R (2019) Low microRNA-622 expression predicts poor prognosis and is associated with ZEB2 in glioma. Onco Targets Ther 12:7387–7397
112. Zhou J, Xu D, Xie H, Tang J, Liu R, Li J, Wang S, Chen X, Su J, Zhou X, Xia K, He Q, Chen J, Xiong W, Cao P, Cao K (2015) miR-33a functions as a tumor suppressor in melanoma by targeting HIF-1α. Cancer Biol Ther 16:846–855
113. Doudna JA (2020) The promise and challenge of therapeutic genome editing. Nature 578:229–236
114. Zhan T, Rindtorff N, Betge J, Ebert MP, Boutros M (2019) CRISPR/Cas9 for cancer research and therapy. Semin Cancer Biol 55:106–119
115. He XY, Liu BY, Peng Y, Zhuo RX, Cheng SX (2019) Multifunctional vector for delivery of genome editing plasmid targeting β-catenin to remodulate cancer cell properties. ACS Appl Mater Interfaces 11:226–237

116. Qiao J, Sun W, Lin S, Jin R, Ma L, Liu Y (2019) Cytosolic delivery of CRISPR/Cas9 ribonucleoproteins for genome editing using chitosan-coated red fluorescent protein. Chem Commun (Camb) 55:4707–4710
117. Zhang H, Bahamondez-Canas TF, Zhang Y, Leal J, Smyth HDC (2018) PEGylated chitosan for nonviral aerosol and mucosal delivery of the CRISPR/Cas9 system in vitro. Mol Pharm 15:4814–4826
118. Liu BY, He XY, Xu C, Xu L, Ai SL, Cheng SX, Zhuo RX (2018) A dual-targeting delivery system for effective genome editing and in situ detecting related protein expression in edited cells. Biomacromolecules 19:2957–2968
119. Liu BY, He XY, Zhuo RX, Cheng SX (2018) Tumor targeted genome editing mediated by a multi-functional gene vector for regulating cell behaviors. J Control Release 291:90–98
120. Chen JS, Dagdas YS, Kleinstiver BP, Welch MM, Sousa AA, Harrington LB, Sternberg SH, Joung JK, Yildiz A, Doudna JA (2017) Enhanced proofreading governs CRISPR-Cas9 targeting accuracy. Nature 550:407–410

Adv Polym Sci (2021) 288: 251–270
https://doi.org/10.1007/12_2021_94

Published online: 12 May 2021

Chitosan Polymeric Micelles as Oral Delivery Platform of Hydrophobic Anticancer Drugs

Andreia Almeida and Bruno Sarmento

Contents

A. Almeida
i3S – Institute for Research and Innovation in Health/INEB – Institute of Biomedical Engineering (INEB), University of Porto, Porto, Portugal

ICBAS – Institute of Biomedical Sciences Abel Salazar, University of Porto, Porto, Portugal
e-mail: andreia.almeida@ineb.up.pt

B. Sarmento (✉)
i3S – Institute for Research and Innovation in Health/INEB – Institute of Biomedical Engineering (INEB), University of Porto, Porto, Portugal

CESPU, IINFACTS – Institute for Research and Advanced Training in Health Sciences and Technologies, Porto, Portugal
e-mail: bruno.sarmento@i3s.up.pt

Abstract Oral drug administration is the most common and preferred route of administration for both patients and doctors. However, some drugs, as the case of hydrophobic anticancer drugs, cannot be administered orally since they are hampered by the physiological and biological barriers of the gastrointestinal tract. Thus, the use of nanotechnology, particularly, the use of polymeric micelles has received great attention to overcome these limitations and achieve a better therapeutic outcome. Polymeric micelles have the ability to protect the drug from the harsh environment of the gastrointestinal tract, improving the stability and solubility of the drug for an enhanced oral bioavailability. Chitosan is a natural polymer with very interesting biological properties and has received growing interest to produce polymeric micelles to encapsulate hydrophobic drugs. The limitations of the oral administration and the mechanisms of the intestinal absorption, as well as the chitosan modifications to produce polymeric micelles are reviewed in this chapter. Finally, recently developed chitosan-based polymeric micelles were included in the review, with a focus on the delivery of anticancer drugs for oral chemotherapy.

Keywords Cancer chemotherapy · Drug delivery · Micellar systems · Nanomedicines

1 Introduction

Cancer is the second cause of death worldwide, after cardiovascular diseases. The World Health Organization (WHO) has declared that 10 million deaths per year are due to cancer and around 70% of the deaths arise from low- and middle-income countries [1]. Cancer can be defined by the uncontrolled cell division that grows beyond their normal limits and can invade other parts of the body and/or spread to other organs. The latter is the process called metastasis and is the main cause of death from cancer [2]. The treatment usually is based on chemotherapy, radiotherapy, and/or surgery, and if cancer is detected at the early stage, the likelihood of patients being cured is very high. Thus, the pharmaceutical industry and researchers have been searching for new forms of chemotherapy, as the oral chemotherapy, for cancer treatment with better therapeutic effects and less aggressive, as the conventional chemotherapy.

The oral route has remained the preferred drug administration route due to the ease and lack of need for venous access, good patient compliance, and improved patient life quality since patients can receive treatment in the comfort of their own homes. However, the number of new drugs administered orally has not grown much in recent years. This may be due to the difficulties that are found in the gastrointestinal tract (GIT) as pH differences, enzymes, and other physiological barriers. After passing through the GIT and into the intestine, drugs must cross the epithelium and enter the liver to undergo first-pass metabolism, which is a critical issue. Most drugs that are taken orally reduce their concentration even before they reach the intestine, and when they are subjected to intestinal and hepatic metabolism, their bioavailability becomes very low. Particularly anticancer drugs, despite being very effective against several types of cancer, are very hydrophobic, which means low solubility in aqueous solution, poor oral availability, and low intestinal permeability [3].

To overcome all of these inadequacies, hydrophobic drugs can be encapsulated or associated with polymeric micelles in order to be protected from the biological fluids, to preserve its stability until the epithelium and at the same time improving its absorption due to the increased solubility. Polymeric micelles are able to increase the circulation time of the drugs, lessen the P-glycoprotein (P-gp) efflux effect, and therefore improve its therapeutic efficacy. These systems are based on amphiphilic polymers, which are constituted by hydrophobic and hydrophilic segments and have been proving to have a substantial potential in the enhancement of the bioavailability of highly hydrophobic anticancer drugs [4–6]. Due to hydrophobic interactions, hydrophobic drugs are encapsulated into the micelle core and thus protected after oral administration.

There are several types of polymers used to produce polymeric micelles. Particularly in this review, we will focus on chitosan, a natural polymer widely used on the delivery of hydrophobic anticancer drugs.

2 Oral Delivery of Drugs

Oral chemotherapy is a cost-effective treatment preferable for patients and doctors. Differently from intravenous administration that exposes a high dose of a cytotoxic agent to the systemic circulation after administration but still at low concentration for cancer treatment, the oral chemotherapy can prolong the time that a drug is in contact with the cancer cells at a relatively low dose and in a safer concentration for normal cells. To be effective, drugs administered orally have to be travel through the GIT and be absorbed until they reach the blood circulation. However, the oral route has some limitations as the biological and/or physico-chemical interactions of the drug with the GIT, which includes pH variations, presence of various enzymes as well as the bacterial flora in the intestine. Next, we will discuss in more detail why it is not that easy to create new orally administrable drugs.

2.1 Limitations to the Oral Administration of Drugs

The pH differences along the GIT, especially in the stomach, together with the enzymatic production can lead to modifications on drug solubility and stability due to oxidation and hydrolysis processes [7], being highly important to protect drugs from this harsh environment of the GIT with nanosystems, as the case of polymeric micelles.

When a drug is administered orally, it is subjected to mechanisms called efflux of drugs. One of the most common efflux pumps present in the intestinal tract is the P-gp, which is highly expressed in the apical side of the intestinal epithelial cells, particularly in the jejunum [8]. It is an ATP-dependent membrane transporter protein that has been identified as the primary cause of multidrug resistance. Thus drugs that are substrates to this protein are pumped out to the lumen of the intestine and are not absorbed, being responsible for reducing the oral bioavailability of many molecules. Moreover, a drug can undergo first-pass intestinal metabolism [9]. It is a drug metabolization that occurs before the drug enters the bloodstream, via enzymes present in the brush border or inside the enterocytes, the main cellular population of the intestine, which also leads to a decrease in intestinal absorption.

After the drug passes the first-pass intestinal metabolism and reaches the blood capillaries, drugs are subjected to the first-pass hepatic metabolism. This consists in the metabolization of the drug by the liver, where every drug that is absorbed is transported through the portal circulation [9]. In the liver, due to the metabolization by several hepatic enzymes, the amount of drug available to proceed to the systemic circulation is decreased, since the main function of the liver is to clean the body from endogenous compounds. This is the main reason why many drugs have little oral bioavailability [10]. These processes, both intestinal and hepatic metabolisms, involve the presence of the cytochrome P450 3A4 (CYP3A4), a member of cytochrome P450 (CYP450), a super family of enzymes [8].

Furthermore, the GIT is covered by a mucus layer that extends from the gastric mucosa to the colon, becoming thicker at the ileum and colon regions. At the colon, the mucus layer can reach up to 800 μm, which can be a physical barrier for the intestinal absorption of molecules [11]. Nowadays, this limitation can be overcome by the use of nanoformulations, which protects the drug until the intestine and has the surface modified to enhance its mucodiffusion for a better drug absorption. In addition, the use of mucoadhesive polymers such as chitosan can improve bioadhesion and increase drug intake into the systemic circulation [12].

2.2 *Intestinal Absorption Mechanisms*

During intestinal absorption, molecules or nanosystems can be transported by the intestinal wall to reach the blood circulation by several mechanisms depending on the molecule/system used (Fig. 1). Other parameters very important in the intestinal absorption, which will dictate the mechanism of transport, are the type of material (in case of nanosystems), the size of the molecule/system, the surface charge, and the hydrophobicity/hydrophilicity balance [13, 14]. Each mechanism of transport will be explained below in more detail.

The paracellular transport involves the diffusion of molecules by the spaces between the cells through the opening of the tight junctions until it reaches the bloodstream (Fig. 1c). Usually, this type of transport involves smaller molecules and

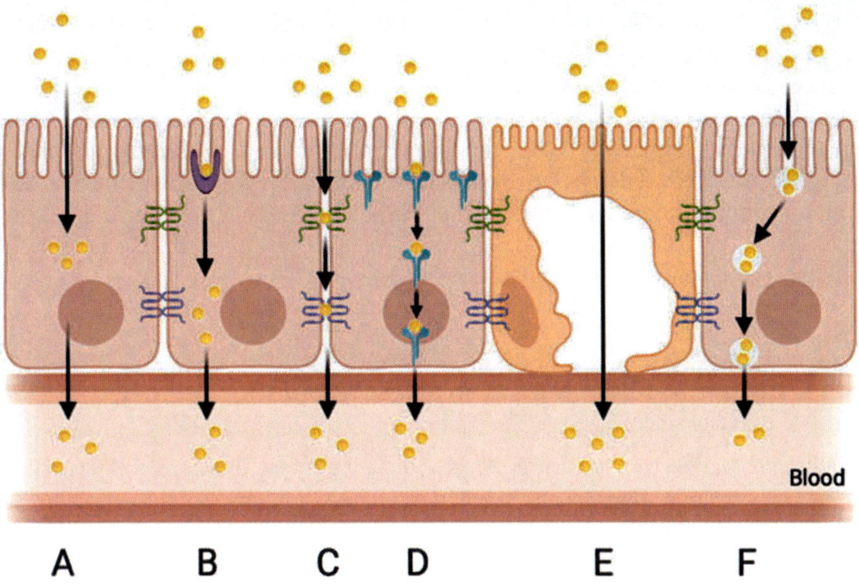

Fig. 1 Different mechanisms involved in the small intestine absorption. (**a**) passive diffusion; (**b**) active transport; (**c**) paracellular transport; (**d**) receptor-mediated endocytosis; (**e**) endocytosis via M cells; (**f**) pinocytosis

it is known that chitosan has the ability to temporarily open the tight junctions of the epithelium [15, 16].

Transcellular transport is a process in which substances can cross the epithelial cell membrane from the apical to the basolateral side, and it includes passive or active transport as well as endocytosis [17]. The passive diffusion is the simple diffusion of substances through the cell, a process more frequent for lipophilic molecules (Fig. 1a). However, the active transport is energy-dependent by the use of ATP to selectively transfer substances through the cellular membrane (Fig. 1b). Endocytosis is defined by the transport of substances from the cellular membrane to the cytoplasm and is the most frequent mechanism of transport of nanosystems. There are three types of endocytosis: (1) the receptor-mediated endocytosis, which involves the combination of a ligand with a receptor at the apical side of the cell, which crosses the cell to the basolateral side and delivers the molecule to the blood capillaries (Fig. 1d). It is based on a targeted delivery, and for the case of nanosystems, they are decorated with a ligand for a specific type of cells that have the receptor for this specific ligand; (2) the phagocytosis via M cells, where macromolecules and nanosystems are thought to be transported, despite the fact that these cells are very rare in the intestine (Fig. 1e); (3) the pinocytosis, which involves the invagination of the cellular membrane forming a vesicle, which will transport the molecules until the basolateral side of the cell and, consequently, reach the blood circulation (Fig. 1f). This latter can even be divided into four major mechanisms:

macropinocytosis, clathrin-mediated endocytosis, caveolae-mediated endocytosis, clathrin-independent endocytosis and caveolae-independent endocytosis [18].

The use of nanotechnology, as polymeric micelles, can be an advantage when formulating new oral delivery dosage forms. Chitosan is a biocompatible and biodegradable polymer with ease of forming derivatives due to its chemical constitution. These chitosan derivatives have been widely used to form polymeric micelles for oral administration due to its amphiphilic ability. In the next section, it will be explained the main chemical modifications of chitosan to produce amphiphilic polymers and, consequently, generate polymeric micelles.

3 Chitosan Modifications into an Amphiphilic Polymer

Chitosan is derived from chitin, the most abundant polymer in nature, and has called a lot of attention due to its properties. Being composed by ß-(1,4)-linked D-glucosamine and *N*-acetyl-D-glucosamine units, chitosan is known by its low toxicity [19], biodegradability [20], biocompatibility [21], antimicrobial activity [22], ability to temporarily open the tight junctions of the epithelium [16], and mucoadhesiveness [12]. Also, chitosan has an evident antitumor activity due to the positive surface charge that can neutralize the negatively charged tumor surface and thus improve the intestinal absorption [23]. Of course, all these advantages are highly influenced by the molecular weight and degree of acetylation (DA), which will affect its chemical and biological features. Moreover, chitosan solubility is dependent on its DA, ionic concentration, and pH.

Other important characteristic of chitosan is the presence of reactive groups on its backbone, which can facilitate chitosan modifications into new chitosan derivatives. The presence of C2-anime, C3-hydroxyl, and C6-hydroxyl makes possible several chemical reactions as alkylation, acylation, hydroxyalkylation, or carboxyalkylation. Here, we briefly describe the main properties of these modifications.

3.1 *Alkylation*

Alkylation can occur with both functional groups of chitosan (-NH_2 or -OH); however, the reactions using the amino groups are more frequent. This is one of the most frequent modifications of chitosan and is characterized by the grafting of alkyl chains in the chitosan backbone. Since alkylation occurs mainly in the amino groups by the reductive amination of chitosan, *N*-alkylated chitosan derivatives are generated [24]. As in other substitutions, the hydrophobic character of chitosan will depend on the length of the alkyl chain and the substitution degree. As higher the alkyl chain and/or the substitution degree, a more hydrophobic chitosan derivative will be obtained. Several works have been developed based on this reaction and, typically, new chitosan derivatives are obtained to improve chitosan properties.

Chen et al. [25] demonstrated better hemostatic activity in the *N*-alkylated chitosan as compared with unmodified chitosan. Better properties make chitosan more attractive for biomedical applications.

3.2 Acylation

In this type of derivatives, both -NH_2 and -OH groups are used in the reaction. Thus, we can have *N*-acylation or *O*-acylation, depending on the site of the grafting. This modification is characterized by the incorporation of acyl groups, which is typically a carbodiimide reaction mediated by 1-ethyl-3-(3-dimethylaminopropyl) carbodiimide (EDC) and *N*-hydroxysuccinimide (NHS) [24]. The majority of these derivatives are nanoparticles or micelles with self-assembly ability. Also, the chitosan solubility in aqueous or organic solvents is improved, depending on the molecular weight, acyl chain length, and substitution degree. Piegat et al. [26] demonstrated that chitosan grafted with linoleic acid had better antibacterial activity, compared to unmodified chitosan. Several other studies used acylation to make chitosan an amphiphilic derivative to produce polymeric micelles [6, 27–30].

3.3 Hydroxylation and Carboxylation

Both reactions are also common when modifying chitosan mainly because they improve the solubility of chitosan in water. Hydroxyalkylated chitosan is obtained by the reaction of chitosan with epoxides, while carboxyalkylated chitosan with acidic groups. The reaction may occur in both reactive groups of chitosan and, besides improving its water solubility, other biological properties are improved, as the antibacterial and antioxidant activity [31, 32].

An ideal carrier should be biocompatible and biodegradable, present no cytotoxicity, and form stable complexes able to cross the GIT and deliver the drug for an improved intestinal permeability. Chitosan is able to form those complexes through its chemical modification into an amphiphilic polymer and consequently produce polymeric micelles. In the next section, it will be discussed in detail the main advantages of using polymeric micelles and its main properties.

4 Polymeric Micelles as Delivery Platform

Polymeric micelles are known by the core–shell structure formed spontaneously by amphiphilic copolymers. The self-assembly process is due to the gain of entropy of the solvent molecules since the hydrophobic segments of the polymer aggregate and leave the aqueous medium [33]. The micelle core is hydrophobic and is covered by a

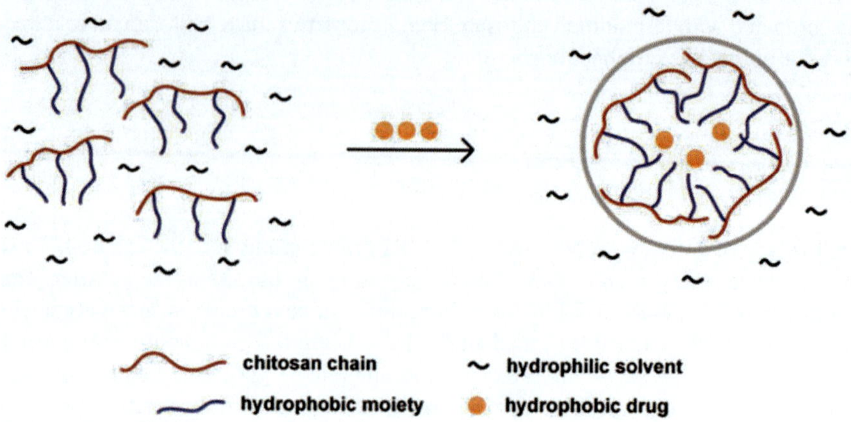

Fig. 2 Schematic representation of hydrophobically modified chitosan self-assembly. Aggregates can entrap hydrophobic drugs in their hydrophobic core. Reprinted with permission from [34]

hydrophilic corona, forming the core–shell structure (Fig. 2). In aqueous solution, the micelle core is used as a reservoir for hydrophobic biomolecules, while the shell offers the colloidal stability of these systems. The payloads of these systems can range from drugs, proteins, peptides, DNA, or siRNA. However, in this chapter, we will focus exclusively on hydrophobic anticancer drugs.

One important characteristic of the polymeric micelles is the critical micelle concentration (CMC) that is the concentration of the polymer above which it is possible to form micelles spontaneously [35]. The lower the CMC value, the more stable the micelles are. Imagining in in vivo scenario, low CMC prevents micelle dissociation upon dilution in the biological fluids and blood circulation, as the case of oral administration. Actually, the balance between hydrophilic/hydrophobic moieties will dictate several properties as the CMC, stability, solubility, or encapsulation efficiency. Other important parameter to take into consideration is the micelle production method. There are five different main ways of producing them as the direct dissolution, the evaporation solvent, dialysis, freeze-drying, or the oil-in-water emulsion [36]. The production method is crucial for having a good drug loading and encapsulation efficiency and is determinant in the final micelle size.

It is possible to enumerate several advantages of using polymeric micelles as a drug delivery vehicle. Polymeric micelles can improve the aqueous solubility of hydrophobic drugs due to the hydrophobic core; provide a controlled deliver due to the polymer used in the micellar structure and target the deliver by using specific ligands to the cancer cells receptors; can protect the drug from the biological fluids improving its stability; can encapsulate a great amount of drug, increasing the drug dose that reaches the targeted cells and thus promoting a higher treatment efficacy; can decrease the side effects by the fact of the drug be encapsulated through the GIT and be specifically delivered; and have longer life-time and lower aggregation concentration, which means good thermodynamic stability, better than those low-molecular weight surfactants [33, 37, 38].

There are several copolymers very often used for micelles production as the case of the biodegradable poly(lactic acid) (PLA), poly(glycolic acid) (PGA), poly(lactic-co-glycolic acid) (PLGA) and poly(ε-caprolactone) (PCL), and non-degradable copolymers of poly(ethylene oxide) (PEO), poly(ethylene glycol) (PEG), and poly (propylene oxide) (PPO). However, natural polymers as chitosan and chitosan derivatives have gained much attention on micelles production for delivery of hydrophobic anticancer drugs. As previously mentioned, chitosan is mucoadhesive and, when talking about oral formulations, it is great if we can use mucoadhesive polymers to enhance the adhesion to the intestinal epithelium and thus improve the residence time of the system for a better therapeutic outcome.

Although a significant number of polymeric micelles are in clinical trials, as the case of NK012 micelles, based on PEG-poly(L-glutamic acid)-SN-38 conjugated [39, 40], NK105 micelles, based on PEG-poly(L-aspartic acid) loaded with paclitaxel (PTX) [41] or IT-101, based on polymer–cyclodextrin–camptothecin conjugated [42, 43], there is no micellar system based on chitosan at this phase. However, in the next section, we describe the latest polymeric systems based on chitosan to deliver anticancer hydrophobic drugs with promising results for oral chemotherapy.

5 Hydrophobic Anticancer Drugs Loaded into Polymeric Micelles

Chemotherapy is, in most of the cases, crucial for cancer treatment. There are several drugs effective against cancer; however, they have low solubility and low intestinal permeability, which turns the treatment ineffective. Moreover, these drugs are cytotoxic for normal cells, despite being efficient in stopping the cell division process. Thus, the scientific community started to think about adjuvants or other forms of delivery that could overcome these drawbacks. Actually, the improvements of oral dosage forms are still the major concerns of oral chemotherapy [44].

According to the biopharmaceutical classification system (BCS), most of the hydrophobic anticancer drugs are classified as BCS 4, which means low solubility and low permeability. Thus, if administered orally, these drugs will present low bioavailability. In that regard, polymeric micelles appeared as a promising strategy to efficiently delivery anticancer drugs, particularly hydrophobic drugs. Moreover, chitosan-based micelles have been frequently used for oral delivery of anticancer drugs due to its capability to protect the drug through the GIT, enhancing the solubility and absorption of poor water-soluble drugs. Several hydrophobic anticancer drugs encapsulated or associated with polymeric micelles based on chitosan for cancer treatment will be presented in this section.

5.1 Paclitaxel Delivery

PTX is a well-known anticancer drug commonly used in chitosan-derived systems. Its mechanism of action influences the cell cycle through the hyperstabilization of the microtubules, which ceases the mitosis until cell dead [45]. PTX is approved by U.S. Food and Drug Administration (FDA) for the treatment of several types of cancer such as breast, ovarian, and lung cancer, and it is classified by the BCS as class IV drug, i.e., poorly soluble and poorly permeable. Thus, PTX is commercially formulated as Taxol™ for intravenous administration; however, its excipients lead to serious side effects as neurotoxicity, nephrotoxicity, and hypersensitivity [46].

Other alternatives using nanomedicine are being applied to overcome these drawbacks. This is the case of Abraxane®, albumin nanoparticle-based system FDA approved, Lipusu®, a liposome system approved by the State Food and Drug Administration of China, Paclical®, a micellar system in clinical phase III, or even Genexol-PM®, already in the market. However, all these nanosystems are being tested for intravenous administration, which is not the ideal route of administration due to the drug instability in the bloodstream and lack of specificity, the high probability of infections or thrombosis, and also, the high cost of these formulations for the patients [47]. Thus, there is an urgent need in developing new oral drug delivery systems with high efficacy and low toxicity and cost for the patients.

Almeida et al. developed a new chitosan-based system through the introduction of PCL into the chitosan backbone [27]. It is still a preliminary work, but the authors found a twofold increase in the in vitro intestinal permeability of PTX compared with the free drug. In a similar work, chitosan was modified with myristoyl chloride as hydrophobic segment to produce micelles. PTX was efficiently encapsulated into the system and showed no cytotoxicity against colorectal cancer cells, just as the PTX permeability that was doubly increased compared to the free drug [28].

In other study, Xu et al. used Pluronic F127-chitosan copolymer modified with cysteine and sodium cholate to prepare polyion complex micelles. The in vivo pharmacokinetic study in rats revealed an increase of the oral bioavailability of PTX from around 9% to 42%, when the drug is administered freely or encapsulated in the micelles, respectively. Moreover, it was observed an area under the concentration–time curve (AUC) fivefold higher and a maximal plasma concentration (C_{max}) threefold higher when the drug was administered encapsulated into the micelles compared to the free drug [46]. The authors attributed these results to the cysteine modified micelles, which can improve the mucoadhesiveness and so increase the contact of the micelles with the intestinal mucosa and consequently provide a higher drug concentration. More recently, chitosan hydrophobically modified with stearic acid and conjugated with L-carnitine for the targeting of the intestinal organic cation/carnitine transporter 2 (OCTN2) was successfully developed and PTX encapsulated with a high drug loading (16%). The pharmacokinetic studies performed in rats demonstrated twofold increase of the C_{max} and a relative oral bioavailability of 166% of PTX encapsulated into the micelles, compared to the commercial form of PTX, Taxol™ [48]. No differences in the C_{max} were observed for the micelles without conjugation with L-carnitine and a relative oral bioavailability of 84% was observed, which suggest the active targeting of the micelles and transport by the OCTN2 had great impact in the increase of the oral bioavailability of PTX. Also, internalization studies performed in Caco-2 cells using coumarin-6 as model drug exhibited a significantly higher uptake by the micelles conjugated with L-carnitine, compared to the freely coumarin-6 or the micelles without L-carnitine conjugation [48].

Recently, chitosan was modified into a multifunctional amphiphilic polymer using polylysine and cysteine. The in vitro antitumor efficacy studies performed in Caco-2 cells showed higher cytotoxicity after 48 h of incubation for the micelles, compared with Taxol, and higher cellular uptake compared with non-conjugated micelles. Furthermore, the in vivo data revealed a sixfold higher AUC and oral absorption of PTX by the micelles, ninefold increase of PTX in the tumor, and twofold tumor reduction, when comparing with the free drug [49].

In another study, trimethyl chitosan (TMC) was conjugated with PTX (TMC-PTX) and folic acid (FA-TMC-PTX) and tested for oral and intravenous administration, containing approximately 11% PTX. An increase in the oral absorption of PTX was firstly observed in vitro, where both systems presented a twofold increase on the intestinal mucoadhesion, and a 16- and 19-fold increase of PTX intestinal permeability for TMC-PTX and FA-TMC-PTX micelles, respectively, compared to PTX solution [4]. The in vivo studies demonstrated that both micellar systems had an impact in the tumor suppression, where FA-TMC-PTX was the most efficient in both routes of administration (Fig. 3). The animals body weight was maintained during the study, and survival experiment showed a significantly increase in the life span for the animals treated with both micellar systems, compared with saline or PTX. Thus, both systems showed promising results for cancer treatment by oral or intravenous administration.

Later, Wang et al. used the same tumor model to assess the efficacy of the carboxymethyl chitosan derivative micelles (200 nm), highly encapsulated with PTX with a drug loading of 35%. These micelles were able to increase the AUC and the C_{max} of 16 and 26-fold, respectively, compared with Taxol™, and also, the animals treated with micelles showed an antitumor inhibition effect around 46%, twofold higher compared with the free drug [5].

A summary of the chitosan derivatives used to encapsulate and deliver paclitaxel is depicted in Table 1.

5.2 *Curcumin Delivery*

Curcumin (CUR) is a natural active drug with applicability in several fields due to its antioxidant, antidepressant, anti-inflammatory, antimicrobial, chemopreventive, and anticancer properties [50, 51]. Its mechanism of action is related with the cell signaling pathway as cell cycle progression inhibition, reduction of angiogenesis and cell proliferation, and inducing cell apoptosis [52, 53]. CUR is known for being

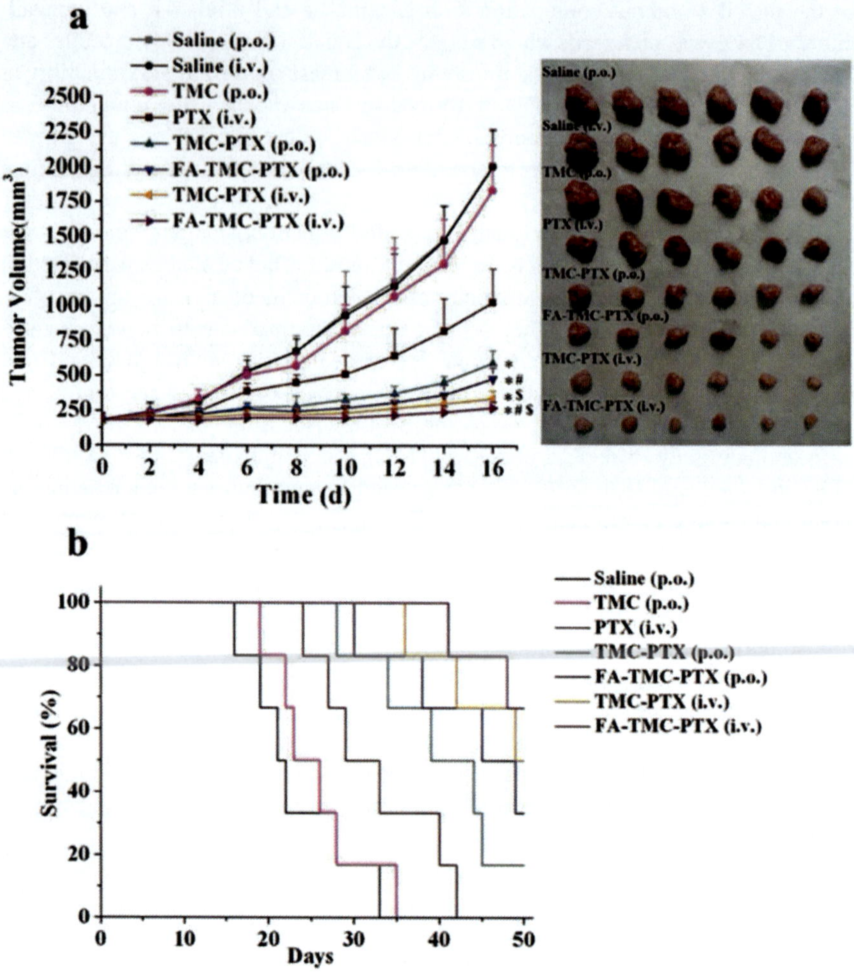

Fig. 3 Antitumor efficacy of PTX, TMC-PTX, and FA-TMC-PTX on H22 tumor-bearing mice. Five administrations were conducted every alternate day. (a) Tumor growth curves following i.v. or p.o. administration of saline and distinct PTX formulations. Indicated values were mean $\pm$ SD ($n = 6$). $^{*}P < 0.05$ versus PTX (i.v.); $^{\#}P < 0.05$ versus TMC-PTX via the same administration route; \$ $P < 0.05$ versus p.o. administration. (b) Survival curves of H22 tumor-bearing mice receiving i.v. or p.o. administration of saline and distinct PTX formulations ($n = 6$). Reprinted with permission from [4]

listed on the FDA's GRAS (generally regarded as safe) since an oral dose of 12 g was reported as tolerable in human clinical trials [53]. However, as hydrophobic drug, its oral availability (<1%) is hampered by the low aqueous solubility and intestinal absorption. Therefore, several nanoformulations are being explored in order to improve the oral availability of CUR and thus improve the therapeutic efficacy of the drug.

Table 1 Summary of the chitosan derivatives used to encapsulate and deliver paclitaxel

System	Average size (nm)	CMC	Loading efficiency	Evaluation	References
Chitosan-*g*-polycaprolactone	~400	9.55×10^{-4} mg/mL	82%	In vitro	[27]
3,6-*O*,*O*′-dimyristoyl chitosan	~480	8.90×10^{-3} mg/mL	100%	In vitro	[28]
F127-chitosan modified with cysteine and bile salt	~60	2.50×10^{-3} mol/L	85%	In vitro and in vivo	[46]
L-carnitine conjugated chitosan-stearic acid	~160	14.31 μg/mL	84%	In vitro and in vivo	[48]
Polylysine and cysteine functionalized chitosan	~165	–	91%	In vitro and in vivo	[49]
Trimethyl chitosan and folic acid	~175	0.08 mg/mL	(Drug loading of 11%)	In vitro and in vivo	[4]
Carboxymethyl chitosan-rhein	~217	31.25 μg/mL	87%	In vitro and in vivo	[5]

In a work developed by Silva et al., chitosan was modified with myristoyl chloride to obtain an amphiphilic polymer with self-assembly properties. CUR was efficiently encapsulated into micelles and no cytotoxicity was found in colorectal cancer cells, demonstrating to be a safe carrier of the drug [29]. Regarding its intestinal permeability, CUR encapsulated into chitosan micelles showed a significant increase in the intestinal permeability, compared with the free drug. The authors believe the permeability enhancement was due to the mucoadhesive character of chitosan and its ability to open the tight junctions of the epithelium [29]. In another study developed by Woraphatphadung et al., polymeric micelles based on a new chitosan derivative with pH-sensitive properties for CUR delivery in the colon was achieved. The CUR release in the simulated gastric fluid (SGF) was relatively low around 20%, increasing to 50–55% in the simulated intestinal fluid (SIF) and 60–70% in the simulated colonic fluid (SCF), which, compared to the free drug, was significantly higher (~20%). Also, micelles showed higher cell inhibition (IC_{50} 6.18 ± 0.18 μg/mL) than the free drug (IC_{50} 11.38 ± 3.07 μg/mL) [54]. In the same year, chitosan was modified with arginine and thiolated fucoidan aiming to improve the tight junctions opening ability and to enhance the mucoadhesion of the micelles and permeation of the drug through the P-gp inhibition. Data showed that micelles were able to increase the apparent permeability coefficient (Papp) of CUR (2×10^{-6}) compared with the free drug (1×10^{-7}), as well as, micelles loaded with CUR were able to significantly inhibit the growth of Caco-2 cells [55]. A similar work developed by Raja et al. mentioned that chitosan was modified with

Table 2 Summary of the chitosan derivatives used to encapsulate and deliver curcumin

System	Average size (nm)	CMC	Loading efficiency (%)	Evaluation	References
Ammonium myristoyl chitosan	~372	9.38×10^{-6} mg/mL	69	In vitro	[29]
Arginine-modified chitosan and thiolated fucoidan	~167	–	62	In vitro	[55]
N-naphthyl-*N*,*O*-succinyl chitosan	~321	0.07 mg/mL	30	In vitro	[54]
Chitosan modified with acrylonitrile and arginine	~199	–	76	In vitro and in vivo	[56]

acrylonitrile and arginine to develop a system with improved mucoadhesion and solubility for CUR delivery. CUR was highly encapsulated into the system and its aqueous solubility was significantly enlarged in comparison with the free drug, and also a mucoadhesion enhancement was observed due to the arginine and acrylonitrile incorporation as well as due to the use of chitosan.

A summary of the chitosan derivatives used to encapsulate and deliver curcumin is depicted in Table 2.

5.3 *Other Hydrophobic Anticancer Drugs*

Camptothecin (CPT) is an anticancer drug with strong activity against several types of tumors. Its mechanism of action is based on the inhibition of the nuclear enzyme topoisomerase I, ending cell division, leading to apoptosis [57]. Despite being very powerful on stopping cell division, its therapeutic efficacy is hampered by the low solubility and bioavailability and due to the high toxicity. Also, CPT is pH-dependent, which can suffer hydrolysis in aqueous media easily. As a result of being encapsulated in polymeric micelles, drugs can be protected from biological fluids and delivered to the target site in higher doses for better treatment.

Silva et al. [2] developed a new chitosan amphiphilic derivative based on 3,6-*O*,*O′*-dimyristoyl to encapsulate and efficiently deliver CPT. Empty micelles presented a safe profile against colorectal cancer cell lines and CPT-loaded micelles exhibited a low release rate, showing no cytotoxic effect against the same cell lines. However, when tested the CPT permeability, polymeric micelles showed to improve sevenfold the CPT transport across Caco-2/HT29-MTX co-culture model, result that can be explained by the ability of chitosan in opening the tight junctions of the epithelium [2]. Also, in preliminary study, Almeida et al. modified chitosan with *O*-methyl-*O′*-succinylpolyethylene glycol and oleic acid to produce polymeric micelles to encapsulate CPT aiming to improve its solubility and bioavailability as potential candidate for oral chemotherapy [30].

Doxorubicin (DOX) has been used against several types of cancer as chemotherapy. However, DOX can be easily pumped out of the cells due to the P-gp efflux pump. Thus, being encapsulated in polymeric micelles can avoid efflux transport and improve oral absorption.

Stearic acid-*g*-chitosan was used to encapsulate and improve the oral availability of DOX through the formation of polymeric micelles by self-assembly [6]. This system revealed good permeability as compared with free DOX, increasing the transport in the Caco-2 monoculture model from 20- to 40-fold, depending on the SD. More interesting, in vivo studies performed in rats demonstrated a threefold improvement in the oral availability of DOX encapsulated into micelles compared with free DOX and an AUC 4-times superior. Also, these micelles had an elimination half-life of 11 h, much more prolonged than the control, which may favor the cancer treatment [6]. In another study, Mu et al. [58] used quercetin-chitosan conjugated (QT-CS) to produce polymeric micelles and improve the oral bioavailability and solubility od DOX. Results showed a safe profile for the polymeric micelles against Caco-2 cells and a twofold higher DOX uptake from the loaded micelles, as compared with free DOX. The apparent permeability coefficient (Papp) performed in Caco-2 monolayers was tenfold higher for the DOX-loaded chitosan micelles than for free DOX, demonstrating the effect of chitosan opening the tight junctions of the epithelium and thus increasing the intestinal permeability.

Docetaxel (DTX) is an antimicrotubular agent approved for the treatment of several types of cancer. However, this drug is administered intravenously and no oral form is available until now. Kumar et al. [59] recently developed a new chitosan amphiphilic derivative through the grafting of oleic acid on carboxymethyl chitosan backbone for the delivery of DTX. Caco-2 model was used to investigate the Papp, which revealed to be up to sixfold higher for the DTX-loaded micelles than the free DTX. The authors believe this increase on the intestinal permeability was achieved by paracellular transport, mainly due to the use of chitosan to open the tight junctions of the epithelium. These micelles also demonstrated to be stable in gastrointestinal fluids, where their size and PdI were nearly maintained. Moreover, the in vivo pharmacokinetic study after oral administration revealed an almost twofold and threefold increase on the C_{max} and AUC, respectively, for the DTX polymeric micelles as compared with the free DTX, leading to an increase of DTX oral absorption.

A summary of the chitosan derivatives used to encapsulate and deliver camptothecin, doxorubicin, and docetaxel is depicted in Table 3.

Table 3 Summary of the chitosan derivatives used to encapsulate and deliver camptothecin, doxorubicin, and docetaxel

Drug	System	Average size (nm)	CMC	Loading efficiency (%)	Evaluation	References
Camptothecin	3,6-*O*,*O*′-dimyristoyl chitosan	~290	–	70	In vitro	[2]
	O-methyl-O′-succinylpolyethylene glycol and oleic acid-*g*-chitosan	~146	0.08 mg/mL	78	In vitro	[30]
Doxorubicin	Stearic acid-*g*-chitosan	33.4–130.9	0.16 to 0.25 mg/mL	67–76	In vitro and in vivo	[6]
	Quercetin–chitosan conjugated	~137	0.03 mg/mL	88	In vitro	[58]
Docetaxel	Oleic acid-*g*-carboxymethyl chitosan	~213	1.00 μg/mL	57	In vitro and in vivo	[59]

6 Conclusions

The advances of nanotechnology have enabled the delivery of anticancer drugs or chemotherapeutic agents through nanoparticles, micelles, or liposomes. These transport vehicles and, especially detailed here, polymeric micelles based on chitosan have a great potential since they can potentially overcome some of the biological and physiological barriers of the gastrointestinal tract as well as reduce systemic toxicity and improve the solubility and bioavailability of many anticancer drugs. Particular attention was given to chitosan, a biocompatible and biodegradable polymer with mucoadhesive properties and ability to temporarily open the tight junctions of the epithelium for the production of polymeric micelles. Thus, this chapter summarizes the main limitations of the oral administration and identifies various anticancer therapeutic agents using polymeric micellar systems as a carrier vehicle to delivery and improve the therapeutic efficacy and safety of this drugs.

Acknowledgments This work was financed by FCT – Fundação para a Ciência e a Tecnologia/ Ministério da Ciência, Tecnologia e Ensino Superior in the framework of the project "Institute for Research and Innovation in Health Sciences" (UID/BIM/04293/2019). Andreia Almeida (SFRH/ BD/118721/2016) acknowledges Fundação para a Ciência e a Tecnologia (FCT), Portugal, for financial support.

Declaration of Interest The authors declare that they have no known competing financial interests or personal relationships that could have appeared to influence the work reported in this work.

Credit Authorship Contribution Statement Andreia Almeida: Conceptualization, Writing – original draft; Bruno Sarmento: Writing – review & editing.

References

1. Organization W.W.H. (2021) Cancer. [cited 2021]
2. Silva DS et al (2017) Self-aggregates of 3, 6-O,O′-dimyristoylchitosan derivative are effective in enhancing the solubility and intestinal permeability of camptothecin. Carbohydr Polym 177:178–186
3. Khafagy E-S, Morishita M (2012) Oral biodrug delivery using cell-penetrating peptide. Adv Drug Deliv Rev 64(6):531–539
4. He R, Yin C (2017) Trimethyl chitosan based conjugates for oral and intravenous delivery of paclitaxel. Acta Biomater 53:355–366
5. Wang X et al (2019) Preparation and evaluation of carboxymethyl chitosan-rhein polymeric micelles with synergistic antitumor effect for oral delivery of paclitaxel. Carbohydr Polym 206:121–131
6. Yuan H et al (2011) Stearic acid-g-chitosan polymeric micelle for oral drug delivery: in vitro transport and in vivo absorption. Mol Pharm 8(1):225–238
7. Sood A, Panchagnula R (2001) Peroral route: an opportunity for protein and peptide drug delivery. Chem Rev 101(11):3275–3304
8. Dahmani FZ et al (2012) Enhanced oral bioavailability of paclitaxel in pluronic/LHR mixed polymeric micelles: preparation, in vitro and in vivo evaluation. Eur J Pharm Sci 47(1):179–189

9. Pond SM, Tozer TN (1984) First-pass elimination basic concepts and clinical consequences. Clin Pharmacokinet 9(1):1–25
10. Thanki K et al (2013) Oral delivery of anticancer drugs: challenges and opportunities. J Control Release 170(1):15–40
11. Ensign LM, Cone R, Hanes J (2012) Oral drug delivery with polymeric nanoparticles: the gastrointestinal mucus barriers. Adv Drug Deliv Rev 64(6):557–570
12. Palazzo C et al (2017) Mucoadhesive properties of low molecular weight chitosan-or glycol chitosan-and corresponding thiomer-coated poly (isobutylcyanoacrylate) core-shell nanoparticles. Eur J Pharm Biopharm 117:315–323
13. des Rieux A et al (2006) Nanoparticles as potential oral delivery systems of proteins and vaccines: a mechanistic approach. J Control Release 116(1):1–27
14. Kammona O, Kiparissides C (2012) Recent advances in nanocarrier-based mucosal delivery of biomolecules. J Control Release 161(3):781–794
15. Su F-Y et al (2012) Protease inhibition and absorption enhancement by functional nanoparticles for effective oral insulin delivery. Biomaterials 33(9):2801–2811
16. Wang J et al (2017) Mechanism of surface charge triggered intestinal epithelial tight junction opening upon chitosan nanoparticles for insulin oral delivery. Carbohydr Polym 157:596–602
17. Shakweh M, Ponchel G, Fattal E (2004) Particle uptake by Peyer's patches: a pathway for drug and vaccine delivery. Expert Opin Drug Deliv 1(1):141–163
18. Conner SD, Schmid SL (2003) Regulated portals of entry into the cell. Nature 422(6927):37–44
19. Yu H et al (2021) A novel nanohybrid antimicrobial based on chitosan nanoparticles and antimicrobial peptide microcin J25 with low toxicity. Carbohydr Polym 253:117309
20. Yu Z et al (2018) Silica in situ enhanced PVA/chitosan biodegradable films for food packages. Carbohydr Polym 184:214–220
21. Dimida S et al (2017) Effects of genipin concentration on cross-linked chitosan scaffolds for bone tissue engineering: structural characterization and evidence of biocompatibility features. Int J Polymer Sci 2017
22. Sahariah P, Masson M (2017) Antimicrobial chitosan and chitosan derivatives: a review of the structure–activity relationship. Biomacromolecules 18(11):3846–3868
23. Trickler W, Nagvekar A, Dash AK (2008) A novel nanoparticle formulation for sustained paclitaxel delivery. AAPS PharmSciTech 9(2):486–493
24. Silva D, Almeida A, Azevedo C, Campana-Filho SP, Sarmentob B (2018) Synthesis and applications of amphiphilic chitosan derivatives for drug delivery applications. In: Nanoparticles in life sciences and biomedicine, p 45
25. Chen Z et al (2017) Blood coagulation evaluation of N-alkylated chitosan. Carbohydr Polym 173:259–268
26. Piegat A et al (2021) Antibacterial activity of N,O-acylated chitosan derivative. Polymers 13 (1):107
27. Almeida A et al (2018) Synthesis and characterization of chitosan-grafted-polycaprolactone micelles for modulate intestinal paclitaxel delivery. Drug Deliv Transl Res 8(2):387–397
28. Silva DS et al (2017) Synthesis and characterization of 3, 6-O,O′-dimyristoyl chitosan micelles for oral delivery of paclitaxel. Colloids Surf B Biointerfaces 152:220–228
29. Silva DS et al (2018) N-(2-hydroxy)-propyl-3-trimethylammonium, o-mysristoyl chitosan enhances the solubility and intestinal permeability of anticancer curcumin. Pharmaceutics 10 (4):245
30. Almeida A et al (2020) Novel amphiphilic chitosan micelles as carriers for hydrophobic anticancer drugs. Mater Sci Eng C:110920
31. Mi Y et al (2020) New synthetic chitosan derivatives bearing benzenoid/heterocyclic moieties with enhanced antioxidant and antifungal activities. Carbohydr Polym 249:116847
32. Zhao D et al (2018) Biomedical applications of chitosan and its derivative nanoparticles. Polymers 10(4):462

33. Bromberg L (2001) Hydrophobically modified polyelectrolytes and polyelectrolyte block copolymers. In: Handbook of surfaces and interfaces of materials. Elsevier, Amsterdam, pp 369–404
34. Quiñones JP, Peniche H, Peniche C (2018) Chitosan based self-assembled nanoparticles in drug delivery. Polymers 10(3):235
35. Deshmukh AS et al (2017) Polymeric micelles: basic research to clinical practice. Int J Pharm 532(1):249–268
36. Simões SM et al (2015) Polymeric micelles for oral drug administration enabling locoregional and systemic treatments. Expert Opin Drug Deliv 12(2):297–318
37. Gaucher G et al (2005) Block copolymer micelles: preparation, characterization and application in drug delivery. J Control Release 109(1–3):169–188
38. Bromberg L (2008) Polymeric micelles in oral chemotherapy. J Control Release 128(2):99–112
39. Raefsky E, Spigel DR, Infante JR, Bendell JC, Jones SF, Lipman AJ, Trent D, Kawamura S, Greco FA, Hainsworth JD, Burris HA (2011) Phase II study of NK012 in relapsed small cell lung cancer. J Clin Oncol 29(15_suppl):7079–7079
40. Schneider M et al (2016) A paired comparison between glioblastoma "stem cells" and differentiated cells. Int J Cancer 138(7):1709–1718
41. Fujiwara Y, Mukai H, Saeki T, Ro J, Lin YC, Nagai SE, Lee KS, Watanabe J, Ohtani S, Kim SB, Kuroi K, Tsugawa K, Tokuda Y, Iwata H, Park YH, Yang Y, Nambu Y (2019) A multinational, randomised, open-label, parallel, phase III non-inferiority study comparing NK105 and paclitaxel in metastatic or recurrent breast cancer patients. Br J Cancer 120(5):475–480
42. Schluep T, Hwang J, Cheng J, Heidel JD, Bartlett DW, Hollister B, Davis ME (2006) Preclinical efficacy of the camptothecin-polymer conjugate IT-101 in multiple cancer models. Clin Cancer Res 12(5):1606–1614
43. Schluep T, Cheng J, Khin KT, Davis ME (2006) Pharmacokinetics and biodistribution of the camptothecin–polymer conjugate IT-101 in rats and tumor-bearing mice. Cancer Chemother Pharmacol 57(5):654–662
44. Feng S-S, Chien S (2003) Chemotherapeutic engineering: application and further development of chemical engineering principles for chemotherapy of cancer and other diseases. Chem Eng Sci 58(18):4087–4114
45. Spencer CM, Faulds D (1994) Paclitaxel. Drugs 48(5):794–847
46. Xu W et al (2015) Cysteine modified and bile salt based micelles: preparation and application as an oral delivery system for paclitaxel. Colloids Surf B Biointerfaces 128:165–171
47. Fonte P et al (2015) Polymer-based nanoparticles for oral insulin delivery: revisited approaches. Biotechnol Adv 33(6):1342–1354
48. Yang T et al (2020) L-carnitine conjugated chitosan-stearic acid polymeric micelles for improving the oral bioavailability of paclitaxel. Drug Deliv 27(1):575–584
49. Du X et al (2020) Polylysine and cysteine functionalized chitosan nanoparticle as an efficient platform for oral delivery of paclitaxel. Carbohydr Polym 229:115484
50. Maheshwari RK et al (2006) Multiple biological activities of curcumin: a short review. Life Sci 78(18):2081–2087
51. Lopresti AL, Hood SD, Drummond PD (2012) Multiple antidepressant potential modes of action of curcumin: a review of its anti-inflammatory, monoaminergic, antioxidant, immune-modulating and neuroprotective effects. J Psychopharmacol 26(12):1512–1524
52. Guo LD et al (2013) Curcumin inhibits proliferation and induces apoptosis of human colorectal cancer cells by activating the mitochondria apoptotic pathway. Phytother Res 27(3):422–430
53. Aggarwal BB, Kumar A, Bharti AC (2003) Anticancer potential of curcumin: preclinical and clinical studies. Anticancer Res 23(1/A):363–398
54. Woraphatphadung T et al (2018) Development of chitosan-based pH-sensitive polymeric micelles containing curcumin for colon-targeted drug delivery. AAPS PharmSciTech 19 (3):991–1000

55. Chen C-H et al (2018) Mutlifunctional nanoparticles prepared from arginine-modified chitosan and thiolated fucoidan for oral delivery of hydrophobic and hydrophilic drugs. Carbohydr Polym 193:163–172
56. Raja MA et al (2016) Self-assembled nanoparticles based on amphiphilic chitosan derivative and arginine for oral curcumin delivery. Int J Nanomedicine 11:4397
57. Liu LF et al (2000) Mechanism of action of camptothecin. Ann N Y Acad Sci 922(1):1–10
58. Mu Y et al (2019) Multifunctional quercetin conjugated chitosan nano-micelles with P-gp inhibition and permeation enhancement of anticancer drug. Carbohydr Polym 203:10–18
59. Kumar R et al (2020) Polymeric micelles based on amphiphilic oleic acid modified carboxymethyl chitosan for oral drug delivery of bcs class iv compound: intestinal permeability and pharmacokinetic evaluation. Eur J Pharm Sci 153:105466

Adv Polym Sci (2021) 288: 271–292
https://doi.org/10.1007/12_2021_96

Published online: 6 May 2021

Chitosan-Based Theranostics for Cancer Therapy

A. S. Soubhagya and M. Prabaharan

Contents

Abstract Chitosan, a natural-based cationic polysaccharide, has a great potential to be utilized as drug delivery systems, tissue engineering scaffolds, and wound dressings due to its biocompatibility, bioactivity, biodegradability, antibacterial property, gelling behavior, cell adhesion, and proliferation abilities. Due to the presence of primary amino and hydroxyl groups, chitosan can be chemically modified or functionalized with other bioactive molecules easily to improve its physico-chemical and biological properties required for advanced biomedical applications. In this context, considerable efforts have been made to conjugate chitosan and its derivatives with different types of photosensitizers/photothermal agents, quantum dots (QDs), bioactive molecules, metals, and metal oxides to develop theranostic agents for concurrent imaging and treatment of tumors. The chitosan-based theranostics were found to have better cellular imaging capability, tumor-targeted drug release, and multimodal therapeutic efficiencies. This chapter reviews the recent progress of chitosan-based theranostics, their properties, and applications in advanced cancer therapy.

A. S. Soubhagya and M. Prabaharan (✉)
Department of Chemistry, Hindustan Institute of Technology and Science, Chennai, India

Keywords Bioactive molecules · Cancer therapy · Chitosan · Quantum dots · Theranostic agents

1 Introduction

Cancer is a most divergent and lethal disorder that remains the second-leading cause of mortality worldwide [1]. The most commonly used cancer treatment methods are surgery, radiation therapy, and chemotherapy. These methods can be utilized individually or in a combined manner based on the type, stage, and location of cancer present in the body and the age, general health, and other factors of the patient. Although these methods have experienced many advancements in recent years, still their practical applications are not completely satisfied to suppress the cancer cells more effectively with minimum risk factors. Most of these treatment methods are non-specific to deliver the drug in the tumor site and hence show cytotoxicity to the normal cells and poor drug biodistribution in the diseased area. Moreover, these approaches are generally not sufficient to overcome biological barriers, to treat metastatic disease, and for monitoring, treating, and imaging cancer simultaneously [2].

In recent years, nanotechnology has been regarded as a promising tactic to deal with cancer and has been broadly exploited to expand the traditional cancer treatment methods [3, 4]. Nanotechnology plays a vital role in developing the theranostics for diagnosis, early detection, monitoring, and therapy of cancer [5]. These theranostics have therapeutic, diagnostic, and imaging properties. Moreover, they can carry an additional payload of drugs. Because of the large surface area to volume ratio, theranostic nanoparticles (NPs) can comprise various targeting ligands that provide specificity and high affinity for target cells. Due to the smaller size (<200 nm), theranostic NPs can preferentially accumulate in the cancer cells due to the enhanced permeation and retention (EPR) effect and may exhibit long blood circulation effects by avoiding recognition by the reticuloendothelial system (RES). Therefore, theranostic NPs with adequate drug loading and releasing ability would considerably enhance the efficacy of cancer therapy and potentially overcome the drawbacks related to conventional treatment methods.

So far, different types of NPs based on polymeric materials, ceramics, carbon nanotubes, and inorganic materials have been established as theranostics for cancer diagnosis and therapy [6]. Using surface modification, these NPs have been conjugated with a hydrophobic anticancer drug to improve its bioavailability. Moreover, they have been coated with hydrophilic and biocompatible polymers to improve their circulation in the bloodstream for a longer period and avoid recognition by the RES [7]. The theranostics NPs with a suitable size and surface modification can pass through the blood–brain barrier (BBB) and deliver therapeutic concentration of drugs in the brain for the treatment of central nervous system cancers [8]. Moreover, these NPs may have the benefits of overcoming multidrug resistance [9].

Chitosan is a cationic biopolymer, which is derived from chitin by the alkaline deacetylation process. In recent years, chitosan has been extensively considered as a promising biomaterial for the development of advanced theranostics for cancer therapy due to its desired characteristics such as biocompatibility, biodegradability, gel-forming ability, the capability to conjugate imaging agents, biomarkers, ligands, and therapeutic agents. The chitosan-based NPs with imaging, targeting, and therapeutic efficiencies have been proved to be a potential material for cancer theranostics. In this chapter, the recent improvements of chitosan-based theranostics such as glycol chitosan loaded with photosensitizers/photothermal agents, chitosan loaded with QDs, chitosan-noble metal conjugates, chitosan-based magnetic NPs, chitosan-based hybrid NPs, and chitosan-multimodal nanocomposites in cancer therapy have been reviewed in detail.

2 Theranostics Based on Chitosan and Its Derivatives

An ideal theranostic material must have a diagnosis, imaging, and therapeutic abilities instantaneously. A variety of multifunctional NPs containing therapeutic molecules, imaging agents, and targeting ligand have been developed for the early detection and inhibition of tumors. Due to the occurrence of targeting molecule, such NPs can reach the target-site more specifically and release the drug in a controlled manner and thereby avoid the unwanted side effects. The imaging agent loaded with NPs can be utilized to image the tumor cells and thereby monitor the progress of the treatment. In recent years, the NPs based on polymers, noble metals, metal oxides, and ceramics have been largely considered as theranostics for the simultaneous imaging and therapy of various diseases [10, 11]. Among the polymers, chitosan-based materials have been shown the potential to be used for the developments of advanced theranostic systems with various other functional materials for the diagnosis, imaging, and therapy of cancer because of their favorable characteristics such as biocompatibility, nontoxicity, biodegradability, bioactivity, non-toxicity, and lower adverse effects [12].

2.1 *Chitosan Loaded with Photosensitizers/Photothermal Agents*

Photodynamic therapy (PDT) is one of the important approaches for cancer treatment, where photosensitizers play a major role to create reactive oxygen species (ROS), namely singlet oxygen and hydroxyl radicals in presence of light with a suitable wavelength [13]. These ROS can kill various biological objects in the surrounding medium and hence are measured as a promising candidate for the abolition of tumor cells. The photosensitizers accumulated in the tumor sites can also produce an intense fluorescence signal under light irradiation, which can be used as imaging agents for imaging and diagnostic applications [14]. However, the usage of photosensitizers as a therapeutic agent is highly limited due to their nonspecific phototoxicity, less water solubility,and ineffective tumor-targeting ability. The limitations of photosensitizers can be overcome by encapsulating them into chitosan-based tumor-targeted drug delivery carriers.

Glycol chitosan is a water-soluble derivative of chitosan. It is prepared by reacting chitosan with hydrophobic glycol molecules. Due to its amphiphilic nature, glycol chitosan can be self-assembled to NPs for the delivery of therapeutic and imaging agents in the tumor site. The glycol chitosan NPs loaded with photosensitizer and anti-cancer drugs have been considered as a promising candidate for imaging and therapy of cancer because of their biocompatibility, cationic nature, improved EPR effect, and functional groups for surface modification and bioconjugation. Lee et al. [15] developed glycol chitosan-based NPs loaded with protoporphyrin IX (PpIX) as a photosensitizer for PDT and imaging of cancer. The PpIX-loaded glycol chitosan NPs presented the improved tumor-targeting ability compared to free photosensitizer when treated with SCC7 tumor-bearing mice. Also, amphiphilic self-assembled NPs were prepared using the glycol chitosan conjugated with chlorin e6 (Ce6), a photosensitizer for PDT of cancer [16]. Since Ce6 generated ROS more effectively in the NPs, the Ce6-loaded glycol chitosan showed enhanced PDT in MDA-MB-231 breast cancer cells. In another approach, FITC-labeled glycol chitosan-Ce6 NPs and chitosan-Ce6 NPs were prepared and analyzed their biodistribution in vivo in the cancer cells [17]. Compared to free Ce6, chitosan-Ce6 NPs presented reduced in vitro cell viability and improved in vivo tumor phototoxicity, which confirms its potential for the PDT and imaging of cancer.

Shrestha and Kishen prepared rose Bengal-loaded chitosan and studied its antibiofilm efficiency on gram-positive and gram-negative bacteria using PDT [18]. Under the treatment of rose bengal-loaded chitosan, the shape of bacterial biofilm was disintegrated. This result indicated that rose bengal-loaded chitosan had improved antibiofilm activity compared to the control photosensor. Sun et al. [19] developed methylbenzene blue-loaded carboxymethyl chitosan NPs as a theranostic agent for cancer therapy. These NPs exhibited pH-dependent drug release behavior due to the presence of carboxymethyl chitosan. The prolonged controlled drug release from the NPs was observed at pH 7.4 medium compared to pH 5.5 medium. Under a laser light treatment, these NPs showed antibiofilm activity and inhibition of cancer cell growth in acidic conditions. Moreover, the developed NPs exhibited self-imaging ability and antitumor activity against the tumor mice model. Recently, Pandya et al. [20] prepared chitosan-tetraphenyl chlorin conjugate NPs loaded with mertansine or cabazitaxel as imaging contrast agents and for PDT of cancer. Due to the strong interaction between the drugs and photosensitizer, the chitosan-tetraphenyl chlorin conjugate NPs showed the improved drug loading capacity. These NPs presented the enhanced cytotoxicity and PDT effect on breast cancer cell lines, suggests their suitability for cancer theranostics.

Photothermal therapy (PTT) is another promising approach to destruct cancer cells by thermal energy. In PTT, near-infrared (NIR) light is engrossed and transformed to confined heat by the NPs, resulting in the demolition of the tumor cells [21]. Kumar et al. [22] developed polycaprolactone coated and IR

820 dye-loaded glycol chitosan composite for imaging and PTT of cancer cells. These composites retained their structural stability even after the exposure to laser light and presented the sustained release profile. Also, they exhibited the imaging capability, improved cellular uptake, and enhanced photothermal effect on breast cancer cells. Manivasagan et al. [23] fabricated chitosan-polypyrrole composite NPs for imaging-guided PTT of cancer. Because of the synergistic effects of chitosan and polypyrrole, these composites had improved biocompatibility, stability, and NIR absorbance. The in vivo study demonstrated that the prepared composite NPs had better antitumor activity against the tumor-bearing mice under NIR radiation. Recently, Lee et al. [24] prepared chitosan oligosaccharide lactate conjugated with ZW800–1 NIR fluorophore for imaging and targeted PTT of cancer. The HT-29 tumor-bearing mice treated with developed conjugate presented a considerable decrease in the volume of tumor after NIR laser irradiation. Further, chitosan oligosaccharide lactate-ZW800–1 conjugate showed light-triggered PTT and strong fluorescence in tumor sites.

2.2 *Chitosan Loaded with QDs*

The inorganic nanocrystals that belong to groups III–V and II–VI are known as QDs. Due to the fluorescent properties, QDs are considered an imaging agent in the field of cancer theranostics. The fluorescent properties of QDs can be altered by changing their size and configuration [25]. So far, considerable works have been done to develop multifunctional theranostics based on chitosan and its derivatives encapsulated with QDs for cancer therapy.

Tan et al. [26] prepared chitosan-QD composite surface modified with human epidermal growth factor receptor 2 (HER2) antibodies for the imaging and targeted delivery of si-RNA to MCF-7 and SKBR3 human breast cancer cell lines. Due to the presence of a greater number of HER2 receptors, more internalization of the chitosan-QD composite was found in SKBR3 cells compared to MCF-7 cells. When chitosan-QD was utilized to release HER2 si-RNA, HER2 gene expression levels were decreased to 80%. Also, Yuan et al. [27] prepared chitosan-ZnO QDs for imaging and tumor-targeted drug delivery. Due to the presence of chitosan, the prepared materials showed the improved drug loading efficiency and biocompatibility. Also, these materials presented the imaging ability by emitting the blue light due to the existence of ZnO QDs. In another approach, curcumin-loaded *O*-carboxymethyl chitosan grafted with ZnO QDs was prepared for cancer therapy [28]. Since ZnO QDs were conjugated with chitosan, these materials presented enhanced stability for better drug release, imaging, and therapeutic efficiencies.

Ma et al. [29] developed carboxymethyl chitosan loaded with CdTe QDs. Due to the strong binding ability of carboxymethyl chitosan to CdTe QDs with Zn^{2+}, the developed materials presented an improved fluorescence property. These NPs accumulated more effectively at the tumor site of tumor-bearing mice after intravenous inoculation. Due to the biocompatibility and imaging ability, the developed NPs showed great potential for the imaging and therapy of cancer. In another study, a film

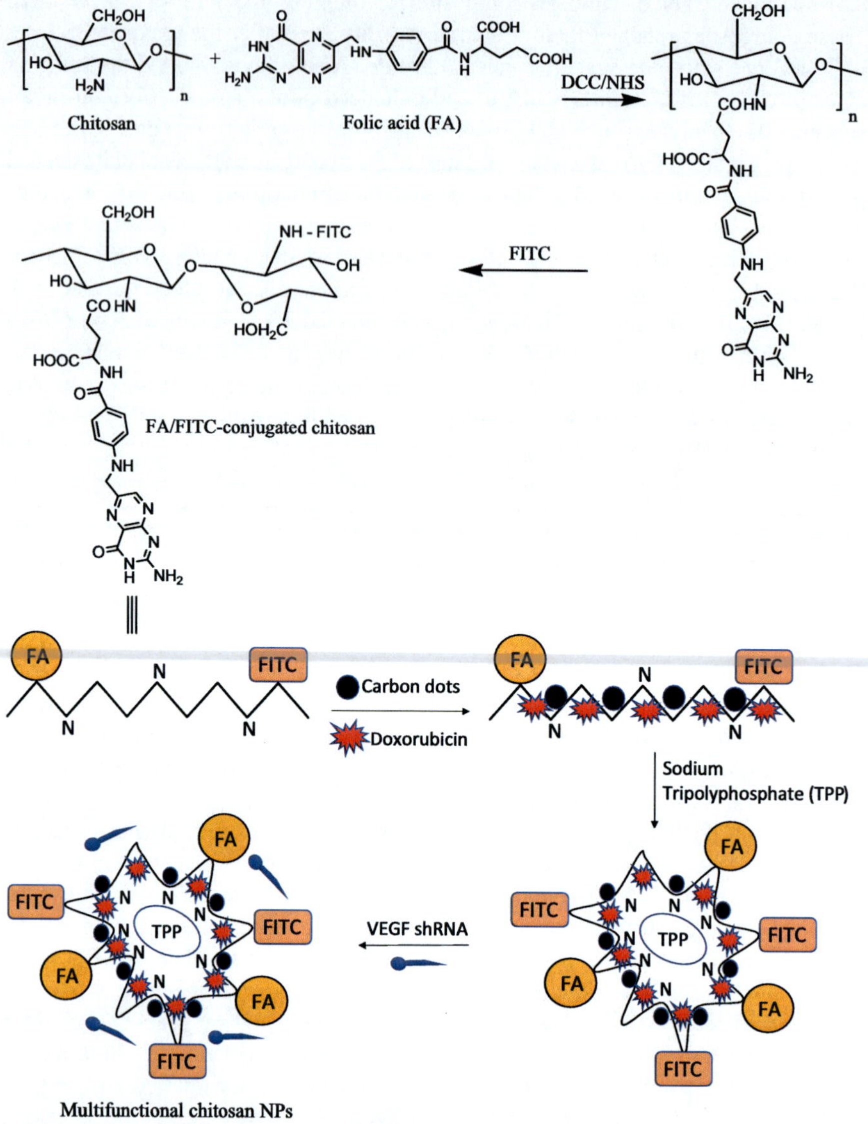

Fig. 1 Schematic representation of the preparation of multifunctional chitosan NPs

based on CdTe QDs functionalized with chitosan-*L*-cysteine was prepared. This film showed better imaging ability and antibacterial property due to the synergistic effect of CdTe QDs and chitosan-L-cysteine [30].

Multifunctional chitosan NPs containing folic acid (FA), FITC, doxorubicin (DOX), carbon QDs, and vascular endothelial growth factor (VEGF) shRNA were prepared as a theranostic agent for cancer therapy (Fig. 1) [31]. The average particle size and charge of these NPs were found to be 154 nm and 23.2 mV, respectively.

These NPs effectively delivered the VEGF shRNA into HeLa cells by shielding shRNA from degradation. The NPs presented the pH-responsive anti-cancer drug release profile. Due to the presence of FA, the prepared NPs exhibited improved cellular uptake by HeLa cells through folate-receptor-mediated endocytosis. The HeLa cells treated with multifunctional chitosan NPs presented reduced VEGF expression, minimal cell proliferation, and augmented cell apoptosis. Besides, these NPs confirmed outstanding fluorescence cellular imaging due to the presence of carbon QDs. In another study, Ding et al. [32] developed FA-conjugated carboxymethyl chitosan NPs containing QDs and Fe_3O_4 NPs for cellular imaging and tumor-targeted drug delivery. Due to the presence of FA, these NPs were successfully uptaken by tumor cells via the FA-receptor-mediated endocytosis mechanism. Moreover, they exhibited simultaneous fluorescence imaging and magnetic properties due to the synergistic effects of QDs and Fe_3O_4 NPs. Further, Lin et al. [33] fabricated nano-micelles based on pluronic-containing anticancer drug and ZnO/CdTe QDs. In this study, the surface of nano-micelles was altered by FA-conjugated chitosan to improve their tumor-targeting ability and blood circulation time. The in vivo study results disclosed that these NPs had controlled drug release and fluorescence cellular imaging capabilities.

Recently, Janus et al. [34] prepared chitosan-based carbon QDs doped with nitrogen for diagnosis, cellular imaging, and controlled drug delivery. The developed QDs presented photoluminescence property in visible light. The chitosan-based carbon QDs modified with amino acids such as lysine and glutamic acid showed a higher quantum yield. Biocompatibility studies demonstrated that the developed NPs had no cytotoxicity on human dermal fibroblasts.

In another approach, Yu et al. [35] designed FA-conjugated copper sulfide (CuS)-chitosan QDs for imaging-guided cancer therapy. These NPs presented adequate biocompatibility, increased cellular uptake, strong NIR light absorption, and improved photothermal efficiency. They also effectively targeted and amassed in the cancer cells within 60 min. Under laser light irradiation, these NPs potentially inhibited the growth rate of tumor cells.

2.3 *Chitosan-Noble Metal Conjugates*

In recent years, noble metals such as gold (Au), silver (Ag), and palladium (Pd) have received much importance as cancer theranostics due to their exceptional optical, electrical, and photothermal behaviors. The metal NPs encapsulated with chitosan and its derivatives have a large potential to be used for cancer imaging and therapy due to their excellent stability and multifunctional effects [36]. Chitosan can form complexation with metal ions through hydroxyl and amino groups present in the polymer chain and hence can improve the stability of metal NPs [37].

Thangam et al. [38] developed the multifunctional chitosan NPs encapsulated with Ag and natural fluorescent protein, R-phycoerythrin for cancer therapy. The theranostic ability of these NPs was studied against MDA-MB-231 breast cancer

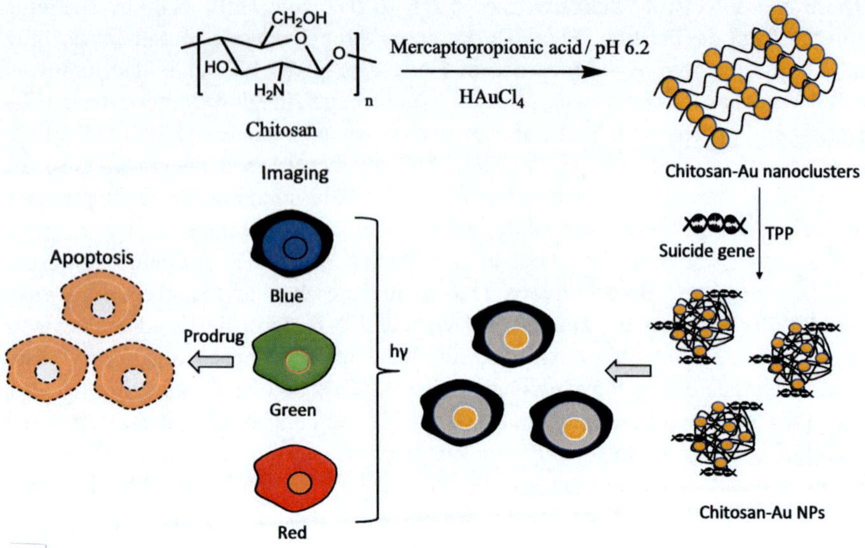

Fig. 2 Schematic representation of the preparation of fluorescence chitosan-Au NPs

cells. The developed NPs presented noticeably high cytotoxicity in the MDA-MB-231 cells, but less cytotoxicity in the normal cells. In recent years, Au NPs are largely utilized as a theranostic agent for cancer treatment because of their non-toxicity, high surface-to-volume ratio, excellent optical and photothermal properties [39]. The Au NPs encapsulated chitosan can have the improved drug loading and tumor-targeted drug delivery ability and imaging property.

Zhang et al. [40] developed chitosan-modified Au NPs for the radiotherapy of cancer. Due to the improved radiation sensitivity, the prepared NPs showed remarkable damage to tumor cells under X-ray irradiation. In another strategy, Sahoo et al. [41] developed chitosan-Au NPs for the apoptosis of cervical cancer cells (Fig. 2). Due to the fluorescence property of Au NPs, the developed NPs presented the optical imaging of cells without the usage of additional dyes. Also, as a radiosensitizer, the chitosan-Au NPs enhanced the efficiency of the radiotherapy. Yan et al. [42] prepared poly(vinyl alcohol)/chitosan nanofibers incorporated with Au NPs for delivery of the drug to cancer cells and imaging of cancer cells. Due to the surface plasmon resonance (SPR) property of Au NPs, the developed nanofibers can be used to attain image-guided therapies.

The Pd NPs coated with RGD peptide-linked chitosan oligosaccharide were developed for imaging and PTT of cancer cells (Fig. 3) [43]. These particles presented the selective accumulation in MDA-MB-231 breast cancer cells and improved photothermal effects under irradiation of 808-nm laser light at 2 W/cm^2 power density. Due to the coating of chitosan oligosaccharide, the NPs showed adequate biocompatibility, colloidal stability, and water dispersity. Since the developed NPs exhibit a good amplitude of photoacoustic signals, they facilitated the imaging of cancer cells using a photoacoustic tomography system.

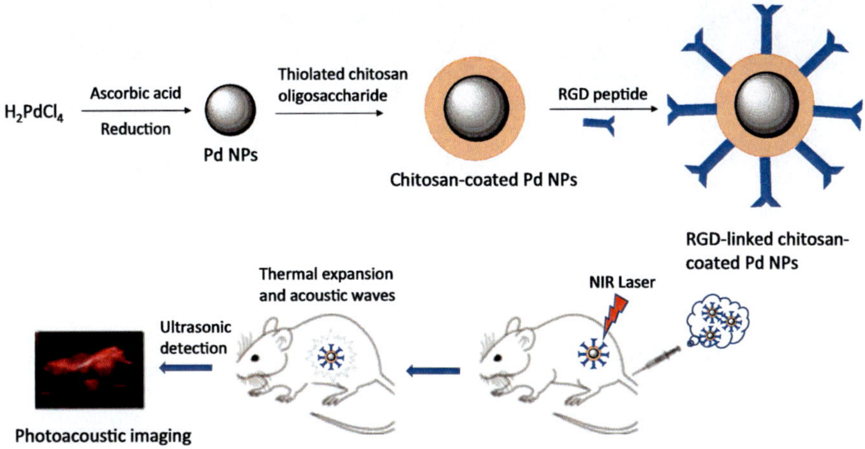

Fig. 3 Schematic illustration of the preparation of RGD-linked chitosan-coated Pd NPs, photothermal extirpation, and photoacoustic imaging of cancer tissue using them

Recently, Sun et al. [44] fabricated Au NPs coated with glycol-chitosan for cancer cell imaging and therapy. These NPs showed relatively improved cellular uptake by breast cancer cell lines and good photoacoustic signals in the in vitro studies. The improved photoacoustic signals could be due to the plasmon coupling effect of glycol-chitosan-coated Au NPs in cancer cells. Further, these NPs presented photoacoustic cancer cellular imaging without using any antibodies or surface alteration. Wang et al. [45] prepared core-shell composite NPs based on Fe_3O_4, chitosan, and Au NPs as cancer theranostics. These NPs had improved cytotoxicity against different types of human carcinoma cell lines. The developed core-shell composite NPs could be used for the imaging and therapy of pancreatic, gastric, and colon cancers. Further, Ma et al. [46] developed Au encapsulated chitosan NPs for cancer therapy. These NPs presented outstanding biocompatibility, improved colloidal stability, pH-dependent drug release property, high drug loading, and fluorescence imaging abilities.

2.4 *Chitosan-Based Magnetic NPs*

Among the inorganic nanomaterials, magnetic NPs have been extensively used in hyperthermia, drug delivery, magnetic resonance imaging (MRI), drug delivery, bioseparation, and catalysis due to their exceptional multifunctional characteristics [47, 48]. The magnetic NPs exhibit excellent superparamagnetic behavior when their size is reduced to <15 nm. In recent years, much interest has been given to chitosan-magnetic NPs conjugates for the tumor-targeted drug delivery and imaging modalities using an externally applied magnetic field [49]. Among the magnetic NPs, superparamagnetic iron oxide (Fe_3O_4) NPs are widely considered as a promising

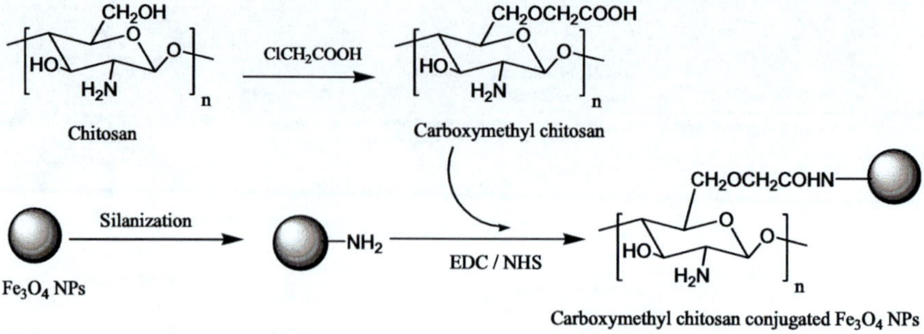

Fig. 4 Preparation of carboxymethyl chitosan conjugated Fe_3O_4 NPs

MRI contrast agent for cancer therapy because of their lesser toxicity and increased proton relaxation that results in a lesser detection limit [50]. Hence, superparamagnetic Fe_3O_4 NPs have been combined with chitosan as cancer theranostic agents. For instance, Lee et al. [51] prepared self-assembled chitosan-linoleic acid NPs encapsulated with Fe_3O_4 NPs as a contrast agent. These NPs were found to increase the colloidal stability of the Fe_3O_4 NPs. Moreover, they reduced the cytotoxicity and showed the targeted MRI of the cancer cells without imaging the normal cells. Maria et al. [52] prepared Fe_3O_4-loaded chitosan NPs for the sustained delivery of docetaxel and imaging of cancer cells. The results showed that the drug retains its efficiency during its loading into the NPs and hence NPs carried the drug into tumor cells effectively. Further, Shi et al. [53] prepared carboxymethyl chitosan conjugated Fe_3O_4 NPs as a contrast agent in MRI (Fig. 4). These NPs were efficiently taken up by the human mesenchymal stem cells (hMSCs) through endocytosis and showed minimal cytotoxicity as compared to the control medium. Due to high labeling competence, the developed Fe_3O_4 NPs detected <100 labeled cells by MRI. Santos et al. [54] synthesized multifunctional Fe_3O_4/chitosan-*L*-glutamic acid core-shell NPs for tumor-targeted drug delivery and hyperthermia usages. These NPs presented the improved drug loading and controlled drug release profile with a localized hyperthermia ability.

Fan et al. [55] developed multifunctional Fe_3O_4 NPs surface-modified with carboxymethyl chitosan and FA (Fig. 5). The results showed that the surface-modified Fe_3O_4 NPs decrease not only the cytotoxicity against the normal cells but also the detention of Fe_3O_4 NPs by macrophages. Due to the presence of FA, the established NPs were effectively taken up by FA-receptor positive cancer cells for MRI, drug targeting, and hyperthermia. Balan et al. [56] reported magnetic NPs comprised of N-palmitoyl chitosan and Fe_3O_4 NPs by ionic gelation method. The developed NPs had a mean size of ~150 nm and a zeta potential of 16.78 mV. In addition, they offered the drug loading ability, high value of magnetic saturation (54.59 emu/g), and superparamagnetic behavior required for cancer theranostics. In another approach, Wang et al. [57] developed magnetic graphene functionalized with chitosan for the simultaneous delivery of therapeutics and magnetic NPs to the tumor site. Because of the presence of chitosan, the developed magnetic graphene

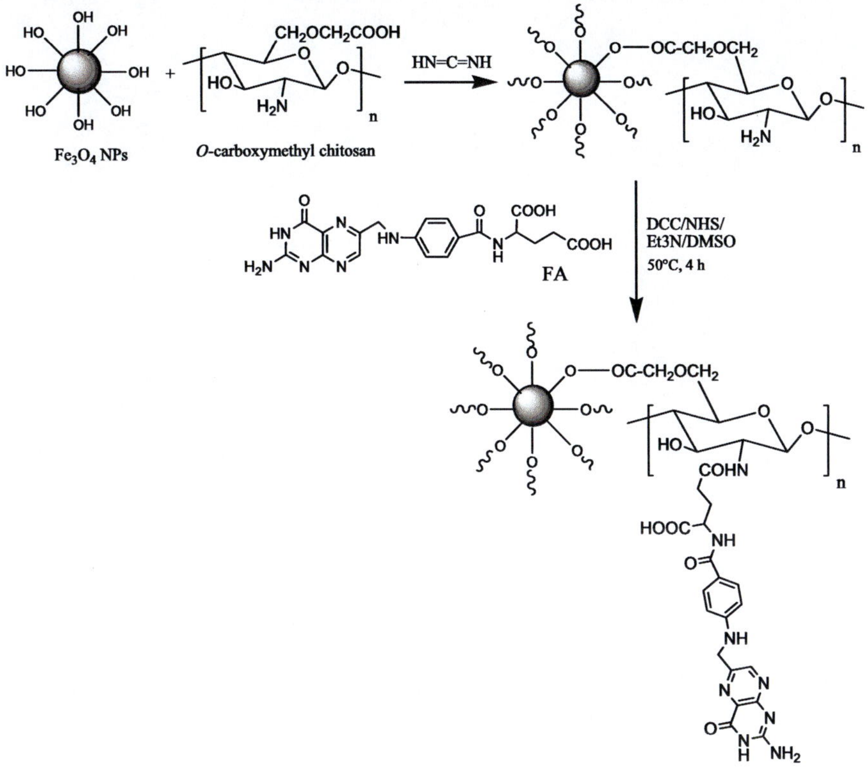

Fig. 5 Preparation of carboxymethyl chitosan and FA coated Fe_3O_4 NPs

sheets were found to be biocompatible, stable, and dispersed in the aqueous medium. These NPs presented the improved drug loading ability, pH-dependent drug release profile, cellular uptake by the A549 human lung cancer cells. Due to the enhanced cellular uptake, the cytotoxicity of the drug-loaded NPs towards the tumor cells was measured to be larger than that of free drug.

Also, Lim et al. [58] developed magnetic NPs based on *N*-naphthyl-*O*-dimethymaleoyl chitosan and Fe_3O_4 for effective cancer therapy. These NPs were found to have the capability of pH-dependent drug release and MRI abilities (Fig. 6). They showed an increased rate of drug release at acidic medium than that at alkaline medium, which could be desired for effective delivery of therapeutics in cancer sites.

Zhou et al. [59] prepared chitosan-coated Fe_3O_4 NPs with a collective ability of drug distribution and hyperthermia effect using a coacervation followed by a chemical cross-linking technique. In this study, FA-conjugated poly(ethylene glycol) (PEG) was functionalized on the surface of the chitosan-coated Fe_3O_4 NPs to expand the long blood circulation and targeting ability. Due to the presence of targeting ligand, the developed NPs showed higher cellular uptake by HeLa cells through folate-receptor-mediated endocytosis. Guanghui et al. [60] fabricated Fe_3O_4 and DOX-loaded chitosan NPs and found that the synthesized NPs could be suitable

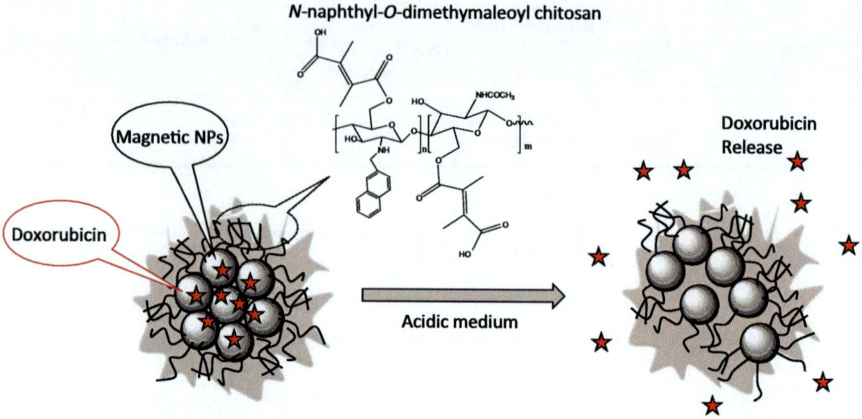

Fig. 6 Illustration of pH-dependent drug release behavior of magnetic *N*-naphthyl-*O*-dimethymaleoyl chitosan NPs

for MRI and delivery of therapeutic agents due to their imaging contrast ability and pH-sensitive properties, respectively. Further, Wang et al. [61] prepared multifunctional DOX-loaded Fe_3O_4–CdTe@SiO_2–carboxymethyl chitosan core-shell NPs. These NPs exhibited good biocompatibility, outstanding magnetic-guided tumor-targeting, and fluorescence labeling abilities. The nanobubbles encapsulated with magnetic NPs have also been considered as a promising candidate for cancer theranostics due to their long blood circulation ability and tumor-targetability. For instance, Yang et al. [62] developed carboxymethyl hexanoyl chitosan nanobubbles loaded with camptothecin and Fe_3O_4 NPs for the imaging and therapy of cancers. Due to the synergistic effect of carboxymethyl hexanoyl chitosan and Fe_3O_4 NPs, the developed nanobubbles demonstrated the increased cytotoxicity against the breast tumor cells and tumor-specific accumulation.

Li et al. [63] developed the multifunctional DOX-loaded Fe_3O_4 NPs blended with chitosan and graphene oxide by electrospray method (Fig. 7). These NPs had the particle size in the range of 100–1,100 μm. Due to the presence of Fe_3O_4 and graphene oxide, the multifunctional Fe_3O_4 NPs showed magnetic responsive behavior with improved drug loading efficiency and stimuli-responsive DOX release profile under NIR irradiation for the chemotherapy of cancer. In another study, chitosan NPs coated with Fe_3O_4 were reported for cancer theranostics [64]. These NPs presented the enhanced MRI signals, imaging of tumor cells, and tumor-specific drug delivery. The biodistribution study conducted using BALB/c mice demonstrated that the developed NPs mainly accumulated in kidneys and liver as they are the important organs that participated in the removal of Fe_3O_4 NPs. These results confirmed that Fe_3O_4-coated chitosan NPs are a promising material for the diagnosis, imaging, and therapy of liver cancer.

In recent years, gene silencing and RNA interference practices have received much interest in cancer therapy. However, due to the rapid disintegration of siRNA by endonucleases before cellular uptake, the usage of these methods is restricted. To

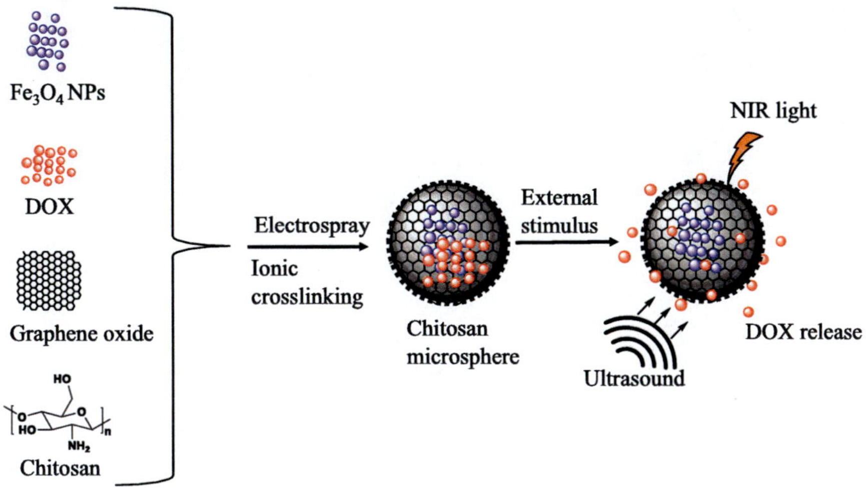

Fig. 7 Schematic representation for the preparation of DOX-loaded Fe3O4/chitosan/graphene oxide NPs and stimuli-responsive DOX release

overcome these drawbacks, Bruniaux et al. [65] developed chitosan functionalized with Fe_3O_4 NPs and siRNA for RNA interference and MRI contrast abilities. In this study, the coating of PEG and poly-L-arginine on the surface of NPs was found to improve their biocompatibility and siRNA transfection ability. Because of the existence of chitosan, the developed NPs presented the pH-dependent delivery of siRNA into the cells. Likewise, Israel et al. [66] reported chitosan-coated Fe_3O_4 NPs as nanocarriers for siRNA for gene silencing therapy.

Olono et al. [67] fabricated the core-shell of $Fe_3C/\gamma\text{-}Fe_2O_3$ NPs encapsulated micelles based on sodium dodecyl sulfate and oleic acid stabilized with chitosan using the sonochemical synthesis method. These micelles showed an average size of 19.71 nm, a polydispersity index of 0.134 and a zeta potential of −41.5 mV. Due to their drug loading efficiency and imaging property, these nanocarriers could be an ideal theranostic agent for cancer therapy. Kumar and Srivastava [68] prepared FITC-linked polycaprolactone glycol chitosan IR 820 NPs for hyperthermia-induced cancer cell death. These NPs showed outstanding photostability for 5 min. The results confirmed that FITC-linked polycaprolactone glycol chitosan IR 820 NPs could be an effective theranostic material for image-guided PTT of cancer. Recently, Baktash et al. [69] prepared hybrid NPs based on chitosan grafted graphene oxide combined with Fe_3O_4 NPs as a pH-sensitive theranostic for cancer therapy (Fig. 8). The DOX-loaded NPs showed the increased rate of drug release at the acidic medium that could be ideal for tumor-targeted drug delivery. Due to the presence of Fe_3O_4 NPs and graphene oxide, the T2 contrast efficacy of the developed NPs was found to be improved. MTT study conducted with L929 cell lines showed the good biocompatibility of NPs, suggesting less contact of GO with the cell membrane due to the presence of chitosan. The results confirmed that the drug-loaded NPs

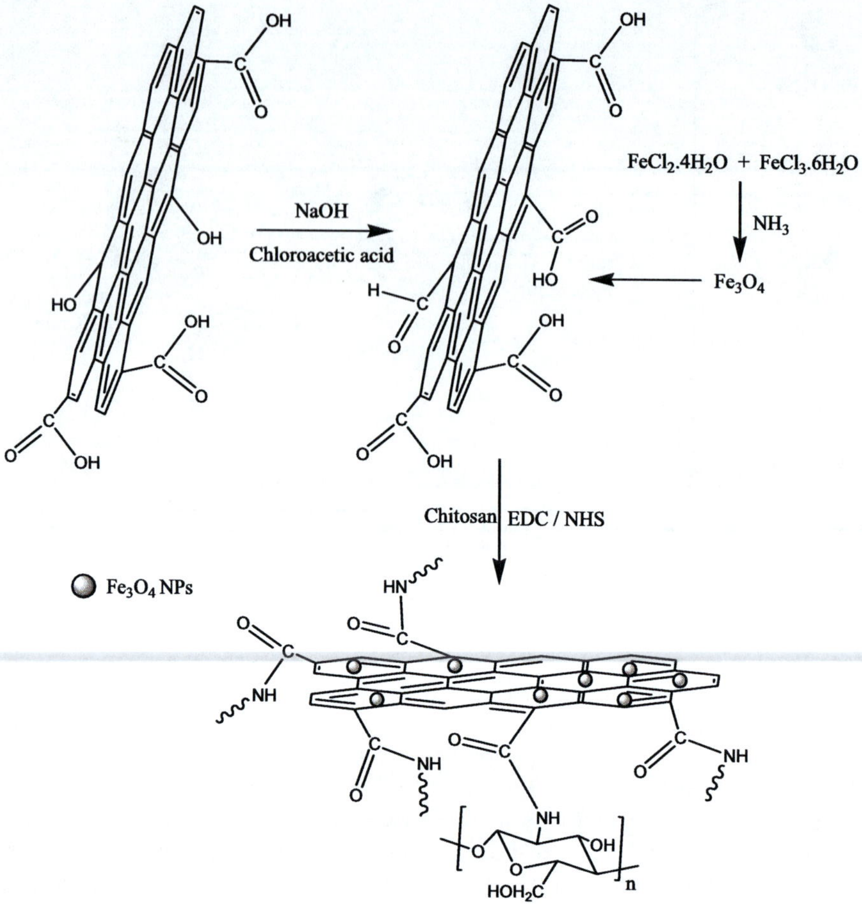

Fig. 8 Preparation of chitosan grafted graphene oxide loaded with Fe_3O_4 NPs

developed using low molecular weight chitosan presents improved cytotoxicity against the MCF7 cancer cells.

2.5 *Chitosan-Based Hybrid NPs*

In recent years, glycol chitosan-5*β*-cholanic acid conjugate has received much interest for its outstanding biocompatibility and tumor-specific biodistribution. The NPs based on glycol chitosan-5β-cholanic acid conjugate presented the selective accumulation in liver tumor tissues, suggesting that these NPs can evade the recognition by RES in the liver [70]. Further, these NPs were found to be extremely dispersed in the brain tumor, indicating the possibility of using them as cancer

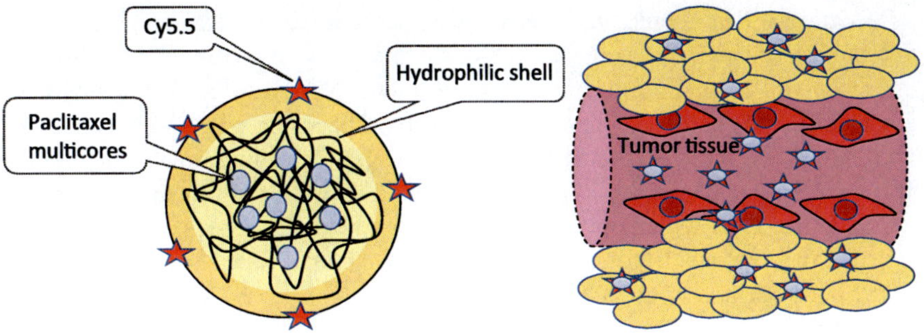

Fig. 9 Schematic diagram of paclitaxel-loaded chitosan NPs linked with Cy5.5 for imaging and therapy of cancer

theranostics [71]. Considering together, glycol chitosan-5β-cholanic acid NPs could be ideal for cancer theranostics irrespective of the location and type of the cancers. Similarly, Yoon et al. [72] developed glycol chitosan-based NPs as a carrier for si-RNA and chemotherapeutic agents and studied their physicochemical characteristics such as particle size, surface, and pH-sensitive properties.

Kim et al. [73] developed paclitaxel-loaded chitosan NPs linked with Cyanine-5.5 (Cy5.5) for imaging and therapy of cancer (Fig. 9). The developed NPs at the optimal amount of administration resulted in the increased rate of tumor cell death compared to free drugs and showed the imaging capability. Na et al. [74] also reported chitosan NPs loaded with paclitaxel and Cy5.5 for imaging and therapy of cancer. In another approach, Srinivasan et al. [75] developed chitosan-IR820 conjugates for imaging and hyperthermia of cancer cells. These conjugates showed hyperthermic cell growth inhibition in human sarcoma cancer cells, SKOV-3, and antibody upon exposure with laser light. The extent of cell growth inhibition due to hyperthermia was found to be higher in chitosan-IR820 conjugates than IR820 alone exist in cancer cells/antibody.

Carbon-based materials such as graphene and graphene oxide have the potential to be used in drug delivery, bioimaging, wound healing, and tissue engineering [76]. These materials can be suitable for the PTT of cancer cells due to their biocompatibility and functionalization ability [77]. Since carbon-based materials have layered sheet-like structure, they offer a high surface area for the increased adsorption of therapeutic agents. By considering this merit, Fu et al. [78] prepared GO-modified carboxymethyl chitosan and linked it to hyaluronic acid and FITC for tumor-targeted controlled delivery of DOX. The developed NPs presented a high drug loading ability along with the increased rate of drug release at acidic conditions (pH 5.8). Also, these NPs showed increased cellular uptake and thereby cancer cell apoptosis. Baghbani et al. [79] prepared the nanodroplets based on curcumin-loaded chitosan/perfluorohexane for ultrasound (US) imaging contrast and cancer therapy. Under US exposure, the nanodroplets presented the increased cytotoxicity against the 4 T1 human breast cancer cell lines, which confirmed their potential to be used in image-guided cancer therapy.

Recently, a nanodroplet has been developed for siRNA delivery using the microbubble transmission technique [80]. In this study, the core of nanodroplets was made up of perfluoropentane and magnetic NPs with a shell containing deoxycholic acid conjugated chitosan and siRNA. The nanodroplets were converted into microbubbles after exposure to US for a longer circulation time in the bloodstream. The magnetic NPs within the microbubbles were utilized to improve the localization via an externally applied magnetic field. The developed materials presented a four-fold decrease in the viability of human lung and breast cancer cells. In another study, DOX-loaded chitosan-Ag hybrid NPs were prepared and analyzed their physicochemical properties [81]. Due to the existence of chitosan, these NPs showed the pH-responsive controlled drug release behavior. Moreover, they showed enhanced cytotoxicity against subcutaneous tumors and cancer cell lines. Further, Zhang et al. [82] fabricated methylene blue-loaded hybrid PEG-chitosan/Fe_3O_4 NPs for imaging and combined PDT/PTT of cancer. Due to the formation of singlet oxygen, the prepared NPs effectively killed the cancer cells under NIR radiation. Also, they exhibited controlled photoexcitation to eradicate the tumor cells without affecting normal cells. These hybrid NPs could be considered for MRI and collective PDT/PTT of cancer.

2.6 *Chitosan-Based Multimodal Nanocomposites*

Chitosan has been utilized as a matrix for encapsulating two or more theranostic NPs to obtain nanocomposites with improved theranostic capability. In this context, Lin et al. [83] developed the multi-modal nanocarriers based on chitosan functionalized with Fe_3O_4 NPs, methotrexate-PEG, and Cy5.5. In this study, methotrexate-PEG was used as a prodrug, which can target the tumor cells. These NPs presented higher accumulation in the target site, decreased adverse effects, sustained drug release, enhanced therapeutic efficacy, and higher cellular uptake compared to the control. Due to the presence of Fe_3O_4 NPs and Cy5.5, these nanocarriers could be used in multi-modal imaging and cancer therapy. Wang et al. [84] prepared multifunctional core-shell NPs based on Fe_3O_4/Au encapsulated chitosan as a photothermal and dual-imaging agent. The NPs showed outstanding magnetic properties, plasmonic activities, and fewer hemolytic effects. Due to the exceptional photothermal behavior of Fe_3O_4/Au NPs, the developed NPs could be ideal for PTT and simultaneous MRI and field imaging of cancer cells. Kim et al. [85] fabricated Cy5.5-labeled Au NPs coated with glycol chitosan and linked with fibrin-targeting peptides for CT imaging and therapy of cerebrovascular thrombi using a tissue plasminogen activator. Key et al. [86] prepared peptide-conjugated glycol chitosan NPs loaded with Fe_3O_4 and Cy5.5 for MRI/NIR fluorescence imaging and therapy of cancer. These NPs specifically accumulated in the tumor site without gathering in the normal tissues and presented multimodal cellular imaging capability.

Recently, Liu et al. [87] prepared DOX-loaded polyoxometalate NPs surface coated with mesoporous silica and FA–chitosan for multimodal imaging and chemotherapy of cancer. These NPs presented upconverting luminescence under NIR light and CT imaging characteristics. Due to the generation of heat under laser light irradiation and the release of DOX, the prepared NPs showed their potential to be used for the combined PPT/chemotherapy of cancer. Because of the presence of FA-chitosan, they also exhibited the tumor-targeted pH-responsive drug release. Further, Choi et al. [88] developed glycol chitosan NPs containing iodine for US and computed tomography (CT) imaging of cancer tissues. In this study, diatrizoic acid-containing iodine was linked to glycol chitosan and then perfluoropentane was loaded into the resulting product to form a multimodal imaging agent. The prepared NPs showed effective accumulation and US/CT signals in the tumor site.

3 Summary

Over the last few decades, considerable efforts have been taken to utilize chitosan and its derivatives for the development of theranostics for effective imaging and therapy of cancer due to their desirable physiochemical and biological properties. Chitosan and its derivatives have been largely combined with photosensitizers/ photothermal agents, QDs, noble metals, magnetic NPs, and other functional materials to fabricate the advanced cancer theranostics. Since photosensitizers-loaded chitosan selectively accumulates in the tumor site, they can be effectively utilized for PTT and fluorescence imaging of cancer cells. Due to the presence of tumor-targeting ability and imaging capability, the targeting ligand-linked chitosan NPs encapsulated with QDs and anticancer drugs demonstrated better cellular imaging and therapeutic efficacy at the tumor site. In recent years, chitosan-Au NPs conjugates are largely considered for the combined PTT and imaging of tumors. In this system, the Au NPs provided the optical and photothermal properties to chitosan-Au NPs conjugates, while chitosan offered better drug loading, controlled release, and tumor-targeting abilities. Due to the exceptional superparamagnetic and photothermal properties of Fe_3O_4 NPs, chitosan and its derivatives have been conjugated with Fe_3O_4 NPs to fabricate different types of chitosan-based magnetic NPs as theranostics for cancer therapy. These magnetic NPs could be a promising material for simultaneous MRI, PTT, and chemotherapy of cancer because of the synergistic effects of Fe_3O_4 NPs and chitosan-based materials. Further, chitosan-based materials conjugated with different fluorescent dyes and functional materials, namely Fe_3O_4 NPs, Au NPs, therapeutic agents, 5β-cholanic acid, PEG, and GO established their suitability for multimodal imaging and therapy of cancer. Overall, it is apparent that chitosan-based theranostics have a large potential to be utilized in real-time cancer therapy after the systematic evaluation of their cytotoxicity and immunotoxicity and completion of preclinical studies.

References

1. Nair JB, Joseph MM, Mohapatra S, Safeera M, Ghosh S, Sreelekha TT, Maiti KK (2016) A dual-targeting octaguanidine–doxorubicin conjugate transporter for inducing caspase-mediated apoptosis on folate-expressing cancer cells. Chem Med Chem 11(7):702–712
2. Peng N, Wu B, Wang L, He W, Ai Z, Zhang X, Wang Y, Fan L, Ye Q (2016) High drug loading and pH-responsive targeted nanocarriers from alginate-modified SPIONs for anti-tumor chemotherapy. Biomater Sci 4(12):1802–1813
3. Gonciar D, Mocan T, Matea CT, Zdrehus C, Mosteanu O, Mocan L, Pop T (2019) Nanotechnology in metastatic cancer treatment: current achievements and future research trends. J Cancer 10(6):1358–1369
4. Li Y, Zhang H (2019) Nanoparticle-based drug delivery systems for enhanced tumor-targeting treatment. J Biomed Nanotech 15(1):1–27
5. Teijeiro-Valino C, Novoa-Carballal R, Borrajo E, Vidal A, Alonso-Nocelo M, de la Fuente FM, Lopez-Casas PP, Hidalgo M, Csaba N, Alonso MJ (2019) A multifunctional drug nanocarrier for efficient anticancer therapy. J Control Release 294:154–164
6. Jahangirian H, Kalantari K, Izadiyan Z, Rafiee-Moghaddam R, Shameli K, Webster TJ (2019) A review of small molecules and drug delivery applications using gold and iron nanoparticles. Int J Nanomedicine 14:1633
7. Nunes SS, De Barros ALB (2015) The use of coating agents to enhance liposomes blood circulation time. J Mol Pharm Org Proc Res 3(1):e120
8. He Q, Liu J, Liang J, Liu X, Li W, Liu Z, Ding Z, Tuo D (2018) Towards improvements for penetrating the blood–brain barrier - recent progress from a material and pharmaceutical perspective. Cell 7(4):24
9. Jiang H, Guo D, Chen D, Wu Y, Jin X, Zhu X (2019) A new insight into the reversal of multidrug resistance in cancer by nanodrugs. Biomater Sci 7(8):3489–3496
10. Mirrahimi M, Abed Z, Beik J, Shiri I, Dezfuli AS, Mahabadi VP, Kamrava SK, Ghaznavi H, Shakeri-Zadeh A (2019) A thermo-responsive alginate nanogel platform co-loaded with gold nanoparticles and cisplatin for combined cancer chemo-photothermal therapy. Pharmacol Res 143:178–185
11. An J, Yang XQ, Cheng K, Song XL, Zhang L, Li C, Zhang XS, Xuan Y, Song YY, Fang BY, Hou XL (2017) In vivo computed tomography/photoacoustic imaging and NIR-triggered chemo–photothermal combined therapy based on a gold nanostar, mesoporous silica, and thermosensitive liposome-composited nanoprobe. ACS Appl Mater Interfaces 9 (48):41748–41759
12. Wang Y, Liu X, Deng G, Wang Q, Zhang L, Wang Q, Lu J (2017) Multifunctional PS@CS@Au–Fe_3O_4–FA nanocomposites for CT, MR and fluorescence imaging guided targeted-photothermal therapy of cancer cells. J Mater Chem B 5(22):4221–4232
13. Calavia PG, Bruce G, Pérez-García L, Russell DA (2018) Photosensitiser-gold nanoparticle conjugates for photodynamic therapy of cancer. Photochem Photobiol Sci 17(11):1534–1552
14. Simões JCS, Sarpaki S, Papadimitroulas P, Therrien B, Loudos G (2020) Conjugated photosensitizers for imaging and PDT in cancer research. J Med Chem 63(23):14119–14150
15. Lee SJ, Park K, Oh YK, Kwon SH, Her S, Kim IS, Choi K, Lee SJ, Kim H, Lee SG, Kim K, Kwon IC (2009) Tumor specificity and therapeutic efficacy of photosensitizer-encapsulated glycol chitosanbased nanoparticles in tumor-bearingmice. Biomaterials 30:2929–2939
16. Lim CK, Shin J, Kwon IC, Jeong SY, Kim S (2012) Iodinated photosensitizing chitosan: self-assembly into tumor-homing nanoparticles with enhanced singlet oxygen generation. Bioconjug Chem 23:1022–1028
17. Lee SJ, Koo H, Jeong H, Huh MS, Choi Y, Jeong SY, Byun E, Choi K, Kim K, Kwon IC (2011) Comparative study of photosensitizer loaded and conjugated glycol chitosan nanoparticles for cancer therapy. J Control Release 152:21–29
18. Shrestha A, Kishen A (2012) Polycationic chitosan-conjugated photosensitizer for antibacterial photodynamic therapy. Photochem Photobiol 88:577–583

19. Sun L, Jiang W, Zhang H, Guo Y, Chen W, Jin Y, Chen H, Du K, Dai H, Ji J, Wang B (2019) Photosensitizer-loaded multifunctional chitosan nanoparticles for simultaneous in situ imaging, highly efficient bacterial biofilm eradication and tumor ablation. ACS Appl Mater Interfaces 11:2302–2316
20. Pandya AD, Øverbye A, Sahariah P, Gaware VS, Høgset H, Masson M, Høgset A, Mælandsmo GM, Skotland T, Sandvig K, Iversen TG (2020) Drug-loaded photosensitizer-chitosan nanoparticles for combinatorial chemo- and photodynamic therapy of cancer. Biomacromolecules 21:1489–1498
21. Doughty ACV, Hoover AR, Layton E, Murray CK, Howard EW, Chen WR (2019) Nanomaterial applications in photothermal therapy for cancer. Materials (Basel) 2(5):779
22. Kumar P, Srivastava R (2015) IR 820 stabilized multifunctional polycaprolactone glycol chitosan composite nanoparticles for cancer therapy. RSC Adv 5:56162–56170
23. Manivasagan P, Bui NQ, Bharathiraja S, Moorthy MS, Oh YO, Song K, Seo H, Yoon M, Oh J (2017) Multifunctional biocompatible chitosan-polypyrrole nanocomposites as novel agents for photoacoustic imaging-guided photothermal ablation of cancer. Sci Rep 7:43593
24. Lee S, Jo G, Jung JS, Yang DH, Hyun H (2020) Near-infra-red fluorescent chitosan oligosaccharide lactate for targeted cancer imaging and photothermal therapy. Artif Cells Nanomed Biotech 48:1144–1152
25. Kulkarni NS, Guererro Y, Gupta N, Muth A, Gupta V (2019) Exploring potential of quantum dots as dual modality for cancer therapy and diagnosis. J Drug Del Sci Technol 49:352–364
26. Tan WB, Jiang S, Zhang Y (2007) Quantum-dot based nanoparticles for targeted silencing of HER2/neu gene via RNA interference. Biomaterials 28:1565–1571
27. Yuan Q, Hein S, Misra RD (2010) New generation of chitosan encapsulated ZnO quantum dots loaded with drug: synthesis, characterization and in vitro drug delivery response. Acta Biomater 6:2732–2739
28. Upadhyaya L, Singh J, Agarwal V, Pandey AC, Verma SP, Das P, Tewari RP (2015) Efficient water soluble nanostructured ZnO grafted O-carboxymethyl chitosan/curcumin-nanocomposite for cancer therapy. Process Biochem 50:678–688
29. Ma Q, Lin ZH, Yang N, Li Y, Su XG (2014) A novel carboxymethyl chitosan–quantum dot-based intracellular probe for Zn^{2+} ion sensing in prostate cancer cells. Acta Biomater 10:868–874
30. Kumar H, Srivastava R, Dutta PK (2013) Highly luminescent chitosan-L-cysteine functionalized CdTe quantum dots film: synthesis and characterization. Carbohydr Polym 97:327–334
31. Yang H, Xu M, Li S, Shen X, Li T, Yan J, Zhang C, Wu C, Zeng H, Liu Y (2016) Chitosan hybrid nanoparticles as a theranostic platform for targeted doxorubicin/VEGF shRNA codelivery and dual-modality fluorescence imaging. RSC Adv 6:29685
32. Ding Y, Yin H, Chen R, Bai R, Chen C, Hao X, Shen S, Sun K, Liu F (2017) Carboxymethyl chitosan based nanocomposites containing chemically bonded quantum dots and magnetic nanoparticles. Appl Surf Sci:433
33. Lin YJ, Huang KB, Wu YC, Rani P, Lin HR (2019) Pluronic-chitosan-folate nano-micelles incorporated with quantum dots for anti-cancer drug therapy. Int J Polym Mater Polym Biomater 68:16
34. Janus L, Piatkowski M, Pragłowska JR, Bogdał D, Matysek D (2019) Chitosan-based carbon quantum dots for biomedical applications: synthesis and characterization. Nano 9:274
35. Yu W, Yu N, Wang Z, Li X, Song C, Jiang R, Geng P, Li M, Yin S, Chen Z (2019) Chitosan-mediated green synthesis and folic-acid modification of CuS quantum dots for photoacoustic imaging guided photothermal therapy of tumor. J Colloid Inter Sci 555:480–488
36. Mohammed MA, Syeda JTM, Wasan KM, Wasan EK (2017) An overview of chitosan nanoparticles and its application in non-parenteral drug delivery. Pharmaceutics 9(4):53
37. Detsi A, Kavetsou E, Kostopoulou I, Pitterou I, Rozaria A, Pontillo N, Tzani A, Christodoulou P, Siliachli A, Zoumpoulakis P (2020) Nanosystems for the encapsulation of natural products: the case of chitosan biopolymer as a matrix. Pharmaceutics 12:669

38. Thangam R, Sundarraj S, Vivek R, Suresh V, Sivasubramanian S, Paulpandi M, Karthick SV, Ragavi AS, Kannan S (2015) Theranostic potentials of multifunctional chitosan-silver-phycoerythrin nanocomposites against triple negative breast cancer cells. RSC Adv 5:12209–12223
39. Wang Z, Dong J, Zhao Q, Ying Y, Zhang L, Zou J, Jiang S (2020) Gold nanoparticle-mediated delivery of paclitaxel and nucleic acids for cancer therapy. Mol Med Rep 22:4475–4484
40. Zhang C, Huang P, Bao L, He M, Luo T, Gao G, Cui D (2011) Enhancement of gastric cell radiation sensitivity by chitosan-modified gold nanoparticles. J Nanosci Nanotechnol 11:9528–9535
41. Sahoo AK, Banerjee S, Ghosh SS, Chattopadhyay A (2013) Simultaneous RGB emitting Au nanoclusters in chitosan nanoparticles for anticancer gene theranostics. ACS Appl Mater Interfaces 6:712–724
42. Yan E, Cao M, Wang Y, Hao X, Pei S, Gao J, Wang Y, Zhang Z, Zhang D (2016) Gold nanorods contained polyvinyl alcohol/chitosan nanofiber matrix for cell imaging and drug delivery. Mater Sci Eng C 58:1090–1097
43. Bharathiraja S, Bui NQ, Manivasagan P, Moorthy MS, Mondal S, Seo H, Phuoc NT, Phan TTV, Kim H, Lee KD, Oh J (2018) Multimodal tumor-homing chitosan oligosaccharide coated biocompatible palladium nanoparticles for photo-based imaging and therapy. Sci Rep 8:500
44. Sun IC, Ahn CH, Kim K, Emelianov S (2019) Photoacoustic imaging of cancer cells with glycol-chitosan-coated gold nanoparticles as contrast agents. J Biomed Opt 24(12):1–5
45. Wang X, Almoallim HS, Cui Q, Alharbi SA, Yang H (2021) In situ decorated Au NPs on chitosan-encapsulated Fe_3O_4-NH_2 NPs as magnetic nanocomposite: investigation of its anti-colon carcinoma, anti-gastric cancer and anti-pancreatic cancer. Int J Biol Macromol 171:198–207
46. Ma K, Cheng Y, Wei X, Chen D, Zhao X, Jia P (2020) Gold embedded chitosan nanoparticles with cell membrane mimetic polymer coating for pH-sensitive controlled drug release and cellular fluorescence imaging. J Biomater Appl 35(7):088532822095259
47. Guo T, Lin M, Huang J, Zhou C, Tian W, Yu H, Jiang X, Ye J, Shi Y, Xiao Y, Bian X, Feng X (2018) The recent advances of magnetic nanoparticles in medicine. J Nanomater 2018:7805147
48. Natarajan S, Harini K, Gajula GP, Sarmento B, Petersen MTN, Thiagarajan V (2019) Multifunctional magnetic iron oxide nanoparticles: diverse synthetic approaches, surface modifications, cytotoxicity towards biomedical and industrial applications. BMC Mater 1:2
49. Farnaz A, Hoda J, Hossein A, Vaghari H, Anarjan N, Ahmadi O, Berenjian A (2017) Chitosan magnetic nanoparticles for drug delivery systems. Crit Rev Biotechnol 37:492–509
50. Nabavinia M, Huarac JB (2020) Recent progress in iron oxide nanoparticles as therapeutic magnetic agents for cancer treatment and tissue engineering. ACS Appl Bio Mater 3 (12):8172–8187
51. Lee CM, Jeong HJ, Kim SL, Jeong HJ, Kim EM, Park EH, Kim DW, Lim ST (2009) SPION-loaded chitosan–linoleic acid nanoparticles to target hepatocytes. Int J Pharm 371:163–169
52. Maria VL, Daniel T, Dolores T, Anxo V, Fernando D, Alonso MJ (2009) Highly efficient system to deliver taxanes into tumor cells: docetaxel-loaded chitosan oligomer colloidal carriers. Biomacromolecules 9:2186–2193
53. Shi Z, Neoh KG, Kang ET, Shuter B, Wang SC, Poh C, Wang W (2009) Carboxymethyl chitosan-modified superparamagnetic iron oxide nanoparticles for magnetic resonance imaging of stem cells. ACS Appl Mater Inter 1(2):328–335
54. Santos DP, Ruiz MA, Gallardo V, Zanoni MVB, Arias JL (2011) Multifunctional antitumor magnetite/chitosan-l-glutamic acid (core/shell) nanocomposites. J Nanopart Res 13:4311–4323
55. Fan C, Gao W, Chen Z, Fan H, Li M, Deng F, Chen Z (2011) Tumor selectivity of stealth multifunctionalized superparamagnetic iron oxide nanoparticles. Int J Pharm 404(1–2):180–190
56. Balan V, Butnaru M, Verestiuc L (2013) Synthesis and characterization of magnetic nanoparticles based on N-palmitoyl-chitosan with potential applications in cancer theranostics. In: The 4th IEEE international conference on E-health and bioengineering
57. Wang C, Ravi S, Garapati US, Das M, Howell M, Mallela JM, Alwarapan S, Mohapatra SS, Mohapatra S (2013) Multifunctional chitosan magnetic-graphene (CMG) nanoparticles: a

theranostic platform for tumor-targeted co-delivery of drugs, genes and MRI contrast agents. J Mater Chem B 1(35):4396–4405
58. Lim EK, Sajomsang W, Choi Y, Jang E, Lee H, Kang B, Kim E, Haam S, Suh JS, Chung SJ, Huh YM (2013) Chitosan-based intelligent theragnosis nanocomposites enable pH-sensitive drug release with MR-guided imaging for cancer therapy. Nanoscale Res Lett 8:467
59. Zhou S, Li Y, Cui F, Jia M, Yang X, Wang Y, Xie L, Zhang Q, Hou Z (2014) Development of multifunctional folate-poly (ethylene glycol)-chitosan-coated Fe3O4 nanoparticles for biomedical applications. Macromol Res 22:58–66
60. Guanghui Z, Jianzhi W, Xiaomen P, Yanfeng L, Xuemei Y, Ma Y (2014) Facile solvo thermal synthesis of mesostructured Fe_3O_4/chitosan nanoparticles as delivery vehicles for pH-responsive drug delivery and magnetic resonance imaging contrast agents. Chem Asian J 9:546–553
61. Wang G, Jin L, Dong Y, Niu L, Liu Y, Ren F, Su X (2014) Multifunctional Fe_3O_4–CdTe@SiO_2–carboxymethyl chitosan drug nanocarriers: synergistic effect toward magnetic targeted drug delivery and cell imaging. New J Chem 38:700–708
62. Yang PS, Tung FI, Chen HP, Liu TY, Lin YY (2014) A novel bubble-formingmaterial for preparing hydrophobic-agent-loaded bubbles with theranostic functionality. Acta Biomater 10:3762–3774
63. Li S, Xiao L, Deng H, Shi X, Cao Q (2017) Remote controlled drug release from multifunctional Fe3O4/GO/chitosan microspheres fabricated by an electrospray method. Colloid Surf B 151:354–362
64. Kania G, Sternak M, Jasztal A, Chlopicki S, Błazejczyk A, Nasulewicz-Goldeman A, Wietrzyk J, Jasinski K, Skórka T, Zapotoczny S, Nowakowska M (2018) Uptake and bioreactivity of charged chitosan-coated superparamagnetic nanoparticles as promising contrast agents for magnetic resonance imaging. Nanomedicine 14:131–140
65. Bruniaux J, Ben Djemaa S, Herve-Aubert K, Marchais H, Chourpa I, David S (2017) Stealth magnetic nanocarriers of siRNA as platform for breast cancer theranostics. Int J Pharm 532:660–668
66. Israel LL, Lellouche E, Greneche JM, Bechor M, Michaeli S, Lellouche JP (2016) Ultrasound-mediated surface engineering of theranostic magnetic nanoparticles: an effective one-pot functionalization process using mixed polymers for siRNA delivery. J Nanomed Nanotechnol 7:385/381–385/314
67. Sauceda-Oloño PY, Lucero-Acuña JA, Zavala-Rivera P (2018) Encapsulation of iron oxide nanoparticles type core-shell in chitosan as possible theranostic agent. Microsc Microanal 24:2018
68. Piyush K, Rohit S (2018) FITC conjugated polycaprolactone-glycol-chitosan nanoparticles containing the longwave emitting fluorophore IR 820 for in-vitro tracking of hyperthermia-induced cell death. BioRxiv. https://doi.org/10.1101/273748
69. Baktash MS, Zarrabi A, Avazverdi E, Reis NM (2020) Development and optimization of a new hybrid chitosan-grafted graphene oxide/magnetic nanoparticle system for theranostic applications. J Mol Liquids 322(24):114515
70. Na JH, Koo H, Lee S, Min KH, Park K, Yoo H, Lee SH, Park JH, Kwon IC, Jeong SY, Kim K (2011) Real-time and non-invasive optical imaging of tumor-targeting glycol chitosan nanoparticles in various tumor models. Biomaterials 32:5252–5261
71. Veiseh O, Sun C, Fang C, Bhattarai N, Gunn J, Kievit F, Du K, Pullar B, Lee D, Ellenbogen RG, Olson J, Zhang M (2009) Specific targeting of brain tumors with an optical/magnetic resonance imaging nanoprobe across the blood–brain barrier. Cancer Res 69:6200–6207
72. Yoon HY, Son S, Lee SJ, You DG, Yhee JY, Park JH, Swierczewska M, Lee S, Kwon IC, Kim SH, Kim K, Pomper MG (2014) Glycol chitosan nanoparticles as specialized cancer therapeutic vehicles: sequential delivery of doxorubicin and Bcl-2 siRNA. Sci Rep 4:6878
73. Kim K, Kim JH, Park H, Kim YS, Park K, Nam H, Lee S, Park JH, Park RW, Kim IS, Choi K, Kim SY, Park K, Kwon IC (2010) Tumor-homing multifunctional nanoparticles for cancer

theragnosis: simultaneous diagnosis, drug delivery, and therapeutic monitoring. J Control Release 146:219–227
74. Na JH, Koo H, Lee S, Min KH, Park K, Yoo H, Lee SH, Park JH, Kwon IC, Jeong SY, Kim K (2011) Real-time and noninvasive optical imaging of tumor-targeting glycol chitosan nanoparticles in various tumor models. Biomaterials 32:5252–5261
75. Srinivasan S, Manchanda R, Fernandez-Fernandez A, Lei T, McGoron AJ (2013) Near-infrared fluorescing IR820-chitosan conjugate for multifunctional cancer theranostic applications. J Photochem Photobiol B 119:52–59
76. Hoseini-Ghahfarokhi M, Mirkiani S, Mozaffari N, Abdolahi Sadatlu MA, Ghasemi A, Abbaspour S, Akbarian M, Farjadian F, Karimi M (2020) Applications of graphene and graphene oxide in smart drug/gene delivery: is the world still flat? Int J Nanomedicine 15:9469–9496
77. Wang C, Wang X, Chen Y, Fang Z (2020) In-vitro photothermal therapy using plant extract polyphenols functionalized graphene sheets for treatment of lung cancer. J Photochem Photobio B 204:111587
78. Fu G, Zhu L, Yang K, Zhuang R, Xie J, Zhang F (2016) Diffusion-weighted magnetic resonance imaging for therapy response monitoring and early treatment prediction of photothermal therapy. ACS Appl Mater Interfaces 8:5137–5147
79. Baghbani F, Chegeni M, Moztarzadeh F, Hadian-Ghazvini S, Raz M (2017) Novel ultrasound-responsive chitosan/perfluorohexane nanodroplets for image-guided smart delivery of an anti-cancer agent: curcumin. Mater Sci Eng C 74:186–193
80. Lee JY, Crake C, Teo B, Carugo D, Victor MS, Seth A, Stride E (2017) Ultrasound-enhanced siRNA delivery using magnetic nanoparticle-loaded chitosan-deoxycholic acid nanodroplets. Adv Healthcare Mater:6
81. Mohamed N (2020) Synthesis of hybrid chitosan silver nanoparticles loaded with doxorubicin with promising anti-cancer activity. BioNanoScience 10(3)
82. Zhang G, Gou H, Liu Y, Xi K, Jiang D, Jia X (2020) pH-responsive PEG-chitosan/iron oxide hybrid nanoassemblies for low-power-assisted PDT/PTT combination therapy. Nanomedicine (Lond) 5(11):1097–1112
83. Lin J, Li Y, Li Y, Wu H, Yu F, Zhou S, Xie L, Luo F, Lin C, Hou Z (2015) Drug/dye-loaded, multifunctional PEG-chitosan-iron oxide nanocomposites for methotraxate synergistically self-targeted cancer therapy and dual model imaging. ACS Appl Mater Interfaces 7:11908–11920
84. Wang X, Liu H, Chen D, Meng X, Liu T, Fu C, Hao N, Zhang Y, Wu X, Ren J, Tang F (2013) Multifunctional Fe_3O_4@ P(St/MAA)@chitosan@ Au core/shell nanoparticles for dual imaging and photothermal therapy. ACS Appl Mater Interfaces 5:4966–4971
85. Kim JY, Ryu JH, Schellingerhout D, Sun IC, Lee SK, Jeon S, Kim J, Kwon IC, Nahrendorf M, Ahn CH, Kim K, Kim DE (2015) Direct imaging of cerebral thromboemboli using computed tomography and fibrin-targeted gold nanoparticles. Theranostics 5:1098
86. Key J, Dhawan D, Cooper CL, Knapp DW, Kim K, Kwon IC, Choi K, Park K, Decuzzi P, Leary JF (2016) Multicomponent, peptide-targeted glycol chitosan nanoparticles containing ferrimagnetic iron oxide nanocubes for bladder cancer multimodal imaging. Int J Nanomedicine 11:4141–4155
87. Liu S, Li W, Gai S, Yang G, Zhong C, Dai Y, He F, Yang P, Suh YD (2019) A smart tumor microenvironment responsive nanoplatform based on upconversion nanoparticles for efficient multimodal imaging guided therapy. Biomater Sci 7(3):951–962
88. Choi D, Jeon S, You DG, Um W, Kim JY, Yoon HY, Chang H, Kim DE, Park JH, Kim H, Kim K (2018) Iodinated echogenic glycol chitosan nanoparticles for X-ray CT/US dual imaging of tumor. Nano 2(2):117–127

Adv Polym Sci (2021) 288: 293–380
https://doi.org/10.1007/12_2021_93

Published online: 19 May 2021

An Overview on Chitosan-Based Adjuvant/ Vaccine Delivery Systems

Selin Parmaksız and Sevda Şenel

Contents

Abstract Vaccination is a highly effective and safe method in preventing life-threatening infections in both human and animals. It is based on the principle of activating the immune system and providing protection against infection by administration of the pathogen (antigen) that causes the disease. In general, vaccines can be classified as live-attenuated, inactivated, toxoid vaccines, subunit and highly purified recombinant vaccines. Use of subunit and highly purified vaccines enhances the safety; however, their immunogenicity is relatively low due to their monomeric nature and the absence of other immunostimulatory components. Among the strategies to improve the immunogenicity of subunit vaccines is use of adjuvants and/or delivery systems. Chitosan has been one of the most attractive materials that has been investigated as vaccine adjuvant/delivery system, due to its promising features such as bioadhesivity, biocompatibility, biodegradability, and penetration enhancing activity, as well as bioactive properties (immunostimulatory, anti-inflammatory). In

S. Parmaksız and S. Şenel (✉)
Department of Pharmaceutical Technology, Faculty of Pharmacy, Hacettepe University, Ankara, Turkey
e-mail: selin.yuksel@hacettepe.edu.tr; ssenel@hacettepe.edu.tr

this review, after giving brief information about chitosan and its role in immunization, recent studies on chitosan-based adjuvant/delivery systems will be reviewed in line with the delivery route.

Keywords Adjuvant · Chitosan · Cutaneous · Mucosal · Parenteral · Vaccine delivery

1 Introduction

Vaccination is a highly effective and safe method in preventing life-threatening infections in both human and animals. It is based on the principle of activating the immune system and providing protection against infection by administration of the pathogen (antigen) that causes the disease [1]. In general, vaccines can be classified as live-attenuated, inactivated, toxoid vaccines, subunit and highly purified recombinant vaccines. Nucleic acid (DNA and RNA)-based vaccines have also attracted attention, especially during the COVID-19 pandemics. Use of subunit and highly purified vaccines contains the specific fragment of the pathogen that induces the immune response, and by this means, they provide enhanced safety. However, the immunogenicity of subunit vaccines is usually relatively low due to their monomeric nature and the absence of other immunostimulatory components [2–5]. Among the strategies investigated to improve the immunogenicity of subunit vaccines are polyvalent antigen display, use of adjuvants and/or delivery systems [6, 7]. Adjuvants have diverse mechanisms of action and should be selected for use on the basis of the route of administration and the type of immune response (antibody, cell-mediated, or mucosal immunity) that is desired for a particular vaccine [3, 8, 9]. As the activity of the adjuvant is the result of multiple factors, the enhanced immune response obtained with the adjuvant for one antigen cannot be extrapolated to another antigen. Furthermore, each antigen differs in its physical, biological, and immunogenic properties. Adjuvants are divided mainly into two groups as immunostimulant and delivery systems according to their mechanism of action [7, 8, 10, 11]. Immunostimulants act by directly stimulating the immune system specifically through pattern-recognition receptors (PRRs), such as toll-like receptors (TLRs), NOD-like receptors (NLR), C-type lectins, and RIG-like receptors of immune cells [12–14]. Delivery systems provide the presentation of the required amount of the antigen and/or adjuvants to the appropriate immune cells. Adsorbent mineral salts (e.g., aluminum salts), particulate systems, and combinations of them are included in this group [14, 15]. Alum is the only compound that has approval as an adjuvant (since 1930), whereas the other adjuvants used in commercially available vaccines have approval with the product. It is noteworthy that although many studies have been carried out to date, the number of adjuvants available on the market is still very limited, mainly due to the safety issues [8, 16–18]. There are numerous studies on new adjuvants which are in clinical trial stage [19, 20].

Particulate delivery systems, especially lipid-based systems such as immune-stimulating complexes (ISCOMs), liposomes, emulsions, virosomes, lipid nanoparticles as well as polymer-based micro- and nanoparticles have been designed to deliver the antigens. The main aim is to exert all the properties of pathogens with the exception of causing disease [21, 22]. Key features of pathogens that can be mimicked by vaccine delivery systems are their size, shape, and surface molecule organization. Besides the above-mentioned features, uptake of these systems by APCs or reaching to lymph nodes depends also on surface charge, hydrophobicity, and hydrophilicity [23]. In general, cationic particles show greater uptake and activation of APCs than neutral or anionic particles. This is attributed to the enhancement of binding to the negatively charged cell surface by the positively charged particle, which subsequently sets off internalization into the cell [24, 25]. Particulate systems are also reported to provide a depot effect and extend the residence time of the antigen [26–28]. Furthermore, depending on the type of the particulate system as well as the ingredients used to prepare this system, mechanism of uptake by the immune system can also differ. Virus size particles (20–200 nm) are taken into cells by endocytosis and directly reach lymph nodes and create immune response by interacting with B cells. Larger size particles (0.5–5 μm) are taken into cells with macropinocytosis, particles larger than 0.5 μm are taken up by macrophages predominantly by phagocytosis [29–31]. In addition to antigen delivery, particulate systems themselves have been shown to enhance the immune responses.

Natural (e.g., chitosan, alginate, hyaluronic acid) and synthetic polymers (e.g., polyanhydrides, polyesters (poly lactic acid (PLA), polyglycolic acid (PGA), poly (lactic-co-glycolic acid) (PLGA), polycaprolactone (PCL)) and their combinations have been investigated for the preparation of polymeric micro- and nanoparticles for antigen delivery [32, 33]. Amongst them, chitosan has been one of the most attractive polymers that has been investigated as vaccine adjuvant/delivery system, due to its promising features such as bioadhesivity, biocompatibility, biodegradability, and penetration enhancing activity, as well as bioactive properties (immunostimulatory, anti-inflammatory) [34–49]. Chitosan-based systems have been prepared both as an adjuvant/antigen-delivery system for systemic and mucosal immunization. The studies showed that chitosan stimulates both humoral and cellular immune responses against various antigens. In the following sections, after giving brief information about chitosan and its role in immunization, recent studies on chitosan-based adjuvant/delivery systems will be reviewed in line with the delivery route.

2 Chitosan and Immunization

Chitosan is obtained by deacetylation of chitin, which is a polysaccharide found in the exoskeleton of crustaceans and insects and in the cell walls of some bacteria and fungi [39, 42, 49–51]. Chitosan is a linear copolymer consisting of poly [ß-(1 → 4)-

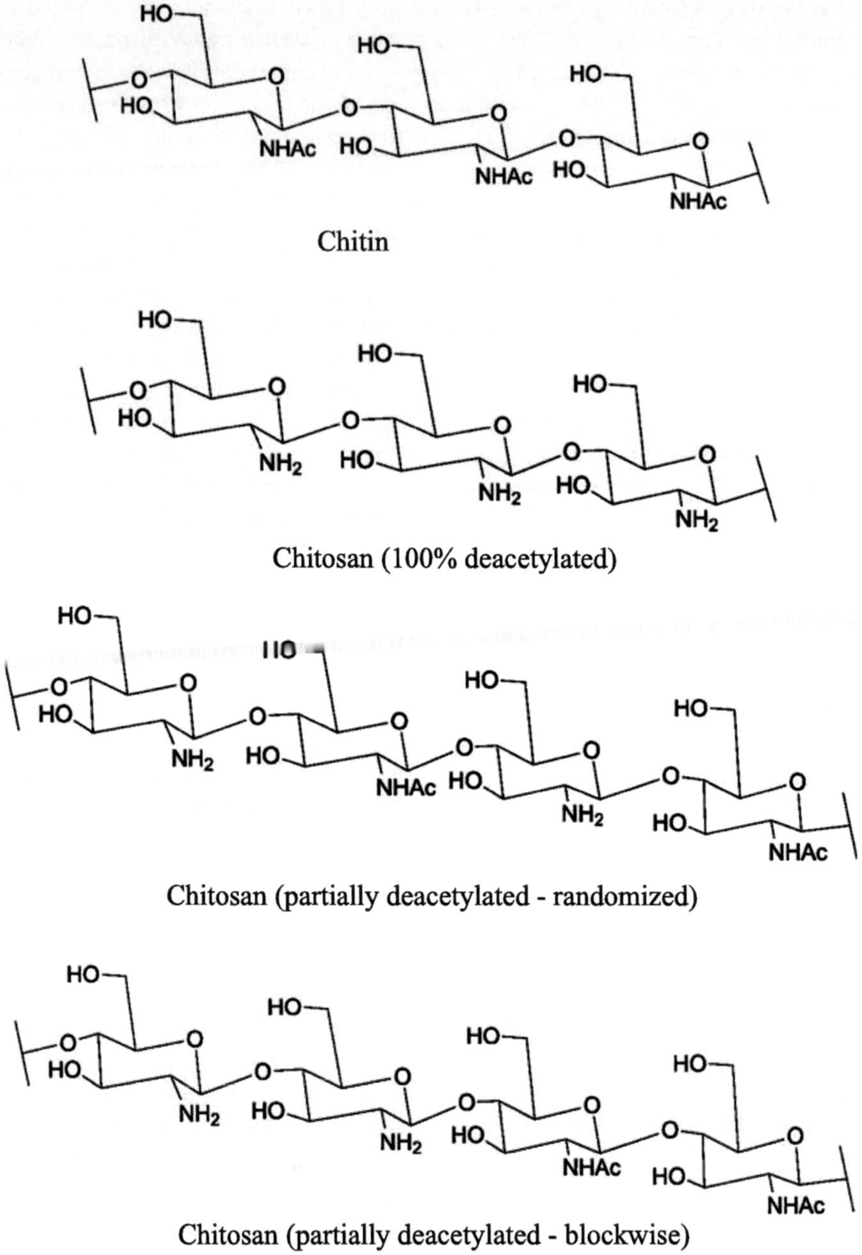

Fig. 1 Chemical structure of chitin and chitosan

2-amino-2-deoxy-D-glucopyranose] with randomly located N-acetylglucosamine groups depending upon the degree of deacetylation (DD) of the polymer (Fig. 1). A monograph for water-soluble chitosan chloride is included in the European Pharmacopeia [52] and a monograph for chitosan in the United States Pharmacopeia/National Formulary [53]. In these monographs, chitosan source is animal, specified as shells of shrimps and crabs, and its degree of deacetylation is required to be above 70%. Other specifications such as appearance, solubility, identification, bacterial endotoxin, microbial limits, loss on drying, residue on ignition (or sulfated ash), limit of heavy metals, limit of protein content, average molecular weight and molecular weight distribution, appearance of solution, matter insoluble in water, pH, viscosity are also stated in the pharmacopeia. Non-animal source (fungal) chitosan is also available for medical purposes; however, it has not been included in the pharmacopeia yet.

Chitosan can be characterized in regard to its intrinsic properties (purity, molecular weight, viscosity, and degree of deacetylation) and physical form. It is commercially available in various grades of purity, molecular weight (low, medium, high), and degree of deacetylation, obtained from either animal or non-animal sources. Chitosan is readily soluble in dilute acidic solutions below pH 6.0. Degree of deacetylation (DD), which represents the proportion of N-acetyl-D-glucosamine units with respect to the total number of units, can be randomized or blockwise (Fig. 1). The DD and deacetylation pattern play an important role in defining the physicochemical and biological properties of chitosan. Chitosan has cationic charge due to its free amino groups, which are also the active sites ready to interact with various molecules enabling to tailor structures and systems with desired properties. Chitosan is degraded by enzymes such as chitosanase and lysozyme. The biodegradation rate of chitosan is controlled by its DD and molecular weight as well as its crystallinity. Up to date, no adverse issues have been reported about its safety in medical applications, both internally and externally [54]. It is important to make sure that it does not cause any allergies or any other side effects as it is mainly animal-sourced.

For decades, chitosan has been widely investigated in medical field both as a therapeutic agent and as a delivery system due to its promising features [55–63]. For immunization, chitosan and its derivatives have been investigated both as an adjuvant and antigen-delivery system, in different forms such as gel, film, fibers, micro- and nanoparticle gel, aqueous dispersions for systemic and mucosal immunization [37, 42, 43]. Chitosan has been shown to induce both humoral and cellular immune responses against various antigens. Zaharoff and his group [36] have shown in C57BL/6 mice that chitosan solution, prepared using water-soluble glutamate salt of chitosan, improved humoral and cell-mediated immune responses to a subcutaneous vaccination with a model protein antigen in the absence of additional adjuvants. They reported that chitosan exhibited mainly two characteristics that may allow it to function as an immune adjuvant. First, the viscous chitosan solution was shown to display an antigen depot. Second, it was found to induce a transient 67% cellular expansion in draining lymph nodes. In presence of chitosan, antigen-specific serum IgG titers were enhanced over fivefold and antigen-specific $CD4^+$

proliferation over sixfold. The nature of the immune responses facilitated by chitosan was reported to be a mixed TH1/TH2 response. Delayed type hypersensitivity (DTH) responses were used to confirm that chitosan also induced a robust cell-mediated immune response. Studies by several groups have continued to elucidate the mechanism underlying the adjuvant activity of chitosan. These proposed mechanisms can be summarized as antigen protection, depot formation, enhanced antigen uptake and presentation and direct modulation of immune responses [24, 64–68]. The positive surface charge of chitosan was shown to enhance the immune response by increasing the interaction of antigen with immune cells that have negative surface charge. In addition, chitosan extends the contact time of the antigen with the surface and increases the uptake by APCs due to its bioadhesive property. In recent years, studies revealing the immunomodulatory effects of chitosan have been reported. Bueter et al. [69] have shown that chitosan activates the NLRP3 inflammasome, and this contributes to Th1 cell polarization. Caroll et al. [44] have described the mechanism by which chitosan promoted DC activation and induction of cellular immunity. Chitosan was shown to stimulate the intracellular release of DNA, which engaged the cGAS-STING pathway to mediate the selective production of type I IFN and interferon-stimulated genes (ISGs). These cytokines were reported to be responsible for mediating the activation of DCs and induction of cellular immunity. The authors concluded that chitosan could engage the STING-cGAS pathway to trigger innate and adaptive immune responses.

There are numerous studies performed in vivo in animal models [36, 37, 48, 70–73] while a small number of studies in human showing the adjuvant effect of chitosan. McNeela et al. [74] have formulated a nasal diphtheria vaccine and showed that chitosan glutamate significantly augmented Th2-type responses, which correlated with protective levels of toxin-neutralizing antibodies in intranasally boosted individuals. One year after this study, it was reported that a remarkably efficient and reliable induction of levels of meningococcal and diphtheria immunity associated with protection against disease was obtained by simple syringe insufflation of Menjugate-C in chitosan glutamate in 36 healthy volunteers [75]. Read et al. [76] have investigated the use of chitosan, as a nasal delivery system with inactivated, subunit influenza vaccine. Subjects received nasally standard inactivated trivalent influenza vaccine with chitosan. Serum hemagglutination inhibition (HI) titers following intranasal vaccination with the nasal chitosan-influenza vaccine was reported to meet the criteria set by the regulatory authority in terms of seroprotection rate. El-Kamary et al. [77] have developed an adjuvanted dry powder intranasal Norwalk Virus-Like Particle (VLP) vaccine formulation including monophosphoryl lipid A (MPL) and chitosan (ChiSys®, Archimedes Development Ltd.) and performed a Phase 1–2 study. It was reported that the intranasal monovalent adjuvanted Norwalk VLP vaccine was well tolerated and highly immunogenic. As a follow-up of this study, intranasally administered adjuvanted Norwalk VLP vaccine was investigated in healthy adults against experimental human Norwalk virus illness [78]. Protection against illness and infection was demonstrated after challenge with a homologous virus. Neimert-Andersson et al. [79] have demonstrated the safety and tolerability of a chitosan-based hydrogel (ViscoGel-based on a low deacetylated medical grade chitosan) in a Phase I/IIa clinical trial in combination with a model antigen (Act-HIB). However, no adjuvant effect was observed on the antibody response after i.m. administration, while ViscoGel was shown to have an impact on the IFN-γ response to the vaccine. Importantly, Act-HIB in combination with ViscoGel did not inhibit a cell-mediated response, different than that Act-HIB alone. The reason for not observing any adjuvant effect can be attributed to the low deacetylation degree of the chitosan used as well as its solubility properties. Scherließ et al. [80] have reported that the adjuvant effect of chitosan would be dependent on its molecular weight, degree of deacetylation and also its particle size and preparation technique that would affect the solubility and viscosity. In this line, when the clinical studies reported so far is examined, it is noteworthy that in general water-soluble chitosan salts such as chitosan glutamate with deacetylation degree above 80% were used in these studies.

3 Chitosan-Based Systems for Immunization

As described above, chitosan itself exerts immunostimulant property. Furthermore, due to its versatile chemical and physical properties structure, it allows to prepare delivery systems in different forms and with desired properties such as particle size, surface charge, conjugation with other molecules, etc. (Fig. 2). Chitosan has been used in aqueous dispersion, gel, micro- and nanoparticle form to study its effect on immune response and for delivery of different antigens.

There are numerous studies reported on advantages of these dosage forms over to each other.

In our previous study, we have compared the aqueous dispersion and nanoparticles of chitosan and its derivatives carboxymethyl chitosan (MCC) and trimethyl chitosan (TMC), for nasal delivery of tetanus toxoid in vivo in Balb/c mice [38]. Chitosan and TMC nanoparticles (300–400 nm), which had positively charged

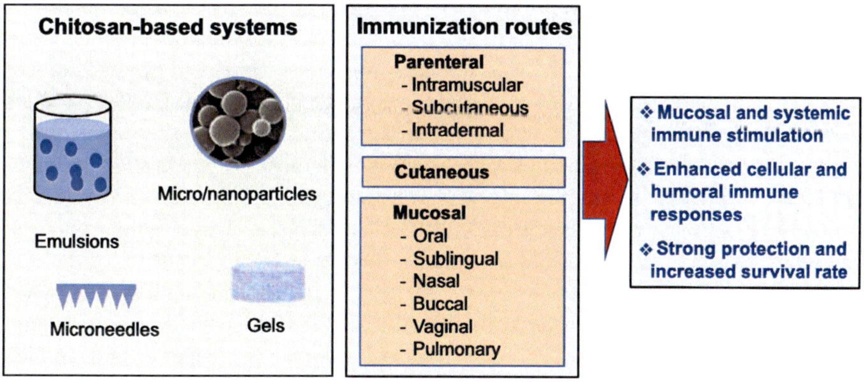

Fig. 2 Chitosan-based adjuvant/vaccine delivery systems and delivery routes

surfaces induced higher serum IgG titers when compared to those prepared with MCC, which were negatively charged and smaller in size (40–90 nm). Both the aqueous dispersion and nanoparticle systems prepared by chitosan derivatives were found to enhance mucosal immune responses and no significant difference was found between the aqueous dispersion and the nanoparticles. Nonetheless, encapsulation of the antigen into nanoparticles would be more advantageous aqueous dispersion, especially in regard to protection of the antigen both in vitro and in vivo before reaching the target M-cells, and especially for mucosal delivery, removal of the chitosan nanoparticles from the application site will be less compared to aqueous dispersion. In another study, we have developed TMC-MCC nanoparticles (280 nm) by complexation between the oppositely charged chitosan derivatives without using any crosslinker and immune responses against intranasal administration of TT loaded systems were compared to that obtained with chitosan, TMC, and MCC nanoparticles [40]. TMC-MCC complex nanoparticles were shown to induce higher mucosal and systemic immune responses. Results of these two studies demonstrated that the surface charge and particle size of the chitosan-based system play an important role in obtaining an enhanced immune response. Gordon et al. [81] have compared in vitro and in vivo the immunostimulatory capacity of chitosan nanoparticles and thermosensitive chitosan hydrogels as particulate delivery systems for a model antigen, ovalbumin (OVA). No significant immunogenicity was obtained with nanoparticles, whereas chitosan gel formulation was shown to stimulate both cell-mediated and humoral immunity in vivo. In a recent paper released from our group, a comparison was performed in regard to cellular uptake and macrophage activation between the gel, nanoparticle and microparticle forms of chitosan-based adjuvant system incorporated with porins [82]. Uptake from J774A.1 macrophage cells was observed to be higher with the gels, following microparticles and nanoparticles. Expression of MHC-II molecules in murine macrophage cells was not observed with chitosan gel or nanoparticles, while high stimulation of MHC II expression was obtained with the microparticles. On the other side, CD80 and CD86 expression was observed to increase with both gel and nanoparticles.

In the vast majority of the studies performed on chitosan, nanoparticles have been the preferred adjuvant/delivery systems [10, 39, 42, 49]. They offer numerous advantages besides being biodegradable, bioadhesive, and biocompatible: chitosan nanoparticles do not require elevated temperature, shear stress, and organic solvent use during preparation, preparation methods are easily applicable and stable particles are obtained, with high loading capacity. Furthermore, the particle size can be tuned according to the desired properties. Combination with other polymers and loading of additional adjuvants is also possible with chitosan nanoparticles. The most distinct property of chitosan nanoparticles is their positive surface charge, which plays an important role in enhancement of the immune responses.

In studies that have been going on for years, it has been shown that when chitosan-based systems are used with different antigens by different administration routes of administration, they increase humoral and cellular response and provide protection against pathogen. Various chemical modifications of chitosan have also been investigated in order to improve the applicability for vaccine delivery [83]. In the following sections, the chitosan-based delivery systems for vaccine will be reviewed for each delivery route after giving a brief introduction to each route.

4 Chitosan for Mucosal Immunization

Mucosal surfaces (e.g., nasal, oral, buccal, sublingual, rectal, vaginal) are the main entrance of many pathogens to the body, thus majority of the infections occur at or take their departure from the mucosal surfaces. To combat infection, mucosal surfaces are equipped with physical, chemical, and immunological defense mechanisms. In particular, mucosal tissues comprise a highly compartmentalized and specialized immune system in the form of the mucosa-associated lymphoid tissue (MALT) [84, 85]. MALT helps to induce pathogen-specific immune responses and in the secretion of immunoglobulin A (IgA) at mucosal surfaces to protect against infection [86–88]. The MALT includes the nasopharynx-associated lymphoid tissue (NALT), the bronchus-associated lymphoid tissue (BALT), and the gut-associated lymphoid tissue (GALT), which comprises Peyer's patches (PPs) and isolated lymphoid follicles (ILFs) [86, 89, 90]. Each mucosal route induces different immune responses in terms of potency and longevity, due to the differences in the organization and cellular make-up of lymphoid structures in different mucosal tissues. Mucosal immunization is considered as an attractive alternative to conventional parenteral immunization, eliciting immune defense in both mucosal and systemic tissue for protecting from pathogen invasion at mucosal surfaces without any necessity of needle-based vaccination [91–95]. However, currently very few mucosal vaccines are available on the market. This is due to the lack of effective delivery systems that are able to protect antigen as well as provide strong adjuvanticity.

4.1 Oral and Sublingual Delivery

Oral immunization has limitations due to degradation of the vaccine in the acidic pH as well as the presence of the digestive enzymes and low intake of antigens in the lymphoid tissue of the gastrointestinal tract [96, 97]. Investigations on design of new adjuvant/delivery systems focus on protecting the antigens from degradation in the intestine and stomach and deliver them efficiently to the gut-associated lymphoid tissue (GALT) [98]. Furthermore, antigens can get diluted before absorption at the mucosa. Hence, a comparatively larger amount of vaccine may be needed for oral vaccination to produce an effective immune response that is comparable to other routes. Preferably, an oral mucosal vaccine is intended to adhere to the intestinal mucosa or selectively target the M cells. Among the approaches to improve the delivery of vaccines orally is the use of delivery systems to encapsulate them in particulate systems such as polymeric micro- and nanoparticles, liposomes, etc. [26, 42, 99–102].

Chitosan-based systems have been investigated both as adjuvant and delivery system also for oral delivery of vaccines both in human and veterinary field. In the early 2000s, van der Lubben et al. [60] have shown that chitosan microparticles (<10 μm) were taken up by the Peyer's patch after intragastric administration to mice. A dose-dependent immune reaction was observed for mice vaccinated with different doses of DT incorporated to chitosan microparticles. Up to present, chitosan micro- and nanoparticles have been investigated for oral delivery of numerous viral and bacterial antigens as well as DNA vaccines (Table 1).

Saraf et al. [103] have developed chitosan nanoparticles for hepatitis B surface antigen (HBsAg) anchored with lipopolysaccharide (LPS) as an adjuvant. The nanoparticles were coated with alginate as a stabilizing agent for oral mucosal immunization. Enhanced production of secretory IgA was observed in all mucosal secretions. Alginate coating was shown to provide higher immune responses when compared to uncoated particles.

Renu et al. [104] have prepared chitosan nanoparticles for delivery of Salmonella subunit vaccine through drinking water or feed in poultry. Cellular uptake of F-protein coated chitosan nanoparticles was found to be higher than F-protein alone. Nanoparticles were reported to trigger the expression of both cell surface and endosomal TLRs, as well as enhanced Th1 and Th2 cytokines mRNA expression in vitro. Moreover, drinking water and feed-delivered nanoparticles were shown to reduce the Salmonella load in chickens, suggesting that reduced bacterial load was mediated by the adjuvant effect as well as efficient vaccine delivery to mucosal immune cells. In another study, carboxymethyl chitosan/chitosan nanoparticles (CMCS/CS-NPs) loaded extracellular products (ECPs) of *Vibrio anguillarum* were developed for oral antigen delivery in fish vaccination. Oral immunization of ECPs-loaded CMCS/CS-NPs group in turbot was shown to elevate specific antibody and higher concentrations of lysozyme activity and complement activity in fish serum than ECPs solution [105].

Chang et al. [106] have evaluated the effect of chitosan hydrolysate (obtained via cellulase degradation of chitosan), low molecular weight chitosan, and a chitooligosaccharide mixture on mitogen-induced and antigen-specific immune responses after oral administration in Balb/c mice. All chitosan products were shown to significantly increase the proliferation of splenocytes and Peyer's patch lymphocytes. Chitosan oligomixture was found to be the most potent chitosan hydrolytic product to elicit IgM, IgG, and IgA production, while low molecular weight chitosan was the most potent one to elevate splenic IFN-γ production and IFN-γ/IL-4 (Th1/Th2) ratio.

Slütter et al. [107] have compared the immune responses obtained by oral administration of nanoparticles prepared using chitosan and its N-trimethylated derivative, TMC. Intraduodenal vaccination with OVA-loaded nanoparticles was shown to induce significantly higher antibody responses than immunization with OVA alone. TMC nanoparticles were found to induce anti-OVA antibodies after only a priming dose. The interaction of nanoparticles with the intestinal epithelium was also explored, using a follicle associated epithelium model. Only TMC nanoparticles were found to exert intrinsic adjuvant effect on dendritic cells. It was

Table 1 Chitosan-based systems for mucosal delivery of vaccines

Antigen	Chitosan type	Dosage form	Immunization route	Adjuvant actadded	In vitro studies	In vivo studies	Immune response	Ref.
Brucella abortus recombinant proteins (rMdh, rOMP 10 and 19)	Chitosan (MW is not stated) (DD: 82.4%) (Jakwang, Ansung, Korea)	Nanoparticle *rMdh-CS NPs: 475.4 ± 124.5* *rOMP 10-CS NPs: 360.8 ± 84.5 nm* *rOMP 19-CS NPs: 439.5 ± 89.0 nm* *Cocktail of three NP formulations*	Nasal	–	–	Immunization in 6-week-old female BALB/c mice *Dose: 30 µg* *Boost: On day 14 and 28* – Determination of IgA, IgG and its subtypes – Determination of cytokine levels	– Significant production of IFN-γ and IL-4 levels suggesting the induction of a mixed Th1–Th2 response – Antigen-specific IgA levels with a Th2-polarized immune response following cocktail antigen loaded CS NPs immunization	[120]
HBsAg	Chitosan (MW 3.8–20 kDa) (DD: >75%) (HiMedia, India)	Nanoparticle *HBsAg-CS NPs: 430.46 ± 1.76 nm* *Alginate coated HBsAg-CS NPs: 440.34 ± 2.84* *LPS anchored alginate coated HBsAg-CS NPs: 605.23 ± 2.81*	Oral	LPS	– Mucoadhesion study (performed by Adamczak et al.'s method [158]) (cellulose nitrate membrane) – Cell viability (RAW264.7 cells)	Immunization in 6-week-old female BALB/c mice *Dose: 20 µg* *Boost: On days 2 and 3* – Determination of IgA in mucosal fluids, IgG and its subtypes	– Higher mucoadhesion with alginate-coated NPs compared to non-coated NPs – Dose- and time-dependent cytotoxic effect with LPS anchored alginate-coated NPs – High cell viability with alginate-coated NPs – Highest IgA and IgG levels with LPS anchored alginate-coated NPs	[103]

(continued)

Table 1 (continued)

Antigen	Chitosan type	Dosage form	Immunization route	Adjuvant actadded	In vitro studies	In vivo studies	Immune response	Ref.
OMPs and flagellin (F) protein of *Salmonella serovar Enteritidis*	Chitosan (MW: 50–190 kDa) (DD: 75–85%) (sigma-Aldrich, Saint Louis, USA)	Nanoparticle *OMPs-F-CS NPs: 514 nm*	Oral	–	– Cellular uptake study (PBMCs of chicken) – Determination of expression of TLRs mRNA (PBMCs of chicken)	Immunization in 5-week-old SPF white Leghorn layer chicks – *Experiment 1 oral immunization with OMPs-F-CS NPs* – *Experiment 2 oral immunization with drinking water and feed-delivered OMPs-F-CS NPs* *Dose: 25, 50, 250 μg* *Boost: Days 21 and 42* *Challenge: Oral with nalidixic acid-resistant virulent S. enteritidis on day 63* – Determination of IgY and IgA	– Increased cellular uptake of NPs – Significantly increased expression of TLRs mRNA with NP formulation – Increased IgY and IgA levels with all doses – Reduction on the Salmonella infection in intestine on day 73 – Significantly enhanced IgY and IgA levels following drinking water or feed-delivered OMPs-F-CS NPs immunization	[104]
Killed/inactivated swine influenza virus (H1N2-OH10) antigen (KAg)	Chitosan (MW: 50–190 kDa) (DD: 75–85%) (sigma-Aldrich, Saint Louis, USA)	Nanoparticle *Chitosan nanovaccine: Combination of KAg-CS NPs (141 nm) and KAg-CS NPs-poly(I:C) (411 nm)*	Nasal	Poly(I:C)	–	Immunization in 5-week-old cesarean-delivered colostrum-derived SwIV antibody free piglets *Dose: 10^7 $TCID_{50}$*	– Higher and similar HI titers with chitosan nanovaccine and commercial vaccine – Induction of	[122]

		KAg-poly(I:C) (for comparison of adjuvant properties with NPs)				*equivalent of KAg Boost: Day 21 challenge: i.n. and intratracheal with SW/OH/24366/ 2007 (H1N1-OH7) influenza virus on day 35* – HI assay – Determination of cytokine gene expressions	cross-reactive systemic antibody and increased both Th1 (IFN-γ, IL-6, and IL-2) and Th2 (IL-10 and IL-13) cytokines gene expression in TBLN associated with increased IFN-γ secreting T-helper/memory cells with chitosan nanovaccine compared to commercial vaccine – Not substantially induction of IgA – Similar reduction on virus load and infection with chitosan nanovaccine and commercial vaccine	
Recombinant MxiH antigen of *Shigella flexneri* produced by *E. coli* BL21	Chitosan (MW: 50–190 kDa) (DD: 75–85%) (sigma-Aldrich, Saint Louis, USA)	Nanoparticle (100 nm) *MxiH-Freund's adjuvant (for comparison of adjuvant properties with CS NPs)*	Nasal	–	–	Immunization in 21-day-old male BALB/c mice *Dose: Not stated* *Boost: On days 20 and 55* *Challenge: i.n. with S. flexneri serotype 1a on day 60* – Determination	– Enhanced and similar IgG and IgA levels with NPs and Freund's adjuvant – Higher IFN-y levels with NPs compared to Freund's adjuvant – Similar IL-4 levels with	[159]

(continued)

Table 1 (continued)

Antigen	Chitosan type	Dosage form	Immunization route	Adjuvant added	In vitro studies	In vivo studies	Immune response	Ref.
						of IgA, IgG and its subtypes	Freund's adjuvant-MxiH compared to CS and MxiH alone – Highest survival rate with NPs (60%) immunized mice compared to Freund's adjuvant (50%)	
Acinetobacter baumannii biofilm-associated protein (bap)	Chitosan (MW: 190–310 kDa) (DD: 75–85%) (sigma-Aldrich, Saint Louis, USA)	Nanoparticle (1,080 nm) *Bap-Freund's adjuvant (for comparison of adjuvant properties with CS NPs)*	Nasal	–	–	Immunization in 4- to 6-week-old female BALB/c mice *Dose: 20 μg* *Boost: Days 14 and 28* *Challenge: i.n. with A. baumannii to passive immunized mice with immune sera of immunized mice* – Determination of IgA, IgG and its subtypes	– Higher IgG and IgA levels with NPs compared to Freund's adjuvant – 100% survival following the passive immunized mice with immune sera obtained from bap-CS NPs immunized mice after 48 h post-challenge	[160]
Native or toxoid extracellular proteins (ECP) from a field strain of *Clostridium perfringens*	Chitosan (MW and DD are not stated) (sigma-Aldrich, Saint Louis, MO)	Nanoparticle – *Native ECP CS NP: 389.7 nm* – *Toxoid ECP CS NP: 352.5 nm*	Oral	–	– Hemolytic effect (red blood cells of chicks)	Immunization in 1-day-old cobbs 500 chicks *Dose: 50 μg* *Boost: On days 3, 7, and 14* – Determination	– Very low hemolytic effect with native ECP CS NPs and no hemolytic effect with toxoid ECP CS NPs – Enhanced IgA	[161]

						of IgA, IgG and its subtypes – Splenic mononuclear cells proliferation assay	levels with toxoid ECP CS NPs on day 17 – Enhanced IgA levels with native ECP CS NPs on day 21 – Significant proliferation of splenic mononuclear cells with NP formulations indicating a fast immune response by T cells as a result of high recall response	
GAS lipopeptides (LP1, LP2, and LP3)	Chitosan (used for TMC synthesis) (MW: 50–190 kDa) (DD: 75–85%) (sigma-Aldrich, Saint Louis, USA)	Nanoparticle *LP1-TMC NPs: 200 ± 10 nm* *LP2-TMC NPs: 424 ± 18 nm* *LP3-TMC NPs: 126 ± 14 nm*	Nasal	–	– Dendritic cell and macrophage activation: Determination of costimulatory molecules expression (splenic cells from C57BL/6 J mice)	Immunization in 6-week-old female C57BL/6 J mice *Dose: 30 μg* *Boost: On days 21 and 42* – Determination of IgA, IgG and its subtypes levels – Indirect bactericidal assay (with sera of immunized mice against GAS strains)	– Significantly higher expression of costimulatory molecules in both dendritic cells and macrophages with NP formulations compared to lipopeptides alone – Highest IgG titers with LP3-TMC NPs following first and boost immunizations – Increased IgG titers with LP1-TMC NPs and LP3 following second boost	[125]

(continued)

Table 1 (continued)

Antigen	Chitosan type	Dosage form	Immunization route	Adjuvant actadded	In vitro studies	In vivo studies	Immune response	Ref.
							immunization – No detectable of mucosal IgA titers with any groups – Highest opsonization with sera of immunized mice with LP3-TMC NPs	
BSA (as a model antigen)	aCS and atCS Chitosan (used for aCS and atCS synthesis) (MW: 50–190 kDa) (DD: 75–85%) (sigma-Aldrich, Saint Louis, USA)	Nanoparticle *BSA loaded aCS NPs: 125.2 ± 5.5 nm* *BSA loaded atCS NPs: 110.0 ± 3.3 nm* – BSA-alum *(for comparison of adjuvant properties with NPs)* – BSA-CpG ODN *(for comparison of adjuvant properties with NPs)*	– Nasal Compared to subcutaneous BSA-alum	–	– Mucoadhesion studies (hydrated cellulose acetate membrane filter with mucin in nasal fluid) – Cytotoxicity assay (Calu-3 and A549 cells)	Immunization in 6- to 8-week-old female BALB/c mice *Dose: 20 μg* *Boost: On day 14* – Determination of IgA, IgG and its subtypes levels – determination of cytokine levels	– Concentration dependent cytotoxic effect in both Calu-3 and A549 cells – Significantly higher IgG and IgA titer and cytokine expression with aCS and atCS NPs compared to CpG ODN and alum – Stimulation of Th1-related cytokine levels and Th2-based antibody levels with i.n. immunization of NPs and stimulation of Th2-related cytokine levels and	[128]

							Th1-based antibody levels with s.c. immunization of NPs indicating the mixed Th1/Th2 immune response	
– Cocktail antigens of *Brucella abortus 544* (sodC, OMP19, BLS, and PrpA) – LPS of *Brucella abortus 544*	Chitosan *(MW and DD are not stated)* (sigma-Aldrich, Saint Louis, USA)	Nanoparticle	Nasal	LPS	–	Immunization in 5-week-old female BALB/c mice *Dose: 5 μg LPS; 25 μg each antigen (sodC, OMP19, BLS, and PrpA)* *Boost: On day 14* *Challenge: i.n. with B. abortus 544 on day 30* – Determination of IgA, IgG and its subtypes levels – Determination of cytokine levels – Splenocyte cell proliferation assay	– Higher IgG and mucosal IgA levels with cocktail antigens loaded CS NPs and LPS+ cocktail antigens loaded CS NPs – Higher levels of IFN-γ and T cell proliferation with cocktail antigens loaded CS NPs and LPS+ cocktail antigens loaded CS NPs – Decreased bacterial load in lungs and spleen with cocktail antigens loaded CS NPs and LPS+ cocktail antigens loaded CS NPs – No significant effect of LPS to enhance humoral/cellular immune response and protection against infection	[121]

(continued)

Table 1 (continued)

Antigen	Chitosan type	Dosage form	Immunization route	Adjuvant actadded	In vitro studies	In vivo studies	Immune response	Ref.
Human influenza A/Puerto Rico/8// 1934 (H1N) virus antigen (PR8)	Chitosan and TMC Chitosan (used for TMC synthesis) (MW: 120 kDa) (DD: 92%) (Primex, Avaldsnes, Norway)	Nanoparticle *PR8 loaded CS NPs: 380.6 nm* *PR8 loaded TMC NPs: 318.2 nm* *Alginate coated PR8 loaded CS NPs: 471.1 nm* *Alginate coated PR8 loaded TMC NPs: 453.2 nm*	Nasal	–	–	Immunization in female BALB/c mice *Dose: 15 μg* *Boost: On days 14 and 28* – Determination of IgG and its subtypes levels	– Enhanced IgG levels with NP formulations – Highest IgG1 levels with mixed Th1/Th2 immune response after vaccination with PR8 loaded CS NPs – Highest IgG2a levels indicating the Th1-based immune response after vaccination with alginate-coated PR8 loaded TMC NPs	[126]
–	Chitosan (used for LMW chitosan, chitosan hydrolysate and oligomixture synthesis) (MW: 300 kDa) (DD: 95%) (synthesis from chitin)	–	Oral	–	– Phagocytic activity of macrophages (murine peritoneal macrophages) – Lymphocyte proliferation assay (murine spleen and Peyer's patches cells)	Immunization in 4-week-old female BALB/c mice *Single dose: 500 mg/kg body weight* – Determination of IgA, IgM, IgG and its subtypes levels – Determination of cytokine levels	– Increased the phagocytic activity of macrophages with chitosan hydrolysate and oligomixture – Significantly increased splenocyte and Peyer's patches lymphocytes proliferation with all chitosan derivatives	[106]

						– Splenocyte proliferation assay	indicating upregulation MHC-II molecules of APCs – Highest IgG, IgA, and IgM levels with chitosan oligomixture compared to LMW chitosan and chitosan hydrolysate – No stimulation of IgE levels with any chitosan indicating no risk to induce type 1 hypersensitivity – Higher Th1 cytokine secretion with chitosan oligomixture, LMW chitosan and with chitosan hydrolysate, respectively	
Recombinant hcp gene from *Campylobacter jejuni* (rhcp)	Chitosan (MW: 3.8–20 kDa) (DD: >75%) (HiMedia, India)	Nanoparticle *(180 ± 6 nm)* *Rhcp-IFA (for comparison of adjuvant properties with CS NPs) (s.c.)*	Oral Compared to subcutaneous *rhcp-IFA*	–	– Cytotoxicity assay (Caco-2, INT407 and primary CEICs)	Immunization in 40-day-old Vencobb chicks *Dose: 50 μg* *Boost: On days 14 and 21 (with 25 μg)* *Challenge: Orally with C. jejuni isolate BCH71 on day*	– No cytotoxic effect in any cell lines – Higher specific IgY and IgA levels with NPs and IFA – Upregulation of Th1 and Th17 cytokines gene expressions with	[162]

(continued)

Table 1 (continued)

Antigen	Chitosan type	Dosage form	Immunization route	Adjuvant actadded	In vitro studies	In vivo studies	Immune response	Ref.
						28 – Determination of IgA, IgG and its subtypes – Assessment of cytokine gene expressions	NPs and IFA – Significantly reduced bacterial load with NPs immunized mice post-challenge compared to IFA	
Urease (produced by *E. coli*)	TMC (MW and DD are not stated) (provided by Dr. Sahebghadam Lotfi, Tarbiat Modares university, Tehran, Iran)	Nanoparticle (300–400 nm)	– Oral – Parenteral (intraperitoneal)	–	–	Immunization in 4- to 6-week-old female SPF BALB/c mice *Dose: 30 μg (i.p.)/ 75 μg (oral)* *Boost: On days 7 and 14 (oral)* *On day 15 (i.p.)* *Challenge: i.p. with virulent Brucella spp. on day 45* – Determination of IgA, IgG and its subtypes – Lymphocyte proliferation assay – Cytokine response	– Highest IgG, IgG1, IgG2a and IFN-γ, IL-12, IL-4 levels with i.p. multi-vaccinated groups indicating potent cellular mixed Th1–Th2 immune response – No detectable IgA levels with any group – Higher IgG2a and IL-17 levels with oral multi-vaccinated groups indicating a high cellular mixed Th1–Th17 immune response – Highest lymphocyte proliferation and protection with i.p. multi-	[163]

							vaccinated groups – Higher induction of cellular immune response and protection against infection with i.p. and multi-vaccination compared to oral and single vaccination	
OVA	Chitosan (Protasan UP CL213) (MW: 150–400 kDa) (DD: 75–90%) (Novamatrix, Norway)	Nanoparticle (196.5 nm) – *OVA-cholera toxin (CT) (for comparison of adjuvant properties with CS NPs)*	Oral	–	– Cytotoxicity studies – Measurement of permeability and tight junction opening effect (by TEER studies) (Caco-2 cells)	Immunization in 6- to 8-week-old female BALB/c mice *Dose: 200 μg* *Boost: On days 7,14, and 21* – Determination of IgA and IgG levels	– Dose-dependent cytotoxic effect – Reduction in TEER and enhanced permeability indicating the tight junction modulating effect and permeability enhancer ability of chitosan – Induction of higher IgG levels with CT compared to NPs – Induction of high IgA levels with CT and NPs, respectively	[164]
Bee (*Apis mellifera*) venom (BV)	Chitosan *(MW, DD, and the source are not stated)*	Nanoparticle – *BV-alginate NPs:* *541.5 ± 50.9 nm* – *BV-alginate-CS*	Nasal	–	–	Immunization in 4-week-old pigs *Dose: 2 mg* *Boost: On day 21 (with 1 and*	– Significant increases in Th cells, Th1-related cytokines, and IgG levels with	[165]

(continued)

Table 1 (continued)

Antigen	Chitosan type	Dosage form	Immunization route	Adjuvant added	In vitro studies	In vivo studies	Immune response	Ref.
		NPs: *434.6 ± 22.1 nm*				*2 mg)* *Challenge: i.n. with PRRSV on day 7* – Determination of PRRS-specific antibodies – Lymphocyte proliferation assay – Determination of PRRSV specific IFN-γ secreting cells	alginate-CS NPs compared to alginate NPs – Significant increase in the $CD4^+$ T cell population, Th1-related cytokines IFN-γ-secreting cells, Th/memory cells, and transcriptional factors with high dose of alginate-CS NPs compared to low dose of alginate-CS NPs – Significant reduction in the microscopic lesion score and viral load with high dose of alginate-CS NPs compared to low dose of alginate-CS NPs	

Inactivated avian infectious bronchitis virus (IBV) vaccine	Chitosan (MW 190–310 kDa) (DD: 75–85%) (sigma-Aldrich, Saint Louis, USA)	Nanoparticle – *IBV loaded CS NPs: 286 nm* – *Live attenuated IBV vaccine + IBV loaded nanoparticles (L + NPs)*	Nasal	–	– Cytotoxicity studies (chicken embryo fibroblast cells)	Immunization in 14-day-old SPF chickens *Dose: 100 μL* *Challenge: Nasal with IBVPR-05 strain on day 17* – Determination of IgA and IgG levels – Quantification of IFN-γ gene expression	– Low cytotoxic effect – No local and systemic reaction and macroscopic lesions after immunization – significantly lower viral load with IBV loaded CS NPs and L + NPs – Higher IFN-y mRNA expressions with L + NPs compared to IBV loaded CS NPs – Enhanced IgA levels with IBV loaded CS NPs compared to L + NPs – Enhanced IgG levels with L + NPs compared to IBV loaded CS NPs	[166]
J8P from GAS M-protein	TMC Chitosan (used for TMC synthesis) (MW: 50–190 kDa) (DD: 75–85%) (sigma-Aldrich, Saint Louis, USA)	Nanoparticle – *PGA-J8P-TMC NPs: 201 ± 8 nm* *-J8P-TMC solution* – *J8P-CTB (for comparison of adjuvant properties with nanogels)*	Nasal	–	– DC and macrophage activation studies: Determination of surface marker expression	Immunization in 6-week-old female inbred C57BL/6 mice *Dose: 10 μg* *Boost: On day 21 and 42* *Challenge: Nasal with streptomycin resistant GAS protein on day 62*	– Strong stimulation of both DCs and macrophages with NPs and CTB – Significantly higher IgG and IgA levels with NPs compared to CTB and TMC solution – Significantly lower bacterial load	[167]

(continued)

Table 1 (continued)

Antigen	Chitosan type	Dosage form	Immunization route	Adjuvant actadded	In vitro studies	In vivo studies	Immune response	Ref.
						– Determination of IgG and IgA levels – Opsonization assay	in mucosal fluids and strong opsonic activity against GAS strain with NPs compared to CTB and TMC solution	
BSA	Mannosylated chitosan (mCS) Chitosan (used for mannosylated chitosan synthesis) (MW: 111 kDa) (DD: 95%) (Shanghai KABO trading co. ltd.)	Nanoparticle *BSA-mCS NPs:* *355.6 ± 25.3 nm* *Eudragit coated BSA-mCS NPs:* *558.2 ± 35.6 nm* *BSA-CS NPs* *Eudragit coated BSA-CS NPs*	Oral	–	– Targeting property of formulations (Ileal region of small intestine of Sprague Dawley rats)	Immunization in 6- to 8-week-old female BALB/c mice *Dose: 100 μg* *Boost: On days 7, 21, 35, and 49* – Determination of IgG and IgA levels	– Highest localization of mCS NPs in Peyer's patches compared to CS NPs and Eudragit-coated mCS NPs – Highest IgG and IgA level with Eudragit-coated mCS NPs followed by the Eudragit-coated CS NPs and then mCS NPs	[109]
Recombinant chlamydia trachomatis fusion antigen CTH522	Glycol chitosan (MW: 100 kDa) (DD is not stated) (sigma-Aldrich, Saint Louis, USA)	Nanoparticle *Glycol chitosan coated DDA/TDB modified PLGA nanoparticles (GC coated LPNs):* *245.1 ± 5.4 nm* *DDA/TDB liposomes:* *227.5 ± 41.4 nm*	Nasal	–	–	Immunization in 7- to 9-week-old C57BL/6 mice *Dose: 5 μg* *Boost: Four times at two-to-three-week intervals* – Determination of IgG and IgA levels – Cellular immune response: Determination of IFN-γ levels	– Significantly higher IgG responses and IgA responses with GC-coated LPNs than DDA/TDB liposomes – Equal levels of systemic T-cell responses with GC-coated LPNs and DDA/TDB liposomes	[129]

HBsAg	Chitosan (MW is not stated) (DD: 95%) (ChitoClear™ Primex BioChemicals AS, Avaldsnes, Norway)	Nanoparticle *CpG-HBsAg loaded CS NPs:* *878 ± 39 nm* *CpG-HBsAg loaded alginate coated CS NPs: 1428 ± 34 nm* *HBsAg loaded glucan particles (GP): 3087 ± 268 nm*	– Oral – Parenteral (subcutaneous)	–	– Cellular uptake studies (PBMCs from healthy donors)	Immunization in 7-week-old female C57BL/6 mice *Dose: 1.5 μg (s.c.)/ 20 μg (oral)* *Boost: On days 14 and 28 (20 μg / oral)* – Determination of IgA, IgG and its subtypes levels	– Internalization of nanoparticles by PBMCs and murine PPs – Highest serum IgG titers with priming s.c. immunization followed by two oral boosts – Higher IgG levels with subcutaneous immunization of CpG-CS NPs and GP NPs compared to oral immunization of alginate-coated CpG-CS NPs and GP NPs – Predominant IgG1 levels with subcutaneous immunization of CpG-CS NPs indicating the Th2-based immune response – Enhanced IgA levels with three oral immunizations for NPs	[110]
Matrix protein-2 (sM2) and fusion peptide of hemagglutinin (HA2)	Chitosan (MW: 50–190 kDa) (DD: 75–85%) (sigma-	Nanoparticle *sM2HA2-PGA-CS NPs*	Nasal	– CTA1	–	Immunization in 5-week-old BALB/c mice *Dose: 2 μg*	– Highest Th1 cytokine expression, IgG and IgA levels with	[168]

(continued)

Table 1 (continued)

Antigen	Chitosan type	Dosage form	Immunization route	Adjuvant actadded	In vitro studies	In vivo studies	Immune response	Ref.
H1N1, H5N1, and H9N2 influenza viruses	Aldrich, Saint Louis, USA)	*sM2HA2-CTA1-PGA-CS NPs*				*Boost: On days 14 and 28* *Challenge: i.n. with pathogenic influenza subtypes on day 35 or 196* – Determination of IgA, IgG and its subtypes titers – Cytokine expression and splenocyte proliferation assay	adjuvanted NPs compared to non-adjuvanted NPs – Highest protection against divergent influenza infections with reduced viral load and enhanced survival rate with adjuvanted NPs (100%) compared to non-adjuvanted NPs (60%) – High cellular and humoral immune response and protection with adjuvanted NPs after 6 months post-immunization indicating stimulation of the long-lasting immune responses and protection	
HBsAg	Chitosan (MW: 3 kDa) (DD: 80%) (Golden-Shell pharmaceutical co. ltd., Zhejiang, China)	Microparticle *HBsAg – PLGA MPs:* *773.6 ± 42.6 nm* *HBsAg-CS modified PLGA (CS-PLGA)*	– Nasal Compared to subcutaneous HBsAg-alum	–	– Cellular uptake studies (peritoneal macrophages from Sprague–Dawley rats)	Immunization in female Sprague–Dawley rats *Dose: 20 µg* *Boost: On day 21* – Determination	– Enhanced cellular uptake and remaining time in nasal cavity with chitosan modified and chitosan-	[131]

		MPs: *850.8 ± 25.3 nm* *HBsAg-Mannan modified PLGA (MN-PLGA) MPs:* *806.0 ± 41.2 nm* *HBsAg-Mannan and CS modified PLGA (MN-CS-PLGA) MPs:* *914.0 ± 52.8 nm* *HBsAg-alum (for comparison of adjuvant properties with microparticles)*				of IgG levels – Determination of cytokine levels	mannan modified PLGA MPs compared to non-modified PLGA MPs – Significantly higher IgG levels with MN-CS-PLGA MPs than PLGA, MN-PLGA, and CS PLGA MPs – Similar IgG levels with MN-CS-PLGA MPs and alum – Significantly higher Th1 cytokines with MN-CS-PLGA MPs than alum, PLGA, MN-PLGA, and CS PLGA MPs	
Monovalent detergent splitinfluenza (A/Victoria/210/2009H3N2) (HA)	Methylglycol chitosan (MW and DD are not stated) (sigma-Aldrich, Saint Louis, USA)	Liposome (DOPA: Chol) *Unmodified liposomes: 112 ± 34 nm* *PEG modified liposomes: 87 ± 13–105 ± 15 nm* *Pluronic modified liposomes: 99 ± 27–174 ± 9 nm* *Methylglycol chitosan coated liposomes*	Sublingual	CRX-601	–	Immunization in female 6- to 8-week-old BALB/c mice *Dose: 3 μg* *Boost: On days 21 and 42* – Determination of IgG and IgA titers	– Significantly increased IgG and IgA levels with surface-modified liposomes compared to unmodified liposomes – Highest antibody titers with methylglycol chitosan-coated liposomes compared to	[115]

(continued)

Table 1 (continued)

Antigen	Chitosan type	Dosage form	Immunization route	Adjuvant actadded	In vitro studies	In vivo studies	Immune response	Ref.
							unmodified liposomes indicating the robust and consistent immune response	
Recombinant env23 and env13 antigens ofenvelope protein gp46	TMC and chitosan Chitosan (used for TMC synthesis) (MW is not stated) (DD: 95%) (ChitoClear™ Primex BioChemicals AS, Avaldsnes, Norway)	Nanoparticle *env23-CS NPs: 478 ± 40 nm* *env13-CS NPs: 390 ± 44 nm* *env23-TMC NPs: 417 ± 17 nm* *env13-TMC NPs: 360 ± 66 nm*	– Nasal – Parenteral (subcutaneous)	–	–	Immunization in male 6- to 8-week-old BALB/c mice *Dose: 7.5 μg* *Boost: On days 14 and 28* – Determination of IgG and its subtypes levels	– Higher IgG and IgG1 levels with s.c. immunization of env23 or env13 loaded CS and TMC NPs than nasally immunized groups – Highest IgG2a/IgG1 ratio with env23-NPs after nasal administration indicating the Th1-based immune response	[169]
OMP31 of *Brucella ovis*	Chitosan (mw: 161 kDa) (DD: 90%) (Parafarm®, Argentina)	Microsphere (8.66 ± 0.03 μm)	Nasal	–	–	Immunization in 5-month-old rams *Dose: 0.5 mL* *Boost: On days 21 and 42* – Determination of IgA and IgG levels	– Significantly enhanced IgA and IgG titers indicating the strong induction of mucosal and systemic immune responses	[170]

HBsAg	Chitosan (MW is not stated) (DD: 95%) (ChitoClear™ Primex BioChemicals AS, Avaldsnes, Norway)	Nanoparticle *Poly-ε-caprolactone/ CS NPs: 208 nm*	Nasal	–	– Cellular uptake and cytotoxicity studies (Caco-2 cells)	Immunization in female 8-week-old C57BL/6 mice *Dose: 1.5, 5, and 10 μg* *Boost: On days 7 and 21* – Determination of IgA and IgG levels	– No cytotoxic effect – Enhanced cellular uptake – Induction of similar IgG titers with nanoparticles containing low and high dose of antigen – Higher IgA levels with nanoparticles containing high dose of antigen	[171]
A monovalent detergent-split flu virus vaccine	Methylglycol chitosan (MGC) (MW and DD are not stated) (sigma-Aldrich, St. Louis, MO)	Solution	Sublingual	CRX-601	–	Immunization in 6- to 8-week-old female BALB/c mice *Dose: 3 μg* *Boost: On days 21 and 42* – Determination of IgA, IgG and its subtypes – Influenza hemagglutination inhibition assay	– Significantly increased serum IgG, HI titers, and mucosal IgA titers with MGC, CRX-601, MGC + CRX-601 – Induction of a predominately IgG1 response, which is indicative of a Th2 skewed response	[172]
Bacillus anthracis protective antigen (PA)	Chitosan (MW is not stated) (DD: 95%) (ChitoClear™ Primex BioChemicals AS, Avaldsnes, Norway)	Nanoparticle *PA-C48/80 loaded Cs NPs: 500.9 ± 65.2 nm* *PA-C48/80 loaded alginate CS NPs: 564.3 ± 201.4 nm*	Nasal	C48/80	– Cytotoxicity studies (spleen cells from C57BL/6 mice) – Cellular uptake studies (RAW264.7 cells) – In vitro	Immunization in female 6- to 8-week-old C57BL/6NCr mice *Dose: 2.5 μg* *Boost: On days 7 and 21* – Determination	– Higher cell viability with NPs than solution – Higher cellular uptake with CS NPs compared to alginate CS NPs – Higher IgA,	[127]

(continued)

Table 1 (continued)

Antigen	Chitosan type	Dosage form	Immunization route	Adjuvant actadded	In vitro studies	In vivo studies	Immune response	Ref.
					anthrax lethal toxin (LeTx) neutralization assay (J774A.1 cells) (with sera of immunized mice)	of IgA, IgE, IgG and its subtypes – Determination of cytokine levels	neutralizing antibodies and balanced Th1/Th2 profile with CS NPs than alginate CS NPs – No stimulation of IgE levels indicating a limited risk of induced type 1 hypersensitivity – Induction of Th17 cytokine expression with CS NPs	
Measles antigen (MA)	CS of three different molecular weight (42, 74, and 106 kDa) Chitosan (used for chitosan derivatives synthesis) (MW: 310–375 kDa) (DD: ≥75%) (sigma-Aldrich, St. Louis, MO)	Nanoparticle *MA loaded-alginate-CS (42 kDa) NPs: 432 ± 52 nm* *MA loaded-alginate-CS (74 kDa) NPs: 651 ± 28 nm* *MA loaded-alginate-CS (106 kDa) NPs: 1003 ± 45 nm*	Oral	–	– Cytotoxicity assay (HT 29human epithelial cell line)	Immunization in 6-week-old SPF male BALB/c mice *Dose: 20 μg* *Boost: On days 7, 14, and 28* – Determination of IgA and IgG levels	– No significant difference in cytotoxicity among nanoparticle formulations with ≥70% cell viability – Highest mucosal IgA and systemic IgG levels with lowest molecular weight of chitosan	[111]
DTx	Chitosan *(MW, DD, and the source are not stated)*	Microparticle *DTx-loaded uncoated microparticles*	– Oral Compared to intramuscular	–	–	Immunization in 6- to 8-week-old female BALB/c mice	– Specific and effective uptake of the particles by the M cells of PPs	[112]

		DTx-loaded alginate-coated microparticles: 5.25 ± 0.3 μm *Alum-adsorbed DTx (for comparison of adjuvant properties with microparticles)*	alum-adsorbed DTx			*Dose: 0.5, 1, and 2 lf* *Boost: On day 21* – Determination of IgA and IgG levels – Uptake study	– Enhanced and dose-dependent IgG and IgA levels with microparticles compared to alum	
BmpB	Chitosan (MW:9600) (DD: 91.8%) (provided by prof. Nah (Sunchon National University, Korea))	Microparticle *BmpB-PLGA MPs: 3.36 ± 0.74 μm* *BmpB-CS-PLGA MPs: 3.57 ± 0.73 μm* *BmpB-CKS9-CS-PLGA MPs: 3.69 ± 0.77 μm*	Oral	–	–	Immunization in 8-week-old female BALB/c mice *Dose: 200 μg* *Boost: On days 1, 7, 8, 14, and 15* – Determination of IgA, IgG and its subtypes levels	– Highest IgG and IgA titers with BmpB-CKS9-CS-PLGA MPs – Induction of both Th1 and Th2 immune response – Slight increase in immune responses with coating of PLGA microparticles with chitosan – Significantly increased systemic and mucosal immune responses with the addition of M cell ligand	[113]
Influenza whole virus (WV)	Chitosan (MW: 50–190 kDa) (DD: 75–85%) (sigma-Aldrich, Saint Louis, USA)	Nanoparticle (581.1 ± 32.6 nm) *WV loaded CS NPs* *WV and CpG loaded CS NPs* *WV and QS loaded CS NPs*	– Nasal – Parenteral (intramuscular)	– CpG ODN – QS	–	Immunization in 7-week-old New Zealand albino female rabbits *Dose: 10 μg* *Boost: On days 45, 60 (i.n.), and*	– Enhanced HI antibody titers with all NPs – Effective induction of cellular and immune response with CpG-NPs	[133]

(continued)

Table 1 (continued)

Antigen	Chitosan type	Dosage form	Immunization route	Adjuvant actadded	In vitro studies	In vivo studies	Immune response	Ref.
						75 (i.m.) – Hemagglutination inhibition assay – Determination of IgA and IgG levels – Determination of cytokine levels	compared to QS-NPs and non-adjuvanted NPs	
OVA	TMC Chitosan (used for TMC synthesis) (MW: 50–190 kDa) (DD: 75–85%) (sigma-Aldrich, Saint Louis, USA)	Solution – *OVA conjugated TMC (OVA-TMC)* – *Physical mixture of OVA and TMC (OVA + TMC)* – *OVA-alum (for comparison of adjuvant properties)*	Nasal	–	– Cellular uptake study (RAW264.7 cells)	– Immunization in 6- to 8-week-old female BALB/c mice *Dose: 20 μg* *Boost: On days 14 and 28* – Nasal ciliotoxicity and transport to cervical lymph nodes studies in male Sprague–Dawley (SD) rats following i.n. immunization – Determination of IgA, IgG and its subtypes levels	– Enhanced and time-dependent cellular uptake with OVA-TMC compared to OVA +TMC – Significantly increased the transport of OVA to both superficial cervical lymph nodes and deep cervical lymph nodes, indicating that chitosan is able to open tight junctions of the nasal mucosa – No toxic effect on the integrity of nasal cilia and epithelium – Higher IgA,	[173]

							IgG, IgG1, and IgG2a levels with OVA-TMC compared to OVA +TMC – No significant difference in antibody levels between OVA-TMC and OVA-alum – Predominantly Th2 type immune response with nasal administrated OVA-TMC and OVA-alum	
Tetanus toxoid (TT)	Chitosan (MW: 45 kDa) (DD: 95%) (sigma-Aldrich, Saint Louis, USA)	Nanoparticle *(CS functionalized gold NPs (CS-Au NPs))* *TT loaded CS-au NPs: 40 ± 3.5 nm* *TT loaded QS-CS-au-NPs: 42.5 ± 4 nm)* *TT-QS (for comparison of adjuvant properties with nanoparticles)*	Oral	QS	–	Immunization in 2- to 3-week-old Swiss albino mice *Dose: 20 lf/mL* *Boost: On days 14 and 28* – Determination of IgG and IgA levels	– Higher IgG and IgA levels with QS-CS-au-NPs and CS-au-NPs, respectively, compared to QS	[114]
HBsAg	– Chitosan (MW: 110–150 kDa) (DD is not stated) (sigma-Aldrich, Saint Louis, USA)	Nanoparticle *PLGA NPs: 164.3 ± 29.2 nm* *CS coated PLGA NPs:*	Nasal	–	– Cytotoxicity study (Calu-3 cells)	Immunization in 8–10 weeks old female BALB/c mice *Dose: 10 μg*	– No cytotoxic effect – Higher nasal retention, mucoadhesive	[130]

(continued)

Table 1 (continued)

Antigen	Chitosan type	Dosage form	Immunization route	Adjuvant actadded	In vitro studies	In vivo studies	Immune response	Ref.
	– Glycol chitosan (MW: 250 kDa) (DD is not stated) (sigma-Aldrich, Saint Louis, USA)	*175.8 ± 38.4 nm GC coated PLGA NPs:* *181.4 ± 35.6 nm* *Alum (for comparison of adjuvant properties with nanoparticles)*				*Boost: On day 21* – Determination of IgA and IgG levels – Determination of cytokine levels	properties and lower mucociliary clearance with GC-coated PLGA NPs than CS-coated PLGA NP – Highest IgG levels with alum – Higher IgG levels with GC-coated PLGA NPs compared to CS-coated PLGA NPs – Highest mucosal IgA levels with GC-coated NPs compared to CS coated NPs – Higher Th1 cytokine expression with GC-coated and CS-coated NPs, respectively, compared to alum	

							– Higher Th2 cytokine expression with alum compared to GC-coated and CS-coated NPs	
WIV	– TMC (MW: 36 kDa) (DD: 17%) – Re-acetylated TMC (TMC-RA) (MW: 43 kDa) (DD: 54%) Chitosan (used for TMC and TMC-RA synthesis) (MW: 28–43 kDa) (DD: 17%) (Primex, Siglufjordur, Iceland)	Nanoparticle *WIV-TMC NPs ≅ 250 nm* *WIV-RA-TMC NPs ≅ 300 nm*	Nasal	–	– Cellular uptake studies (Calu-3 cells) – Activation of dendritic cells: Determination of cell-surface expression of co-stimulatory molecules (BMDCs from C57BL/6 mice)	– Imaging of nasal residence time in female nude BALB/c mice	– Significantly higher association and cellular uptake of RA-TMC NPsthan TMC NPs – No additive effect of TMC or TMC-RA on DCs activation – No increase in nasal residence time with nanoparticle formulations – Higher enzymatic degradation of RA-TMC than TMC	[174]
rHA	Chitosan (MW: 50–190 kDa) (DD: 75–85%) (sigma-Aldrich, Saint Louis, USA)	Nanoparticle *rHA loaded γ-PGA/CS NPs:* ≅900 nm *rHA-CT (for comparison of adjuvant properties with nanoparticles)*	Nasal	–	–	Immunization in 5-week-old female BALB/c mice *Dose: 20 μg* *Boost: On days 14 and 28* *Challenge: i.n. with the HPAI H5N1 virus on day 42* – Determination of IgA, IgG and its	– Higher stimulation of IgG, IgA and neutralizing antibodies with NPs compared to commercialized parenteral influenza vaccines – Balanced IgG2a and IgG1 levels indicating	[175]

(continued)

Table 1 (continued)

Antigen	Chitosan type	Dosage form	Immunization route	Adjuvant actadded	In vitro studies	In vivo studies	Immune response	Ref.
						subtypes levels – Determination of cytokine expression – Hemagglutination inhibition assay	the induction of both humoral and cell-mediated immunity effectively – Higher IFN-γ expression and similar IL-4 expression with NPs compared to CT – Strong protection with 100% survival rate with NPs and CT	
HBsAg	TMC Chitosan (used for TMC synthesis) (MW and DD are not stated) (India Sea foods)	Nanoparticle *HBsAg-TMC NPs: 143 ± 33 nm* *HBsAg-TMC solution* *HBsAg-alum (for comparison of adjuvant properties with nanoparticles)*	– Nasal Compared to intramuscular HBsAg-alum	–	–	Immunization in 6- to 8-week-old female BALB/c mice *Dose: 10 μg* *Boost: On days 14 and 28* – Determination of IgA and IgG levels	– Highest IgG titers with alum – Higher IgG titers with nanoparticles compared to solution – Strong induction of IgA titers with TMC NPs compared to TMC solution and alum	[176]

OVA	TMC Chitosan (used for TMC synthesis) (MW: 120 kDa) (DD is not stated) (Primex, Siglufjordur, Iceland)	Nanoparticle *OVA-TMC NPs: 313 ± 31 nm* *OVA-LPS/TMC NPs: 323 ± 39 nm* *OVA-PAM/TMC NPs: 365 ± 46 nm* *OVA-CpG/TMC NPs: 375 ± 99 nm* *OVA-MDP/TMC NPs: 418 ± 89 nm* *OVA-CTB/TMC NPs: 304 ± 22 nm*	– Nasal – Parenteral (intradermal)	– LPS – PAM_3CSK_4 (PAM) – CpG DNA – MDP – CTB	–	Immunization in 8-week-old female BALB/c mice *Dose: 10 µg* *Boost: On day 21* – Determination of serum IgG, IgG1, IgG2a, and secretory IgA	– Higher IgG, IgG1, and sIgA levels with LPS or MDP adjuvanted NPs than non-adjuvanted NPs after nasal vaccination – No enhanced antibody levels with CTB, PAM, or CpG adjuvanted NPs after nasal vaccination – Higher IgG titers with CpG or LPS adjuvanted NPs than non-adjuvanted NPs after intradermal vaccination	[177]

aCS Aminated chitosan, *Ag* Antigen, *APC* Antigen presenting cell, *atCS* Aminated and thiolated chitosan, *BLS* Brucella lumazine synthase, *BMDC* Bone marrow dendritic cells, *BmpB* Membrane protein B of Brachyspira hyodysenteriae, *BSA* Bovine serum albumin, *C48/80* The mast cell activator compound 48/80, *CEIC* Chicken embryo intestinal cells, *Chol* Cholesterol, *CKS9* M cell homing peptide, *CS* Chitosan, *CT* Cholera toxin, *CTA1* Cholera toxin subunit A1, *CTB* Cholera toxin subunit B, *DC* Dendritic cell, *DD* Deacetylation degree, *DDA* Dimethyldioctadecylammonium bromide, *DOPA* 1,2-dioleoyl-sn-glycero-3-phosphocholine, *DTx* Diphtheria toxoid, *HBsAg* Hepatitis B surface antigen, *HI* Hemagglutinin inhibition assay, *hcp* Hemolysin co-regulated protein, *GAS* Group A streptococcus, *HI* Hemagglutination inhibition, *IFA* Incomplete Freund's adjuvant, *i.m.* Intramuscular, *i. n.* Intranasal, *i.p.* Intraperitoneal, *LMW* Low molecular weight, *LPS* Lipopolysaccharide, *MDP* Muramyl dipeptide, *MHC-I/II* Major histocompatibility complex-I/II, *MW* Molecular weight, *NP* Nanoparticle, *ODN* Oligodeoxynucleotides, *OMP* outer membrane protein, *OVA* Ovalbumin, *PA* Protective antibody, *PBMC* Peripheral mononuclear cells, *PEG* Poly (ethylene glycol), *γ-PGA* Poly-γ-glutamic acid, *PLGA* Poly(d, l-lactide-co-glycolide), *PPs* Peyer's patches, *PrpA* Proline racemase subunit A, *PRRS* Porcine reproductive and respiratory syndrome, *PRRSV* Porcine reproductive and respiratory syndrome virus, *rHA* Recombinant influenza hemagglutinin, *rMdh* Recombinant malate dehydrogenase, *rOMP* Recombinant outer membrane proteins, *QS* Quillaja saponins, *s.c.* Subcutaneous, *sodC* Cu-Zn superoxide dismutase, *SPF* Specific pathogen free, *SwIV* Swine influenza virus, *TBLN* Tracheobronchial lymph nodes, $TCID_{50}$ Fifty-percent tissue culture infective dose, *TDB* Trehalose-6,6′-dibehenate, *TEER* Transepithelial electrical resistance, *TLR* Toll like receptor, *TMC* Trimethyl chitosan, *WIV* whole inactivated influenza virus concluded that nanoparticles can increase the M-cell dependent uptake and enhance the association of the antigen with dendritic cells depending on their composition.

Abkar et al. [108] have investigated the Omp31-loaded *N*-trimethyl chitosan nanoparticles against *Brucella melitensis* infection. Nanoparticles were administered orally and intraperitoneally in mice. Intraperitoneal immunization was shown to induce Th1–Th2 immune responses, while with oral immunization a mixed Th1–Th17 immune response was obtained. Vaccinated groups of mice when challenged with *B. melitensis* 16 M were found to be significantly protected in the orally administered group in comparison with intraperitoneally immunized mice. The authors suggested that administration route plays a main role in determining the immune response type.

Xu et al. [109] have developed mannosylated chitosan nanoparticles (MCS NPs), which were enterically coated with Eudragit® L100, targeting the APCs in the region of Peyer's patches. Oral immunization using BSA-loaded Eudragit® L100-coated MCS NPs was found to elicit strong systemic IgG antibody and mucosal IgA responses. The enhanced transcytotic property across M cells and binding affinity to the APCs of the nanoparticles was attributed to the chemically modified chitosan with mannose, which induced the receptor-mediated endocytosis for targeting into APCs, leading to high transfection efficiency and enhanced immune responses.

Soares et al. [110] have encapsulated recombinant hepatitis B surface antigen into alginate-coated chitosan nanoparticles and compared this system to that prepared with glucan particles following oral and subcutaneous administration in mice. Firstly, they showed in vitro that both systems could be internalized by peripheral blood mononuclear cells and murine Peyer's patches, which was considered as an important indicator for oral immunization. The results of in vivo studies demonstrated that there was a need for a subcutaneous priming prior oral boosting to induce HBsAg seroconversion. It was suggested that with this immunization schedule with only one injectable dose would be favorable for mass vaccination, especially in developing countries. Biswas et al. [111] have prepared sodium alginate-coated nanoparticles using three different molecular weight (42, 74, and 106 kDa) chitosan for delivery of measles antigen and administered orally to mice. Nanoparticles were shown to induce strong immune response and significant correlation was observed between the immune response with chitosan molecular weight, higher IgG and IgA levels with lowest molecular weight (42 kDA).

In another study, alginate-coated chitosan microparticles were prepared for delivery of diphtheria toxoid [112]. The developed microparticles were shown to deliver the antigen effectively to the M cell resulting significant immune responses. Moreover, when compared to Alum, alginate chitosan microparticles were found more effective in stimulating the systemic and mucosal immune responses.

Chitosan has been utilized also in combination with particulate systems prepared with other materials. As an example, Jiang et al. [113] have used chitosan conjugated with M cell ligand, CKS9, to coat the PLGA microparticles loaded with a membrane protein B of *Brachyspira hyodysenteriae* (BmpB) against swine dysentery. Coating of PLGA microparticles with chitosan was reported to result in a slight increase in immune responses, while with the addition of M cell ligand systemic and mucosal immune responses were increased significantly. In another study, Barhate et al. [114] have developed chitosan functionalized-gold nanoparticles a carrier for tetanus toxoid along with immunostimulant. A significant increase was observed in immune response following the oral administration in mice with chitosan functionalized-gold nanoparticles, and in presence of Quillaja saponaria extract, the immunogenicity was increased even more. In another study, liposomes carrying the synthetic toll like receptor-4 agonist, CRX-601 were modified with PEG copolymers (Pluronics) and phospholipid-PEG conjugates and coated with methylglycol chitosan to deliver influenza antigens [115]. These systems were evaluated for their ability to generate an immune response in a sublingual murine influenza vaccine model. Liposomes coated with chitosan were reported to result in the most robust and consistent immune response with sublingual administration.

Buffa et al. [116] have investigated the impact of chitosan as well as a range of TLR ligands as potential adjuvants for different routes of mucosal immunization (sublingual – SL, intranasal – IN, intravaginal – IV) and a parenteral route (subcutaneous – SC) in the murine model in regard to assess their ability to enhance antibody responses to HIV-1 CN54gp140 (gp140) and tetanus toxoid (TT) in systemic and vaginal compartments. Chitosan was found to significantly enhance systemic responses to TT by the sublingual route, providing a strong Th2 biasing effect for SL- and IN-administration with both TT and gp140. IN-immunization with gp140 was shown to provide a more balanced Th1/Th2 profile than SL- or SC-routes. The authors have concluded that in the design of mucosal vaccine strategies route, antigen and adjuvant effects play important role.

4.2 *Nasal Delivery*

Nasal mucosa is another route widely investigated for mucosal immunization. In nasal immunization, antigens can be taken either by the M-cell connected to the nasal-associated lymphoid tissues (NALTs) or antigen-presenting cells present in nasal mucosa, hence intranasal vaccination can initiate via NALT or diffusion into nasal mucosa [117, 118]. On the other hand, in the nasal cavity the mucociliary clearance limits the nasal residence time of the particles and causes removal of the administered vaccine system quickly.

Due to its positive charge as well as its mucoadhesive and penetration enhancing properties, chitosan-based systems become very attractive for nasal immunization, by prolonging the residence time of the vaccine as well as by enhancing the uptake of the antigen. It has been reported that the nasal formulations based on chitosan induce significant serum IgG responses similar to and secretory IgA levels superior to that is induced by a parenteral administration of the vaccine [119]. In most cases, for nasal delivery, chitosan and its derivatives have been used in aqueous dispersion, gel, or especially, particulate forms. As an example, for chitosan gel, our group has investigated base chitosan and water-soluble chitosan in gel form as an adjuvant/ delivery system for intranasal immunization against foot-and-mouth disease (FMD)

[43]. Chitosan-based FMD gels were shown to induce FMD antigen-specific serum IgG and nasal IgA levels, the latter being significantly stronger as compared to that obtained following subcutaneous administration of the FMD antigen in Freund's incomplete adjuvant. No difference in regard to obtained immune responses was found between base chitosan and water-soluble chitosan.

Shim et al. [120] have investigated the induction of mucosal and systemic immunity by chitosan nanoparticles loaded with three *Brucella abortus* recombinant proteins after intranasal immunization of BALB/c mice. Antigen-specific IgA with a Th2-polarized immune responses were elicited by the nanoparticle formulations and combination of the highly immunogenic antigens elicited IgG specific to each type of antigen. Similarly, Senevirethane et al. [121] have investigated a chitosan-based Brucella nasal vaccine using Brucella immunogens (sodC, omp19, BLS, and PrpA) against nasal Brucella challenge in BALB/c mice. Enhanced IgG and mucosal IgA levels and protection were obtained with cocktail antigen loaded nanoparticles.

In general, in order to obtain high immune responses, inclusion of adjuvant is required for nasal immunization. Various research groups have investigated chitosan alone as an adjuvant or in combination with other adjuvants. Renu et al. [122] have used the adjuvant, poly(I:C) to improve the innate and adaptive immune responses elicited by a killed/inactivated swine influenza virus antigen (KAg)-loaded chitosan nanoparticle (CS NPs-KAg) in pigs following the nasal immunization. Unlike chitosan nanoparticles without poly(I:C), incorporation of poly(I:C) was shown to increase both the Th1 and anti-inflammatory cytokines gene expression in TBLN and this effect was associated with a high frequency of IFN-γ secreting T-helper/memory and γδT cells in PBMCs. However, the sIgA response in the airways was not substantially induced. Induced cross-reactive cell-mediated immune response, elevated HI titers, and higher frequency of antigen-specific IFN-γ secreting T-helper/memory and γδT cells obtained in young swine were reported to be superior to commercial vaccine.

Trimethylated chitosan (TMC) has been widely investigated for nasal delivery of vaccines. Hagenaars et al. [123] have investigated the influence of TMC on nasal delivery of whole inactivated influenza virus in mice by comparing the nasal residence time and the specific location in the nasal cavity. A comparable nasal clearance profile was observed with and without TMC, therefore it was concluded that adjuvant effect of TMC in intranasal formulations should not be ascribed to differences in nasal residence time of the bulk vaccine. However, it was stated that presence of TMC resulted in an increased contact area between the antigen and epithelium, which was related to the improved immunogenicity of TMC-adjuvanted formulations. Later, the same group investigated a TMC-antigen nanoconjugate and TMC nanoparticles for nasal vaccine delivery. They suggested that efficient codelivery of antigen and adjuvant to DCs, rather than a particulate form of the antigen/adjuvant combination, is decisive for the immunogenicity of the antigen [124]. Our group has shown that nanoparticles prepared with TMC showed similar immunogenicity to nanoparticles obtained with chitosan for nasal delivery of tetanus toxoid [38]. For both formulations IgG2a/IgG1was found to be smaller than 1, indicating a predominant Th2 response. Nevagi et al. [125] have developed a group A streptococcus (GAS) peptide vaccine based on combined lipidic TLR 2 agonist and self-adjuvanting polymers. Nanoparticles were formed via ionic complexation of the anionic lipopeptide conjugates with trimethyl chitosan (TMC), which has a cationic nature.

Mosafer et al. [126] have prepared alginate-coated TMC and chitosan nanoparticles for delivery of PR8 influenza virus. After nasal immunization in BALB/c mice, chitosan nanoparticles elicited higher IgG2a and IgG1 antibody titers than that with TMC nanoparticles. The antibody titers were decreased with alginate coating and less immune response was induced when compared to coated TMC formulations. Bento et al. [127] have developed a combined adjuvant system based on chitosan and the mast cell activator compound 48/80 (C48/80) in nanoparticle form and coated with alginate to deliver anthrax protective antigen. The adjuvant effect was found to be higher when in combination form, while presence of alginate was found to decrease the immune response.

Sinani et al. [128] have used aminated plus thiolated chitosan (atChi) polymers to develop nanoparticles for nasal immunization. Amination and amination plus thiolation were shown to significantly improve the mucoadhesive property of chitosan. Aminated plus thiolated chitosan nanoparticles showed slightly higher antibody titers compared to aminated chitosan nanoparticles which was attributed to the increased mucoadhesiveness of atChi polymer due to thiol groups in its structure.

Rose et al. [129] have coated the lipid-polymer hybrid nanoparticles (LPNs) with a water-soluble derivative of chitosan, glycol chitosan and investigated mucosal immune responses against recombinant *Chlamydia trachomatis* fusion antigen CTH522. Increased antigen-specific mucosal immune responses were induced in the lungs and the genital tract with the glycol chitosan-coated LPN adjuvant upon nasal immunization of mice. The mucosal responses were characterized by CTH522-specific IgG/IgA antibodies, together with CTH522-specific interferon γ-producing Th1 cells. Similarly, in another study, PLGA nanoparticles coated with chitosan or glycol chitosan were investigated in vivo for nasal delivery of hepatitis B surface antigen [130]. The glycol chitosan-PLGA nanoparticles were found to show lower clearance and better local and systemic uptake when compared to chitosan coated and uncoated PLGA. Furthermore, significantly higher systemic and mucosal immune responses were induced with glycol chitosan-coated nanoparticles.

In another study, surface-functionalized, pH-responsive PLGA microparticles were investigated for nasal delivery of hepatitis B surface antigen [131]. Mannan and chitosan surface modification was found to enhance intracellular microparticle uptake by macrophages. Residence time of the nanoparticles was increased by chitosan coating and chitosan-mannan-PLGA microparticles were observed to induce stronger humoral and cell-mediated immune responses compared to PLGA, mannan-PLGA, and chitosan-PLGA microparticles [131].

In the examples given so far, it has been shown that coating with chitosan increases the immune responses against various antigens. However, in a study reported by Amin et al. [132], it was shown in rabbits that after nasal administration, coating of liposomes by chitosan failed to increase both the residence time of liposomes in nasal cavity and systemic responses and coated liposomes could not induce the mucosal responses as efficiently as non-coated liposomes. The authors concluded that coating of liposomes affected their interaction potential with nasal-associated lymphoid tissue cells. It was interesting that the size of the liposomes both coated and uncoated was found to be around 2.3 μm and no data was given about the surface charge of both uncoated and coated liposomes. Therefore, we believe that further evaluations are needed to confirm this result.

Dehghan et al. [133] have loaded two different adjuvants, CpG oligodeoxynucleotide (CpG ODN) and *Quillaja* saponins (QS) into chitosan nanoparticles and investigated immune responses against influenza virus in rabbits. Induction of humoral and cellular immune responses after nasal administration was found to be highest with CPG ODN-chitosan nanoparticles. IgG titers were found to be similar between chitosan nanoparticles and CPG ODN-chitosan nanoparticles up to 60 days, yet on days 75 and 90, significantly higher IgG levels were obtained with CPG ODN-chitosan based systems.

The reader can find the summary of the recent studies on chitosan-based particulate systems for nasal immunization in Table 1.

5 Chitosan for Parenteral Immunization

For immunization the main parenteral delivery routes are intramuscular, subcutaneous, and intradermal. When the literature is searched for utilization of chitosan for parenteral immunization, it is noteworthy that the vast majority of the examples are from veterinary field. The recent studies on chitosan-based systems for parenteral immunization are summarized in Table 2. In this section, we will mention some of these studies only to highlight the issues that are important in regard to formulation aspect. Li et al. [134] have evaluated the potential and mechanisms of a thermosensitive chitosan hydrogel as an adjuvant for a recombinant protein (rP-HSP90C) containing epitope C from heat shock protein 90 of *Candida albicans* rP-HSP90C vaccine. Chitosan hydrogels were shown to evoke a long-lasting IgG levels and also enhance Th1, Th2, Th17 responses and the ratio of Th1/Th2. In addition, chitosan hydrogel recruited immune cells at the injection site. Chitosan hydrogel was suggested as an efficient adjuvant in fungal vaccines.

A water-soluble chitosan, N-2-Hydroxypropyl trimethyl ammonium chloride chitosan/N-2-HACC) was evaluated as an adjuvant in solution form for immunization against porcine parvovirus (PPV) [135]. It was reported that PPV/N-2-HACC protected PPV infection mainly through enhancing the humoral immunity rather than cellular immunity. In another study, the adjuvant property of the nanoparticles prepared with chitosan, sulfated chitosan and HACC was evaluated in chickens against inactivated Newcastle disease following subcutaneous administration [136]. Chitosan and HACC nanoparticles were shown to induce antibody titers and cytokine levels and provided 100% protection, whereas sulfated chitosan nanoparticles were not effective to stimulate immune response. This difference

Table 2 Chitosan-based systems for parenteral delivery of vaccines

Antigen	Chitosan type	Dosage form	Immunization route	Adjuvant added	In vitro studies	In vivo studies	Immune response	Ref.
OMP and flagellar antigen of *E. coli* O1 and O78 strains (OMP-F)	– Chitosan – Chitosan ascorbate *(MW, DD, and the source are not stated)*	Nanoparticle (81–177 nm) *OMP-F-loaded CS NPs* *OMP-F-encapsulated CS NPs* *OMP-F-loaded ascorbate CS NPs* *OMP-F-encapsulated ascorbate CS NPs*	Subcutaneous	Montanide ISA 71 R VG (adjuvant is mixed with nanovaccines in a ratio (1:1))	–	Immunization in 3-week-old broiler chicken *Dose: 500 μg* *Boost: On day 7* *Challenge: Into the thigh region with E. coli serotype O1 and O78* – Determination of IgG levels	– Highest antibody titer with antigen encapsulated NPs without adjuvant – Similar antibody response to non-adjuvanted and adjuvanted NPs – Strong protection on day 35 – Higher immune activity and protection with chitosan than chitosan ascorbate	[178]
Recombinant protein containing HSP90C of *Candida albicans* (rP-HSP90C)	Chitosan (MW: 50–190 kDa) (DD: 75–85%) (sigma-Aldrich, Saint Louis, USA)	Hydrogel – rP-HSP90C-Quil-A (for comparison of adjuvant properties with hydrogel)	Subcutaneous	–	– Cell viability (RAW264.7 cells) – Cellular uptake and intracellular antigen trafficking – Inhibition of expression of MHC-I by	Immunization in 6- to 8-week-old male C57BL/6 mice *Dose: 25 μg* *Boost: On day 14* *Challenge: i.v. with C. albicans on day 28* – Determination of IgG and its	– Nontoxic – Enhanced cellular uptake – Strong activation of dendritic cells – Escape of antigen from lysosomes to cytoplasm to induce CTL response	[134]

(continued)

Table 2 (continued)

Antigen	Chitosan type	Dosage form	Immunization route	Adjuvant added	In vitro studies	In vivo studies	Immune response	Ref.
					proteasome inhibitor – Activation of dendritic cells (BMDCs of C57BL/6 mice)	subtypes – Determination of cytokine levels – Differentiation of splenocyte T cells – Recruitment of immune cells into a local immune site	– Enhanced IgG1, IgG2a, and IgG2b levels with predominantly IgG2a/IgG1 ratio indicating enhanced Th1-biased response – Highest IgG levels with Quil A and antigen – Increased $CD4^{+}T$ and $CD8^{+}T$ cells, Th1, Th2, and Th17 cytokine levels and CTL response – Enhanced survival rate (86%) – Enhanced recruitment of antigen presenting cells	
Inactivated porcine parvovirus (PPV)	N-2-hydroxypropyl trimethyl ammonium chloride chitosan (N-2-	Emulsion (chitosan was added into the inactivated PPV at ratio of 0.5% (v/v)	Intramuscular	–	–	Immunization in 6- to 7-month-old sows *Dose:* $10^{6.0}$ $(TCID_{50})$	– Highest antibody titers in 8 weeks after immunization – No increase in	[135]

	HACC) *(MW, DD, and the source are not stated)*	and emulsified with mineral oil)				*Boost: On day 14* *Challenge: Intranasal with homologous PPV-H strain on day 70* – Determination of IgG and its subtypes -Histopathological change and PPV control	cellular immune response – No pathological changes in animals after immunization – Significant protection of sows from homologous strain infection	
Inactivated Newcastle disease virus (NDV)	-α-chitosan (CS) (MW: 1856 kDa) (DD: 86.0%) (Qingdao Yunzhou biochemical Corp. (Qingdao, China)) – HACC (MW: 5003 kDa) (DD is not stated) – SCS (MW: 5030 kDa) (DD is not stated)	Nanoparticle *CS-NPs: 200–406 nm* *HACC NPs: 320–391 nm* *SCS NPs: 156–195 nm*	Subcutaneous	–	– Cell viability (RAW264.7 cells)	– Safety test in 10-day-old SPF white Leghorn chicken Immunization in 8-day-old SPF white Leghorn chicken *Dose: Not stated* *Boost: Day 14* *Challenge: Highly virulent NDV (F48E9 strain) on day 70* – Determination of IgG and its subtypes – Proliferation of $CD4^+$ and $CD8^+$ T cells	– Low toxicity – Higher viability with HACC NPs compared to SCS NPs – High humoral immune response with CS and HACC NPs, no activation with SCS-NPs – Lower antibody titers with CS NPs and HACC NPs compared to commercial vaccine – High IL-2 and IFN-γ levels which indicated the elevation of	[136]

(continued)

Table 2 (continued)

Antigen	Chitosan type	Dosage form	Immunization route	Adjuvant added	In vitro studies	In vivo studies	Immune response	Ref.
						– Determination of cytokine levels	cellular immunity with CS NP and HACC NP – Higher lymphocyte proliferation with CS NPs and HACC NPs compared to SCS-NPs and commercial vaccine – Higher $CD4^+$ and $CD8^+$ T cells with CS NPs and HACC NPs – 100% protection with CS and HACC NPs on day 7	
rMdh of *Brucella abortus*	Chitosan (CS) (MW is not stated) (DD: 82.4%) (Jakwang, Ansung, Korea)	Nanoparticle (664.6 ± 161.7 nm)	–	–	– Generation of the in vitro M cell model – Cellular uptake and transport studies in M cells – Stimulation of M cells: Determination	–	– Similar uptake efficiency with rMdh and rMdh-CS NPs and localization of rMdh-CS NPs in M cells – Increased antigen transport with NPs	[179]

					of TLRs related genes Determination of cytokine levels		– Significant expression of TLR2 and no significant expression of TLR4 with NPs – Enhanced Th2-related immune responses with NPs – Significant stimulation of TLR-MyD88 signaling pathway with NPs – Significantly enhanced production of TNF-α, IL-1β, and IL-6 with NPs through MyD88-dependent TLR2 signaling in M cells – Significantly increased and similar expression of the IRF4 gene, which is a transcription factor in lymphocyte	

(continued)

Table 2 (continued)

Antigen	Chitosan type	Dosage form	Immunization route	Adjuvant added	In vitro studies	In vivo studies	Immune response	Ref.
							differentiation and a regulator of TLR signaling, with rMdh alone or NPs	
FliC of *Brucella abortus* and *B. melitensis*	– Chitosan *(MW, DD, and the source are not stated)*	Nanoparticle *Mannosylated CS NPs: 142.8 ± 8.17 nm*	Subcutaneous	–	–	Immunization in 6- to 8-week-old female BALB/c mice: *Dose: 30 μg* *Boost: On days 14 and 28* *Challenge: i.p. with B. abortus and B. melitensis on day 56* – Determination of IgG and its subtypes – Determination of cytokine levels	– Higher IgG and IgG subtypes with NPs – Higher IgG2a levels than IgG1 and IFN-y and IL-2 cytokine levels with NPs which indicates the Th1-based immune response – Low production of IL-10 levels with all groups which indicates low stimulation of Th2-based immune response – Significant protection with NPs	[137]

Neurotoxin II from *Androctonus australis* hector scorpion venom (Aah II)	Chitosan (MW: 50–190 kDa) (DD: 75–85%) (sigma-Aldrich, Saint Louis, USA)	Nanoparticle (208 ± 9 nm) *Aah II-chitosan VacciGrade™ (for comparison of adjuvant properties with CS NPs)* *Aah II-colloidal aluminum hydroxide wet gel adjuvant (VAC 20) (for comparison of adjuvant properties with CS NPs)*	Subcutaneous	–	–	Immunization in 6- to 8-week-old female Swiss mice *Dose: 1.5 μg* *Boost: On day 14* *Challenge: s.c. with increasing lethal doses of Aah II toxin on day 44* – Toxicity studies – Determination of IgG and its subtypes	– Nontoxic – Strong inflammatory response with alum, NPs, and chitosan VacciGrade™ – Highest specific antibody titer with NPs on day 22 compared to alum and chitosan VacciGrade™ – High IgG1/IgG2a ratio indicating the Th2-based immune response with NPs – Significantly higher protection with the Aah II-CS NPs immunized mice compared to Aah II-alum and Aah II-chitosan VacciGrade™	[180]
AVA	FUC-HTCC – Chitosan (used for FUCC-HTCC synthesis) (MW is not	Nanoparticle – *The formation of positively and negatively charged nanoparticles at*	Intraperitoneal	–	– Cell viability and cellular uptake studies (L929 and jaw	Immunization in 6- to 8-week-old female A/J mice *Dose: 10 μL* *Boost: Day 14*	– High cell viability (≥80%) and no plasma membrane damage – Higher	[181]

(continued)

Table 2 (continued)

Antigen	Chitosan type	Dosage form	Immunization route	Adjuvant added	In vitro studies	In vivo studies	Immune response	Ref.
	stated) (DD: 93.8%) (Wako pure chemical industries ltd., Osaka, Japan)	*different FUC/HTCC weight ratios* – *Positively charged AVA-FUC-HTCC NPs: 245.3 ± 13.1 nm)* – *Negatively charged AVA-FUC-HTCC NPs: 312.3 ± 58.8 nm)* *AVA-CpG ODN (for comparison of adjuvant properties with NPs)*			II dendritic cells)	*Challenge:* *i.p. with anthrax lethal toxin on day 70* – Determination of IgA, IgG and its subtypes	cellular uptake with negatively charged NPs – Higher IgG titers and Th2-based immune response with positively charged NPs – Th1-based immune response with CpG ODN – Higher survival rate (100%) with positively charged NPs after 20 days post-challenge	
–	Half N-acetylated chitosan derivatives[a]: – CS_{5-80} (MW: 7.05 kDa) (DD: 76.9%) – CS_{15-80} (MW: 15.01 kDa) (DD: 76.1%) – CS_{5-50} (MW: 6.54 kDa)	Nanoparticle *CS_{5-80} NPs: 101.84 ± 4.49 nm* *CS_{5-50} NPs: 126.1 ± 7.585 nm* *CS_{5-Man} NPs: 146.19 ± 5.637 nm* *CS_{15-80} NPs: 116.26 ± 10.88 nm* *CS_{15-50} NPs: 153.86 ± 8 nm*	–	CpG ODN	– Cell viability – Cellular uptake – Macrophage activation (RAW264.7 cells)	–	– Nontoxic – Increased cellular uptake with NPs compared to CpG ODN alone – Highest cellular uptake with CpG ODN-CS_{15-80} NPs – Highest IL-6 levels with CpG	[182]

	(DD: 49.5%) – $CS_{15–50}$ (MW: 18.36 kDa) (DD: 51.2%) -Mannosylated chitosan (CS_{man}) Chitosan (used for chitosan derivatives synthesis) (MW: 201.8 kDa) (DD: 76.9%) (Millipore sigma, St. Louis, MO, USA)	*$CS_{15–Man}$ NPs: 185.05 ± 12.01 nm*					ODN-$CS_{5–Man}$ NPs and CpG ODN-$CS_{15–Man}$ NPs compared with all groups – Higher IL-6 levels with CpG ODN-$CS_{5–50}$ NPs and CpG ODN-$CS_{15–50}$ NPS compared to naked CpG ODN, CpG ODN-$CS_{5–80}$ NPs and CpG ODN-$CS_{15–80}$ NPs	
HAV	Chitosan (MW is not stated) (DD: ≥90%) (bio basic, Canada)	Nanoparticle *HAV-CS NPs: 461.27 ± 105.93 nm* *HAV-alginate coated CS NPs: 654.13 ± 167.91 nm* *HAV-alum (for comparison of adjuvant properties with NPs)*	Intraperitoneal	–	–	Immunization in 6- to 8-week-old female and male BALB/c mice: *Single dose: 4 IU* – Determination of IgA, IgG and its subtypes – Splenocyte cell proliferation assay – Measurement of seroconversion rate in sera of immunized mice using ELISA	– Highest IgG levels, splenocyte proliferation and protection with alginate-coated NPs compared to non-coated NPs and alum, respectively – Slight tendency towards Th2 immune response with alum and alginate-coated NPs	[138]

(continued)

Table 2 (continued)

Antigen	Chitosan type	Dosage form	Immunization route	Adjuvant added	In vitro studies	In vivo studies	Immune response	Ref.
							– Balanced IgG1/IgG2a ratio with non-coated NPs – Highest IL-10 and IFN-γ levels with alginate-coated NPs and highest IL-2 levels with non-coated CS NPs – 100% sero-conversion rate in the sera of non-coated NPs immunized group indicating the strong protection with specific anti-bodies against hepatitis A	
DENV-2 NS1	TMC Chitosan (used for TMC synthesis) (MW is not stated) (DD: 94%) (Seafresh chitosan lab,	Nanoparticle *TMC NPs: 282.7 ± 1.65 nm*	Intraperitoneal	–	– Cellular uptake and dendritic cell activation studies: Determination of costimulatory	Immunization in 6- to 8-week-old female BALB/c mice *Dose: 8.5 and 17 μg* *Boost: Days*	– Enhanced cellular uptake of antigen – Strong dendritic cell stimulation – Enhanced	[139]

	Bangkok, Thailand)				molecules (human MoDCs)	*15 and 30* *Challenge: With immunized mice sera in DENV-1,2,3, and 4 infected BHK-21 cells* – Determination of IgG levels Determination of cytokine levels – Differentiation and proliferation of splenocyte T cells	upregulation of proinflammatory cytokines, Th1 and Th2 related cytokines, growth factors and IFN-α both in vitro and in vivo – Higher IgG levels – Induced Th1 and Th2 immune responses – Elimination of infected cells – Enhanced splenocyte proliferation – No increase in IFN-y + CD4+ cells proliferation – Enhanced IFN-y + CD8+ cells proliferation	
–	Chitosan (Protasan UP CL213) (MW: 150–400 kDa) (DD: 75–90%) (Novamatrix, Norway)	Gel Nanoparticle Microparticle *CS NPs:* *366.4 ± 4.8 nm* *Porins-CS NPs:* *337.7 ± 1.7 nm* *CS MPs:*	–	– OMP of *Salmonella enterica serovar typhi (S. typhi)* (porins)	– Cytotoxicity studies – Cellular uptake studies – Macrophage activation studies: Determination	–	– Higher cell viability with NPs and gel – Highest cellular uptake with MPs – Expression of MHC-II	[82]

(continued)

Table 2 (continued)

Antigen	Chitosan type	Dosage form	Immunization route	Adjuvant added	In vitro studies	In vivo studies	Immune response	Ref.
		1.30 ± 0.01 μm *Parins-CS MPs: 1.69 ± 0.01 μm*			of surface markers and cytokine levels		molecules with the MPs – Increased CD80 and CD86 expression with gel and NPs	
OVA	Catechol modified chitosan Chitosan (used for catechol modified chitosan synthesis) (MW-50–200 kDa) (DD: ≥ 85%) (Zhejiang Axong biotechnology co. ltd., China)	Nanosheets	Subcutaneous	–	– Cytotoxicity assay (DC2.4 cells) – Cellular uptake study – Maturation and activation of dendritic cells: Determination of surface marker expression (BMDCs from C57BL/6 mice)	Immunization in 6-week-old female C57BL/6 mice *Dose: 100 μg* – In vitro trafficking of antigens	– High biocompatibility with the 100% cell viability – Promoting the dendritic cell proliferation (to 124% from 90%) – Enhanced cellular uptake – Enhanced exogenous antigen uptake and induction of antigen endo/lysosomal escape which indicates the induction of strong $CD8^{+}T$ cells immune response – Significantly	[183]

							higher MHC-II expressions implying the stimulation of $CD4^{+}T$ cells via MHC-II pathway – Increased MHC-I complex level which indicates the effective antigen processing via MHC-I pathway – Upregulation of costimulatory molecules expression revealing the promotion of dendritic cell maturation – Significantly increased expression of chemokine receptor CCR7 which indicates the improved BMDCs migrating to lymph nodes – Enhanced secretion of Th1 cytokines (TNF-a	

(continued)

Table 2 (continued)

Antigen	Chitosan type	Dosage form	Immunization route	Adjuvant added	In vitro studies	In vivo studies	Immune response	Ref.
							and IFN-γ) and IL-6 but not Th2 cytokines (IL-4) – Accumulation in lymph nodes, which indicates the protections of antigens from degradation and increased retention time in the presence of nanosheets	
OMP31 of *Brucella melitensis*	Chitosan *(MW, DD, and the source are not stated)*	Nanoparticle *OMP31 loaded CS NPs: 99.1 ± 15 nm* – *OMP31 loaded AH NPs (for comparison of adjuvant properties with CS NPs)* – *OMP31 loaded CP NPs: 176.5 ± 11 nm (for comparison of adjuvant properties with CS NPs)*	Subcutaneous	–	–	Immunization in 4- to 6-week-old BALB/c mice *Dose: 30 μg* *Boost: On day 15* *Challenge: s.c. with B. melitensis 16 M on day 45* – Determination of IgA, IgG and its subtypes levels Determination of cytokine levels – Splenocyte	– Higher IgG and IgG1 titer after immunization with AH NPs than CS and CP NPs – Induction of IgG2a titer with all NPs – Th2-based immune response with AH NPs and Th1-based immune response with CS and CP NPs	[184]

						cell proliferation assay	– Induced secretion of IL-4, IL-12, IL-10, and IFN-γ with AH and CP NPs – Induced secretion of only IL-12 and IFN-γ with CS NPs – High splenic cell proliferation after immunized groups with all NPs – Equal and higher protection against *B. melitensis* 16 M with CS and CP NPs	
–	-α-chitosan (MW: 1856 kDa) (DD: 86.0%) (Qingdao Yunzhou bio-chemical Corp. (Qingdao, China)). HACC (MW: 5003 kDa) (DD is not stated) (from α-chitosan)	Solution	–	–	– Cell viability assay – Determination of NO and cytokine production (RAW264.7 cells)	–	– Low cytotoxicity with ≥80% cell viability – Higher NO production – Significantly stimulated the expression of IL-6 and TNF-a – Activation of MAPK which promotes the	[185]

(continued)

Table 2 (continued)

Antigen	Chitosan type	Dosage form	Immunization route	Adjuvant added	In vitro studies	In vivo studies	Immune response	Ref.
							production of cytokines that also activate the JAK/STAT signaling pathway and JAK/STAT pathways which modulate the immune response	
–	Chitosan (MW: 150 kDa) (DD: 86%) (FMC biopolymer AS Novamatrix, Sandvika, Norway)	Nanoparticle *PLGA NPs: 127 ± 12 nm* *Chitosan coated PLGA NPs: 151 ± 20 nm*	–	–	– Nanoparticle binding and uptake studies (moDC) – Activation of dendritic cells surface markers expression (moDC)	–	– Higher cellular uptake of chitosan-coated NPs than non-coated NPs – Uptake of NPs into the cells by macropinocytosis – Uptake of chitosan-coated NPS into the cells by clathrin-mediated endocytosis – No significant induction of activation markers on the surface of dendritic cells after incubation	[186]

							with nanoparticle formulations	
pDNA of rabies virus	Chitosan glutamate *(MW, DD, and the source are not stated)*	Nanoparticles in poloxamer 407 gel (hydrogel) *pDNA loaded CS-PLGA NPs: 422 nm*	Intramuscular	– Alum and MF59	– Determination of NO production and antigen presenting molecules (DC2.4 cells)	Immunization in 6- to 8-week-old female Swiss Webster mice *Dose: 100 µg* – Determination of IgG and its subtypes levels – Determination of cytokine levels – Splenocyte proliferation assay	– Higher NO production with adjuvanted NPs than non-adjuvanted NPs – Enhanced MHC II expression, resulting in a balanced MHC I and MHC II response expression with adjuvanted NPs – Th1 biased MHC I response with non-adjuvanted NPs – High IgG2a/IgG1 ratio, IL-4 and IFN-y expression which indicates Th1-based immune response with non-adjuvanted NPs and high IgG1/IgG2a ratio	[140]

(continued)

Table 2 (continued)

Antigen	Chitosan type	Dosage form	Immunization route	Adjuvant added	In vitro studies	In vivo studies	Immune response	Ref.
							which indicates Th2-based immune response with adjuvanted NPs – Significantly higher prolifera-tion of splenocytes with adjuvanted NPs than non-adjuvanted nanoparticle – Enhanced $CD8^+$ T cells population in splenocytes and lymph node cells with adjuvanted and non-adjuvanted NPs	
OVA	TMC Chitosan (used for TMC synthe-sis) (MW: 50 kDa) (DD: 85%)	Nanoparticle in hydrogel (A bioinspired catechol-functionalized hyaluronic acid hydrogel)	Subcutaneous	–	– Cell viabil-ity assay (NIH/3TC fibroblasts) – Cellular uptake and dendritic cell	Immunization in 6- to 8-week-old C57BL/6 mice *Dose: 10 μg* *Boost: On day 28* – Determination of IgG levels	– High cell via-bility (≅100%) – Uptake of for-mulations by endo/lysosomal transport which has important role	[141]

	(Koyo chemical, Hyogo, Japan)	*OVA loaded TMC NPs:* *233.1 ± 17.7 nm* *OVA-alum (for comparison of adjuvant properties with CS NPs)*			maturation studies surface marker expression (BMDCs)	– In vivo degradability and cytotoxicity assay	in antigen delivery and processing – Higher cellular uptake of NPs than NPs in hydrogel – Stimulation of DCs with NPs and NPs in hydrogel – Recruitment of immune cells to the outer layer of NPs in hydrogel – No cytotoxic effects on organs – Highest IgG levels with NPs in hydrogel – Similar and higher IgG levels with TMC NPs and alum	
ptfA gene of avian *Pasteurella multocida*	Chitosan (MW is not stated) (DD: 90%) (Golden Shell pharmaceuticals, Zhejiang, China)	Nanoparticle (≅ 200 nm)	Intramuscular	–	–	Immunization in 4-week-old chickens *Dose: 100 μg* *Boost: On days 14 and 28* *Challenge:* *i.m. with the virulent avian*	– Higher antibody levels and lymphocyte proliferation with NPs compared to attenuated live vaccine – Higher concentrations of the	[187]

(continued)

Table 2 (continued)

Antigen	Chitosan type	Dosage form	Immunization route	Adjuvant added	In vitro studies	In vivo studies	Immune response	Ref.
						P. multocida strain CVCC474 on day 42 – Determination of serum antibody concentrations – Cytokine assay – Lymphocyte proliferation assay	Th2 cytokine IL-4 with attenuated live vaccine compared to NPs – Higher concentrations of the Th2 cytokine IL-10 and Th1 cytokine IFN-γ with NPs compared to attenuated live vaccine – Higher survival rate with attenuated live vaccine (88%) and NPs (68%) immunized mice compared to antigen alone (56%)	
HBsAg	Chitosan (MW is not stated) (DD: 95%) (ChitoClear™ Primex Bio-Chemicals AS, Avaldsnes, Norway)	Nanoparticle *HBsAg loaded CS NPs: 837 ± 22 nm* *HBsAg loaded CS:-β-glucan NPs: 1274 ± 47 nm*	Subcutaneous	–	– Cytotoxicity studies – Macrophage studies (RAW264.7 cells)	Immunization in 7- to 8-week-old female C57BL/6 mice: *Dose: 1 or 1.5 μg* *Boost: On day 14* – Determination of IgG1, IgG2c, IgG3, and IgE	– Nontoxic – Localization of nanoparticles near the macrophage cell membrane – Higher IgG levels with CS:-β-glucan NPs	[188]

						levels -determination of cytokine levels	compared to HBsAg loaded CS NP – Higher Th2-based antibody levels with CS:β-glucan NPs compared to CS NPs – Higher antibody responses with higher dose – No detectable IgE levels indicating no risk to induce type 1 hypersensitivity – No enhanced increase in Th2 type cytokine and Th1 type cytokine with nanoparticles – Increased the Th17 type cytokine with CS:-β-glucan NPs compared to CS NP	
OMP25, BLS, and rHSP60 of	Chitosan (MW and DD are not stated) (sigma-Aldrich,	Nanoparticles *rOMP25-BLS loaded CS NPs rOMP25-BLS-*	Intraperitoneal	–	–	Immunization in 6-week-old female BALB/c mice	– Higher predominantly Th1-based antibody levels with	[189]

(continued)

Table 2 (continued)

Antigen	Chitosan type	Dosage form	Immunization route	Adjuvant added	In vitro studies	In vivo studies	Immune response	Ref.
Brucella melitensis	Saint Louis, USA)	*rHSP60 loaded CS NPS*				Dose: 20 μg Boost: On days 15 (40 μg) and 30 (30 μg) – Determination of IgG and its subtypes – Determination of cytokine levels – Lymphocyte proliferation assay	rOMP25-BLS loaded CS NPs compared to rOMP25-BLS-rHSP60 loaded CS NPs – Higher titer of IFN-γ and TNF-α with rOMP25-BLS-rHSP60 loaded CS NPs indicating the increased protective responses in the hosts – Higher titer of lymphocyte proliferation with both CS NPs	
HBsAg	Chitosan (MW: 500 kDa) (DD: 85%) (Jinan Haidebei marine biotechnology co. ltd., Jinan, China)	Nanogels – *The formation of positively and negatively charged HBsAg loaded nanogels by changing the adding order of chitosan and γ-PGA)* – *Negatively*	Intramuscular	–	–	Immunization in 6- to 8-week-old female C57BL/6 J mice *Dose: 3 μg* *Boost: On days 14 and 28* *Challenge:* *i.v. with plasmid pAAV/HBV1.2*	– Higher antibody levels with single dose of positively charged nanogels compared to single dose of negatively charged nanogels and alum – Enhanced and	[142]

		charged nanogels: 213.6 ± 8.6 nm *-positively charged nanogels: 265.3 ± 7.28 nm* – *Combination of positively and negatively charged nanogels (for only challenge study)* *-alum-HBsAg (for comparison of adjuvant properties with nanogels)*				– Determination of IgG levels – Determination of surface molecules of dendritic cells – T cells proliferation assay	similar expression of co-stimulatory molecules with both positively and negatively charged nanogels – No significant differences in CD40 expression and MHC-II expression – Enhanced memory $CD8^+$, IFN-γ+, and TNF-α + $CD8^+$ T cells with both positively and negatively charged nanogels indicating enhanced cellular immune responses to pAAV/HBV1.2 challenge – Strong resistance to pAAV/HBV1.2 plasmid challenge with single dose of positively charged nanogels and the	

(continued)

Table 2 (continued)

Antigen	Chitosan type	Dosage form	Immunization route	Adjuvant added	In vitro studies	In vivo studies	Immune response	Ref.
							combination of positively and negatively charged nanogels	
AIV	Chitosan (extracted from marine shrimp) (MW: 2122 kDa) (DD: 85%)	Nanoparticle and solution – *AIV loaded CS NPs: 397 nm* – *AIV-CS solution* – *Inactivated AIV vaccine*	Subcutaneous	–	–	Immunization in 8-week-old SPF chickens *Dose: 0.1 mL* *Boost: On day 21* – Determination of HI antibody titers – Lymphocyte proliferation assay – Phagocytic activity of immunized chicken peripheral monocytes	– Highest HI antibody titers with CS solution compared to CS NPs and inactivated AIV vaccine – Significantly higher lymphocyte proliferation and phagocytic activity with CS NPs compared to CS solution and inactivated AIV vaccine	[190]
HBsAg	Chitosan (MW: Not stated) (DD: 95%) (ChitoClear™ Primex Bio-Chemicals AS, Avaldsnes, Norway)	Nanoparticle *HBsAg-alum-CS NPs: 280.9 ± 32.7* *HBsAg-Na-CS NPs: 754.2 ± 114.4 nm*	Subcutaneous	–	– Dendritic cell activation and measurement of surface marker expression (BMDCs)	Immunization in 8 to 16-week-old female C57BL/6 mice *Dose: 1.5 µg* *Boost: On day 14* – Determination of specific IgG, IgG1, and IgG2c	– Higher secretion of IL-1β via NLRP3- and ASC-dependent process with alum-CS NPs compared to Na-CS NPs – Upregulation	[143]

						– Determination of cytokine levels – Determination of innate immune response and cellular uptake studies: PECs of intraperitoneally immunized mice with FITC-labelled formulations	of the expression of surface markers with alum-CS NPs – Enhanced IgG levels and balanced Th1/Th2 response with both NPs	
OMP16 of *Haemophilus parasuis*	Chitosan (MW is not stated) (DD: ≥90%) (Sangon biotech, China)	*Microsphere OMP16-alginate-CS NPs: 1.0 ± 0.25 μm IFA-OMP16 (for comparison of adjuvant properties with nanogels)*	Subcutaneous	–	–	Immunization in 28-day-old BALB/c mice *Dose: 50 μg* *Boost: On day 14* *Challenge: i.p. with H. parasuis strain LY02 on day 29* – Determination of IgG levels – Determination of cytokines – Bacteriological analysis in tissues from liver, lung, and spleen of mice	– Significantly increased antibody titers and IL-2, IL-4, and IFN-y cytokine levels with NPs compared to IFA indicating the enhanced stimulation of humoral and cellular immunity – Significantly lower bacterial load and higher survival rate with NPs (80%) compared to IFA (% 70)	[191]

(continued)

Table 2 (continued)

Antigen	Chitosan type	Dosage form	Immunization route	Adjuvant added	In vitro studies	In vivo studies	Immune response	Ref.
H5N1 split vaccine	HTCC Chitosan (used for HTCC synthesis) (MW: 7.8×10^5 kDa) (DD: 90–95%) (Golden-Shell pharmaceutical co. ltd., Zhejiang, China)	Microgel *H5N1 split vaccine-microgel:* *4.30 ± 0.22 μm, 2.51 ± 0.40 μm, 807 ± 74 nm (obtained by different quaternization degree of chitosan)* *H5N1 split vaccine-alum*	Intramuscular	–	– Hemolysis assay on sheep erythrocytes – Cellular uptake study – Determination of cell surface markers expression (BMDCs of BALB/c mice)	Immunization in 4- to 6-week-old female BALB/c mice *Dose: 3.75 or 15 μg* *Boost: On day 14* – Determination of IgG titers and hemagglutination inhibition assay – Determination of cytokine release	– Negligible and size-dependent hemolytic activity – Significantly enhanced antigen cellular uptake with H5N1 split vaccine-microgel compared to alum – Highest ability to facilitate antigen escape from endosome into the cytoplasm with H5N1 split vaccine-microgel indicating the efficiently enhanced antigen cross-presentation and increased cellular immune responses – Significantly higher IgG levels and Th1 cytokines with H5N1 split vaccine-microgel	[144]

							compared to the alum group – Higher immune response in vitro and in vivo with low quaternization degree and size of microgels compared to high quaternization degree and size of microgels – Equivalent level of antibodies and cytokines with microgels for 3.75 μg antigen dose compared to the alum group for 15 μ g antigen dose indicating the antigen dose spring effect of microgels	
OVA	Chitosan (MW: 50–190 kDa) (DD: 75–85%) (sigma-Aldrich, Saint Louis, USA)	Gel – *Poloxamer 407–25R4 sol–gels* – *Chitosan-methyl cellulose (MC) sol-gels* – *Cubosomes*	Subcutaneous	–	–	Immunization in male and female C57BL/6 mice *Dose: 10 μg* *Challenge: With OVA on day 21* – Determination	– Significantly enhanced IgG levels with inclusion of cubosomes into the CS-MC sol--gels compared to	[192]

(continued)

Table 2 (continued)

Antigen	Chitosan type	Dosage form	Immunization route	Adjuvant added	In vitro studies	In vivo studies	Immune response	Ref.
		(Phytantriol: Poloxamer 407: Propylene glycol: MPL)				of IgG levels – Determination of cytokine levels	poloxamer 407–25R4 sol–gels – No effect of inclusion of cubosomes into the poloxamer 407–25R4 sol–gels	
UVI-DENV-2	Chitosan (MW: 80 kDa) (DD: 82%) (Koyo chemical, Japan)	Nanoparticle (372 ± 11 nm)	Intraperitoneal	–	–	Immunization in 6- to 8-week-old female Swiss albino mice *Dose: 0.3, 1, 3, and 10 μg* *Boost: On days 15 and 30* – Determination of IgG and IgM levels – Determination of cytokine and chemokine expression	– Strong IgM and IgG levels – Enhanced Th1, Th2, pro-inflammatory cytokines and chemokine expression	[193]
OVA	– Chitopharm S (MW: 79.4 ± 4.50 kDa) (DD: 81%) (Chitinor AS, Norway)	Solutions of spray dried chitosans – *Chitopharm S: 4.1 μm* – *ChitoClear FG 95:20.8 μm*	Subcutaneous	–	–	Immunization in 6- to 8-week-old C57BL/6 J mice *Dose: 10 μg* *Boost: On day 28* – Determination	– Highest IgG levels with CS (sigma) and lowest with ChitoClear – Similar IgG	[80]

	– ChitoClear FG 95 (MW: 102 ± 4.99 kDa) (DD: 81%) (Primex, Iceland) – Chitosan (MW: 149 ± 6.60) (DD: 76%) (sigma Aldrich, Germany)	– *Chitosan: 2.2 μm* – *Alum (for comparison of adjuvant properties)*				of IgG levels – Determination of CD4 and CD8T cells expansion – Determination of cytokine levels	levels with alum and Chitopharm S – Highest CD8 and CD4 cells population with CS (sigma) compared to other chitosans and alum – No detectable IL-4 and Il-5 cytokine expression with chitosans – Highest Th1 cytokine (IFN-γ) expression with Chitopharm S	
BSA OVA	Chitosan (MW: 95 ± 20 kDa) (DD is not stated) (sigma-Aldrich, Saint Louis, USA)	Micro- and nanoparticles *332 ± 19 nm, 1,034 ± 74 nm, 2,918 ± 333 nm*	–	–	– Cellular uptake study – Activation of DCs and macrophages: Determination of surface molecules and cytokine expression – Antigen presentation assay (co-cultured	–	– Enhanced pro-inflammatory cytokines in both macrophages and dendritic cells – Expression of greater levels of monocyte/macrophage-specific cytokines on macrophages, expression of higher levels of IL-1β and IL-6 on	[194]

(continued)

Table 2 (continued)

Antigen	Chitosan type	Dosage form	Immunization route	Adjuvant added	In vitro studies	In vivo studies	Immune response	Ref.
					BMDCs and RAW264.7 cells)		dendritic cells – Modest increases in surface marker expression with chitosan alone – Enhanced antigen presenting function with increased particle size	

AH Aluminum hydroxide, *AIV* Avian influenza virus, *AVA* Anthrax vaccine adsorbed, *BLS* Brucella lumazine synthase, *BMDC* Bone marrow dendritic cell, *BSA* Bovine serum albumin, *CP* Calcium phosphate, *CS* Chitosan, *CTL* Cytotoxic T lymphocyte, *DC* Dendritic cell, *DD* Deacetylation degree, *DENV-2 NS1* Dengue virus-2 nonstructural protein 1, *FliC* Flagellar filament structural protein, *FUC-HTCC* Fucoidan-N-(−2-hydroxy-3-trimethylammonium) propylchitosan, *HA* Hyaluronic acid, *HACC* Hydroxypropyl trimethyl ammonium chloride chitosan, *HAV* Hepatitis A vaccine, *HBsAg* Hepatitis B surface antigen, *HI* Hemagglutination inhibition, *HSP60* Heat shock protein 60 kDa, *HSP90C* Epitope C from heat shock protein 90, *HTCC* N-[(2-hydroxy-3-trimethylammonium) propyl] chitosan chloride, *IFA* Incomplete Freund's adjuvant, *i.m.* Intramuscular, *i.p.* Intraperitoneal, *i.v.* Intravenous, *MHC-I/II* Major histocompatibility complex-I/II, *MoDCs* Monocyte derived dendritic cells, *MP* Microparticle, *MPL* Monophosphoryl lipid A, *MW* Molecular weight, *Na* Sodium sulfate, *NP* Nanoparticle, *ODN* Oligodeoxynucleotides, *OMP* Outer membrane protein, *OVA* Ovalbumin, *γ-PGA* Poly-γ-glutamic acid, *PECs* Peritoneal exudate cells, *PLGA* Poly (d, l-lactide-co-glycolide), *rMdh* Recombinant malate dehydrogenase, *OMP* Outer membrane protein, *s.c.* Subcutaneous, *SCS* Sulfated chitosan, *SPF* Specific pathogen free, $TCID_{50}$ Fifty-percent tissue culture infective dose, *TLR* Toll like receptor, *TMC* Trimethyl chitosan, *UVI-DENV-2* UV-inactivated dengue virus serotype 2

[a] C_{15-80} possesses a molecular weight of 15 kDa and deacetylation degree of 80%

was attributed to negative charge and low molecular weight of sulfated chitosan nanoparticles.

Sadeghi et al. [137] have developed mannosylated chitosan nanoparticles loaded with FliC protein and investigated the immunogenicity and protective efficacy against Brucella infection in BALB/c mice. The use of mannosylated chitosan as an adjuvant and antigen-delivery system was reported to increase immunogenic properties of FliC protein. Although it was found to be inferior to live-attenuated antigens, appropriate level of protection was observed.

Abdelallah et al. [138] have compared chitosan solution and alginate-coated chitosan nanoparticles to alum as an adjuvant/delivery for the hepatitis A vaccine (HAV). Enhanced humoral and cellular immune responses were obtained with alginate-coated chitosan nanoparticles, increasing the seroconversion rate (100%). Furthermore, the chitosan solution was reported to show comparable immune responses to that obtained with alum with a balanced Th1/Th2 immune pathway.

TMC-based nanoparticles were used for parenteral delivery of nonstructural protein 1 (NS1) of dengue virus (DENV) [139]. TMC nanoparticles were capable of enhancing antibody production and establishing a mixed Th1-Th2 immune response against the antigen. TMC NPs were shown to act also as an adjuvant for cellular immunity. However, the mechanism behind CTL response adjuvanted by TMC could not be elucidated. It was suggested that TMC NPs could disrupt the lysosome and escape into the cytoplasm.

Bansal et al. [140] have prepared a thermosensitive gel incorporated with nanoparticles by adsorbing the negatively charged pDNA of rabies on the surface of cationic modified PLGA chitosan nanoparticles by electrostatic attraction and finally dispersing in poloxamer 407 gel for immunization against rabies. Further, Alum and MF59 were added as adjuvant. The pDNA nanoparticulate vaccine dispersed in hydrogel was shown to induce both humoral and cell-mediated immune responses. The antibody response was enhanced with the additional adjuvants, skewing the Th1/Th2 immune profile toward Th2. The enhanced immune response was attributed to the increased cellular uptake of the pDNA provided by the particulate system. Retention of the vaccine at the application site was prolonged by the gel formulation.

In another study with chitosan-based nanoparticles in hydrogel formulations [141], cellular uptake of positively charged TMC nanoparticles was shown to be more effectively internalized by dendritic cells compared to nanoparticles in negatively charged hydrogel. It was known that positively charged particles (such as TMC/NPs) can improve particle endocytosis by DCs and induce DC maturation. Histological examination in mice samples showed that immune cells were recruited to the outer layer of the NPs-Gel which indicates the implanted NPs-Gel can serve as an in situ depot that recruits and concentrates immune cells. Moreover, in mice that were immunized by a single administration of TMC nanoparticles in hydrogel, significantly higher IgG titers were generated in week 2 following immunization than TMC nanoparticles. These differences were maintained for at least 12 weeks, suggesting that the NPs-gel can serve as a single-injection platform to strengthen and prolong OVA-specific humoral immune responses.

HBsAg loaded nanogels were prepared using chitosan (CS) and poly-γ-glutamic acid (γ-PGA) [142]. Positively charged nanogel and negatively charged nanogels were prepared by adjusting the CS and γ-PGA proportion. A single dose of positively charged nanogels was shown to induce both humoral immunity and cellular immunity and provide long-term immune protection against hepatitis B virus, while with negatively charged nanogels, cellular immunity was induced. Positively charged nanogels were found to be more stable.

Lebre et al. [143] have investigated the ability of chitosan-aluminum nanoparticles (CH-Al NPs), combining the immunostimulatory effects of chitosan and aluminum salts, to promote dendritic cell activation. They assessed their impact on innate and adaptive immune responses and compared the results to those reported for conventional chitosan particles. All tested CH-NP formulations were capable of modulating cytokine secretion by dendritic cells. CH-Al NPs were shown to promote NLRP3 inflammasome activation, enhancing the release of IL-1β without significantly inhibiting Th1 and Th17 cell-polarizing cytokines, IL-12p70 or IL-23, and induced DC maturation, while pro-inflammatory cytokine production was not promoted. A local inflammatory response comparable to that elicited by the vaccine adjuvant alum was observed in with CH-Al NPs. These nanoparticles were combined with HBsAg and administered subcutaneously in mice. Enhanced antigen-specific IgG titers in serum, nasal, and vagina were obtained. The authors suggested the developed chitosan-aluminum combination nanoparticles as a potential adjuvant to enhance both innate and adaptive immune responses.

Similarly, our group has investigated the combination of chitosan with *Salmonella typhi* porins as a new adjuvant system to enhance the immunogenicity of the highly purified antigens [82]. Porins which are open channels filled with water in protein structure found in the outer membrane of gram-negative bacteria have been demonstrated to exert adjuvant effects. Cationic gels, microparticles, and nanoparticles based on chitosan were prepared with high loading efficiency of porins. Porins-chitosan combination systems were found to induce CD80, CD86, and MHC-II expressions at different levels by different formulations depending on the particle size. Similarly, TNF-α and IL-6 levels were found to increase with porins-chitosan combination.

Wang et al. [144] have developed quaternized chitosan microgels without chemical crosslinking as an adjuvant for H5N1 split vaccine and investigated the effect of quaternization degree and size on the adjuvanticity of microgels. Microgels with relatively smaller size (807 nm) and moderate quaternization degree (41% and 60%) were found to be suitable for a maximum immune response.

6 Chitosan for Cutaneous Immunization

Cutaneous immunization (also referred to as skin vaccination or transcutaneous immunization) is a promising alternative to replace parenteral immunization, being a safe and needle-free delivery. This route of immunization takes advantage of the unique immunological features of the skin immune system [145, 146]. The skin, which is the first line of defense preventing the entry of pathogens into the body is composed of the epidermis, the dermis, and the hypodermis. The viable epidermis is composed of approximately 90% keratinocytes. The epidermis and dermis have a high density of immunocompetent cells such as Langerhans cells (LCs) and dermal dendritic cells (dDCs), which represent the second line of defense and initiate specific immune responses by presenting the processed antigens to other cells of the immune system.

For cutaneous vaccine delivery, it is required that the antigen penetrates through the stratum corneum and reaches to dendritic cells [147, 148]. Hence, for a successful cutaneous immunization, it is essential to understand very well the structure and function of the skin as well as the challenges (e.g., size, charge, and other physicochemical properties) in delivery of antigens. In cutaneous vaccination, the antigen is directly delivered to immunocompetent cells either by needle-free jet and powder injection or by creating microchannels or transient cavities in the upper skin layers, using electroporation, laser poration, sono- or iontophoresis, barrier and ablative methods, microneedle-mediated transport [149]. Furthermore, delivery systems including polymeric nanoparticles and lipid-based systems to be delivered topically have also been investigated to stimulate the immune responses [150]. Variety of adjuvants including alum, Quillaja saponins, bacterial adenosine diphosphate (ADP)-ribosylating exotoxins, and specific ligands of toll-like receptors have been utilized in cutaneous vaccination in order to obtain improved immune responses. Currently, studies are still continuing to find more effective adjuvants as well as to elucidate mechanisms of action of these adjuvants.

Various research groups have investigated chitosan in different forms in cutaneous vaccination. Among these systems are microneedles which can be developed using low-cost preparation methods and can be self-administered. Studies have shown that with these microneedle systems, long-lasting humoral and cellular immune response could be generated through skin with high patient compliance [151]. Chen et al. [152] have designed a microneedle system, composed of embeddable chitosan microneedles and a poly(L-lactide-co-D,L-lactide) (PLA) supporting array for delivery of encapsulated antigens to the skin. They have demonstrated in vivo that when microneedles loaded with a model antigen, OVA was embedded in rat skin, microneedles gradually degraded and prolonged OVA exposure at the insertion sites for up to 14 days. Further, in rats immunized by a single microneedle dose of OVA, a significantly higher OVA-specific antibody response which lasted for at least 6 weeks was obtained when compared to intramuscular immunization.

Li et al. [153] have prepared polyvinylpyrrolidone (PVP)-based fast dissolving microneedle arrays microneedles which the needle tips were loaded with chitosan nanoparticles for coencapsulation of a model antigen, OVA and an adjuvant, CpG oligodeoxynucleotides (CpG). After insertion into the skin, the microneedle arrays were fully dissolved within 3 min and the nanoparticles were released in the epidermal layer. Following vaccination in mice, chitosan nanoparticles, loaded with the antigen and the adjuvant were shown to effectively accumulate in peripheral lymph nodes, thus inducing greatly enhanced immune responses compared to subcutaneous injections.

Siddhapura et al. [154] have investigated the potential of tetanus toxoid loaded chitosan nanoparticles following bare topical and microneedle-assisted immunization. Significantly higher humoral and cell-mediated immune responses were obtained with bare chitosan nanoparticles as well as with hollow microneedle- or soluble microneedle-assisted vaccinations. Higher IgG and IgG2a levels and IFN-y and IL-2 cytokines indicated Th-1-based immune response was obtained with hollow microneedle-assisted nanoparticles administration compared to commercial vaccine and soluble microneedle-assisted administration. In another study, a microneedle system was developed using sodium hyaluronate (HA) as the tip and chitosan as the base, for biphasic antigen release, and was evaluated as a dermal delivery system for single-dose vaccination. Upon skin insertion, the dissolvable HA tip was dissolved within the skin for rapid release of the encapsulated antigens, thus priming the immune system, while the biodegradable chitosan base remained in the dermis for prolonged antigen release for 4 weeks, providing booster. It was shown that single immunization with the HA/chitosan microneedle containing OVA stimulated both T helper type 1 (Th1) and Th2 immune responses in rats and induced considerably higher antibody responses when compared to subcutaneous vaccination [155].

TMC was also investigated for cutaneous vaccination [156]. TMC-adjuvanted diphtheria toxoid (DT) formulations were applied cutaneously with microneedles. Mice were vaccinated with DT-loaded TMC nanoparticles and solution of TMC-DT. The formulations were applied onto the skin before or after microneedle treatment with two different microneedle arrays. Independent of the microneedle array used and the sequence of microneedle treatment and vaccine application, cutaneous immunization with the TMC/DT mixture was shown to elicit eightfold higher IgG titers compared to the TMC nanoparticles or DT solution.

Different than the microneedle system, Yang et al. [157] have developed galactosylated chitosan-modified ethosomes combined with silk fibroin nanofibers for cutaneous vaccine delivery. It was demonstrated that the developed system was able to induce anti-OVA-specific IgG levels and increase the expression of IFN-γ, IL-2, and IL-6 by spleen cells in vivo. It was concluded that the large specific surface area of nanofibers would enlarge the contact area between the ethosomes and skin, resulting in enhanced efficacy.

7 Conclusion

For more than two decades, chitosan and its derivatives are being widely investigated as an adjuvant and vaccine delivery system, inducing both humoral and cellular responses, and only recently its mechanism of action has been elucidated. Mechanistic understanding of its immunostimulatory activity will certainly be a key to success in development of chitosan-based systems for immunization. Besides its immunostimulatory activity, chitosan is a promising material as an adjuvant/vaccine delivery system also due to its favorable features such as positive surface charge, mucoadhesive and bioadhesive properties as well as penetration enhancing property, especially for mucosal and cutaneous immunization. By this means, chitosan enhances the interaction with immune cells, increases the uptake of the antigen, acts as depot, resulting in higher immune responses. Chitosan-based systems can be formulated in different forms (gel, micro- and nanoparticle, etc.) with different shape and size. Versatility in its physicochemical properties allows the formulators to engineer customized systems with this polymer for antigen delivery. Furthermore, it has been demonstrated that also combining chitosan with other vaccine delivery platforms as well as other adjuvants results in more enhanced immune responses. Yet, for a successful vaccine delivery system with enhanced efficacy, the properties of chitosan to be used must be very well versed. In spite of numerous studies performed on chitosan-based systems, up to date, there is no chitosan-based or chitosan containing vaccine reached to the market yet. Reason for this can be explained mainly due to the hurdles in standardization of chitosan and its derivatives, which still remains as the major issue to be solved in order to maximize the possibility of its approval. Another reason may be that it is a material of animal origin. Nevertheless, results are encouraging for the future and chitosan appears to have a great potential in vaccine delivery applications.

References

1. Ogra PL, Faden H, Welliver RC (2001) Vaccination strategies for mucosal immune responses. Clin Microbiol Rev 14(2):430–445
2. Baxter D (2007) Active and passive immunity, vaccine types, excipients and licensing. Occup Med 57(8):552–556
3. Moyle PM, Toth I (2013) Modern subunit vaccines: development, components, and research opportunities. ChemMedChem 8(3):360–376
4. Karch CP, Burkhard P (2016) Vaccine technologies: from whole organisms to rationally designed protein assemblies. Biochem Pharmacol 120:1–14
5. Vetter V, Denizer G, Friedland LR, Krishnan J, Shapiro M (2018) Understanding modern-day vaccines: what you need to know. Ann Med 50(2):110–120
6. Bastola R, Noh G, Keum T, Bashyal S, Seo J-E, Choi J, Oh Y, Cho Y, Lee S (2017) Vaccine adjuvants: smart components to boost the immune system. Arch Pharm Res 40(11):1238–1248
7. O'Hagan DT, Friedland LR, Hanon E, Didierlaurent AM (2017) Towards an evidence based approach for the development of adjuvanted vaccines. Curr Opin Immunol 47:93–102
8. Di Pasquale A, Preiss S, Tavares Da Silva F, Garcon N (2015) Vaccine adjuvants: from 1920 to 2015 and beyond. Vaccines (Basel) 3(2):320–343
9. Weinberger B (2018) Adjuvant strategies to improve vaccination of the elderly population. Curr Opin Pharmacol 41:34–41
10. Şenel S (2020) Current status and future of chitosan in drug and vaccine delivery. React Funct Polym 147:104452
11. Garçon N, Morel S, Didierlaurent A, Descamps D, Wettendorff M, Van Mechelen M (2011) Development of an AS04-adjuvanted HPV vaccine with the adjuvant system approach. BioDrugs 25(4):217–226

12. De Gregorio E, Caproni E, Ulmer JB (2013) Vaccine adjuvants: mode of action. Front Immunol 4
13. Toussi D, Massari P (2014) Immune adjuvant effect of molecularly-defined toll-like receptor ligands. Vaccine 2(2):323–353
14. Christensen D (2016) Vaccine adjuvants: why and how? Hum Vacc Immunother 12 (10):2709–2711
15. De Temmerman ML, Rejman J, Demeester J, Irvine DJ, Gander B, De Smedt SC (2011) Particulate vaccines: on the quest for optimal delivery and immune response. Drug Discov Today 16(13–14):569–582
16. Del Giudice G, Rappuoli R, Didierlaurent AM (2018) Correlates of adjuvanticity: a review on adjuvants in licensed vaccines. Semin Immunol 39:14–21
17. Reed SG, Bertholet S, Coler RN, Friede M (2009) New horizons in adjuvants for vaccine development. Trends Immunol 30(1):23–32
18. Seder R, Reed SG, O'Hagan D, Malyala P, D'Oro U, Laera D, Abrignani S, Cerundolo V, Steinman L, Bertholet S (2015) Gaps in knowledge and prospects for research of adjuvanted vaccines. Vaccine 33(Supp 2):B40–B43
19. Diaz-Arévalo D, Zeng M (2020) Chapter 7 – nanoparticle-based vaccines: opportunities and limitations. In: Shegokar R (ed) Nanopharmaceuticals. Elsevier, Amsterdam, pp 135–150
20. O'Hagan DT, Lodaya RN, Lofano G (2020) The continued advance of vaccine adjuvants – 'we can work it out'. Semin Immunol 50:101426
21. Bachmann M, Jennings G (2010) Vaccine delivery: a matter of size, geometry, kinetics and molecular patterns. Nat Rev Immunol 10:787–796
22. Lebre F, Hearnden CH, Lavelle EC (2016) Modulation of immune responses by particulate materials. Adv Mater 28:5525–5541
23. Gregory AE, Titball R, Williamson D (2013) Vaccine delivery using nanoparticles. Front Cell Infect Microbiol 3:13–13
24. Foged C, Brodin B, Frokjaer S, Sundblad A (2005) Particle size and surface charge affect particle uptake by human dendritic cells in an in vitro model. Int J Pharm 298(2):315–322
25. Thiele L, Merkle HP, Walter E (2003) Phagocytosis and phagosomal fate of surface-modified microparticles in dendritic cells and macrophages. Pharm Res 20(2):221–228
26. Lavelle EC, O'Hagan DT (2006) Delivery systems and adjuvants for oral vaccines. Expert Opin Drug Deliv 3(6):747–762
27. Shah RR, O'Hagan DT, Amiji MM, Brito LA (2014) The impact of size on particulate vaccine adjuvants. Nanomedicine 9(17):2671–2681
28. McKee AS, Marrack P (2017) Old and new adjuvants. Curr Opin Immunol 47:44–51
29. O'Hagan DT, Valiante NM (2003) Recent advances in the discovery and delivery of vaccine adjuvants. Nat Rev Drug Discov 2:727–735
30. O'Hagan DT, Singh M (2003) Microparticles as vaccine adjuvants and delivery systems. Expert Rev Vaccines 2(2):269–283
31. Xiang SD, Scholzen A, Minigo G, David C, Apostolopoulos V, Mottram PL, Plebanski M (2006) Pathogen recognition and development of particulate vaccines: does size matter? Methods 40(1):1–9
32. Sahdev P, Ochyl LJ, Moon JJ (2014) Biomaterials for nanoparticle vaccine delivery systems. Pharm Res 31(10):2563–2582
33. Joshi VB, Geary SM, Salem AK (2013) Biodegradable particles as vaccine delivery systems: size matters. AAPS J 15(1):85–94
34. Şenel S, McClure SJ (2004) Potential applications of chitosan in veterinary medicine. Adv Drug Deliv Rev 56(10):1467–1480
35. Maxwell S, Şenel S, McClure S (2006) Chitosan for delivery of mucosal vaccines in ruminants. In: Paper presented at the 33rd annual meeting & exposition of the controlled release society, Vienna, Austria, 18–22 July

36. Zaharoff DA, Rogers CJ, Hance KW, Schlom J, Greiner JW (2007) Chitosan solution enhances both humoral and cell-mediated immune responses to subcutaneous vaccination. Vaccine 25(11):2085–2094
37. Sayın B, Şenel S (2008) Chitosan and its derivatives for mucosal immunization. In: Jayakumar R, Prabaharan M (eds) Current research and developments on chitin and chitosan in biomaterials science, vol 1. Research Signpost, Kerala, pp 145–165
38. Sayın B, Somavarapu S, Li XW, Thanou M, Sesardic D, Alpar HO, Şenel S (2008) Mono-N-carboxymethyl chitosan (MCC) and N-trimethyl chitosan (TMC) nanoparticles for non-invasive vaccine delivery. Int J Pharm 363(1–2):139–148
39. Arca HC, Gunbeyaz M, Senel S (2009) Chitosan-based systems for the delivery of vaccine antigens. Expert Rev Vaccines 8(7):937–953
40. Sayın B, Somavarapu S, Li XW, Sesardic D, Şenel S, Alpar OH (2009) TMC-MCC (N-trimethyl chitosan-mono-N-carboxymethyl chitosan) nanocomplexes for mucosal delivery of vaccines. Eur J Pharm Sci 38(4):362–369
41. Günbeyaz M, Faraji A, Özkul A, Puralı N, Şenel S (2010) Chitosan based delivery systems for mucosal immunization against bovine herpesvirus 1 (BHV-1). Eur J Pharm Sci 41 (3–4):531–545
42. Şenel S (2011) Chitosan-based particulate systems for non-invasive vaccine delivery. In: Jayakumar R, Prabaharan M, Muzzarelli RAA (eds) Advances in polymer science, vol 243. Springer, Berlin, pp 111–137
43. Çokçalışkan C, Özyörük F, Gürsoy RN, Alkan M, Günbeyaz M, Arca HÇ, Uzunlu E, Şenel S (2014) Chitosan-based systems for intranasal immunization against foot-and-mouth disease. Pharm Dev Technol 19(2):181–188
44. Carroll EC, Jin L, Mori A, Munoz-Wolf N, Oleszycka E, Moran HBT, Mansouri S, McEntee CP, Lambe E, Agger EM, Andersen P, Cunningham C, Hertzog P, Fitzgerald KA, Bowie AG, Lavelle EC (2016) The vaccine adjuvant chitosan promotes cellular immunity via DNA sensor cGAS-STING-dependent induction of type I interferons. Immunity 44(3):597–608
45. Li D, Fu D, Kang H, Rong G, Jin Z, Wang X, Zhao K (2017) Advances and potential applications of chitosan nanoparticles as a delivery carrier for the mucosal immunity of vaccine. Curr Drug Deliv 14(1):27–35
46. Dhakal S, Renu S, Ghimire S, Shaan Lakshmanappa Y, Hogshead BT, Feliciano-Ruiz N, Lu F, Hogenesch H, Krakowka S, Lee CW, Renukaradhya GJ (2018) Mucosal immunity and protective efficacy of intranasal inactivated influenza vaccine is improved by chitosan nanoparticle delivery in pigs. Front Immunol 9
47. Li J, Cai C, Li J, Li J, Li J, Sun T, Wang L, Wu H, Yu G (2018) Chitosan-based nanomaterials for drug delivery. Molecules 23(10)
48. Moran HBT, Turley JL, Andersson M, Lavelle EC (2018) Immunomodulatory properties of chitosan polymers. Biomaterials 184:1–9
49. Şenel S, Yüksel S (2020) Chitosan-based particulate systems for drug and vaccine delivery in the treatment and prevention of neglected tropical diseases. Drug Deliv Transl Res 10:1644–1674
50. Younes I, Rinaudo M (2015) Chitin and chitosan preparation from marine sources. Structure, properties and applications. Mar Drugs 13(3):1133–1174
51. Peniche C, Argüelles-Monal W, Goycoolea FM (2008) Chapter 25 – chitin and chitosan: major sources, aroperties and applications. In: Belgacem MN, Gandini A (eds) Monomers, polymers and composites from renewable resources. Elsevier, Amsterdam, pp 517–542
52. European Pharmacopoeia (2020) Council of Europe, 10th edn., Monograph 1774
53. USP 42/NF37 (2019) United States Pharmacopeial Convention, Rockville
54. Smith A, Perelman M, Hinchcliffe M (2014) Chitosan: a promising safe and immune-enhancing adjuvant for intranasal vaccines. Hum Vaccin Immunother 10(3):797–807
55. Akıncıbay H, Şenel S, Ay ZY (2007) Application of chitosan gel in the treatment of chronic periodontitis. J Biomed Mater Res B Appl Biomater 80(2):290–296

56. Netsomboon K, Bernkop-Schnürch A (2016) Mucoadhesive vs. mucopenetrating particulate drug delivery. Eur J Pharm Biopharm 98:76–89
57. Sandri G, Rossi S, Ferrari F, Bonferoni MC, Muzzarelli C, Caramella C (2004) Assessment of chitosan derivatives as buccal and vaginal penetration enhancers. Eur J Pharm Sci 21 (2–3):351–359
58. Şenel S, Kremer MJ, Kaş S, Wertz PW, Hıncal AA, Squier CA (2000) Enhancing effect of chitosan on peptide drug delivery across buccal mucosa. Biomaterials 21(20):2067–2071
59. Timur SS, Yüksel S, Akca G, Şenel S (2019) Localized drug delivery with mono and bilayered mucoadhesive films and wafers for oral mucosal infections. Int J Pharm 559:102–112
60. van der Lubben IM, Verhoef JC, Borchard G, Junginger HE (2001) Chitosan and its derivatives in mucosal drug and vaccine delivery. Eur J Pharm Sci 14(3):201–207
61. Smith J, Wood E, Dornish M (2004) Effect of chitosan on epithelial cell tight junctions. Pharm Res 21(1):43–49
62. Şenel S, Kaş HS, Squier CA (2000) Application of chitosan in dental drug delivery and therapy. In: Muzzarelli RAA (ed) Chitosan per os: from dietary supplement to drug carrier. Atec, Grottammare, pp 241–256
63. Şenel S (2010) Potential applications of chitosan in oral mucosal delivery. J Drug Del Sci Tech 20(1):23–32
64. Fröhlich E (2012) The role of surface charge in cellular uptake and cytotoxicity of medical nanoparticles. Int J Nanomedicine:5577
65. Alpar H, Somavarapu S, Atuah K, Bramwell V (2005) Biodegradable mucoadhesive particulates for nasal and pulmonary antigen and DNA delivery. Adv Drug Deliv Rev 57 (3):411–430
66. Slütter B, Bal S, Keijzer C, Mallants R, Hagenaars N, Que I, Kaijzel E, van Eden W, Augustijns P, Löwik C, Bouwstra J, Broere F, Jiskoot W (2010) Nasal vaccination with N-trimethyl chitosan and PLGA based nanoparticles: nanoparticle characteristics determine quality and strength of the antibody response in mice against the encapsulated antigen. Vaccine 28(38):6282–6291
67. Agnihotri SA, Mallikarjuna NN, Aminabhavi TM (2004) Recent advances on chitosan-based micro- and nanoparticles in drug delivery. J Control Release 100(1):5–28
68. Riteau N, Sher A (2016) Chitosan: an adjuvant with an unanticipated STING. Immunity 44 (3):522–524
69. Bueter CL, Lee CK, Wang JP, Ostroff GR, Specht CA, Levitz SM (2014) Spectrum and mechanisms of inflammasome activation by chitosan. J Immunol 192(12):5943–5951
70. Wen Z-S, Xu Y-L, Zou X-T, Xu Z-R (2011) Chitosan nanoparticles act as an adjuvant to promote both Th1 and Th2 immune responses induced by ovalbumin in mice. Mar Drugs 9 (6):1038–1055
71. Seferian PG, Martinez ML (2000) Immune stimulating activity of two new chitosan containing adjuvant formulations. Vaccine 19(6):661–668
72. Peluso G, Petillo O, Ranieri M, Santin M, Ambrosio L, Calabro D, Avallone B, Balsamo G (1994) Chitosan-mediated stimulation of macrophage function. Biomaterials 15 (15):1215–1220
73. Fong D, Hoemann CD (2018) Chitosan immunomodulatory properties: perspectives on the impact of structural properties and dosage. Future Sci OA 4(1):Fso225
74. McNeela EA, Jabbal-Gill I, Illum L, Pizza M, Rappuoli R, Podda A, Lewis DJ, Mills KH (2004) Intranasal immunization with genetically detoxified diphtheria toxin induces T cell responses in humans: enhancement of Th2 responses and toxin-neutralizing antibodies by formulation with chitosan. Vaccine 22(8):909–914
75. Huo Z, Sinha R, McNeela EA, Borrow R, Giemza R, Cosgrove C, Heath PT, Mills KHG, Rappuoli R, Griffin GE, Lewis DJM (2005) Induction of protective serum meningococcal bactericidal and diphtheria-neutralizing antibodies and mucosal immunoglobulin a in volunteers by nasal insufflations of the Neisseria meningitidis serogroup C polysaccharide-CRM197 conjugate vaccine mixed with chitosan. Infect Immun 73(12):8256–8265

76. Read RC, Naylor SC, Potter CW, Bond J, Jabbal-Gill I, Fisher A, Illum L, Jennings R (2005) Effective nasal influenza vaccine delivery using chitosan. Vaccine 23(35):4367–4374
77. El-Kamary SS, Pasetti MF, Mendelman PM, Frey SE, Bernstein DI, Treanor JJ, Ferreira J, Chen WH, Sublett R, Richardson C, Bargatze RF, Sztein MB, Tacket CO (2010) Adjuvanted intranasal Norwalk virus-like particle vaccine elicits antibodies and antibody-secreting cells that express homing receptors for mucosal and peripheral lymphoid tissues. J Infect Dis 202 (11):1649–1658
78. Atmar RL, Bernstein DI, Harro CD, Al-Ibrahim MS, Chen WH, Ferreira J, Estes MK, Graham DY, Opekun AR, Richardson C, Mendelman PM (2011) Norovirus vaccine against experimental human Norwalk virus illness. N Engl J Med 365(23):2178–2187
79. Neimert-Andersson T, Binnmyr J, Enoksson M, Langeback J, Zettergren L, Hallgren AC, Franzen H, Lind Enoksson S, Lafolie P, Lindberg A, Al-Tawil N, Andersson M, Singer P, Gronlund H, Gafvelin G (2014) Evaluation of safety and efficacy as an adjuvant for the chitosan-based vaccine delivery vehicle ViscoGel in a single-blind randomised phase I/IIa clinical trial. Vaccine 32(45):5967–5974
80. Scherließ R, Buske S, Young K, Weber B, Rades T, Hook S (2013) In vivo evaluation of chitosan as an adjuvant in subcutaneous vaccine formulations. Vaccine 31(42):4812–4819
81. Gordon S, Saupe A, McBurney W, Rades T, Hook S (2008) Comparison of chitosan nanoparticles and chitosan hydrogels for vaccine delivery. J Pharm Pharmacol 60 (12):1591–1600
82. Yüksel S, Pekcan M, Puralı N, Esendağlı G, Tavukçuoğlu E, Rivero-Arredondo V, Ontiveros-Padilla L, López-Macías C, Şenel S (2020) Development and in vitro evaluation of a new adjuvant system containing Salmonella Typhi porins and chitosan. Int J Pharm 578:119129
83. Xing L, Fan Y-T, Zhou T-J, Gong J-H, Cui L-H, Cho K-H, Choi Y-J, Jiang H-L, Cho C-S (2018) Chemical modification of chitosan for efficient vaccine delivery. Molecules 23(2):229
84. Yuki Y, Kiyono H (2003) New generation of mucosal adjuvants for the induction of protective immunity. Rev Med Virol 13(5):293–310
85. Mestecky J, Blumberg R, Kiyono H, McGhee JR (2003) The mucosal immune system. Fundamental immunology, 5th edn. Lippincott Williams & Wilkins, Lippincott-Raven, Philadelphia
86. Mantis NJ, Rol N, Corthésy B (2011) Secretory IgA's complex roles in immunity and mucosal homeostasis in the gut. Mucosal Immunol 4(6):603–611
87. He B, Xu W, Santini PA, Polydorides AD, Chiu A, Estrella J, Shan M, Chadburn A, Villanacci V, Plebani A, Knowles DM, Rescigno M, Cerutti A (2007) Intestinal bacteria trigger T cell-independent immunoglobulin a(2) class switching by inducing epithelial-cell secretion of the cytokine APRIL. Immunity 26(6):812–826
88. Brandtzaeg P (2010) Function of mucosa-associated lmphoid tissue in antibody formation. Immunol Investig 39(4–5):303–355
89. Kiyono H, Fukuyama S (2004) NALT- versus PEYER'S-patch-mediated mucosal immunity. Nat Rev Immunol 4(9):699–710
90. Holmgren J, Czerkinsky C (2005) Mucosal immunity and vaccines. Nat Med 11(S4):S45–S53
91. Neutra MR, Kozlowski PA (2006) Mucosal vaccines: the promise and the challenge. Nat Rev Immunol 6(2):148–158
92. Walker RI (1994) New strategies for using mucosal vaccination to achieve more effective immunization. Vaccine 12(5):387–400
93. Vajdy M, O'Hagan DT (2001) Microparticles for intranasal immunization. Adv Drug Deliv Rev 51(1):127–141
94. Li M, Wang Y, Sun Y, Cui H, Zhu SJ, Qiu H-J (2020) Mucosal vaccines: strategies and challenges. Immunol Lett 217:116–125
95. Bennett KM, Parnell EA, Sanscartier C, Parks S, Chen G, Nair MG, Lo DD (2016) Induction of colonic M cells during intestinal inflammation. Am J Pathol 186(5):1166–1179

96. van der Lubben IM, Verhoef JC, van Aelst AC, Borchard G, Junginger HE (2001) Chitosan microparticles for oral vaccination: preparation, characterization and preliminary in vivo uptake studies in murine Peyer's patches. Biomaterials 22(7):687–694
97. Brayden DJ, Jepson MA, Baird AW (2005) Keynote review: intestinal Peyer's patch M cells and oral vaccine targeting. Drug Discov Today 10(17):1145–1157
98. Shakya AK, Chowdhury MYE, Tao W, Gill HS (2016) Mucosal vaccine delivery: current state and a pediatric perspective. J Control Release 240:394–413
99. Miquel-Clopés A, Bentley EG, Stewart JP, Carding SR (2019) Mucosal vaccines and technology. Clin Exp Immunol 196(2):205–214
100. Vela Ramirez JE, Sharpe LA, Peppas NA (2017) Current state and challenges in developing oral vaccines. Adv Drug Deliv Rev 114:116–131
101. New RRC (2019) Formulation technologies for oral vaccines. Clin Exp Immunol 198 (2):153–169
102. Gala PR, Bajaj L, Bansal A, Gomes KB, Joshi D, Menon I, Zaman RU, Zughaier SM, D'Souza M, Popescu C, D'Souza N, Knipp GT, D'Souza MJ (2020) Oral vaccine delivery: the coming age of particulate vaccines to elicit mucosal immunity. In: Muttil P, Kunda N (eds) Mucosal delivery of drugs and biologics in nanoparticles, AAPS advances in the pharmaceutical sciences series, vol 41. Springer, Cham
103. Saraf S, Jain S, Sahoo RN, Mallick S (2020) Lipopolysaccharide derived alginate coated hepatitis B antigen loaded chitosan nanoparticles for oral mucosal immunization. Int J Biol Macromol 154:466–476
104. Renu S, Han Y, Dhakal S, Lakshmanappa YS, Ghimire S, Feliciano-Ruiz N, Senapati S, Narasimhan B, Selvaraj R, Renukaradhya GJ (2020) Chitosan-adjuvanted Salmonella subunit nanoparticle vaccine for poultry delivered through drinking water and feed. Carbohydr Polym 243:116434
105. Gao P, Xia G, Bao Z, Feng C, Cheng X, Kong M, Liu Y, Chen X (2016) Chitosan based nanoparticles as protein carriers for efficient oral antigen delivery. Int J Biol Macromol 91:716–723
106. Chang S-H, Wu G-J, Wu C-H, Huang C-H, Tsai G-J (2019) Oral administration with chitosan hydrolytic products modulates mitogen-induced and antigen-specific immune responses in BALB/c mice. Int J Biol Macromol 131:158–166
107. Slütter B, Plapied L, Fievez V, Alonso Sande M, Des Rieux A, Schneider Y-J, Van Riet E, Jiskoot W, Préat V (2009) Mechanistic study of the adjuvant effect of biodegradable nanoparticles in mucosal vaccination. J Control Release 138(2):113–121
108. Abkar M, Fasihi-Ramandi M, Kooshki H, Sahebghadam Lotfi A (2017) Oral immunization of mice with Omp31-loaded N-trimethyl chitosan nanoparticles induces high protection against Brucella melitensis infection. Int J Nanomedicine 12:8769–8778
109. Xu B, Zhang W, Chen Y, Xu Y, Wang B, Zong L (2018) Eudragit® L100-coated mannosylated chitosan nanoparticles for oral protein vaccine delivery. Int J Biol Macromol 113:534–542
110. Soares E, Jesus S, Borges O (2018) Oral hepatitis B vaccine: chitosan or glucan based delivery systems for efficient HBsAg immunization following subcutaneous priming. Int J Pharm 535 (1–2):261–271
111. Biswas S, Chattopadhyay M, Sen KK, Saha MK (2015) Development and characterization of alginate coated low molecular weight chitosan nanoparticles as new carriers for oral vaccine delivery in mice. Carbohydr Polym 121:403–410
112. Shukla A, Mishra V, Bhoop BS, Katare OP (2015) Alginate coated chitosan microparticles mediated oral delivery of diphtheria toxoid. (part a). Systematic optimization, development and characterization. Int J Pharm 495(1):220–233
113. Jiang T, Singh B, Li H-S, Kim Y-K, Kang S-K, Nah J-W, Choi Y-J, Cho C-S (2014) Targeted oral delivery of BmpB vaccine using porous PLGA microparticles coated with M cell homing peptide-coupled chitosan. Biomaterials 35(7):2365–2373

114. Barhate G, Gautam M, Gairola S, Jadhav S, Pokharkar V (2013) Quillaja saponaria extract as mucosal adjuvant with chitosan functionalized gold nanoparticles for mucosal vaccine delivery: stability and immunoefficiency studies. Int J Pharm 441(1–2):636–642
115. Oberoi HS, Yorgensen YM, Morasse A, Evans JT, Burkhart DJ (2016) PEG modified liposomes containing CRX-601 adjuvant in combination with methylglycol chitosan enhance the murine sublingual immune response to influenza vaccination. J Control Release 223:64–74
116. Buffa V, Klein K, Fischetti L, Shattock RJ (2012) Evaluation of TLR agonists as potential mucosal adjuvants for HIV gp140 and tetanus toxoid in mice. PLoS One 7(12):e50529
117. Nagamoto T, Hattori Y, Takayama K, Maitani Y (2004) Novel chitosan particles and chitosan-coated emulsions inducing immune response via intranasal vaccine delivery. Pharm Res 21 (4):671–674
118. Pabst R (2015) Mucosal vaccination by the intranasal route. Nose-associated lymphoid tissue (NALT)—structure, function and species differences. Vaccine 33(36):4406–4413
119. Illum L, Jabbal-Gill I, Hinchcliffe M, Fisher AN, Davis SS (2001) Chitosan as a novel nasal delivery system for vaccines. Adv Drug Deliv Rev 51(1):81–96
120. Shim S, Soh SH, Im YB, Park H-E, Cho C-S, Kim S, Yoo HS (2020) Elicitation of Th1/Th2 related responses in mice by chitosan nanoparticles loaded with Brucella abortus malate dehydrogenase, outer membrane proteins 10 and 19. Int J Med Microbiol 310(1):151362
121. Senevirathne A, Hewawaduge C, Hajam IA, Lalsiamthara J, Lee JH (2019) Intranasally administered anti-Brucella subunit vaccine formulation induces protective immune responses against nasal Brucella challenge. Vet Microbiol 228:112–118
122. Renu S, Feliciano-Ruiz N, Ghimire S, Han Y, Schrock J, Dhakal S, Patil V, Krakowka S, Renukaradhya GJ (2020) Poly(I:C) augments inactivated influenza virus-chitosan nanovaccine induced cell mediated immune response in pigs vaccinated intranasally. Vet Microbiol 242:108611
123. Hagenaars N, Mania M, De Jong P, Que I, Nieuwland R, Slütter B, Glansbeek H, Heldens J, Van Den Bosch H, Löwik C (2010) Role of trimethylated chitosan (TMC) in nasal residence time, local distribution and toxicity of an intranasal influenza vaccine. J Control Release 144 (1):17–24
124. Slütter B, Bal SM, Que I, Kaijzel E, Löwik C, Bouwstra J, Jiskoot W (2010) Antigen−adjuvant nanoconjugates for nasal vaccination: an improvement over the use of nanoparticles? Mol Pharm 7(6):2207–2215
125. Nevagi RJ, Dai W, Khalil ZG, Hussein WM, Capon RJ, Skwarczynski M, Toth I (2019) Structure-activity relationship of group a streptococcus lipopeptide vaccine candidates in trimethyl chitosan-based self-adjuvanting delivery system. Eur J Med Chem 179:100–108
126. Mosafer J, Sabbaghi A-H, Badiee A, Dehghan S, Tafaghodi M (2019) Preparation, characterization and in vivo evaluation of alginate-coated chitosan and trimethylchitosan nanoparticles loaded with PR8 influenza virus for nasal immunization. Asian J Pharm Sci 14(2):216–221
127. Bento D, Staats HF, Gonçalves T, Borges O (2015) Development of a novel adjuvanted nasal vaccine: C48/80 associated with chitosan nanoparticles as a path to enhance mucosal immunity. Eur J Pharm Biopharm 93:149–164
128. Sinani G, Sessevmez M, Gök MK, Özgümüş S, Alpar HO, Cevher E (2019) Modified chitosan-based nanoadjuvants enhance immunogenicity of protein antigens after mucosal vaccination. Int J Pharm 569:118592
129. Rose F, Wern JE, Gavins F, Andersen P, Follmann F, Foged C (2018) A strong adjuvant based on glycol-chitosan-coated lipid-polymer hybrid nanoparticles potentiates mucosal immune responses against the recombinant chlamydia trachomatis fusion antigen CTH522. J Control Release 271:88–97
130. Pawar D, Mangal S, Goswami R, Jaganathan KS (2013) Development and characterization of surface modified PLGA nanoparticles for nasal vaccine delivery: effect of mucoadhesive coating on antigen uptake and immune adjuvant activity. Eur J Pharm Biopharm 85 (3):550–559

131. Li Z, Xiong F, He J, Dai X, Wang G (2016) Surface-functionalized, pH-responsive poly(lactic-co-glycolic acid)-based microparticles for intranasal vaccine delivery: effect of surface modification with chitosan and mannan. Eur J Pharm Biopharm 109:24–34
132. Amin M, Jaafari MR, Tafaghodi M (2009) Impact of chitosan coating of anionic liposomes on clearance rate, mucosal and systemic immune responses following nasal administration in rabbits. Colloids Surf B: Biointerfaces 74(1):225–229
133. Dehghan S, Tafaghodi M, Bolourieh T, Mazaheri V, Torabi A, Abnous K, Tavassoti Kheiri M (2014) Rabbit nasal immunization against influenza by dry-powder form of chitosan nanospheres encapsulated with influenza whole virus and adjuvants. Int J Pharm 475(1–2):1–8
134. Li X, Yang Y, Yang F, Wang F, Li H, Tian H, Wang G (2021) Chitosan hydrogel loaded with recombinant protein containing epitope C from HSP90 of Candida albicans induces protective immune responses against systemic candidiasis. Int J Biol Macromol 173:327–340
135. Zhou M, Qu W, Sun Y, Liang L, Jin Z, Cui S, Zhao K (2020) Water-soluble N-2-Hydroxypropyl trimethyl ammonium chloride chitosan enhanced the immunogenicity of inactivated porcine parvovirus vaccine vaccination on sows against porcine parvovirus infection. Immunol Lett 223:26–32
136. Yang Y, Xing R, Liu S, Qin Y, Li K, Yu H, Li P (2020) Chitosan, hydroxypropyltrimethyl ammonium chloride chitosan and sulfated chitosan nanoparticles as adjuvants for inactivated Newcastle disease vaccine. Carbohydr Polym 229:115423
137. Sadeghi Z, Fasihi-Ramandi M, Azizi M, Bouzari S (2020) Mannosylated chitosan nanoparticles loaded with FliC antigen as a novel vaccine candidate against Brucella melitensis and Brucella abortus infection. J Biotechnol 310:89–96
138. Abdelallah NH, Gaber Y, Rashed ME, Azmy AF, Abou-Taleb HA, Abdelghani S (2020) Alginate-coated chitosan nanoparticles act as effective adjuvant for hepatitis a vaccine in mice. Int J Bio Macromol 152:904–912
139. Jearanaiwitayakul T, Sunintaboon P, Chawengkittikul R, Limthongkul J, Midoeng P, Warit S, Ubol S (2020) Nanodelivery system enhances the immunogenicity of dengue-2 nonstructural protein 1, DENV-2 NS1. Vaccine 38(43):6814–6825
140. Bansal A, Gamal W, Wu X, Yang Y, Olson V, D'Souza MJ (2019) Evaluation of an adjuvanted hydrogel-based pDNA nanoparticulate vaccine for rabies prevention and immunocontraception. Nanomedicine: NBM 21:102049
141. Korupalli C, Pan W-Y, Yeh C-Y, Chen P-M, Mi F-L, Tsai H-W, Chang Y, Wei H-J, Sung H-W (2019) Single-injecting, bioinspired nanocomposite hydrogel that can recruit host immune cells in situ to elicit potent and long-lasting humoral immune responses. Biomaterials 216:119268
142. Wang H, Han Q, Zhao H, Xu D, Zhang J (2018) Single dose HBsAg CS-γ-PGA nanogels induce potent protective immune responses against HBV infection. Eur J Pharm Biopharm 124:82–88
143. Lebre F, Pedroso De Lima MC, Lavelle EC, Borges O (2018) Mechanistic study of the adjuvant effect of chitosan-aluminum nanoparticles. Int J Pharm 552(1–2):7–15
144. Wang Y-Q, Fan Q-Z, Liu Y, Yue H, Ma X-W, Wu J, Ma G-H, Su Z-G (2016) Improving adjuvanticity of quaternized chitosan–based microgels for H5N1 split vaccine by tailoring the particle properties to achieve antigen dose sparing effect. Int J Pharm 515(1–2):84–93
145. Glenn GM, Taylor DN, Li X, Frankel S, Montemarano A, Alving CR (2000) Transcutaneous immunization: a human vaccine delivery strategy using a patch. Nat Med 6(12):1403–1406
146. Babiuk S, Baca-Estrada M, Babiuk LA, Ewen C, Foldvari M (2000) Cutaneous vaccination: the skin as an immunologically active tissue and the challenge of antigen delivery. J Control Release 66(2–3):199–214
147. Li N, Peng L-H, Chen X, Nakagawa S, Gao J-Q (2011) Transcutaneous vaccines: novel advances in technology and delivery for overcoming the barriers. Vaccine 29(37):6179–6190
148. Chen Z, Lv Y, Qi J, Zhu Q, Lu Y, Wu W (2018) Overcoming or circumventing the stratum corneum barrier for efficient transcutaneous immunization. Drug Discov Today 23 (1):181–186

149. Engelke L, Winter G, Hook S, Engert J (2015) Recent insights into cutaneous immunization: how to vaccinate via the skin. Vaccine 33(37):4663–4674
150. Pielenhofer J, Sohl J, Windbergs M, Langguth P, Radsak MP (2020) Current progress in particle-based systems for transdermal vaccine delivery. Front Immunol:11
151. Leone M, Mönkäre J, Bouwstra JA, Kersten G (2017) Dissolving microneedle patches for dermal vaccination. Pharm Res 34(11):2223–2240
152. Chen M-C, Huang S-F, Lai K-Y, Ling M-H (2013) Fully embeddable chitosan microneedles as a sustained release depot for intradermal vaccination. Biomaterials 34(12):3077–3086
153. Li Z, He Y, Deng L, Zhang Z-R, Lin Y (2020) A fast-dissolving microneedle array loaded with chitosan nanoparticles to evoke systemic immune responses in mice. J Mater Chem B 8 (2):216–225
154. Siddhapura K, Harde H, Jain S (2016) Immunostimulatory effect of tetanus toxoid loaded chitosan nanoparticles following microneedles assisted immunization. Nanomedicine: NBM 12(1):213–222
155. Chiu Y-H, Chen M-C, Wan S-W (2018) Sodium hyaluronate/chitosan composite microneedles as a single-dose intradermal immunization system. Biomacromolecules 19 (6):2278–2285
156. Bal SM, Ding Z, Kersten GFA, Jiskoot W, Bouwstra JA (2010) Microneedle-based transcutaneous immunization in mice with N-trimethyl chitosan adjuvanted diphtheria toxoid formulations. Pharm Res 27(9):1837–1847
157. Yang X, Wang X, Hong H, Elfawal G, Lin S, Wu J, Jiang Y, He C, Mo X, Kai G, Wang H (2020) Galactosylated chitosan-modified ethosomes combined with silk fibroin nanofibers is useful in transcutaneous immunization. J Control Release 327:88–99
158. Adamczak MI, Hagesaether E, Smistad G, Hiorth M (2016) An in vitro study of mucoadhesion and biocompatibility of polymer coated liposomes on HT29-MTX mucus-producing cells. Int J Pharm 498(1):225–233
159. Gilavand F, Marzban A, Ebrahimipour G, Soleimani N, Goudarzi M (2020) Designation of chitosan nano-vaccine based on MxiH antigen of Shigella flexneri with increased immunization capacity. Carbohydr Polym 232:115813
160. Darzi Eslam E, Darvish Alipour Astaneh S, Rasooli I, Nazarian S, Jahangiri A (2020) Passive immunization with chitosan-loaded biofilm-associated protein against Acinetobacter baumannii murine infection model. Gene Rep 20:100708
161. Akerele G, Ramadan N, Renu S, Renukaradhya GJ, Shanmugasundaram R, Selvaraj RK (2020) In vitro characterization and immunogenicity of chitosan nanoparticles loaded with native and inactivated extracellular proteins from a field strain of Clostridium perfringens associated with necrotic enteritis. Vet Immunol Immunopathol 224:110059
162. Singh A, Nisaa K, Bhattacharyya S, Mallick AI (2019) Immunogenicity and protective efficacy of mucosal delivery of recombinant hcp of campylobacter jejuni type VI secretion system (T6SS) in chickens. Mol Immunol 111:182–197
163. Abkar M, Fasihi-Ramandi M, Kooshki H, Lotfi AS (2018) Intraperitoneal immunization with urease loaded N-trimethyl chitosan nanoparticles elicits high protection against Brucella melitensis and Brucella abortus infections. Immunol Lett 199:53–60
164. Cole H, Bryan D, Lancaster L, Mawas F, Vllasaliu D (2018) Chitosan nanoparticle antigen uptake in epithelial monolayers can predict mucosal but not systemic in vivo immune response by oral delivery. Carbohydr Polym 190:248–254
165. Lee J, Kim Y-M, Kim J-H, Cho C-W, Jeon J-W, Park J-K, Lee S-H, Jung B-G, Lee B-J (2018) Nasal delivery of chitosan/alginate nanoparticle encapsulated bee (Apis mellifera) venom promotes antibody production and viral clearance during porcine reproductive and respiratory syndrome virus infection by modulating T cell related responses. Vet Immunol Immunopathol 200:40–51
166. Lopes PD, Okino CH, Fernando FS, Pavani C, Casagrande VM, Lopez RFV, Montassier MDFS, Montassier HJ (2018) Inactivated infectious bronchitis virus vaccine encapsulated in

chitosan nanoparticles induces mucosal immune responses and effective protection against challenge. Vaccine 36(19):2630–2636
167. Nevagi RJ, Khalil ZG, Hussein WM, Powell J, Batzloff MR, Capon RJ, Good MF, Skwarczynski M, Toth I (2018) Polyglutamic acid-trimethyl chitosan-based intranasal peptide nano-vaccine induces potent immune responses against group a streptococcus. Acta Biomater 80:278–287
168. Chowdhury MYE, Kim T-H, Uddin MB, Kim J-H, Hewawaduge CY, Ferdowshi Z, Sung M-H, Kim C-J, Lee J-S (2017) Mucosal vaccination of conserved sM2, HA2 and cholera toxin subunit A1 (CTA1) fusion protein with poly gamma-glutamate/chitosan nanoparticles (PC NPs) induces protection against divergent influenza subtypes. Vet Microbiol 201:240–251
169. Amirnasr M, Fallah Tafti T, Sankian M, Rezaei A, Tafaghodi M (2016) Immunization against HTLV-I with chitosan and tri-methylchitosan nanoparticles loaded with recombinant env23 and env13 antigens of envelope protein gp46. Microb Pathog 97:38–44
170. Díaz AG, Quinteros DA, Llabot JM, Palma SD, Allemandi DA, Ghersi G, Zylberman V, Goldbaum FA, Estein SM (2016) Spray dried microspheres based on chitosan: a promising new carrier for intranasal administration of polymeric antigen BLSOmp31 for prevention of ovine brucellosis. Mater Sci Eng C 62:489–496
171. Jesus S, Soares E, Costa J, Borchard G, Borges O (2016) Immune response elicited by an intranasally delivered HBsAg low-dose adsorbed to poly-ε-caprolactone based nanoparticles. Int J Pharm 504(1–2):59–69
172. Spinner JL, Oberoi HS, Yorgensen YM, Poirier DS, Burkhart DJ, Plante M, Evans JT (2015) Methylglycol chitosan and a synthetic TLR4 agonist enhance immune responses to influenza vaccine administered sublingually. Vaccine 33(43):5845–5853
173. Liu Q, Zhang C, Zheng X, Shao X, Zhang X, Zhang Q, Jiang X (2014) Preparation and evaluation of antigen/N-trimethylaminoethylmethacrylate chitosan conjugates for nasal immunization. Vaccine 32(22):2582–2590
174. Verheul RJ, Hagenaars N, Van Es T, Van Gaal EVB, De Jong PHJLF, Bruijns S, Mastrobattista E, Slütter B, Que I, Heldens JGM, Van Den Bosch H, Glansbeek HL, Hennink WE, Jiskoot W (2012) A step-by-step approach to study the influence of N-acetylation on the adjuvanticity of N,N,N-trimethyl chitosan (TMC) in an intranasal nanoparticulate influenza virus vaccine. Eur J Pharm Sci 45(4):467–474
175. Moon H-J, Lee J-S, Talactac MR, Chowdhury MYE, Kim J-H, Park M-E, Choi Y-K, Sung M-H, Kim C-J (2012) Mucosal immunization with recombinant influenza hemagglutinin protein and poly gamma-glutamate/chitosan nanoparticles induces protection against highly pathogenic influenza a virus. Vet Microbiol 160(3–4):277–289
176. Subbiah R, Ramalingam P, Ramasundaram S, Kim DY, Park K, Ramasamy MK, Choi KJ (2012) N,N,N-Trimethyl chitosan nanoparticles for controlled intranasal delivery of HBV surface antigen. Carbohydr Polym 89(4):1289–1297
177. Bal SM, Slütter B, Verheul R, Bouwstra JA, Jiskoot W (2012) Adjuvanted, antigen loaded N-trimethyl chitosan nanoparticles for nasal and intradermal vaccination: adjuvant- and site-dependent immunogenicity in mice. Eur J Pharm Sci 45(4):475–481
178. Mohammed GM, Elzorkany HE, Farroh KY, Abd El-Aziz WR, Elshoky HA (2021) Potential improvement of the immune response of chickens against E. coli vaccine by using two forms of chitosan nanoparticles. Int J Biol Macromol 167:395–404
179. Shim S, Park H-E, Soh SH, Im YB, Yoo HS (2020) Induction of Th2 response through TLR2-mediated MyD88-dependent pathway in human microfold cells stimulated with chitosan nanoparticles loaded with Brucella abortus Mdh. Microb Pathog 142:104040
180. Rebbouh F, Martin-Eauclaire M-F, Laraba-Djebari F (2020) Chitosan nanoparticles as a delivery platform for neurotoxin II from Androctonus australis hector scorpion venom: assessment of toxicity and immunogenicity. Acta Trop 205:105353
181. Chuang C-C, Tsai M-H, Yen H-J, Shyu H-F, Cheng K-M, Chen X-A, Chen C-C, Young J-J, Kau J-H (2020) A fucoidan-quaternary chitosan nanoparticle adjuvant for anthrax vaccine as an alternative to CpG oligodeoxynucleotides. Carbohydr Polym 229:115403

182. Babii O, Wang Z, Liu G, Martinez EC, Van Drunen Littel-Van Den Hurk S, Chen L (2020) Low molecular weight chitosan nanoparticles for CpG oligodeoxynucleotides delivery: impact of molecular weight, degree of deacetylation, and mannosylation on intracellular uptake and cytokine induction. Int J Biol Macromol 159:46–56
183. Pei M, Liang J, Zhang C, Wang X, Zhang C, Ma G, Sun H (2019) Chitosan/calcium phosphates nanosheet as a vaccine carrier for effective cross-presentation of exogenous antigens. Carbohydr Polym 224:115172
184. Abkar M, Alamian S, Sattarahmady N (2019) A comparison between adjuvant and delivering functions of calcium phosphate, aluminum hydroxide and chitosan nanoparticles, using a model protein of Brucella melitensis Omp31. Immunol Lett 207:28–35
185. Yang Y, Xing R, Liu S, Qin Y, Li K, Yu H, Li P (2019) Hydroxypropyltrimethyl ammonium chloride chitosan activates RAW 264.7 macrophages through the MAPK and JAK-STAT signaling pathways. Carbohydr Polym 205:401–409
186. Durán V, Yasar H, Becker J, Thiyagarajan D, Loretz B, Kalinke U, Lehr C-M (2019) Preferential uptake of chitosan-coated PLGA nanoparticles by primary human antigen presenting cells. Nanomedicine NBM 21:102073
187. Gong Q, Kong LY, Niu MF, Qin CL, Yang Y, Li X, Ruan MD, Tian Y, Li ZL (2018) Construction of a ptfA chitosan nanoparticle DNA vaccine against Pasteurella multocida and the immune response in chickens. Vet J 231:1–7
188. Soares E, Jesus S, Borges O (2018) Chitosan:β-glucan particles as a new adjuvant for the hepatitis B antigen. Eur J Pharm Biopharm 131:33–43
189. Yousefi S, Abbassi-Daloii T, Sekhavati MH, Tahmoorespur M (2018) Evaluation of immune responses induced by polymeric OMP25-BLS Brucella antigen. Microb Pathog 115:50–56
190. Mohamed SH, Arafa AS, Mady WH, Fahmy HA, Omer LM, Morsi RE (2018) Preparation and immunological evaluation of inactivated avian influenza virus vaccine encapsulated in chitosan nanoparticles. Biologicals 51:46–53
191. Zheng X, Yang X, Li X, Qiu G-H, Dai A, Huang Q, Huang C, Guo X (2017) Omp16-based vaccine encapsulated by alginate-chitosan microspheres provides significant protection against Haemophilus parasuis in mice. Vaccine 35(10):1417–1423
192. Kojarunchitt T, Baldursdottir S, Dong Y-D, Boyd BJ, Rades T, Hook S (2015) Modified thermoresponsive Poloxamer 407 and chitosan sol–gels as potential sustained-release vaccine delivery systems. Eur J Pharm Biopharm 89:74–81
193. Hunsawong T, Sunintaboon P, Warit S, Thaisomboonsuk B, Jarman RG, Yoon I-K, Ubol S, Fernandez S (2015) A novel dengue virus serotype-2 nanovaccine induces robust humoral and cell-mediated immunity in mice. Vaccine 33(14):1702–1710
194. Koppolu B, Zaharoff DA (2013) The effect of antigen encapsulation in chitosan particles on uptake, activation and presentation by antigen presenting cells. Biomaterials 34(9):2359–2369

Adv Polym Sci (2021) 288: 381–410
https://doi.org/10.1007/12_2021_105

Published online: 6 July 2021

Enhanced Topical Delivery of Drugs to the Eye Using Chitosan Based Systems

Subramanian Natesan, Ravichandiran Vellayutham, Venkateshwaran Krishnaswami, Chandrasekar Ponnusamy, Saranya Thekkilaveedu, Dhilin Pathayappurakkal Mohanan, and Ruckmani Kandasamy

Contents

S. Natesan (✉) and R. Vellayutham
Department of Pharmaceutics, National Institute of Pharmaceutical Education and Research Kolkata, Kolkata, India

V. Krishnaswami, C. Ponnusamy, S. Thekkilaveedu, and R. Kandasamy
Department of Pharmaceutical Technology, Centre for Excellence in Nanobio Translational Research (CENTRE), University College of Engineering, Anna University, Tiruchirappalli, Tamil Nadu, India

D. Pathayappurakkal Mohanan
Department of Pharmaceutics, DM WIMS College of Pharmacy, DM Wayanad Institute of Medical Science, Wayanad, Kerala, India

Abstract Topical administration is considered to be the favored route for ocular drug delivery by virtue of drug penetration into the underlying layers of eye. Targeted delivery of drugs to the ocular tissues especially at the posterior segment of eye is hindered predominantly. Long-term maintenance of therapeutic level of drugs in the ocular tissue (both anterior/posterior segments) is the major specific challenge faced by researchers. Development of novel pharmaceutical formulations may overcome these barriers and imparts therapeutic levels in eye. In this regimen chitosan due to its avoidance of toxicity, biocompatibility, bioadhesion, and permeability-enhancing properties is widely used in ocular topical drug delivery. Chitosan facilitates the transport of drugs to the inner eye or accumulation into the corneal/conjunctival epithelia due to its mucoadhesion and permeation-enhancing properties. In this chapter will be provided insights toward the different ocular barriers, conventional and novel chitosan based ocular topical drug delivery systems. Special emphasis will be made toward the development of chitosan based topical delivery systems for ocular diseases.

Keywords Chitosan · Eye · Mucoadhesion · Ocular diseases · Pharmaceutical · Topical delivery

1 Eye

Human eye is a complex organ partitioned into anterior and posterior segments. The anterior segment of eye comprises cornea, aqueous humour, iris, ciliary body, and crystalline lens. Retina and vitreous humour occupy the posterior segment of the eye. Corneal lamellar stroma supports the structural integrity of the cornea. The tissue lining of the inner surface of the eye is the retina surrounded by vitreous cavity. The neurons, photoreceptors, bipolar cells, horizontal cells, amacrine cells and ganglion cells occupy the neural retina.

Vision loss (both reversible/irreversible) is one of the critical complications of human diseases. The major causes for vision loss or the diseases affecting the eye include age-related macular degeneration (dry/wet), cataract, glaucoma, diabetic retinopathy (proliferative/non-proliferative), retinoblastoma, conjunctivitis (bacterial/viral), and ocular trauma. The diseases affecting the anterior and posterior segments of eye are shown in Fig. 1.

Eye is composed of varied protective barriers which protect the eye from exogenous substances and external stress. The anterior pole (cornea, conjunctiva, aqueous humour, iris, ciliary body, and lens) of the eye occupies approximately one-third

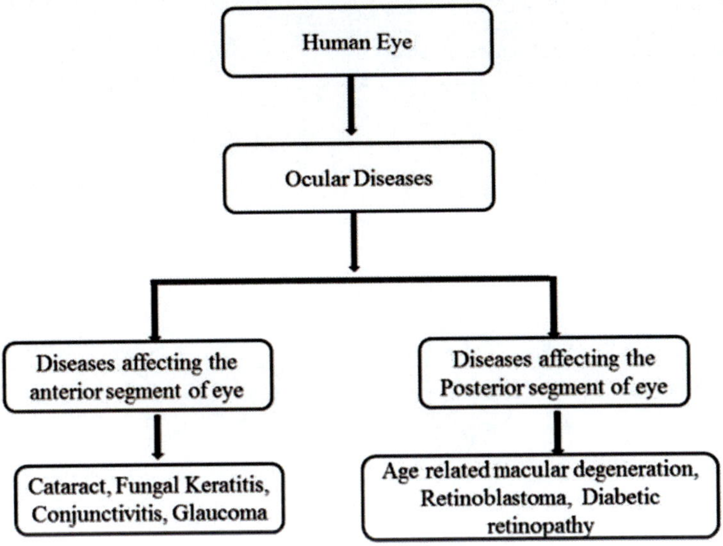

Fig. 1 Diseases affecting the anterior and posterior segments of eye

with the rest of the part as the posterior segment. The posterior pole of the eye includes sclera, choroid, retinal pigment epithelium, neural retina, optic nerve, and vitreous humor.

The three layers present in the human eye can be distinguished with the outer region which comprises cornea and the sclera. The function of cornea is to refract and transmit the light to the lens and the retina followed by protection against infection and structural damage to the deeper parts [1]. The multilayered tissue cornea acts as the principal barrier for the eye. The paracellular resistance of 12–16 kΩ cm is created by 5–6 layers of columnar epithelial cells holding up of tight junctions. The sclera protects the eye from internal and external forces by forming a connective tissue coat and maintains its shape. Conjunctiva is the visible part of the sclera covered by a transparent mucous membrane. The iris, the ciliary body, and the choroid form the middle layer of the eye. The size of the pupil is controlled by the iris. Retina forms the inner layer of the eye which is a complex, layered structure of neurons that capture and process light. The three transparent structures surrounded by the ocular layers are called the aqueous, the vitreous, and the lens. The two-thirds of the eye comprise the posterior segment.

The various diseases affecting the eye both at anterior and posterior segments are shown in Fig. 2. The challenging task for the ocular drug permeation includes tear turnover, nasolacrimal drainage, reflex blinking, and ocular static and dynamic barriers. During topical administration the ocular bioavailability is very low (<5%) [2]. The complicated anatomical features of the eye along with its physiological activity are responsible for more than 90% of drug loss.

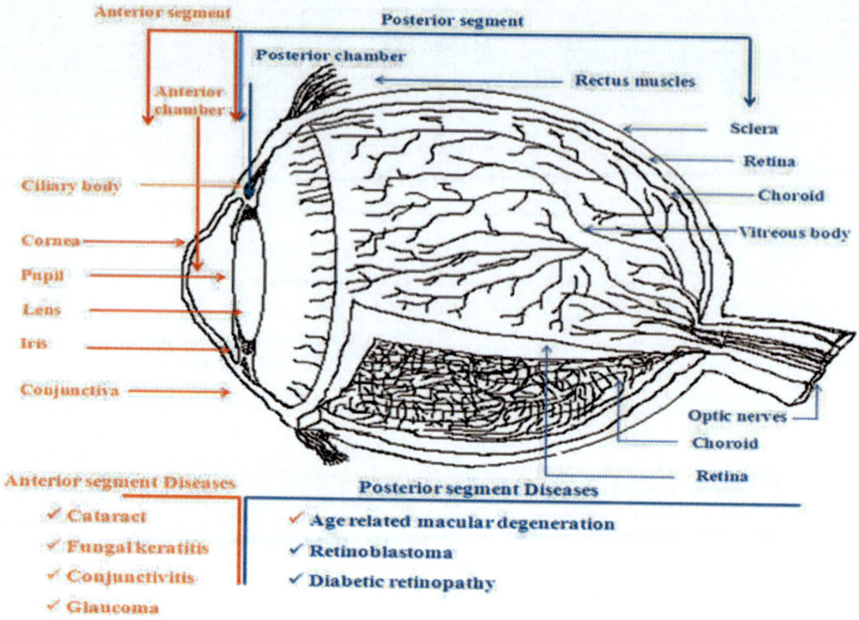

Fig. 2 Anatomical layers of eye showing the various anterior and posterior segment diseases

2 Ocular Barriers

The delivery of the drugs to the eye is a challenging task due to the presence of multiple barriers before reaching the site of action both topically and systemically; it faces multiple obstacles before. As a result, ocular bioavailability from topically administered drug gets reduced to a lower extent. The various barriers (anatomical barriers and physiological barriers) associated with the delivery of drugs to the eye are shown in Fig. 3.

2.1 Anatomical Barriers

The topically administered dosage form enters into the eye with/without cornea support. The hydrophilic drug transport through intercellular spaces is the major challenge; hence, nano-based drug delivery systems were attempted by researchers in order to improve the permeability of administered drugs present in the formulation. The stroma comprising multiple layers of hexagonally arranged collagen fibers containing aqueous pores or channels permits the hydrophilic drugs with restriction for lipophilic drugs [3]. Optimum bioavailability depends upon the lipophilicity and hydrophilicity balance. Wherein, no significant barriers are observed in other layers which are leakier in nature.

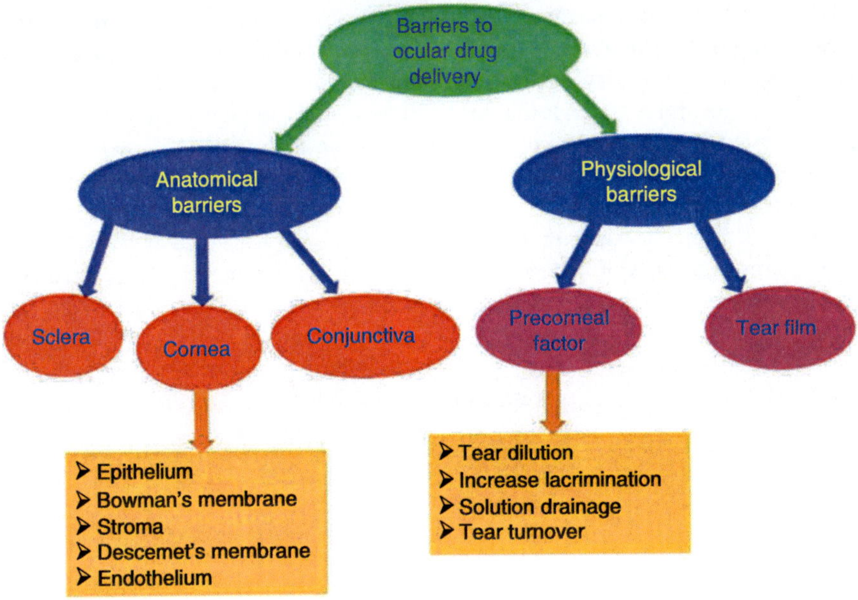

Fig. 3 Barriers associated with ocular drug delivery

The large and small molecular weight drugs enter via the non-corneal route which moves through conjunctiva and sclera. The conjunctiva was found to be more permeable for hydrophilic molecules than cornea due to the presence of tight junction proteins. The limbal area was associated with high vascularity which hinders the delivery by permitting only a small fraction of the dose.

2.2 *Physiological Barriers*

Among the different layers in the cornea the squamous epithelial cells act as a barrier for intercellular drug penetration. Cornea is occupied by only 5% of the eye and the remaining 95% is occupied by the conjunctiva. The pre-corneal factors such as solution drainage, tear dilution, tear turnover, and increased lacrimation reduce the bioavailability of topically administered drugs [4]. The administered dose gets reduced/diluted by lacrimation which leads to poor absorption and enhanced clearance. This in turn significantly reduces the absorption and ultimately ends up with low bioavailability. Thus significant pre-corneal loss of drugs upon topical administration may be prevented by maintaining isotonicity and nonirritating of the formulations.

The tight junctions comprising anastomotic strands create resistance to the paracellular drug absorption The complex nature of stratified corneal epithelium along with the tight junction proteins acts as the barrier for drug permeation

[5]. Extracellular and intracellular calcium levels in tight junctions also reduce the permeability of the administered drugs. The integrity of tight junctions is maintained by the cellular calcium levels and actin filaments present on the cytoskeleton. The permeability of drugs through the tight junctions may be practically enhanced by disrupting the tight junction complexity or by chemical approach. The permeation at corneal epithelium depends upon the charge of the molecules permeating [6].

Sclera occupies 80% of the posterior eye globe. The defense barrier against penetration from undesired molecules is maintained by the ocular surface. The scleral stroma consists of collagen, fibroblasts, etc. with high vascularity occupied by rigid scleral collagenous shell that is lined internally by the uveal tract. The thickness of sclera (0.3–1.0 mm) and choroid (averages 0.25 mm) also acts as a barrier toward the entry. Retinal barriers (including retinal pigment epithelium), static and dynamic barriers prevent entry of toxic substances and drugs.

3 Properties of Chitosan as a Topical Ocular Delivery System

The properties of chitosan favoring the topical delivery of drugs are shown in Fig. 4.

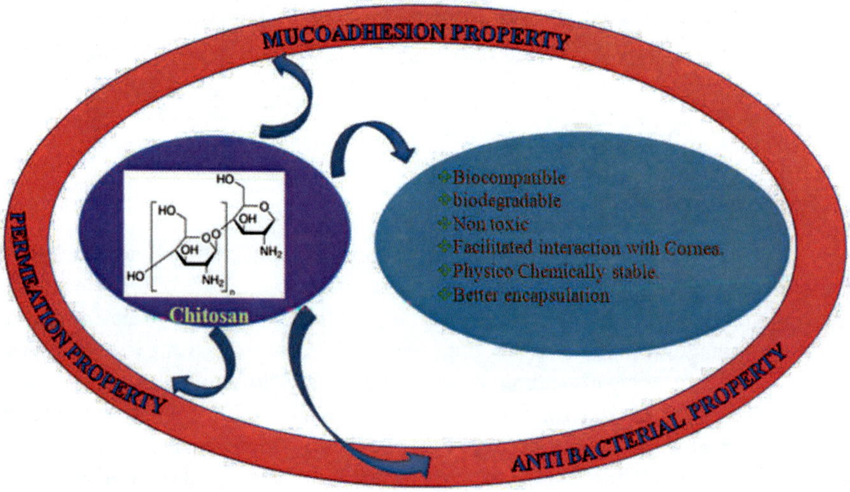

Fig. 4 Physicochemical properties of chitosan

3.1 Mucoadhesive Potential

The mucoadhesion is considered as an essential property for topical ocular drug delivery systems for sufficient retention and penetration of therapeutics to the eye. The mucus, with its anionic substructures such as sialic and sulfonic acids, will efficiently interact with cationic primary amino groups of chitosan through electrostatic and hydrophobic interactions, which is depicted in Fig. 5 [7–9].

The mucoadhesion potential is accelerated by increases in the degree of deacetylation, molecular weight, and decrease in cross-linking [10, 11]. However, it is greatly dependent on the pH of the environment. At low pH of 1.2, weak mucoadhesion is observed between uncharged –COOH and $-SO_3H$ groups of mucin and protonated $-NH_3$ group of chitosan. While at pH nearer to 4.5, the strong electrostatic forces arise between protonated amino groups of chitosan and deprotonated carboxylic and sulfonate groups of mucin. But at neutral pH and above, deprotonated amino groups limited the electrostatic interactions, while the weak van der Waal's forces and hydrogen bonding predominate [12, 13]. In addition to reduced mucoadhesion potential, the low hydrophilicity and swellability at neutral to basic pH challenge the chitosan for delivery to the ophthalmic environment of pH 7.8 [14, 15].

To surmount these problems, several modifications of chitosan have been derived that attach monomers with a functional group promoting mucoadhesion at neutral pH [16–18]. Thiolated chitosan possesses more degree of mucoadhesion due to the strong covalent disulfide bonding between the thiol group (chitosan) and cysteine

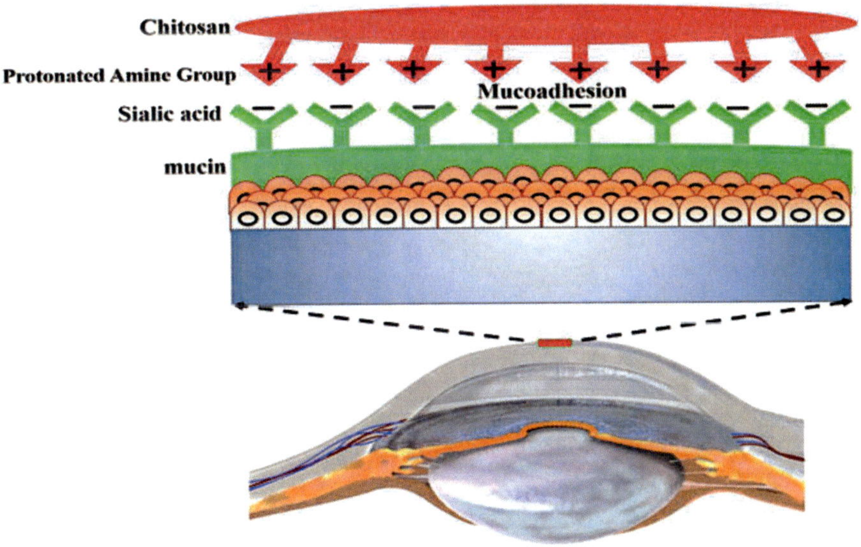

Fig. 5 The mucoadhesive interaction between the protonated amino group of chitosan and negatively charged mucin in the ophthalmic mucus membrane

amino acid (mucus glycoprotein) [9, 19–21]. The trimethyl chitosan (TMCS), formed by the reductive methylation of chitosan is another modified form with improved mucoadhesion potential and solubility at neutral to basic pH [11, 22–24]. Other chitosan derivatives formed with methyl acrylate and acrylic acid [25], N-hydroxysuccinimide [26] and poly(ethylene glycol) diacrylate [27, 28], and catechol [29] offer enhanced mucoadhesive property through hydrogen bonding, covalent bonding, and complexation, respectively, between functional groups of derivatives and mucus membrane.

Even though mucoadhesion is a promising approach, it has some limitations for the drug delivery due to short life-span, high viscosity, hydrophilicity, multiple-barrier property, and saturation of mucosal surface [30–32]. The mucosal layer is dynamically reformed with an ocular clearance period of 5–7.7 min [33]. Thus it can be concluded that, along with mucoadhesive potential, sufficient penetrability of polymeric material across mucosal barrier also plays a vital role in ophthalmic drug delivery [34].

3.2 *Penetration Enhancement Property*

Chitosan is an excellent candidate with enhanced penetration property also. Chitosan interacts with extraocular structures such as the cornea and conjunctiva, offering an increased localized concentration and residence time for therapeutics [35, 36]. The cationic nature will support breaking the epithelial cells tight junctions through structural reorientation of associated proteins [37, 38]. The penetration potential of chitosan increases with an increase in molecular weight and degree of deacetylation. The TMC exhibited excellent penetration enhancement by reducing the trans-epithelial electrical resistance (TEER) up to 25–85% when compared to other chitosan salts such as chitosan glutamate (CG) and chitosan hydrochloride (CH) that showed 60% and 84% reduction in TEER, respectively [39–42]. But TMC is more preferable due to its versatile solubility in any pH.

3.3 *In Situ Gelling Property*

Chitosan is a versatile biopolymer possessing in situ gelling property through various mechanisms such as phase separation and covalent cross-linking. The stimuli-responsive characteristics of chitosan will facilitate rapid sol to gel phase conversion of ophthalmic formulations upon change in temperature and pH of the environment. This will surpass the drawback of traditional ophthalmic delivery solutions by preventing rapid lacrimal clearance, low penetration capacity, and pre-corneal loss, etc. [43–45]. The interaction of conventional ophthalmic formulation and chitosan based in situ hydrogels with the ocular mucosal surface is illustrated in Fig. 6.

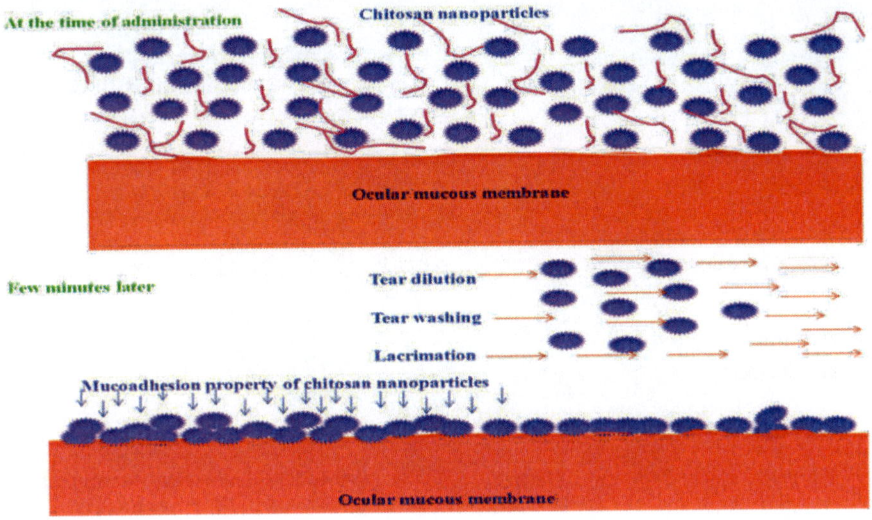

Fig. 6 The fate of conventional eye drops (**a**) and chitosan based in situ hydrogels (**b**) on ocular application

Thermoresponsive gelation is achieved by using low molecular weight polyol phosphates that neutralize the acidic solution of chitosan and convert the solution to gel at body temperature while maintaining isotonicity at ophthalmic pH [46]. Similarly, derivatization of chitosan like thiolation and grafting with other polymers such as poloxamer, gelatin, etc. also promotes thermoresponsive in situ gelation [43, 47–50]. In the case of pH-responsive gelation, the functional groups of chitosan get deprotonated at ophthalmic pH resulting in the expulsion of the loaded drug from the compressed gel matrix [51].

The covalent cross-linking using chemical cross-linkers like diethyl squarate (DES), ethylene glycol diglycidyl ether (EGDE), genipin, blocked diisocyanate, etc. offers excellent in situ gelling potential to develop mechanically strong controlled release formulations when compared to physical gelation methods [52–54].

3.4 Antimicrobial Potential

Chitosan is reported to have attractive broad-spectrum antimicrobial activity against broad-spectrum organisms like bacteria, viruses, and fungi [55, 56]. The antimicrobial potential of chitosan is influenced by the degree of deacetylation, molecular weight, concentration, cationic charge density, and physical state of chitosan; types of the microorganism; pH and temperature of the environment, etc. There are several proposed mechanisms for the antimicrobial action of chitosan such as bacterial cell membrane disruption, inhibition of bacterial nucleic acid synthesis as well as trace

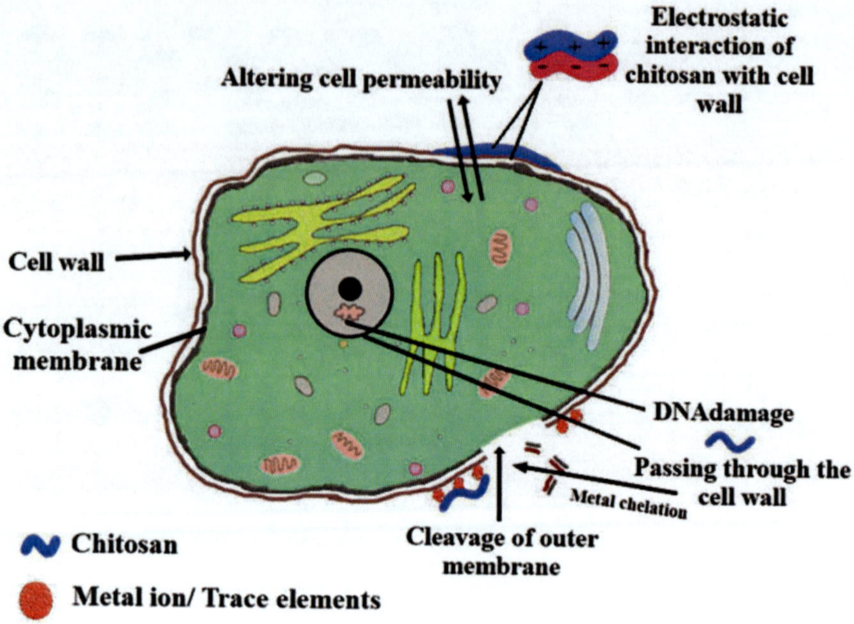

Fig. 7 Various mechanisms responsible for the antimicrobial action of chitosan

metal ion chelating activity that inhibits microbial growth [57–59]. The underlined mechanisms of antimicrobial action of chitosan are depicted in Fig. 7.

The outer membrane of gram-negative bacteria is composed of lipopolysaccharide stabilized by the interaction between anionic functional groups and divalent metal ions. The polycationic chitosan will compete with the divalent metal ions like Mg^{2+} and Ca^{2+} present in the cell membrane, as a result, destabilization of the outer membrane or alteration in the activity of degradative enzymes happens (Fig. 8) [60, 61].

Similarly, interaction of lipoteichoic acids with chitosan in the outer surface of gram-positive bacteria will also cause membrane disruption, leakage of cellular components, and cell death (Fig. 9) [62, 63]. Interestingly, chitosan possesses an antifungal effect by inhibiting sporulation and spore formation. The antimicrobial potential could be ordered as fungi> gram-negative bacteria >gram-positive bacteria [64, 65].

3.5 *Transfection Enhancement Property*

In gene therapy chitosan is recognized as a simple non-viral vector owing to its polycationic nature. It allows efficient encapsulation of anionic genetic materials (DNA, siRNA, etc.), sufficient mucoadhesion, and further penetration through the

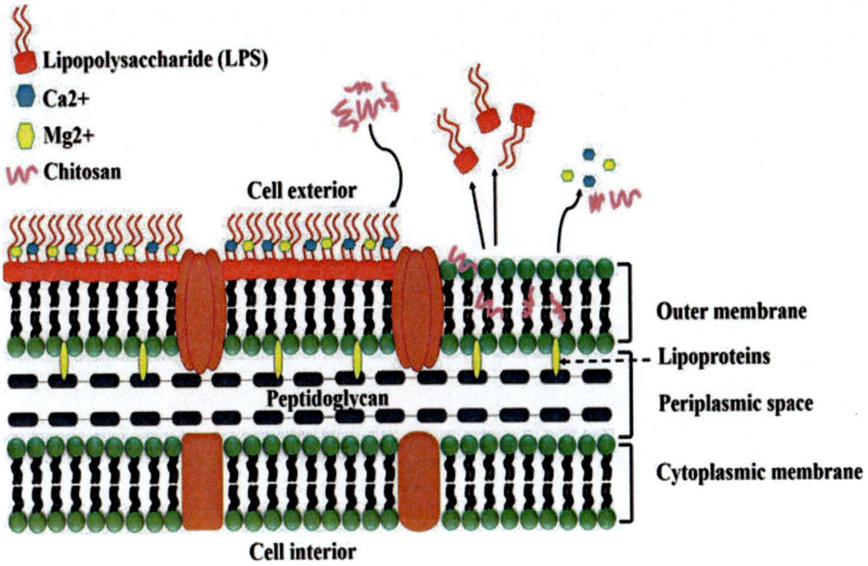

Fig. 8 Antibacterial mechanism of chitosan against gram-negative bacteria

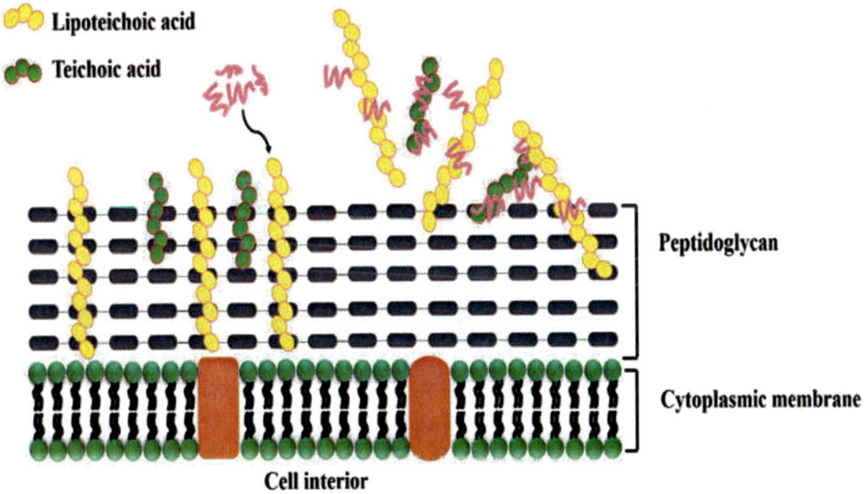

Fig. 9 Antibacterial mechanism of chitosan against gram-positive bacteria

tight cell junctions by structural reorientation of proteins to transport active payload [43]. The net positive charge and nano-size range of lower than 100 nm facilitate the easier endocytosis of fabricated conjugates. Moreover, its non-toxicity, non-immunogenicity, and biocompatibility suggest its plausible role as a non-viral vector in gene therapy when compared to other gold standards such as poly(lysine), poly(ethyleneimine), or poly(arginine) [66, 67]. The molecular weight and degree of deacetylation also influence the transfection efficiency of chitosan. High molecular weight chitosan shows low transfection efficiency as it forms highly viscous strong complexes with chitosan. NOVAFECT is a commercially available low molecular weight, ultrapure chitosan oligomer in deacetylated form having the ability to transfect COS-7 cells when fabricated as chitosan-DNA nanoparticles to treat corneal diseases [68]. Chitosan-DNA conjugate will also protect the payload from intracellular DNAase. The transfection potential was also enhanced with thiolation and trimethylation [69, 70].

However, the transfection efficiency of classical chitosan is comparatively low; hence, various modifications are reported and compared elsewhere. Martien et al. fabricated a nanoparticle using thiolated chitosan for plasmid delivery, proposed a targeted release and higher stability toward nucleases. Here intra-chain disulfide bonds are formed within the complex, which will cleave exclusively under the reducing condition of cytoplasm, leading to site-specific release of genetic material. The results demonstrated a five times increase in transfection efficiency when compared to conventional chitosan/pDNA nanoparticles [71]. In another study, Malmo et al. suggested a higher transfection efficiency of self-branched and self-branched trisaccharide substituted chitosan when compared to linear one [72]. Similarly, thiolated chitosan with trimethylation, chitosan/cyclodextrin, and PEGylated chitosan have been proved its comparative potential in gene transfer in several studies [73–75].

3.6 Controlled Drug Delivery Potential

Chitosan is the best choice for the controlled delivery of anionic drugs. Chitosan will form strong ionic bonding or complexation with polyanionic drugs, offering a prolonged and sustained release of therapeutics effectively [76, 77]. It can also be homogenized with polyanionic excipients (hyaluronic acid, polyacrylates, alginate, carrageenan, pectin, etc.) to form a stable, complex polymeric matrix favorable for the controlled release of active moieties through polymeric erosion and diffusion [78]. Similar nature of controlled release pattern is also observed when chitosan is combined with inorganic multivalent anions (sulfates, tripolyphosphates) [79].

4 Chitosan and Its Derivatives Based Topical Ocular Drug Delivery Systems

4.1 Hydrogel Based Systems

Chitosan and its derivatives are considered well-known stimuli-sensitive drug delivery system, that forms in situ gels based on pH, temperature, and ionic strength of the applied environment. It allows tunable site-specific and sustained delivery therapeutics to the ocular region that surpasses pre-corneal drug loss, lacrimal clearance, and low penetration potential of conventional ophthalmic [43]. Hydrogels are biocompatible structures prepared through physical or chemical cross-linking without the use of any harmful chemicals [80]. Chitosan based hydrogels are reported mainly by using chitosan alone or from modified chitosan or chitosan blends.

4.1.1 Chitosan

Generally, chitosan alone is not used for the preparation of hydrogel as much due to its low viscosity nature. But its intact pH-responsive nature could promote its in situ gelling property in the ophthalmic region as well. However, the formulation prepared via physical (tripolyphosphate, β-glycerophosphate, glucose-1-phosphate, glucose-6-phosphate, etc.) or chemical (blocked diisocyanate, diethyl squarate, genipin, ethylene glycol diglycidyl ether, etc.) cross-linking will efficiently improve its in situ gelling potential and sustained release nature most favorable for ocular application [43].

Fouda et al. developed a proniosome loaded chitosan gel of dorzolamide HCl and compared it with marketed dorzolamide HCl eye drops for reduction of IOP for glaucoma patients. The in vivo study revealed that proniosomal gel has the fastest onset of action along with a drastic reduction of IOP for a prolonged period even up to 8 h when compared with marketed. This suggested the retention of chitosan gel in the ocular surface due to high mucoadhesive potential along with combined penetration enhancement of proniosome and chitosan [81].

Similarly, when the nanovesicle of acetazolamide loaded in tripolyphosphate cross-linked chitosan hydrogel, normalization of IOP was achieved within 45 min and prolonged for 24 h, while the conventional tablet took 1.5 h for action and prolonged only for 5 h. The high viscosity and mucoadhesion of hydrogel support the progressive release of medicament through diffusion mechanism [82].

A thermoresponsive chitosan hydrogel using β-glycerolphosphate was developed by Mohammed et al. for antimicrobial action that exhibits sol to gel transition at 29–37°C. They observed zone of inhibition against *S. aureus* for blank hydrogel also. But complete microbial inhibition was achieved in antibiotics (gentamicin, moxifloxacin) loaded hydrogel in a sustained manner of 4 h [83, 84].

4.1.2 From Modified Chitosan

Several covalently cross-linked modified chitosan hydrogels were reported using carboxymethyl chitosan (CMCS), glycol-modified chitosan (GlyCS), azide-modified GlyCS, L-cysteine modified chitosan, etc. Also modified chitosan with in situ gelling properties such as chitosan-grafted poly(N-isopropyl acrylamide), quaternary ammonium CS, hexanoyl glycol chitosan, hydroxybutyl chitosan, etc. have also been reported elsewhere in ophthalmic applications.

A voriconazole inclusion complex of poly(β-cyclodextrin) loaded chitosan hydrogel was prepared by Yang et al. for fungal endophthalmitis, by spontaneous mixing of separate solutions of carboxymethyl chitosan (CMCS) and Odex (Schiff base reaction). The drug was released from an intertwined complex system of CMCS/Odex network and poly(β-cyclodextrin) linear chain, through a combined mechanism of chemical degradation, diffusion, and swelling [85]. Schiff base reaction was carried out between amino groups of glycol-modified chitosan (GlyCS) and aldehyde group of 4-arm PEG to form a sustained-release porous hydrogel loaded with levofloxacin [86].

The thiolated chitosan has an attractive role in hydrogel preparation owing to its cross-linking potential through disulfide bonds. In one study, anti-VEGF protein (ranibizumab and aflibercept) loaded hydrogel was prepared using thiolated chitosan prepared via coupling of L-cysteine, or 6-mercaptonicotinic acid with chitosan through air oxidation. They observed a minimum burst release, enhanced cross-linking, and controlled release through low swelling, at the same time intact nature of protein drug was also preserved [87].

4.1.3 From Chitosan Blends

Chitosan can form hydrogel with other anionic or non-ionic macromolecules to improve various properties than when used alone. It can be combined with non-ionic polymers such as poloxamers, synthetic polyanions like PAA; polysaccharides like hyaluronic acid, dextran sulfate, xanthan, alginate; as well as proteins like collagen, gelatin, etc.

Gelatin is a protein used to improve thermo gelling property and mechanical strength of chitosan widely. Several studies were reported elsewhere using chitosan/gelatin hydrogel cross-linked with β-glycerol phosphate [88], genipin [89], oxidized sucrose [90], etc. incorporating various therapeutic agents. The hydrogel is formed through electrostatic, hydrophobic interactions between various groups. High molecular weight and hydrogen bond interactions formed due to hydrophilic groups of gelatin backbone and chitosan impart strong bioadhesive strength of hydrogel blend [50, 91, 92].

Poloxamer, a thermoreversible hydrogel at 37°C, is attracted to various biomedical applications nowadays. But weak mechanical strength of the same insists its combination with other polymers like chitosan widely. An ophthalmic in-situ gel of poloxamer 407/chitosan was developed by Irimia et al. for the delivery of topical anesthetic bupivacaine hydrochloride [93].

Poly(vinyl alcohol) (PVA) is another versatile non-ionic polymer that can combine with chitosan. Unlike poloxamer, it offers pH-sensitive gelation rather than thermosensitive at alkaline pH of 6–7.5. The formulation exhibited a swelling controlled release of medicaments [94]. A triple combination of chitosan, PVA, and poly(vinylpyrrolidone) also suggested an improved mucoadhesion with prolonged release of 24 h when compared to marketed formulations of levofloxacin and besifloxacin [95].

4.2 *Colloidal Nanocarriers*

Apart from the development of responsive hydrogel systems, chitosan also holds its hallmark in the development of various colloidal nanoparticulate systems such as nanoparticles, liposomes, micelles, etc. for myriads of applications. Nanocarriers could be effectively applied as liquid dispersions, which adhere for a prolonged time, or surpass effectively the ocular physiological barriers and reach the intraocular region.

4.2.1 Nanoparticles

The nanoparticulate systems nowadays clutch their advantages over conventional drug delivery systems owing to their targeting potential, narrow size range, prolonged-release pattern, sustained activity, and reduced dose size and dosing intervals [96, 97]. Chitosan nanoparticles are considered an excellent system for ocular delivery due to their higher specific surface area and mucoadhesive potential [98, 99]. The chitosan nanoparticles with a size range of 50–500 nm will effectively cross the physiological barriers of eye and mucin mesh of the ocular surface, while the higher particles will adhere only to the mucosal surface and release drugs topically [32, 100–102]. The preparation methods employed for the production of chitosan nanoparticles (CS NP) are ionotropic gelation, spray drying, reverse micelle formation, emulsion droplet coalescence, nanoprecipitation, water-in-oil emulsion cross-linking, etc. [103].

Ionotropic Gelation

It is the simplest and relatively most convenient method for the production of CS NP. The mechanism involved the intra- and inter-molecular ionic cross-linking between the cationic amino group of chitosan and anionic groups of polymers (chondroitin sulfate, hyaluronic acid, sodium cellulose sulfate, alginate, etc.) or small molecules (sodium tripolyphosphate (TPP), sodium sulfate) [104].

The protonated chitosan in acidic solution will experience repulsive forces and attractive forces due to cationic amino group and hydrogen bonding. This will form a cross-linked structure with anionic phosphate groups of aqueous TPP when added to chitosan solution [105, 106]. The ratio of CS/TPP will greatly affect the zeta potential of nanoparticles since both are of opposite charge. Generally, CS NP possesses a surface charge of 20–60 mV due to the protonation of the amino group of chitosan. The higher positive zeta potential of CS NPs is crucial for their prolonged ocular residence through mucoadhesion. Several studies demonstrated that cationic surface charge is reduced with an increase in TPP concentrations [107–110].

But the concentration of CS and TPP showed a different effect in size of CS NP in different studies. The majority of the studies suggested that particle size is directly proportional to the CS concentration [109, 111–113]. Some studies established a decrease in particle size [107, 114], while some showed an initial increase followed by a decrease or vice versa [108, 115]. Similarly, the TPP concentration also suggested a wide range of effects on particle size [110, 116, 117] in a concentration-dependent manner. When considering the drug loading/entrapment efficiency, the concentration of CS and TPP showed an augmentation with an increase in concentration. This is because, the denser the polymer could entrap more drugs and more of TPP promotes efficient cross-linking to lock maximum drugs [108, 109, 114].

Pectin is an anionic polysaccharide, which could undergo ionotropic gelation with chitosan to form polyelectrolyte complexes. Dubey et al. fabricated brinzolamide loaded CS/Pectin nanocapsule through polyelectrolyte complex coacervation method and compared with commercial eye drop for its anti-glaucoma potential. The developed nanocapsule holds a smooth spherical nano-sized structure with improved corneal permeation, hence higher bioavailability [118]. Similar interactions were observed between chitosan and lecithin (L), which is a naturally occurring phospholipid blend, with an abundant phosphate group. The CS/L NP formed facilitated the encapsulation of lipophilic drug Amphotericin B, an antibiotic for fungal keratitis [119].

By Spray Drying

Spray drying is a simple, method for the pilot scale production of microparticles. It involved only drying of drug-loaded polymeric liquid in a fine stream of hot air to produce particles preferably microscale ranges. Ornidazole and pilocarpine are loaded in the chitosan microsphere and evaluated in various studies [120, 121]. Zhou et al. developed a double cross-linked chitosan microparticle loaded with levofloxacin. Here, CS/TPP gel was initially spray dried, which was then further cross-linked by glutaraldehyde during solidification. The developed formulation showed sustained release over 1 day with controllable-swelling potential [122].

From Modified Chitosan

Despite myriads of advantages, plane chitosan limits its application due to certain characteristics such as low aqueous solubility at neutral and alkaline pH, low mucoadhesiveness, etc. When a chitosan nanoparticle is prepared via ionic gelation with TPP, the density of the cationic amino group decreased due to the linkage between CS and TPP. This will attenuate the mucoadhesive property of chitosan, by hindering the interaction between mucin and $–NH_3^+$ group. Furthermore, the cationic surface charge of CS NP is reduced based on TPP concentration [9, 123, 124]. Here the modification of chitosan is required to boost up various properties through functionalization at hydroxyl and amino groups.

The trimethylation of chitosan (TMCS) will form quartered chitosan that exhibits aqueous solubility at a wide range of pH. TMCS/TPP nanoparticle loaded with diclofenac sodium showed improved mucoadhesion and the minimum effective concentration was maintained in aqueous humour even after 12 h of administration [125]. In another study, the flurbiprofen/cyclodextrin inclusion complex was encapsulated in TMCS/TPP system to enhance the ocular residence, aqueous solubility, and mucoadhesion. The in vitro drug release pattern exhibited an initial burst release followed by a sustained pattern [126]. Other derivatives such as 2-carboxybenzaldehyde chitosan [127], succinic anhydride [191] chitosan, methyl methacrylate chitosan [128], chitosan-grafted poly(ethylene glycol) methacrylate [129], thiolated chitosan [130], etc., prepared via modifications at the amino group through Michael addition reaction were also reported elsewhere for nanoparticles synthesis via ionic cross-linking. Also, two free hydroxyl end of chitosan is another potential provision for modification. Glycol chitosan is a successful water-soluble chitosan reported for drug encapsulation (cerium oxide/GLYCS) [131] or drug conjugation (dexamethasone-GLYCS) [132].

4.2.2 Liposomes

Liposomes are lipid vesicular systems comprising aqueous core stabilized with surfactant bilayer that can encapsulate both hydrophilic and lipophilic drugs. The structure has a close resemblance to the cell membrane, hence holds prominent biocompatibility [133]. Abd-Elsalam et al. developed agomelatine-loaded olaminosomes using chitosan hydrochloride for ocular hypertension via the thin-film hydration method. They observed a sustained along with controlled release pattern in in-vitro studies concerning an increased amount of chitosan. The in vivo studies of rabbit suggested that enhanced mucoadhesion and ocular retention favored alleviation of ocular hypertension without any signs of toxicity [134].

Similarly, TMCS coated nanoliposome of pilocarpine hydrochloride for glaucoma demonstrated a cationic charge and increased viscosity in releasing medium. It followed an initial burst release pattern followed by sustained release of medicament. The in vivo study revealed the non-irritant behavior of formulation on the ocular topical application [135]. Octadecyl-quaternized carboxymethyl CS is another derivative, which is amphiphilic in nature employed for the ocular delivery of highly hydrophobic drugs like curcumin and tetrandrine loaded in nanostructured lipid carriers (NLC) and liquid crystalline NPs, respectively. The aqueous solubility, mucoadhesion, and pre-corneal residence time of both drugs were improved with a biphasic drug release pattern [136, 137].

4.2.3 Micelles

Micelles are self-assembled nanostructures (5–30 nm) of amphiphilic polymers that are oriented into the stiff core of non-polar tail and an outer surface of polar head groups spontaneously arranged in an aqueous medium. It has a high loading capacity of hydrophobic drugs, while its susceptibility toward environmental changes like pH, temperature, ionic strength, etc. questions its stability and drug precipitation against storage [138–141]. The plane chitosan does not have the capacity to form micelles of its own, but several forms of modified chitosan are reported elsewhere nowadays. Chitosan grafted methoxy poly(ethylene glycol)-poly(ε-caprolactone) blocks polymer formed diclofenac loaded nanomicelles spontaneously, to form a cationic suspension. The formulation showed an increased bioavailability, corneal penetration, and pre-corneal retention in vivo with an initial burst followed with sustained pattern [142].

In another study, chitosan oligosaccharide-valylvaline-stearic acid derivative-based nanomicelles were synthesized and encapsulated dexamethasone for macular edema. The nanomicelles formed were cationic with sustained-release patterns in simulated tear fluid. The topically applied formulation demonstrated an improved corneal permeability and reached the posterior portion through the conjunctival route in the in-vivo study on rabbits [143].

4.2.4 Other Combined Delivery Systems

The chitosan nanoparticles have been combined with other drug delivery systems to tune up the targeting potential of nanoparticles by combining the advantages of dual delivery vehicles. They were incorporated into various systems such as gels, lenses, ocuserts, coatings, etc. Fabiano et al. developed a 5-fluorouracil-loaded CS NPs of quaternary ammonium chitosan, quaternary ammonium protected thiolated chitosan, and sulfobutyl-chitosan and incorporated them in a hydrogel base of quaternary ammonium chitosan. The first and third formulation exhibited prolonged retention of drugs in aqueous humour in in-vivo studies. This may be due to the high mucoadhesive potential of the cationic hydrogel of the first derivative, while for the third one the ionic interaction between negatively charged sulfobutyl-chitosan NPs and cationic hydrogel together delayed drug diffusion [144]. Also chitosan/alginate (CS/ALG) based core/shell microparticle of bovine serum albumin and timolol maleate nanoparticle were developed and incorporated in HPMC gel which showed delayed release of a medicament for 9 days and 24 h, respectively

[145, 146]. Hyaluronic acid (HA) coated dexamethasone loaded chitosan nanoparticles were developed for improved cellular targeting and compared with simple dexamethasone solutions. The surface charge of CS NPs reduced and size is increased due to coating with HA. The formulation showed better corneal retention and sustained release when compared to the solution [147, 148].

4.3 Ocular Lenses and Ocuserts

Ocular lenses and ocuserts are novel non-implantable ophthalmic devices that create their places in topical ocular delivery of medicaments nowadays. They are advantageous over conventional delivery systems by surpassing fast lacrimal clearance of the drug, improved localized delivery to the lens, and patient compliance. Chitosan is a suitable polymer for the development of such devices due to their inherent mucoadhesiveness, biocompatibility, non-toxicity, and flexible physico-chemical properties.

The flexible films and membranes have been developed mainly as ocuserts by solvent casting method. Chitosan based ocular inserts of bimatoprost and diminazene aceturate were developed where a homogenized dispersion of medicament is obtained by solvent casting method [149, 150]. Similarly, dorzolamide loaded chitosan-hydroxyethyl cellulose films were also developed for long-term delivery of drugs [151]. The dual delivery of timolol and latanoprost was achieved for 9 days when incorporated into flexible nanosheets of chitosan and alginate [152].

A coated lens of chitosan and alginate was developed by Silva et al. for the delivery of diclofenac sodium. The multiple coating of (Alginate-CaCl2)/(Chitosan +Glyoxal) was ultimately coated with Alginate-CaCl2 in a sandwich model to prevent the lacrimal fluid degradation of chitosan. The multiple coated lenses demonstrated a sustained release of an anti-inflammatory drug for 7 days [153]. A biocompatible antibacterial coating of chitosan/gallic acid/zinc oxide NPs developed by Hoyo et al. demonstrated an improved antioxidant activity of 95%. The hybrid lens developed showed better biocompatibility and comfort in application [154].

5 Chitosan Based Ocular Films

In prolonging the ocular delivery of drugs chitosan plays a vital role. In a recent study the efficacy and safety of chitosan-coated timolol maleate (TM) mucoadhesive film were studied. Based on casting and solvent evaporation technique the chitosan-coated timolol maleate (TM) mucoadhesive film was developed and characterized for its in-vitro release and swelling characteristics. They found an enhanced lowering of IOP compared to timolol maleate commercial formulation with 85% of the TM gets released over the first 2 weeks [155]. Brimonidine tartrate (BT) loaded chitosan film prepared by solution casting and air-drying has been reported to possess high corneal permeability with fast drug release kinetics for potential ocular drug delivery. They reported that the developed film is transparent, structurally stable, and mucoadhesive [156].

Ciprofloxacin loaded ocular films using xyloglucan (2%) were developed using polysaccharide by the solvent casting method by Mahajan et al. They observed a cumulative % drug release from the developed film of 98.85% at the end of 24 h. They found a higher amount of ciprofloxacin in cornea and conjunctiva with higher penetration through cornea. Herein xyloglucan can act as a potential film forming polymer for ocular delivery applications [157].

6 Chitosan Based Ocular Inserts

The film-like solid or semi-solid devices of medicated polymers intended for insertion into the conjunctival sac to liberate the medicament to the ocular surface are the ocular inserts. Ocular inserts provide a reservoir for controlled drug delivery of drugs and chitosan ocular inserts were developed by researchers for various therapeutic applications. The advantages of ocular inserts over conventional dosage forms are increased ocular residence, possibility of releasing drugs at a slow and constant rate, increased shelf life, accurate dosing, and avoidance of toxicity. For the treatment of glaucoma Julio Fernandes et al. developed brimonidine tartrate loaded chitosan based ocular inserts by improving the drug bioavailability and found that the inserts are biocompatible and non-toxic in nature with the potential to reduce the intraocular pressure in glaucomatous patients [157, 158].

Chitosan based inserts for sustained release of angiotensin-converting enzyme 2 (ACE2) activator diminazene aceturate (DIZE) were developed and characterized by Foureaux et al. [150]. They observed that the developed DIZE is chemically stable and has the capacity to decrease the IOP for up to 1 month (last week: 11.0 ± 0.7 mmHg) without producing any toxic effects to the eye.

Franca et al. developed and characterized Bimatoprost (BIM) loaded novel sustained-release drug inserts using chitosan by adopting the solvent casting method. The BIM release was observed for 8 h with a biodistribution of ^{99m}Tc-BIM with IOP reduction for 4 weeks, upon single application in glaucoma management [159].

Chitosan grafted titanium alloy was developed by covalent link via a two-step process using TriEthoxySilylPropyl Succinic Anhydride (TESPSA) as the coupling agent which affords effective antibacterial activity. Herein better antibacterial activity was observed against *Escherichia coli* and *Staphylococcus aureus* strains. These developments may be also utilized for ocular drug delivery focusing toward antibacterial approaches [160].

Latanoprost loaded thermosensitive chitosan based hydrogel was developed by Yung et al. for the treatment of glaucoma in order to overcome the frequent and long-term applications. In order to offer a sustained release effect, they developed this topical eye drop formulation to control ocular hypertension. Upon topical administration latanoprost-loaded hydrogel, triamcinolone acetonide-induced elevated IOP gets significantly decreased within 7 days. They concluded that the newly developed chitosan based hydrogel may offer a non-invasive alternative to traditional anti-glaucoma eye drops for glaucoma treatment [88].

7 Chitosan Based Ocular Gels

Chitosan based ocular gels are widely utilized in ocular drug delivery especially the in situ gels are liquid upon instillation and converted into viscoelastic gels in response to environmental changes such as pH, temperature, and specific ions. The acceptance criteria for in situ gels indicates that they are instilled into the eye as a solution and immediately converted into a gel when it contacts with the eye with pre-corneal residence of several hours.

Chen et al. [161] developed and characterized in situ thermosensitive hydrogel based on chitosan by loading levocetirizine dihydrochloride (LD) in combination with disodium α-d-glucose 1-phosphate (DGP) for ocular drug delivery system using sol gel transition. They observed a rapid release at the initial period followed by a sustained release with sol-to-gel phase transition based on chitosan concentration. Their ocular irritation results demonstrated excellent ocular tolerance of the hydrogel with significant prolonged residence time compared to eye drops. The effective anti-allergic conjunctivitis effects compared to LD aqueous solution were observed via sol–gel transition mechanism [161].

Divyesh H. Shastri et al. [162] developed thiolated chitosan based nanoparticulate mucoadhesive ophthalmic in situ gelling system of prulifloxacin for the treatment of bacterial keratoconjunctivitis which is non-irritant to eye. They observed that gelation pH was found to be 7.2 ± 0.2 with entrapment efficiency of more than 80% with sustained release behavior for a period of 12 h. Aikaterini Karava et al. developed chitosan derivatives in order to deliver dexamethasone sodium phosphate and chloramphenicol and observed low cytotoxicity, enhanced mucoadhesive properties, and better antimicrobial activity [163].

8 Conclusion

A considerable amount of research is going on for the chitosan based delivery systems to eye. In particular, the chitosan based ocular films, ocular inserts, ocular gels are gaining recently with more interest especially for challenging water insoluble/poorly permeable drugs. Chitosan based self-assembled systems and chitosan derivatives based ocular delivery systems may have great potential, as well as multiple applications for the future in the design of novel topical systems for eye.

References

1. Willoughby CE, Ponzin D, Ferrari S, Lobo A, Landau K, Omidi Y (2010) Anatomy and physiology of the human eye: effects of mucopolysaccharidoses disease on structure and function- a review. Clin Experiment Ophthalmol 2010(38):2–11
2. Nayak K, Misra M (2018) A review on recent drug delivery systems for posterior segment of eye. Biomed Pharmacother 107:1564–1582
3. Lin S, Ge C, Wang D, Xie Q, Wu B, Wang J, Nan K, Zheng Q, Chen W (2019) Overcoming the anatomical and physiological barriers in topical eye surface medication using a peptide-decorated polymeric micelle. ACS Appl Mater Interfaces 11(43):39603–39612
4. Kwatra D, Mitra AK (2013) Drug delivery in ocular diseases: barriers and strategies. World J Pharmacol 2(4):78–83
5. Bachu RD, Chowdhury P, Al-Saedi ZHF, Karla PK, Boddu SHS (2018) Ocular drug delivery barriers-role of nanocarriers in the treatment of anterior segment ocular diseases. Pharmaceutics 10:28
6. Palani S, Joseph NM, Goda CC, Zachariah A, Ayenew Z (2010) Ocular drug delivery a review. IJPSR 1(3):1–11
7. Rodrigues S, Dionísio M, López CR, Grenha A (2012) Biocompatibility of chitosan carriers with application in drug delivery. J Funct Biomater 3:615–641
8. He P, Davis SS, Illum L (1998) In vitro evaluation of the mucoadhesive properties of chitosan microspheres. Int J Pharm 166:75–88
9. Dhawan S, Singla AK, Sinha VR (2004) Evaluation of mucoadhesive properties of chitosan microspheres prepared by different methods. AAPS PharmSciTech 5:122–128
10. Lehr CM, Bouwstra JA, Schacht EH, Junginger HE (1992) In vitro evaluation of mucoadhesive properties of chitosan and some other natural polymers. Int J Pharm 78:43–48
11. George M, Abraham TE (2006) Polyionic hydrocolloids for the intestinal delivery of protein drugs: alginate and chitosan – a review. J Control Release 114:1–14
12. Barbu E, Verestiuc L, Nevell TG, Tsibouklis J (2006) Polymeric materials for ophthalmic drug delivery: trends and perspectives. J Mater Chem:3439–3443
13. Jacobsen J, Meng-Lund E, Muff-Westergaard C et al (2014) A mechanistic based approach for enhancing buccal mucoadhesion of chitosan. Int J Pharm 461:280–285
14. Nafee NA, Ismail FA, Boraie NA, Mortada LM (2004) Mucoadhesive delivery systems. I. Evaluation of mucoadhesive polymers for buccal tablet formulation. Drug Dev Ind Pharm 30:985–993
15. Greaves JL, Wilson CG (1993) Treatment of diseases of the eye with mucoadhesive delivery systems. Adv Drug Deliv Rev 11:349–383
16. Bhavsar C, Momin M, Gharat S, Omri A (2017) Functionalized and graft copolymers of chitosan and its pharmaceutical applications. Expert Opin Drug Deliv 14:1189–1204
17. Ways TMM, Lau WM, Khutoryanskiy VV (2018) Chitosan and its derivatives for application in mucoadhesive drug delivery systems. Polymers 10:267–304
18. Chopra S, Mahdi S, Kaur J et al (2006) Advances and potential applications of chitosan derivatives as mucoadhesive biomaterials in modern drug delivery. J Pharm Pharmacol 58:1021–1032
19. Werle M, Bernkop-Schnürch A (2008) Thiolated chitosans: useful excipients for oral drug delivery. J Pharm Pharmacol 60:273–281
20. Bernkop-Schnürch A, Guggi D, Pinter Y (2004) Thiolated chitosans: development and in vitro evaluation of a mucoadhesive, permeation enhancing oral drug delivery system. J Control Release 94:177–186
21. Bernkop-Schnürch A (2005) Thiomers: a new generation of mucoadhesive polymers. Adv Drug Deliv Rev 57:1569–1582
22. Jintapattanakit A, Junyaprasert VB, Kissel T (2009) The role of mucoadhesion of trimethyl chitosan and PEGylated trimethyl chitosan nanocomplexes in insulin uptake. J Pharm Sci 98:4818–4848

23. Muzzarelli RAA, Tanfani F (1985) The N-permethylation of chitosan and the preparation of N-trimethyl chitosan iodide. Carbohydr Polym 5:297–307
24. Karavasili C, Katsamenis OL, Bouropoulos N et al (2014) Preparation and characterization of bioadhesive microparticles comprised of low degree of quaternization trimethylated chitosan for nasal administration: effect of concentration and molecular weight. Langmuir 30:12337–12344
25. Agarwal S, Aggarwal S (2015) Mucoadhesive polymeric platform for drug delivery; a comprehensive review. Curr Drug Deliv 12:139–156
26. Menzel C, Hauser M, Frey A et al (2019) Covalently binding mucoadhesive polymers: N-hydroxysuccinimide grafted polyacrylates. Eur J Pharm Biopharm 139:161–167
27. Eshel-Green T, Bianco-Peled H (2016) Mucoadhesive acrylated block copolymers micelles for the delivery of hydrophobic drugs. Colloids Surf B Biointerfaces 139:42–51
28. Eliyahu S, Aharon A, Bianco-Peled H (2018) Acrylated chitosan nanoparticles with enhanced mucoadhesion. Polymers 10:1–17
29. Ryu JH, Choi JS, Park E et al (2020) Chitosan oral patches inspired by mussel adhesion. J Control Release 317:57–66
30. Bernkop-Schnürch A (2005) Mucoadhesive systems in oral drug delivery. Drug Discov Today Technol 2:83–87
31. Sigurdsson HH, Kirch J, Lehr CM (2013) Mucus as a barrier to lipophilic drugs. Int J Pharm 453:56–64
32. Ponchel G, Montisci MJ, Dembri A et al (1997) Mucoadhesion of colloidal particulate systems in the gastro-intestinal tract. Eur J Pharm Biopharm 44:25–31
33. Khutoryanskiy VV (2011) Advances in mucoadhesion and mucoadhesive polymers. Macromol Biosci 11:748–764
34. Zhang X, Cheng H, Dong W et al (2018) Design and intestinal mucus penetration mechanism of core-shell nanocomplex. J Control Release 272:29–38
35. Nagarwal RC, Kant S, Singh PN et al (2009) Polymeric nanoparticulate system: a potential approach for ocular drug delivery. J Control Release 136:2–13
36. Eljarrat-Binstock E, Orucov F, Aldouby Y et al (2008) Charged nanoparticles delivery to the eye using hydrogel iontophoresis. J Control Release 126:156–161
37. De M, Raviña M, Paolicelli P et al (2010) Chitosan-based nanostructures: a delivery platform for ocular therapeutics. Adv Drug Deliv Rev 62:100–117
38. Schipper NG, Olsson S, Hoogstraate JA et al (1997) Chitosans as absorption enhancers for poorly absorbable drugs 2: mechanism of absorption enhancement. Pharm Res 14:923–929
39. Kotzé AF, Thanou MM, Lueben HL et al (1999) Enhancement of paracellular drug transport with highly quaternized N- trimethyl chitosan chloride in neutral environments: in vitro evaluation in intestinal epithelial cells (Caco-2). J Pharm Sci 88:253–257
40. Hamman JH, Schultz CM, Kotzé AF (2003) N-trimethyl chitosan chloride: optimum degree of quaternization for drug absorption enhancement across epithelial cells. Drug Dev Ind Pharm 29:161–172
41. Kotzé AR, Lueβen HL, de Leeuw BJ et al (1997) N-trimethyl chitosan chloride as a potential absorption enhancer across mucosal surfaces: in vitro evaluation in intestinal epithelial cells (Caco-2). Pharm Res 14:1197–1202
42. Kotzé AF, Lueßen HL, De Leeuw BJ et al (1998) Comparison of the effect of different chitosan salts and N-trimethyl chitosan chloride on the permeability of intestinal epithelial cells (Caco-2). J Control Release 51:35–46
43. Natesan S, Krishnaswami V, Thekkila Veedu S, Pathayappurakkal Mohanan D et al (2019) Chitosan-based ocular drug delivery systems. In: Jana S, Jana S (eds) Functional chitosan. Springer, Singapore, pp 107–134
44. Kirchhof S, Goepferich AM, Brandl FP (2015) Hydrogels in ophthalmic applications. Eur J Pharm Biopharm 95:227–238

45. Otero-Espinar FJ, Fernández-Ferreiro A, González-Barcia M et al (2018) Stimuli sensitive ocular drug delivery systems. In: Drug targeting and stimuli sensitive drug delivery systems. William Andrew Publishing, San Diego, pp 211–270
46. Wu Y, Liu Y, Li X et al (2019) Research progress of in-situ gelling ophthalmic drug delivery system. Asian J Pharm Sci 14:1–15
47. Gratieri T, Martins G, Melani E et al (2010) A poloxamer/chitosan in situ forming gel with prolonged retention time for ocular delivery. Eur J Pharm Biopharm 75:186–193
48. Gratieri T, Martins G, De Freitas O et al (2011) Enhancing and sustaining the topical ocular delivery of fluconazole using chitosan solution and poloxamer/chitosan in situ forming gel. Eur J Pharm Biopharm 79:320–327
49. Cheng YH, Hung KH, Tsai TH et al (2014) Sustained delivery of latanoprost by thermosensitive chitosan–gelatin-based hydrogel for controlling ocular hypertension. Acta Biomater 10:4360–4366
50. Tsai CY, Woung LC, Yen JC et al (2016) Thermosensitive chitosan-based hydrogels for sustained release of ferulic acid on corneal wound healing. Carbohydr Polym 135:308–315
51. Ratemi E (2018) pH-responsive polymers for drug delivery applications. In: Stimuli responsive polymeric nanocarriers for drug delivery applications. Woodhead publishing series in biomaterials. Elsevier, Amsterdam, pp 121–141
52. Berger J, Reist M, Mayer JM et al (2004) Structure and interactions in covalently and ionically crosslinked chitosan hydrogels for biomedical applications. Eur J Pharm Biopharm 57:19–34
53. Hennink WE, van Nostrum CF (2002) Novel crosslinking methods to design hydrogels. Adv Drug Deliv Rev 54:13–36
54. Bhattarai N, Gunn J, Zhang M (2010) Chitosan-based hydrogels for controlled, localized drug delivery. Adv Drug Deliv Rev 62:83–99
55. Rabea EI, Badawy MET, Stevens CV et al (2003) Chitosan as antimicrobial agent: applications and mode of action. Biomacromolecules 4:1457–1465
56. Campaniello D, Corbo MR (2010) Chitosan: a polysaccharide with antimicrobial action. In: Application of alternative food-preservation technologies to enhance food safety and stability. Bentham Science Publishers Ltd, Sharjah, pp 92–113
57. Nagy A, Harrison A, Sabbani S et al (2011) Silver nanoparticles embedded in zeolite membranes: release of silver ions and mechanism of antibacterial action. Int J Nanomedicine 6:1833–1852
58. Divya K, Vijayan S, George TK, Jisha MS (2017) Antimicrobial properties of chitosan nanoparticles: mode of action and factors affecting activity. Fibers Polym 18:221–230
59. Kong M, Chen XG, Xing K, Park HJ (2010) Antimicrobial properties of chitosan and mode of action: a state of the art review. Int J Food Microbiol 144:51–63
60. Helander I, von Wright A, Mattila-Sandholm TM (1997) Potential of lactic acid bacteria and novel antimicrobials against gram-negative bacteria. Trends Food Sci Technol 8:146–150
61. Kong M, Chen XG, Liu CS et al (2008) Antibacterial mechanism of chitosan microspheres in a solid dispersing system against *E. coli*. Colloids Surf B Biointerfaces 65:197–202
62. Raafat D, Von Bargen K, Haas A, Sahl HG (2008) Insights into the mode of action of chitosan as an antibacterial compound. Appl Environ Microbiol 74:3764–3773
63. Hassan MA, Omer AM, Abbas E et al (2018) Preparation, physicochemical characterization and antimicrobial activities of novel two phenolic chitosan Schiff base derivatives. Sci Rep 8:1–4
64. Sahariah P, Masson M (2017) Antimicrobial chitosan and chitosan derivatives: a review of the structure–activity relationship. Biomacromolecules 18:3846–3868
65. Helander IM, Nurmiaho-Lassila EL, Ahvenainen R et al (2001) Chitosan disrupts the barrier properties of the outer membrane of gram-negative bacteria. Int J Food Microbiol 71:235–244
66. Mao S, Sun W, Kissel T (2010) Chitosan-based formulations for delivery of DNA and siRNA. Adv Drug Deliv Rev 62:12–27
67. Yu H, Chen X, Lu T et al (2007) Poly(L-lysine)-graft-chitosan copolymers: synthesis characterization, and gene transfection effect. Biomacromolecules 8:1425–1435

68. Klausner EA, Zhang Z, Chapman RL et al (2010) Biomaterials ultrapure chitosan oligomers as carriers for corneal gene transfer. Biomaterials 31:1814–1820
69. Lee D, Mohapatra SS (2008) Chitosan nanoparticle-mediated gene transfer. Methods Mol Biol 433:127–140
70. Bernkop-Schnürch A, Dünnhaupt S (2012) Chitosan-based drug delivery systems. Eur J Pharm Biopharm 81:463–469
71. Martien R, Loretz B, Thaler M et al (2007) Chitosan–thioglycolic acid conjugate: an alternative carrier for oral nonviral gene delivery? J Biomed Mater Res A 82:1–9
72. Malmo J, Vårum KM, Strand SP (2011) Effect of chitosan chain architecture on gene delivery: comparison of self-branched and linear chitosans. Biomacromolecules 12:721–729
73. Varkouhi AK, Lammers T, Schiffelers RM et al (2011) Gene silencing activity of siRNA polyplexes based on biodegradable polymers. Eur J Pharm Biopharm 77:450–457
74. Malhotra M, Lane C, Tomaro-Duchesneau C et al (2011) A novel method for synthesizing PEGylated chitosan nanoparticles: strategy, preparation, and in vitro analysis. Int J Nanomedicine 6:485–494
75. Teijeiro-Osorio D, Remuñán-López C, Alonso MJ (2009) Chitosan/cyclodextrin nanoparticles can efficiently transfect the airway epithelium in vitro. Eur J Pharm Biopharm 71:257–263
76. Bhise KS, Dhumal RS, Paradkar AR, Kadam SS (2008) Effect of drying methods on swelling, erosion and drug release from chitosan-naproxen sodium complexes. AAPS PharmSciTech 9:1–12
77. Sun W, Mao S, Wang Y et al (2010) Bioadhesion and oral absorption of enoxaparin nanocomplexes. Int J Pharm 386:275–281
78. Tapia C, Corbalán V, Costa E et al (2005) Study of the release mechanism of diltiazem hydrochloride from matrices bases on chitosan-alginate and chitosan-carrageenan mixtures. Biomacromolecules 6:2389–2395
79. Shavi GV, Nayak UY, Reddy MS (2011) Sustained release optimized formulation of anastrozole-loaded chitosan microspheres: in vitro and in vivo evaluation. J Mater Sci Mater Med 22(4):865–878
80. Kumar A, Vimal A, Kumar A (2016) Why chitosan? From properties to perspective of mucosal drug delivery. Int J Biol Macromol 91:615–622
81. Fouda NH, Abdelrehim RT, Hegazy DA, Habib BA (2018) Sustained ocular delivery of dorzolamide-HCL via proniosomal gel formulation: in-vitro characterization, statistical optimization, and in-vivo pharmacodynamic evaluation in rabbits. Drug Deliv 25:1340–1349
82. Abdel-Rashid RS, Helal DA, Omar MM, El Sisi AM (2019) Nanogel loaded with surfactant based nanovesicles for enhanced ocular delivery of acetazolamide. Int J Nanomedicine 14:2973–2983
83. Mohammed S, Chouhan G, Anuforom O et al (2017) Thermosensitive hydrogel as an in situ gelling antimicrobial ocular dressing. Mater Sci Eng C 78:203–209
84. Chenite A, Buschmann M, Wang D et al (2001) Rheological characterisation of thermogelling chitosan/glycerol-phosphate solutions. Carbohydr Polym 46:39–47
85. Xu W, Liu K, Li T et al (2019) An in situ hydrogel based on carboxymethyl chitosan and sodium alginate dialdehyde for corneal wound healing after alkali burn. J Biomed Mater Res A 107:742–754
86. Lei L, Li X, Xiong T et al (2018) Covalently cross-linked chitosan hydrogel sheet for topical ophthalmic delivery of levofloxacin. J Biomed Nanotechnol 14:371–378
87. Moreno M, Pow PY, Tabitha TST et al (2017) Modulating release of ranibizumab and aflibercept from thiolated chitosan-based hydrogels for potential treatment of ocular neovascularization. Expert Opin Drug Deliv 14:913–925
88. Cheng Y, Tsai T, Jhan Y et al (2016) Thermosensitive chitosan-based hydrogel as a topical ocular drug delivery system of latanoprost for glaucoma treatment. Carbohydr Polym 144:390–399
89. Song Y, Nagai N, Saijo S et al (2018) In situ formation of injectable chitosan-gelatin hydrogels through double crosslinking for sustained intraocular drug delivery. Mater Sci Eng C 88:1–12

90. El-Feky GS, Zayed GM, Elshaier YAMM, Alsharif FM (2018) Chitosan-gelatin hydrogel crosslinked with oxidized sucrose for the ocular delivery of timolol maleate. J Pharm Sci 107:3098–3104
91. Cheng YH, Ko YC, Chang YF et al (2019) Thermosensitive chitosan-gelatin-based hydrogel containing curcumin-loaded nanoparticles and latanoprost as a dual-drug delivery system for glaucoma treatment. Exp Eye Res 179:179–187
92. Cheng YH, Chang YF, Ko YC, Liu-ling CJ (2020) Sustained release of levofloxacin from thermosensitive chitosan-based hydrogel for the treatment of postoperative endophthalmitis. J Biomed Mater Res B Appl Biomater 108:8–13
93. Irimia T, Muşat GC, Prisada RM et al (2019) Contributions on formulation and preliminary evaluation of ocular colloidal systems of chitosan and poloxamer 407 with bupivacaine hydrochloride. Farmacia 67:702–708
94. Arvind LHS, Ali A (2017) In-situ gel system based on temperature and pH activation for sustained ocular delivery. Indo Am J P Sci 4:558–561
95. Gade SK, Shivshetty N, Sharma N et al (2018) Effect of mucoadhesive polymeric formulation on corneal permeation of fluoroquinolones. J Ocul Pharmacol Ther 34:570–578
96. Das S, Suresh PK (2010) Drug delivery to eye: special reference to nanoparticle. Int J Drug Deliv 2:12–21
97. Giarmoukakis A, Labiris G, Sideroudi H et al (2013) Biodegradable nanoparticles for controlled subconjunctival delivery of latanoprost acid: in vitro and in vivo evaluation. Preliminary results. Exp Eye Res 112:29–36
98. De Campos AM, Diebold Y, Carvalho ELS et al (2004) Chitosan nanoparticles as new ocular drug delivery systems: in vitro stability, in vivo fate, and cellular toxicity. Pharm Res 21:803–810
99. Boddu SHS (2012) Polymeric nanoparticles for ophthalmic drug delivery: an update on research and patenting activity. Recent Pat Nanomed 2:96–112
100. Almeida H, Amaral MH, Lobão P et al (2014) Applications of polymeric and lipid nanoparticles in ophthalmic pharmaceutical formulations: present and future considerations. J Pharm Pharm Sci 17:278–293
101. Lai SK, O'Hanlon DE, Harrold S et al (2007) Rapid transport of large polymeric nanoparticles in fresh undiluted human mucus. Proc Natl Acad Sci U S A 104:1482–1487
102. Lai SK, Wang YY, Hida K et al (2010) Nanoparticles reveal that human cervicovaginal mucus is riddled with pores larger than viruses. Proc Natl Acad Sci U S A 107:598–603
103. Quiñones JP, Peniche H, Peniche C (2018) Chitosan based self-assembled nanoparticles in drug delivery. Polymers (Basel) 10:1–32
104. Bodmeier R, Chen H, Paeratakul O (1989) A novel approach to the oral delivery of micro- or nanoparticles. Pharm Res An Off J Am Assoc Pharm Sci 6:413–417
105. Koukaras EN, Papadimitriou SA, Bikiaris DN, Froudakis GE (2012) Insight on the formation of chitosan nanoparticles through ionotropic gelation with tripolyphosphate. Mol Pharm 9:2856–2862
106. Koukaras EN, Papadimitriou SA, Bikiaris DN, Froudakis GE (2014) Properties and energetics for design and characterization of chitosan nanoparticles used for drug encapsulation. RSC Adv 4:12653–12661
107. Morsi N, Ghorab D, Refai H, Teba H (2017) Nanodispersion-loaded mucoadhesive polymeric inserts for prolonged treatment of post-operative ocular inflammation. J Microencapsul 34:280–292
108. Sabbagh HAK, Abudayeh Z, Abudoleh SM et al (2019) Application of multiple regression analysis in optimization of metronidazole-chitosan nanoparticles. J Polym Res 26:1–4
109. Abdelrahman AA, Salem HF, Khallaf RA, Ali AMA (2015) Modeling, optimization, and in vitro corneal permeation of chitosan-lomefloxacin HCl nanosuspension intended for ophthalmic delivery. J Pharm Innov 10:254–268

110. Abul Kalam M, Khan AA, Khan S et al (2016) Optimizing indomethacin-loaded chitosan nanoparticle size, encapsulation, and release using box-Behnken experimental design. Int J Biol Macromol 87:329–340
111. Manchanda S, Sahoo PK (2017) Topical delivery of acetazolamide by encapsulating in mucoadhesive nanoparticles. Asian J Pharm Sci 12:550–557
112. Ameeduzzafar ISS, Abbas Bukhari SN et al (2018) Formulation and optimization of levofloxacin loaded chitosan nanoparticle for ocular delivery: in-vitro characterization, ocular tolerance and antibacterial activity. Int J Biol Macromol 108:650–659
113. Fathalla ZMA, Khaled KA, Hussein AK et al (2016) Formulation and corneal permeation of ketorolac tromethamine-loaded chitosan nanoparticles. Drug Dev Ind Pharm 42:514–524
114. Sharma S, Sharma A, Singh Sara U, Singh S (2018) Chitosan loaded ketorolac tromethamine nanoparticles for improved ocular delivery in eye inflammation. Indian J Pharm Educ Res 52: S202–S209
115. Barwal I, Kumar R, Dada T, Yadav SC (2019) Effect of ultra-small chitosan nanoparticles doped with brimonidine on the ultra-structure of the trabecular meshwork of glaucoma patients. Microsc Microanal:1–15
116. Chiesa E, Greco A, Riva F et al (2019) Staggered herringbone microfluid device for the manufacturing of chitosan/TPP nanoparticles: systematic optimization and preliminary biological evaluation. Int J Mol Sci 20:1–22
117. Silva NC, Silva S, Sarmento B, Pintado M (2015) Chitosan nanoparticles for daptomycin delivery in ocular treatment of bacterial endophthalmitis. Drug Deliv 22:885–893
118. Dubey V, Mohan P, Dangi JS, Kesavan K (2020) Brinzolamide loaded chitosan-pectin mucoadhesive nanocapsules for management of glaucoma: formulation, characterization and pharmacodynamic study. Int J Biol Macromol 152:1224–1232
119. Chhonker YS, Prasad YD, Chandasana H et al (2015) Amphotericin-B entrapped lecithin/chitosan nanoparticles for prolonged ocular application. Int J Biol Macromol 72:1451–1458
120. Başaran E, Şenel B, Kirimlioğlu GY et al (2015) Ornidazole incorporated chitosan nanoparticles for ocular application. Lat Am J Pharm 34:1180–1188
121. Marques Costa C, Coli Louvisse de Abreu L, Pereira dos Santos E et al (2015) Preparation and evaluation of chitosan submicroparticles containing pilocarpine for glaucoma therapy. Curr Drug Deliv 12:491–503
122. Zhou J, Chen Y, Luo M et al (2019) Dual cross-linked chitosan microspheres formulated with spray-drying technique for the sustained release of levofloxacin. Drug Dev Ind Pharm 45:568–576
123. Fefelova NA, Nurkeeva ZS, Mun GA, Khutoryanskiy VV (2007) Mucoadhesive interactions of amphiphilic cationic copolymers based on [2-(methacryloyloxy)ethyl]trimethylammonium chloride. Int J Pharm 339:25–32
124. Hejjaji EMA, Smith AM, Morris GA (2018) Evaluation of the mucoadhesive properties of chitosan nanoparticles prepared using different chitosan to tripolyphosphate (CS:TPP) ratios. Int J Biol Macromol 120:1610–1617
125. Asasutjarit R, Theerachayanan T, Kewsuwan P et al (2015) Development and evaluation of diclofenac sodium loaded-N-trimethyl chitosan nanoparticles for ophthalmic use. AAPS PharmSciTech 16:1013–1024
126. Shinde UA, Joshi PN, Jain DD, Singh K (2019) Preparation and evaluation of N-trimethyl chitosan nanoparticles of flurbiprofen for ocular delivery. Curr Eye Res 44:575–582
127. Koutroumanis KP, Avgoustakis K, Bikiaris D (2010) Synthesis of cross-linked N-(2-carboxybenzyl)chitosan pH sensitive polyelectrolyte and its use for drug controlled delivery. Carbohydr Polym 82:181–188
128. Jaiswal S, Dutta PK, Kumar S et al (2019) Methyl methacrylate modified chitosan: synthesis, characterization and application in drug and gene delivery. Carbohydr Polym 211:109–117
129. Savin CL, Popa M, Delaite C et al (2019) Chitosan grafted-poly(ethylene glycol) methacrylate nanoparticles as carrier for controlled release of bevacizumab. Mater Sci Eng C 98:843–860

130. Rajawat GS, Shinde UA, Nair HA (2016) Chitosan-N-acetyl cysteine microspheres for ocular delivery of acyclovir: synthesis and in vitro/in vivo evaluation. J Drug Deliv Sci Technol 35:333–342
131. Yu F, Zheng M, Zhang AY, Han Z (2019) A cerium oxide loaded glycol chitosan nano-system for the treatment of dry eye disease. J Control Release 315:40–54
132. Yu A, Shi H, Liu H et al (2020) Mucoadhesive dexamethasone-glycol chitosan nanoparticles for ophthalmic drug delivery. Int J Pharm 575:118943–111870
133. Carvalho SG, Araujo VHS, dos Santos AM et al (2020) Advances and challenges in nanocarriers and nanomedicines for veterinary application. Int J Pharm 580:119214–119229
134. Abd-Elsalam WH, ElKasabgy NA (2019) Mucoadhesive olaminosomes: a novel prolonged release nanocarrier of agomelatine for the treatment of ocular hypertension. Int J Pharm 560:235–245
135. Zhao F, Lu J, Jin X et al (2018) Comparison of response surface methodology and artificial neural network to optimize novel ophthalmic flexible nano-liposomes: characterization, evaluation, in vivo pharmacokinetics and molecular dynamics simulation. Colloids Surf B Biointerfaces 172:288–297
136. Liu R, Wang S, Sun L et al (2016) A novel cationic nanostructured lipid carrier for improvement of ocular bioavailability: design, optimization, in vitro and in vivo evaluation. J Drug Deliv Sci Technol 33:28–36
137. Liu R, Wang S, Fang S et al (2016) Liquid crystalline nanoparticles as an ophthalmic delivery system for tetrandrine: development, characterization, and in vitro and in vivo evaluation. Nanoscale Res Lett 11:1–12
138. Husseini GA, Pitt WG (2008) Micelles and nanoparticles for ultrasonic drug and gene delivery. Adv Drug Deliv Rev 60:1137–1152
139. Gaucher G, Marchessault RH, Leroux JC (2010) Polyester-based micelles and nanoparticles for the parenteral delivery of taxanes. J Control Release 143:2–12
140. Majidinia M, Mirza-Aghazadeh-Attari M, Rahimi M et al (2020) Overcoming multidrug resistance in cancer: recent progress in nanotechnology and new horizons. IUBMB Life 72:855–871
141. Fang X, Cao J, Shen A (2020) Advances in anti-breast cancer drugs and the application of nano-drug delivery systems in breast cancer therapy. J Drug Deliv Sci Technol 57:101662–101678
142. Shi S, Zhang Z, Luo Z et al (2015) Chitosan grafted methoxy poly (ethylene glycol)-poly (ε-caprolactone) nanosuspension for ocular delivery of hydrophobic diclofenac. Nat Publ Gr 5:1–12
143. Xu X, Sun L, Zhou L et al (2020) Functional chitosan oligosaccharide nanomicelles for topical ocular drug delivery of dexamethasone. Carbohydr Polym 227:115356–115367
144. Fabiano A, Piras AM, Guazzelli L et al (2019) Impact of different mucoadhesive polymeric nanoparticles loaded in thermosensitive hydrogels on transcorneal administration of 5-fluorouracil. Pharmaceutics 11:623–638
145. Cui CL, Gan L, Lan XY et al (2019) Development of sustainable carrier in thermosensitive hydrogel based on chitosan/alginate nanoparticles for in situ delivery system. Polym Compos 40:2187–2196
146. Ahdyani R, Novitasari L, Martien R, Danarti R (2019) Formulation and characterization of timolol maleate-loaded nanoparticles gel by ionic gelation method using chitosan and sodium alginate. Int J Appl Pharm 11:48–54
147. Kalam MA (2016) Development of chitosan nanoparticles coated with hyaluronic acid for topical ocular delivery of dexamethasone. Int J Biol Macromol 89:127–136
148. Kalam MA (2016) The potential application of hyaluronic acid coated chitosan nanoparticles in ocular delivery of dexamethasone. Int J Biol Macromol 89:559–568
149. Franca JR, Batista LD, Ribeiro TG et al (2015) Development and validation of a high performance liquid chromatographic method for determination of bimatoprost in chitosan-based ocular inserts. Anal Lett 48:531–540

150. Foureaux G, Franca JR, Nogueira JC et al (2015) Ocular inserts for sustained release of the angiotensin-converting enzyme 2 activator, diminazene aceturate, to treat glaucoma in rats. PLoS One 10:1–18
151. Franca JR, Foureaux G, Fuscaldi LL et al (2019) Chitosan/hydroxyethyl cellulose inserts for sustained-release of dorzolamide for glaucoma treatment: in vitro and in vivo evaluation. Int J Pharm 570:118662–118672
152. Wang L, Jiang YY, Lin N (2020) Promise of latanoprost and timolol loaded combinatorial nanosheet for therapeutic applications in glaucoma. J King Saud Univ Sci 32:1042–1047
153. Silva D, Pinto LFV, Bozukova D et al (2016) Chitosan/alginate based multilayers to control drug release from ophthalmic lens. Colloids Surf B Biointerfaces 147:81–89
154. Hoyo J, Ivanova K, Guaus E, Tzanov T (2019) Multifunctional ZnO NPs-chitosan-gallic acid hybrid nanocoating to overcome contact lenses associated conditions and discomfort. J Colloid Interface Sci 543:114–121
155. Fulgêncio Gde O, Viana FA, Ribeiro RR, Yoshida MI, Faraco AG, Cunha-Júnior AS (2012) New mucoadhesive chitosan film for ophthalmic drug delivery of timolol maleate: in vivo evaluation. J Ocul Pharmacol Ther 28(4):350–358
156. Liab B, Wangbc J, Guib Q, Yang H (2020) Drug-loaded chitosan film prepared via facile solution casting and air-drying of plain water-based chitosan solution for ocular drug delivery. Bioact Mater 5(3):577–583
157. Mahajan HS, Deshmukh SR (2015) Development andevaluation of gel-forming ocular films based on xyloglucan. Carbohydr Polym 122:243–247
158. De Souza JF, Maia KN, Patrício PSDO, Fernandes-Cunha GM, Da Silva MG, Jensen CEDM, Da Silva GR (2016) Ocular inserts based on chitosan and brimonidine tartrate: development, characterization and biocompatibility. J Drug Deliv Sci Technol 32:21–30
159. Franca JR, Foureaux G, Fuscaldi LL, Ribeiro TG, Rodrigues LB, Bravo R et al (2014) Bimatoprost-loaded ocular inserts as sustained release drug delivery systems for glaucoma treatment: *in vitro* and *in vivo* evaluation. PLoS One 9(4):e95461
160. D'Almeida M, Attik N, Amalric J, Brunon C, Renaud F, Abouelleil H et al (2017) Chitosan coating as an antibacterial surface for biomedical applications. PLoS One 12(12):e0189537
161. Chen X, Li X, Zhou Y et al (2012) Chitosan-based thermosensitive hydrogel as a promising ocular drug delivery system: preparation, characterization, and in vivo evaluation. J Biomater Appl 27(4):391–402
162. Shastri DH, Oza PM, Dodiya HD, Shelat PK (2017) Sustained release thiolated chitosan based nanoparticulate in situ gel for ocular delivery of prulifloxacin. Curr Nanomed 7:3
163. Karava A, Lazaridou M, Nanaki S, Michailidou G, Christodoulou E, Kostoglou M, Iatrou H, Bikiaris DN (2020) Chitosan derivatives with mucoadhesive and antimicrobial properties for simultaneous nanoencapsulation and extended ocular release formulations of dexamethasone and chloramphenicol drugs. Pharmaceutics 12:594–614

Adv Polym Sci (2021) 288: 411–446
https://doi.org/10.1007/12_2021_102

Published online: 11 June 2021

Application of Chitosan and Its Derivatives in Transdermal Drug Delivery

Rajitha Panonnummal, Vrinda S. Kumar, R. Jayakumar, and M. Sabitha

Contents

Abstract Chitosan is a natural polymer having much use in drug delivery applications. The polymer is characterized by its biocompatibility, biodegradability, mucoadhesiveness, antimicrobial activity, non-toxicity and positive charge. Insolubility of chitosan in water and other common organic solvents limits its utility. This problem can be solved by chemical modification of its reactive functional groups; NH_2 and –OH to form derivatives with improved solubility. The transdermal drug delivery

Rajitha Panonnummal and Vrinda S. Kumar contributed equally to this work.

R. Panonnummal, V. S. Kumar, and M. Sabitha (✉)
Amrita School of Pharmacy, Amrita Institute of Medical Science and Research Centre, Amrita Vishwa Vidyapeetham, Kochi, India
e-mail: rajitha@aims.amrita.edu; vrindask@aims.amrita.edu; sabitham@aims.amrita.edu

R. Jayakumar
Centre for Nanosciences and Molecular Medicine, Amrita Vishwa Vidyapeetham, Kochi, India
e-mail: rjayakumar@aims.amrita.edu

systems which can control the drug release process have got substantial importance in drug delivery research areas. Chitosan and its derivatives are reported to enhance the skin permeability via a variety of molecular mechanisms. Low molecular weight chitosan is often employed as permeability enhancer in transdermal drug delivery system (TDDS). Various chitosan based transdermal drug delivery systems like composite films, vesicular systems, dendrimers, microspheres, microneedles and other nanocarriers like nanoparticles, nanogel, nanoemulsion, nanolipid carriers with benefits achieved are detailed in this review.

Keywords Chitosan · Microneedles · Microsphere · Nanocarriers · Transdermal drug delivery

1 Introduction

The development of newer drugs is more challenging, time consuming and highly expensive compared to reinventing existing drugs through novel drug delivery systems with an objective to correct the biopharmaceutical issues associated with them. Novel drug delivery approaches are aimed to reduce dose, improve drug bioavailability, reduce toxicity, improve the stability and safety profile of existing drugs along with improved patient compliance. Designing of drug delivery devices is based on various factors like disease condition, nature of drug and its existing issue, route of administration and target site. These drug delivery systems play a critical role in providing a desired therapeutic effect by delivering maximum concentration of drug to the required bio-zone. Researches in the past few decades are focused on the development of delivery systems using polymers, which decides the rate of drug delivery to a particular site of action thereby reducing the side effects. Even though various synthetic polymers with well-established uses in the biomedical field have been reported recently, most of the researchers prefer natural polymers due to their high level safety and biocompatibility. Sodium alginate, chitin, chitosan, starch and pectin are few examples for polymers of natural origin. Among these natural polymers, cellulose derivatives are the most widely accepted for use in drug delivery systems [1–4]. Chitosan is the second most natural polysaccharide, obtained by *deacetylation* of *chitin,* which is derived from *exoskeleton* of crabs, shrimp and cell walls of *fungi* [5–7]. Over the past few decades, novel nano-devices developed successfully using chitosan have been increasingly reported [8–11].

Chitosan is a polymer much preferred for drug delivery applications due to its features like biocompatibility, biodegradability, muco-adhesiveness, antimicrobial activity, non-toxicity, tenability and positive charge [12, 13]. Due to these advantages chitosan has widely been used in various biomedical and pharmaceutical streams, such as drug/gene/vaccine delivery, tissue engineering, wound healing and for manufacturing cosmetic products. The molecular weight of chitosan varies according to the source. The positive charge of chitosan enhances the permeability

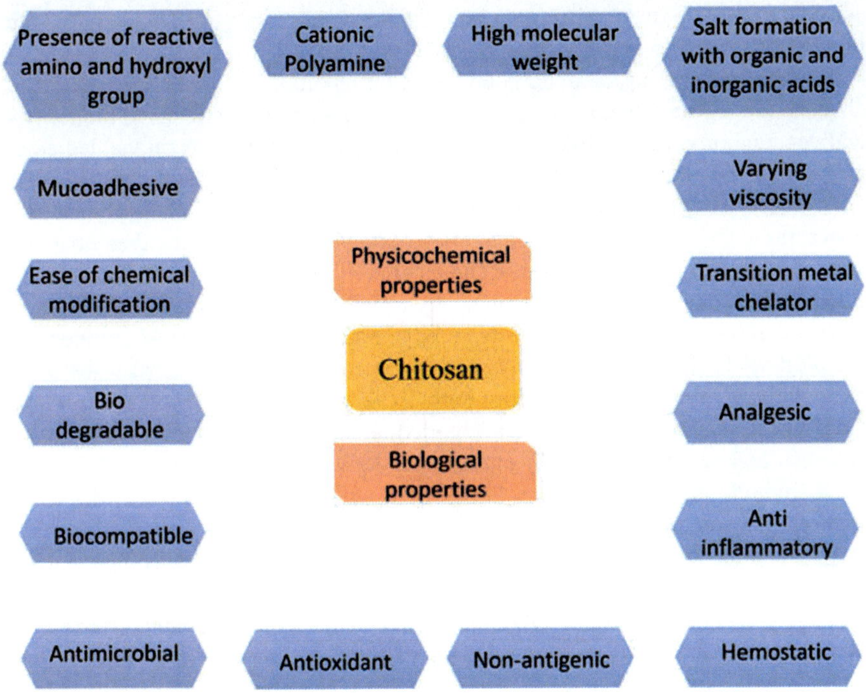

Fig. 1 Various biological and physico-chemical properties of chitosan

by electrostatic interaction with anionic bio-macromolecules such as cells, DNA, etc., which favours the use of chitosan as a permeation enhancer. Also, chitosan has low pH and thus it behaves as a cationic polyelectrolyte which can easily link with negatively charged cross linking agents such as sodium tri-polyphosphate (TPP). Chitosan possesses antibacterial property by altering the bacterial cell wall permeability, resulting in structural leakage and lack of nutrient absorption. Another reason for antibacterial activity of chitosan is the inhibition of transcription due to its binding with DNA [14–19]. Various biological and physico-chemical properties of chitosan are listed in Fig. 1.

1.1 Chitosan Derivatives

Insolubility of chitosan in water and common organic solvents limits its utility. So attempts have been made to overcome the limited aqueous solubility of chitosan by chemical modification of its reactive functional groups; $-NH_2$ and $-OH$ [20, 21]. Modified chitosan derivatives are reported to improve the physico-chemical properties of chitosan by retaining its antimicrobial, antibacterial, anticancer and antiviral properties. Derivatives are often reported to have improved bioactivity with biocompatibility and biodegradability [21].

Acylation is the most common derivatization reported and N-acyl derivatives of chitosan are reported to enhance the solubility of chitosan with retained bioactivity and degradability. Actual solubility of N-acyl derivatives depends on the degree of acylation and is increased on increasing the number of acyl substitution. Whereas O-acylation of chitosan improves the hydrophobic property of the polymer and hence these derivatives are reported to be soluble in lipids and other nonpolar solvents [21]. Alkylation is another chemical modification of chitosan, reported to improve the solubility. However, alkyl group is hydrophobic in nature and hence insertion of too long alkyl groups is reported to improve the hydrophobic property of the polymer [21]. Carboxylated chitosan is reported to have improved water solubility with vast variety of applications in medical, agricultural and bio-chemical field [21]. Quaternization of chitosan is reported to improve its water solubility and antimicrobial activity. The major quaternized derivatives include N,N,N-Trimethyl ethyl ammonium chitosan, N-4, N,N-Dimethyl amino benzyl chloride, N-4-Pyridyl Methyl chitosan, etc. Several N-Aryl derivatives of chitosan are reported to have improved aqueous solubility with enhanced antimicrobial and anti-fungal activity [20]. Graft co-polymerization with hydrophilic polymers such as polyethylene glycol (PEG) is reported to improve the solubility of chitosan without affecting its biocompatibility. In this case the degree of increase in solubility achieved will depend upon the molecular weight of PEG used for grafting. The solubility can be further improved by N-Acetylation of chitosan residues in copolymerized product using acetic un-hydride [22]. Another method to improve the solubility of chitosan is to convert it into carboxymethyl derivative. The resulted carboxymethyl chitosan is reported to have unique physico-chemical properties and biocompatibility; ideal for use in food and cosmetic industry. Grafting of carboxymethyl chitosan with suitable groups like vinyl imidazole or acetyl or benzoyl or chloroacetyl group is reported to enhance the antimicrobial property of chitosan [20]. Thiourea and thiosemicarbazone derivatives of chitosan with improved anti-fungal activity against *S. solani* and *R. solani* are reported [23].

Derivatization of chitosan is also reported to improve its muco-adhesiveness, pH sensitivity and tissue specific targetability [21]. Thiolated chitosan derivatives are reported to have 1.7 times more muco-adhesiveness when compared with pure chitosan. Similarly, quaternary ammonium derivatives are reported to improve the intra-nasal delivery of loaded drugs useful for the treatment of respiratory tract infections [22–27]. pH sensitivity of chitosan can be achieved by grafting chitosan with pH sensitive polymers. The methyl chitosan derivatives are reported to swell at pH 5, whereas carboxymethyl chitosan derivatives exhibit swelling and drug release at intestinal pH and hence suitable to treat conditions like Crohn's disease.

Fatty acid modified quaternary ammonium chitosan is reported to target to liver and is useful to treat various clinical conditions, where a targeted drug delivery to liver is desired [24]. Similarly, glycyrrhizinic acid conjugated chitosan is reported to exhibit liver targeting efficacy, which is proved through the targeted delivery of doxorubicin in a mice model of hepato-cellular carcinoma [25]. Folic acid grafted

Table 1 Derivatives of chitosan with benefits achieved

Sl. No.	Derivatization	Reagents used for derivatization	Benefits achieved	Ref
1	Acylation	Acetic acid and hydrochloric acid Two coupling systems – i.e., 1-ethyl-3-(3-dimethyl-aminopropyl)-1-carbodiimide hydrochloride (EDC) and *N*-hydroxysulfosuccinimide (NHS)	N-acylation improved water solubility O-acylation improved hydrophobic property	[21, 30]
2	Quaternization	N-(3-chloro-2-hydroxypropyl) Trimethyl ammonium chloride	Improved water solubility and antimicrobial activity	[20]
3	Thiolation	Thioglycolic acid Cysteine CS-4-thio-butyl-amdine	Improved mucoadhesiveness	[24]
4	Glycyrrhetinic acid PEG	Glycyrrhetinic acid poly(ethylene glycol)	Liver targeting	[25]
5	Folic acid and PLGA	Poly [D, L-lactide-co-glycolide] (PLGA), folic acid, methanol, dichloromethane, acetonitrile, dimethyl sulfoxide	Lung targeting	[26]
6	Fatty acid and quaternary ammonium	Glycidyltrimethylammonium chloride, N-(3-(di-methyl amino) propyl)-N′-ethyl carbodiimide hydrochloride N-hydroxysuccinimide (NHS), lauric acid, oleic acid	Liver targeting	[22]

PEG chitosan copolymer is reported to exhibit lung targeting behaviour and is reported to enhance the pulmonary delivery of anticancer drug, taxol in a murine model of lung cancer [27–29]. Various derivatives of chitosan with achieved benefits are shown in Table 1.

2 Chitosan Based Drug Delivery Approaches

Chitosan has wide range of applications in drug delivery area. The development of drug delivery systems like nanoparticles, nanogel, nanolipid carriers and nanoemulsion using this natural polymer has been extensively studied and reported. Also, the use of chitosan for various pharmaceutical applications like nasal, ocular, intravenous, oral as well as transdermal drug delivery has been well documented. The safety, cost effectiveness, biocompatibility, biodegradability and non-allergic nature enhance the application of chitosan in research as well as in biomedical field. Despite its unique physico-chemical and biological properties, lower mechanical strength and poor solubility of chitosan have limited its application [31–38]. But these limitations can be overcome by derivatization or by combination with other polymers. The active hydroxyl and amino groups present in chitosan undergo various chemical reactions (hydroxylation, carboxylation, alkylation, acylation and

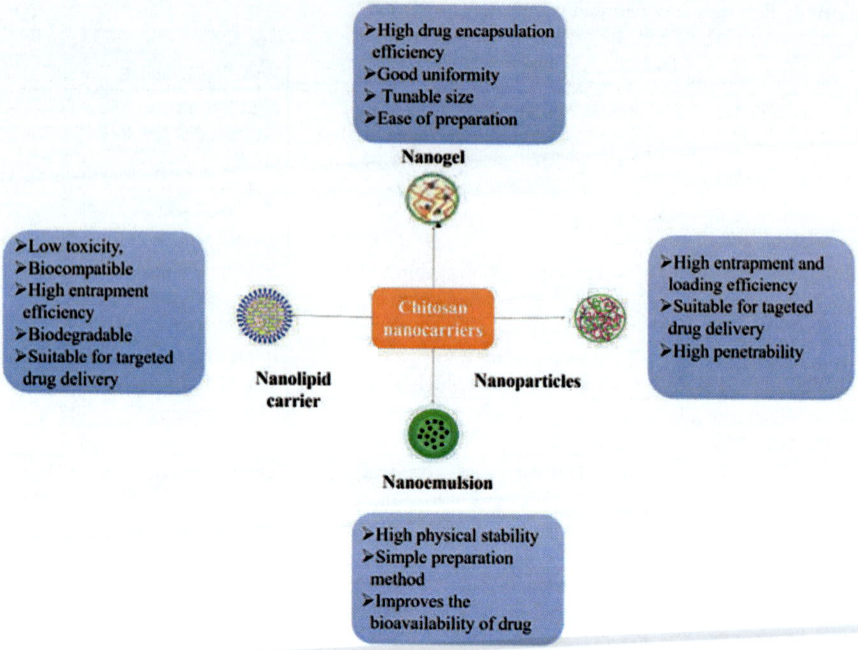

Fig. 2 Advantages of different chitosan based nanocarriers

esterification) and thereby introduce pendant groups to chitosan. This structural modification leads to improved solubility by destroying the crystal structure of chitosan. These chitosan derivatives with improved physico-chemical and biological properties are more suitable as carriers for biomedical and pharmaceutical processes, such as drug/vaccine/gene delivery, tissue engineering, wound healing and manufacture of cosmetic products. In the recent years, pharmaceutical researchers have focused on the development of nanotechnology based systems for effective drug delivery. The use of biopolymers as nanoparticles (NPs) made a tremendous change in the field of medicine with an improved pharmacokinetics and efficacy of existing pharmaceutical compounds. Various studies of polymeric nanocarriers like nanogel, nanoparticles, nanolipid carrier, etc. are reported with an aim to reduce drug dose, enhance drug bioavailability and reduce drug toxicity. There are many studies related to chitosan based nanocarriers developed for different fields of application. The schematic representation of structure of different chitosan based nanocarriers with advantages is depicted in Fig. 2.

3 Chitosan and Chitosan Derivatives in Transdermal Drug Delivery

Transdermal drug delivery systems (TDDS) are one among the most effective drug delivery systems, having tremendous advantages over conventional oral and parenteral routes of drug delivery. The most important advantage is the ability to achieve systemic drug delivery through the application of the formulation into the skin (Fig. 3). It is considered as one of the most effective drug delivery approaches as the systemic delivery of the drugs can be achieved with minimum invasiveness and discomfort. It can avoid the problems of GI degradation, irritability, food and pH interferences associated with the oral drug delivery systems. In addition, adequate bioavailability with minimum fluctuations in plasma drug level and low systemic toxicity can be achieved through the use of TDDS. There is no risk of pain, which is a major problem associated with parenteral dosage forms. If desired rapid termination of action on removal of the system can be achieved while using TDDS. The normal site of application of transdermal patch includes outer arm and buttock and the actual rate of drug permeation depends on the site of application, skin integrity, hydration state, pH, etc. Other drug dependent factors which affect the permeability of TDDS include pKa, log P, solubility and molecular weight of drug [39–41].

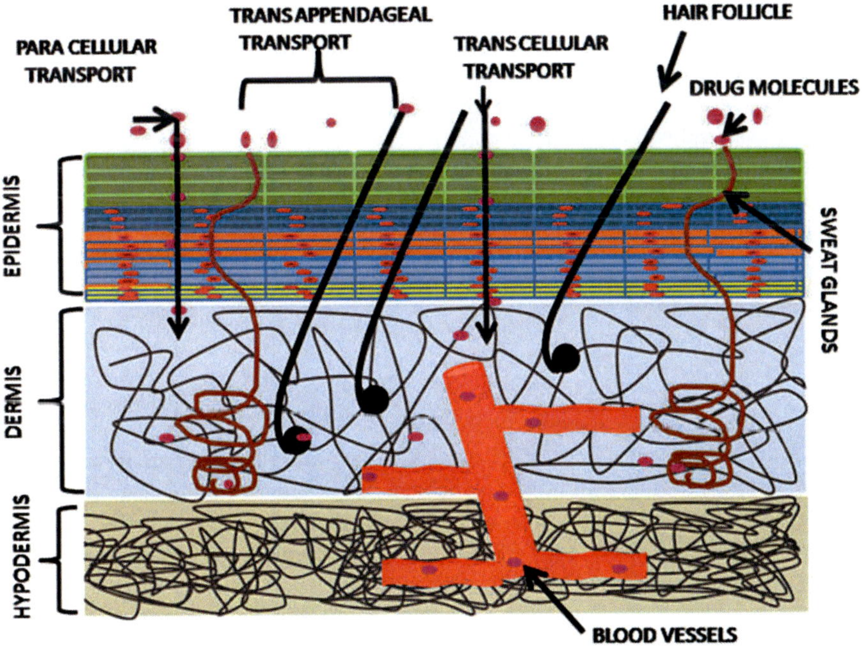

Fig. 3 Schematic representation of the structure of skin with pathways involved in transdermal drug penetration

The biggest challenge in transdermal drug delivery is the development of an effective TDDS with adequate permeability through the stratum corneum layer of skin. The effective permeability can be achieved by the use of suitable polymers having proper hydrophilic-lipophilic balance. Researches in the field of TDDS are mainly focused on the use of natural polysaccharides to achieve the effective stratum corneum permeability. Natural polymers offer lot of advantages including non-toxicity, non-irritability, biocompatibility, biodegradability and are available at low cost. Most of these agents are having a renewable nature and offer the advantage of sustained release property. Among various natural polymers chitosan remains in the foremost position. Chitosan exhibits intrinsic skin adhesiveness and is reported to improve transdermal drug delivery. Its solubility can be modified by derivatization methods and its versatility is useful to conjugate it with moieties that can offer considerable degree of targeting ability to the desired anatomical sites.

Chitosan and its derivatives are reported to enhance the skin permeability via a variety of molecular mechanisms. Low molecular weight chitosan is often employed as permeability enhancer in TDDS. The enhancement in permeability achieved by chitosan significantly depends on pH at the site of application and is reported to be maximum at pH 7.4. It is reported that at this pH its amino group remains as in an optimal level of protonated and unionized form, which allows its interaction with carboxyl group of glycoprotein and keratin of cell membranes. At lower pH (<6) most of the amino group get protonated and a proper interaction with –COOH group is not achieved. But at higher pH (>8) the achieved interaction is too weak to contribute to enhancement in permeation [38]. The carboxymethyl chitosan is reported to enhance the stratum corneum (SC) hydration by allowing the water molecules to enter into cell layers because of its characteristic 3D structure. The hydrated state of SC is retained for long period, causes enhancement in cell membrane fluidity and trans-membrane permeability. In addition, chitosan and its derivatives are reported to alter the secondary structure of keratin protein and cause the loosening of well-organized cell layers and improve the permeability. The gel prepared by using chitosan is reported to increase the skin contact time, which may also contribute to the enhanced permeability achieved by the chitosan [40].

3.1 *Chitosan Based Transdermal Patches*

Transdermal patches are defined as medicated patches capable to deliver specific dose of drug into the systemic circulation on application into the skin. There are different types of transdermal patches; as those having a single layer of drug or multiple layers of drug or drug within a reservoir compartment or drug within a polymeric matrix. The simplest one is single-layered patch, in which the medicine is incorporated directly into the adhesive layer, which itself controls the drug release process. In a multi-layered patch there are multiple polymeric layers loaded with drug molecules, incorporated within the adhesive layer, one of the layers may be for immediate release of drug and the other layers can act as drug reservoir compartment, through which controlled drug release occurs. In reservoir type transdermal patch system, there is a drug reservoir compartment covered with a porous membrane structure, through which drug can be released in a controlled manner. The system has a liner to protect the product during storage and has to be removed before use. Second is the drug reservoir compartment which contains the drug often with excipients like permeability enhancers. Then there may be a membranous structure to control the drug release from the reservoir compartment and a backing film to provide environment protection. Adhesives are used to ensure proper skin adhesiveness of the system. Chitosan based transdermal patches are reported for systemic delivery of variety of drugs and are discussed in the following sections [39–42].

In a study anti-diabetic drug metformin loaded matrix type transdermal patch of chitosan is prepared by solvent evaporation technique. The prepared patch is observed to be having good compatibility, optimum flatness, tensile strength, thickness and significant folding endurance. The patch exhibited sustained release property for 20 h. The prepared patch is expected to be a good option for the transdermal delivery of the selected drug, as it can deliver the drug in a sustained manner and is expected to avoid the rapid first pass effect and GI toxic effects associated with the oral use of the drug. However, the study is limited by in vitro data only and further studies are essential to get more conclusive remarks in this aspect [42]. In another study transdermal patch of chitosan loaded with insulin is developed for use in DM. In the study initially insulin loaded chitosan nanoparticles are prepared by ionotropic gelation technique, which is then loaded into monolithic matrix type transdermal patch composed of HPMC and PVP K300 using solvent casting method. The patch exhibited smooth appearance, with good swelling index (45–50), thickness and higher folding endurance. The system also exhibited good in vitro permeability with 80% of loaded insulin being permeated within 10 h, when tested using goat skin. The patch is seemed to be useful to overcome the discomfort associated with the conventional S.C. use of insulin and is expected to control blood sugar level in a more defined way [43]. The advantage of limited inter-individual variation offered by TDDS is utilized for the transdermal delivery of gliclazide, a second generation sulfonylurea for the treatment of DM. The drug is reported to have poor solubility, bioavailability and exhibited marked inter-individual variability in response. In a study, a transdermal patch of the drug is formulated using sulfoxy amino chitosan, so as to attain a sustained release pattern. The formulation exhibited higher folding endurance (351–359) value, sustained drug release property and adequate skin permeability. The use of sulfoxy amino chitosan controls the drug release and is expected to reduce the inter-individual variability in drug response; as it is reported that if the rate of drug release is lower than the rate of drug diffusion across stratum corneum layer; the chances of inter individual variability is less [44]. However, further studies are essential to confirm it.

In a study lornoxicam loaded chitosan transdermal patch is prepared with an aim to improve the patient compliance of the selected drug. Lornoxicam is a potent NSAID, which is in use for rheumatoid arthritis (RA) treatment, but it needs frequent dosing, due to its short half-life and causes severe gastric irritations and ulceration as

Table 2 List of various drugs formulated as chitosan transdermal patches with benefits achieved

Sl. No.	Loaded drugs and intended diseases	Polymers used	Method of preparation	Achieved benefits	Ref
1	Metformin – diabetes	Chitosan, PVA, HPMC, D-butyl phthalate	Solvent evaporation	Sustained drug release	[42]
2	Insulin – diabetes	Chitosan, HPMC, PEG, PVP	Ionotropic gelation followed by solvent casting	Improved skin permeability and transdermal drug delivery	[43]
3	Gliclazide – diabetes	Sulfoxy amino chitosan, HPMC	Solvent casting	Sustained release and optimum skin permeability	[44]
4	Lornoxicam	Chitosan, propylene glycol, tween 20	Solvent casting	Controlled release and permeability enhancement	[45]

ADRs. A matrix type of transdermal patch loaded with lornoxicam is prepared by solvent casting method using chitosan. Ideal property of chitosan resulted in a patch with higher folding endurance, weight uniformity, optimum thickness, moisture content, etc. The use of tween 20 as permeability enhancer ensures adequate permeability and patch exhibited controlled release pattern with 70% of drug getting released at 12 h time. Unfortunately, the studies are limited by in vitro data and need further studies to prove its proposed hypotheses [45]. Chitosan transdermal patches developed for various drugs with benefits achieved are listed in Table 2.

3.2 *Chitosan Based Composite Films*

The development of transdermal drug delivery systems which can control the drug release process has got substantial importance in drug delivery research areas. Composite films are one among such systems, can act as a reservoir for loaded drug molecules, control its release and thereby improve its therapeutic performance. Ideally composite films are biodegradable polymeric films incorporated with drug molecules. The basic composition of composite films includes the drug molecules, the polymer which forms the film and the solvent. Ideally the drug and the polymer are dissolved or dispersed in a volatile solvent. Upon application into the skin solvent evaporates, leaving a polymeric film with loaded drug molecules over the skin surface [46]. Usually the polymers used for the fabrication of composite films are biodegradable in nature and include gelatine, lignin, cellulose, starch, chitosan, etc. [47].

Among various polymeric materials, chitosan is considered as excellent film forming material. Since chitosan is derived from chitin by deacetylation, its amide groups are deacetylated to ammonium ions, which provide good film forming property to chitosan and improve its water solubility. However, its solubility further

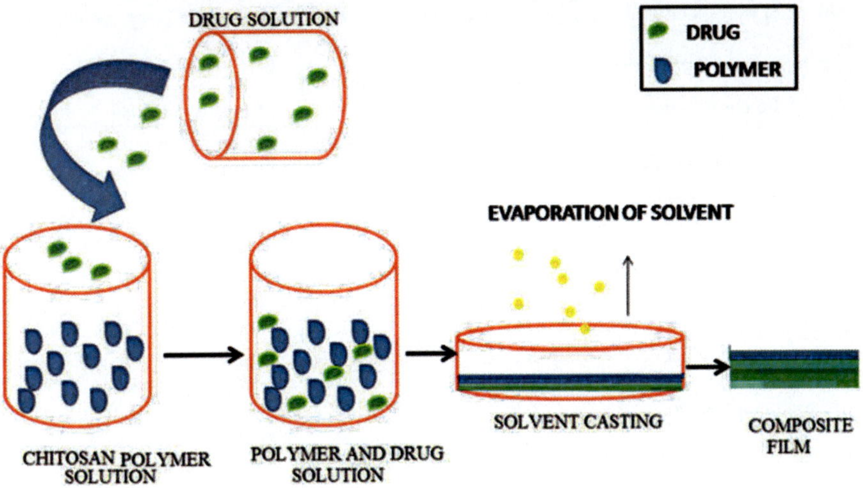

Fig. 4 Preparation of chitosan composite film by solvent casting method

depends on the molecular weight used, pH, the acid used for solubilization, etc. [48]. Chitosan films are characterized by flexibility, durability, biodegradability, biocompatibility, optimum strength and toughness, which closely matches with other commercially available films. In addition, chitosan films are reported to have anti-fungal and antimicrobial property [49–53]. Moreover, the desired level of water and oxygen permeability of chitosan films can be achieved by selecting chitosan having suitable molecular weight, solvent system, etc. [54, 55]. In addition, chitosan can be combined with other polymeric materials such as gelatine, lignin, keratin, starch and cellulose, so as to obtain composite films with desired physico-chemical properties. The most common method used for the fabrication of chitosan composite film is solvent casting/evaporation technique (Fig. 4) [54–59]. Chitosan films are also found application in food industry such as to form a composite film and thereby act as a natural biodegradable packing material for preserving the food stuffs from physical, chemical and biological agents. Here the use of synthetic polymeric materials is reported to often release toxic gases and moreover these are non-biodegradable and are associated with high cost [47, 60, 61]. Films made up of chitosan are also found use in purification of water from heavy metals. In a study activated carbon obtained from palm oil is used to increase the mechanical property of chitosan. The films constructed by the modified chitosan exhibit greater cadmium absorption power from drinking water [62].

Chitosan based composite films have been tried for transdermal delivery of a variety of drug molecules (Table 3) and few are discussed in the following section of the article. Chitosan composite film loaded antiemetic drug ondansetron is developed for its transdermal delivery. Ondansetron is an agent used for the treatment of vomiting associated with cancer chemotherapy. The drug is reported to have poor oral bioavailability and i.v./i.m. injection is reported to cause pain, redness and

Table 3 List of chitosan based composite films for transdermal drug delivery with benefits achieved

Sl. No.	Film composition	Loaded moieties	Therapeutic activity	Benefits achieved	Ref
1	Chitosan polyvinylpyrrolidone, montmorillonite	Curcumin	Antioxidant, anti-inflammatory, anticancer	Optimum mechanical strength and other physico-chemical properties of film, controlled release property and enhanced transdermal penetration	[63]
2	Chitosan, gelatine, glycerol, glutaraldehyde	Silver nanoparticles	Antimicrobial	Optimum mechanical strength and other physico-chemical properties of film, enhanced antimicrobial activity, controlled release, enhanced skin penetration, better cellular uptake	[64]
3	Chitosan, lactic acid, oleic acid, propylene glycol	Ketoprofen	Analgesic, anti-inflammatory	Significant enhancement in skin penetration with in vivo analgesic and anti-inflammatory property	[65]
4	Chitosan, polypyrrole, montmorillonite	Doxorubicin	Anticancer	Optimum mechanical strength and other physico-chemical properties, controlled drug release	[66]
5	Chitosan, HPMC, propionaldehyde	Etoricoxib	Analgesic, anti-inflammatory	Optimum physico-chemical properties, controlled release, enhanced skin penetration	[67]
6	Chitosan, gelatine, ethylene glycol	Ciprofloxacin	Wound healing	Optimum physico-chemical properties, enhanced antimicrobial activity, significant wound healing activity when tested in vivo	[68]
7	Chitosan, poly vinyl alcohol, isopropyl alcohol	Ampicillin	Antimicrobial	Optimum mechanical, physico-chemical properties, controlled drug release	[69]
8	Chitosan, cholesterol, phosphatidylcholine	Griseofulvin loaded liposomes	Anti-fungal	Optimum physico-chemical properties, controlled drug	[70]

(continued)

Table 3 (continued)

Sl. No.	Film composition	Loaded moieties	Therapeutic activity	Benefits achieved	Ref
				release, enhanced skin penetration and improved anti-fungal activity	
9	Chitosan, acetic acid	*Aloe vera* and *Calendula officinalis*	Wound healing	Ideal swelling and optimum physico-chemical properties, enhanced antimicrobial activity	[71]
10	Chitosan, propylene glycol, glutaraldehyde, citric acid	Metronidazole and levofloxacin	Antimicrobial	Optimum physico-chemical properties, sustained drug release, enhanced antibacterial efficacy (in vitro) and significant improvement in periodontitis when tested clinically	[72]
11	Chitosan, polyvinyl alcohol	Nitrofurazone	Antimicrobial	Optimum film properties, controlled drug release with enhanced in vitro antibacterial efficacy	[73]
12	Chitosan, lactic acid, glutaraldehyde, propylene glycol	Mupirocin	Antimicrobial wound healing	Optimum film and ideal swelling property, improved in vitro bio-adhesiveness and enhanced in vivo wound healing property	[74]

burning sensation at the injection site. So in order to make it more effective and patient compatible, a composite film of the drug is constructed using chitosan as polymer. The film exhibited optimum thickness, uniformity in weight, drug content and is bio-adhesive in nature. Excellent skin penetrability, contributed by the use of eucalyptol as a permeability enhancer, suggested the possibility for use of the system for effective transdermal delivery of the selected drug [75].

Propranolol hydrochloride is a beta adrenergic receptor blocker drug, which is in clinical use for various conditions such as hypertension, migraine, anxiety, etc. The drug has got extensive and highly variable first pass effect with systemic availability of only 15–23%. In addition, the reported half-life of drug is about 4 h and it requires multiple dosing. So attempts are made to make the drug therapy more effective and to improve the patient compliance through transdermal route. In a study propranolol loaded chitosan acetate composite film is developed by solvent casting method. Chitosan forms Schiff's base with acetaldehyde resulting in increase in space within the polymeric network, which helps to trap the loaded drug molecules in these spaces and hence contribute to superior drug loading efficacy. Further the film exhibited sustained drug release property with a maximum release of 80% of drug over 24 h, suggesting the possibility for effective transdermal drug delivery [76].

3.3 Chitosan Based Vesicular Drug Delivery Systems

Emergence of vesicular drug delivery system has created revolution in the field of transdermal drug delivery. Vesicular delivery system contains vesicles made up of amphiphilic molecules, usually surrounded by an aqueous layer. These systems are capable of delivering both hydrophilic and lipophilic drug molecules. Hydrophobic drugs can be incorporated into lipid bilayer, whereas hydrophilic drugs can be inserted into central aqueous core. Vesicular drug delivery systems comprise vast variety of delivery carriers including liposomes, niosomes, ethosomes, transferosomes, pharmacosomes, enzymosomes, virosomes, aquasomes, sphingosomes, etc. [77–79]. Out of these a few having importance in transdermal drug delivery have been discussed in the following section of the article.

Liposomes are considered as the most prominent drug delivery carrier among vesicular delivery approaches. As the name indicates liposomes are having vesicular body made up of lipids. It may be composed of either a single lipid or may include a combination of many lipids. Since the major composition of liposomes is lipids, it can interact well with lipid bilayer of skin cells and hence exhibit greater transdermal permeability. In addition, as these are spherical particles having size in the range of 20 nm to 10 μm, the particles can squeeze in between the cell layers, contributing to paracellular transport. In order to improve the transdermal delivery through liposomes, its surface may often be coated with polymeric materials, so as to favour the interaction with skin cells. Among these polymeric materials chitosan has important role, as it is having a positive charge, it can interact well with negatively charged skin cells and thereby improve the skin penetrability. This type of interaction is utilized for delivery of drugs like quercetin, resveratrol, etc. [77–79]. Niosomes come under the class of vesicular delivery approach, which resembles closely with liposomes, but the difference here is that the lipid bilayer is composed of mainly cholesterol and non-ionic surfactants. The non-ionic surfactants mainly include diacyl poly glycol, span 20, etc., which is stabilized by steroids usually cholesterol. Niosomes are capable of enhancing the transdermal delivery of loaded drugs by interacting with lipid bilayer of keratinocytes [80–86]. Another vesicular system with enhanced transdermal permeability is ethosomes which contain ethanol as one of the major ingredients. The other components include phospholipid, steroids like cholesterol and sometimes a permeability enhancer like polygalol. High concentration of ethanol makes the system unique with enhanced transdermal penetrability [87–90]. Transferosomes are newer lipid based vesicular system that can cross the intact skin layers instinctively because of its extremely flexible and ultra-deformable nature. The system has got unique ability to transfer even high molecular weight drug molecules through the inter-cellular transport process across the skin. The

Table 4 Chitosan gel based vesicular systems developed for transdermal drug delivery

Sl. No.	Composition	Drugs loaded	Vesicular system	Ref
1	Phosphatidylcholine, cholesterol, dihexadecyl phosphate (DHP), triton X, chitosan	Quercetin	Liposome	[77]
2	Chitosan, soybean phospholipids, tween-80 ethanol	Baicalein	Liposome	[78]
3	l-a-phosphatidylcholine, chitosan	Resveratrol	Liposome	[79]
4	Lecithin, cholesterol, chitosan	Substance P	Liposome	[95]
5	Span 40, Span 60, cholesterol, chitosan	Urea	Niosome	[81]
6	Span 40, Span 60, cholesterol, chitosan	Celecoxib	Niosome	[82]
7	Span 60, cholesterol, chitosan	Methotrexate	Niosome	[83, 84]
8	Span 20, span 60, tween 20, tween 40, tween 60, tween 80, cholesterol, chitosan	Moxifloxacin	Niosome	[86]
9	Soya lecithin, chitosan Chremophor-A25, sodium tri-polyphosphate Carbopol-934	Ferrous chlorophyllin	Ethosome	[89]
10	Phospholipon G, span 60, span 80 Chitosan, tween 20, tween 80, sodium lauryl sulphate	Rifampicin	Transferosome	[91]

system is composed of phospholipids like phosphatidylcholine, capable to assemble as lipid bilayer in an aqueous environment. In order to enhance its flexibility and to improve the permeability through lipid bilayer, edge activators are often incorporated in its composition. Edge activators are non-ionic surfactants capable to cause destabilization of lipid bilayer, enhance the fluidity and elasticity and hence contribute to increased transdermal penetrability [91–94]. Most of these vesicular systems like niosomes, ethosomes, transferosomes, etc. are often needed to be incorporated into a proper gel matrix so as to develop a final formulation suitable for application into skin for effective transdermal delivery. A number of polymers like carbopol, chitosan, gelatin, HPMC, etc. are in use for this purpose. Among these chitosan has got significant importance as it can provide long residence time, bio-adhesiveness and optimum rheological properties, control the drug release process and improve the transdermal penetrability. Chitosan gel based vesicular systems developed for transdermal drug delivery are listed in Table 4.

3.4 *Chitosan Based Microspheres*

Microspheres are micro-particles having spherical shape, capable to act as carriers for drug molecules. Usually it is composed of biodegradable polymers. Microspheres can be defined as small spherical particles having size in between 1 and 100 μm. Due to greater surface area achieved with smaller particle size, better rate of drug diffusion can be achieved with this system. It is possible to divide the

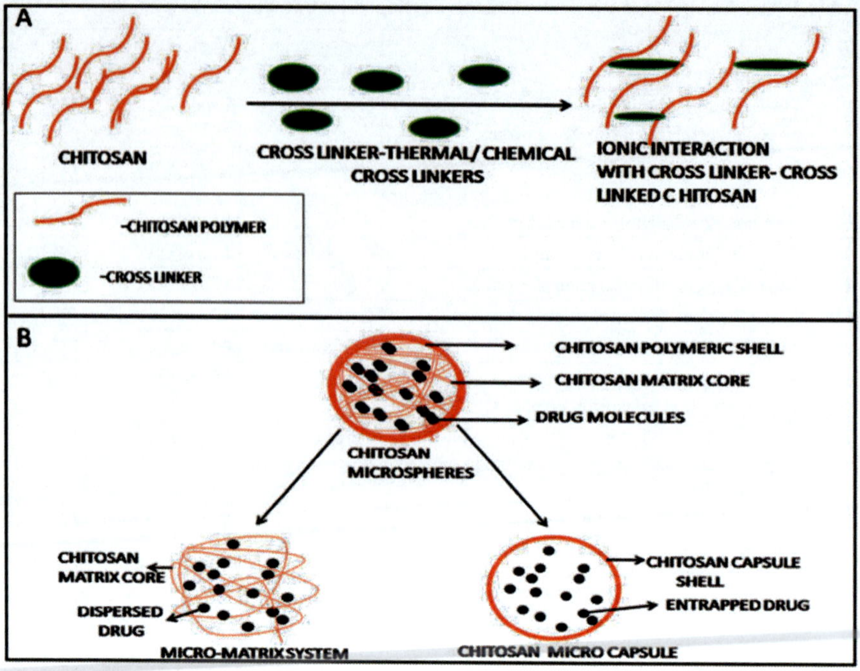

Fig. 5 Schematic representation of chitosan cross linking achieved using cross linkers for microsphere preparation (**a**) and types of chitosan microspheres (**b**)

microspheres into two: as microcapsules if the entrapped material is surrounded by a distinct capsule wall and as micromatrix if the entrapped material is dispersed in a matrix system (Fig. 5b). Usually the shell of microspheres is made up of polymers and is filled with drug materials. The release of loaded drug particles occurs when the shell is degraded. Bio-adhesive microspheres are characterized by the prolonged mucosal residence time and are suitable for intra-nasal drug delivery applications. Floating microspheres are those having bulk density lower than that of the gastrointestinal fluids so as to float in the GI tract. It has been used for NSAIDs. Polymeric microspheres are composed of biocompatible polymers and have been tried for vaccine and other macromolecule delivery. Radioactive microspheres contain radioactive isotopes and are suitable for diagnostic applications. Magnetic microspheres are those which contain supramolecular particles composed of material like magnetite with a desired magnetic moment, which can circulate through capillaries and can deliver the drug molecules to the specific site in a magnetically controlled way. That is, placement of a magnet with sufficient field strength externally at the target site helps to concentrate the magnetic microspheres internally to this site and offers a targeted drug delivery with a reduction in dose [96–98].

Among polymeric microspheres, chitosan microspheres have got considerable attention because of its ability to improve the systemic availability of the loaded drugs, to provide optimum drug release at the site of action, suitability for oral and parenteral administration, ability to achieve controlled and sustained drug delivery and thereby minimizing the frequency of dosing. The physical and chemical properties of chitosan enable the formation of microspheres having unique properties. Chitosan microspheres can be prepared by covalent bonding of its side chain with functional groups of cross linking agents such as glutaraldehyde, poly ethylene glycol, etc. (Fig. 5a). By adjusting the concentration of the polymer and the cross linking agents it is possible to achieve microspheres of desired properties [96–100]. Various methods of preparation of chitosan microspheres include emulsion cross linking, thermal cross linking, ionotropic gelation, co-acervation precipitation, emulsion-droplet coalescence, spray drying, reverse micellar and sieving methods [101, 102].

Cefixime is an antibiotic belonging to the class of third generation cephalosporins having poor oral bioavailability. In order to improve the therapeutic effectiveness of the drug chitosan microspheres composed of inter-penetrating network (IPN) of acrylamide grafted PEG polymer loaded with cefixime is developed. Hydrophilic nature of the grafted PEG helps to form chitosan microspheres through cross linking and IPN provides sustained release property to the system. However further studies are essential to reveal the suitability of the system for its therapeutic effectiveness on topical application [97]. An anti-fungal drug itraconazole loaded chitosan microspheres is prepared by solvent evaporation technique. The cross linking provided by chitosan controls the drug release in a sustained manner, suggested the possibility to use through transdermal route in comparison with the oral, which is limited due to side effects. However further studies are needed to evaluate and confirm the actual anti-fungal potential of the system [98–100].

3.5 *Chitosan Based Microneedles*

Among various transdermal methods of drug delivery, microneedles based drug delivery systems have made revolutionary progress because of the high drug delivery efficacy with safety and painless nature of the system. Microneedles are needles having approximately 1,000 μm length and can assemble at one side of the supporting patch. When it pierces through the stratum corneum in a non-invasive way, it creates micro-channels through which drugs and other fluids can pass to reach into systemic circulation. Due to its micron length the maximum distance that it can pass is up to epidermal layers, which lack nociceptive nerves, making it a painless drug delivery approach. By adjusting the composition of microneedles it is possible to adjust the release profile of loaded drugs. It is possible to classify microneedles as inorganic (silica or any other inorganic materials), metallic (stainless steel or titanium) or polymeric (any polymers like chitosan) based on the materials of construction. Brittleness and tendency to fracture on application to skin limited the use of inorganic microneedles, whereas the uses of metallic microneedles are limited by its low drug loading capacity. Polymeric microneedles are of interest as they have lot of advantages like ease to fabricate, high drug loading

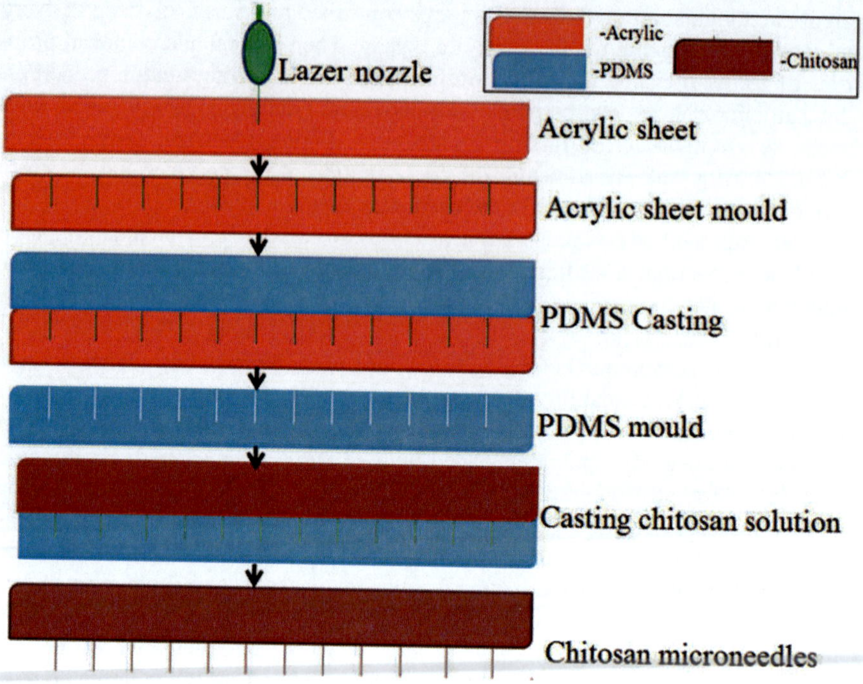

Fig. 6 Preparation of chitosan microneedles by moulding process

capacity, stimuli responsive nature and availability of vast variety of polymeric materials. Among these, chitosan has important role as a polymeric material to fabricate the microneedles of desired properties [101–105]. Biocompatibility, biodegradability, versatility, easy availability and inexpensive nature make chitosan along with its derivatives as excellent polymeric material of choice for the construction of microneedles based delivery system for various drugs for transdermal application [106]. Moulding method is the most common method used for the fabrication of chitosan microneedles (Fig. 6). The method includes the initial engraving of microneedles pattern on acrylic sheets by using CO_2 laser cutter, so that now the acrylic sheet can be called as acrylic mould. Then the surface of the acrylic mould need to be activated by treating with polydimethylsiloxane (PDMS) to form a mould coated with PDMS (PDMS mould). Chitosan solution is then coated over the PDMS mould, dried the system at 40°C so as to facilitate the deposition of chitosan as a shell over the PDMS. Cool down the system at −4°C for 30 min to shrink PDMS, which further hardens the chitosan, so that it is easily detachable from PDMS, resulted in chitosan microneedles formation [107].

Limited SC permeability, often limited the use of conventional TDDS patches, can be easily solved by the use of microneedles, which is reported to improve the transdermal permeation in non-invasive way as shown in Fig. 7.

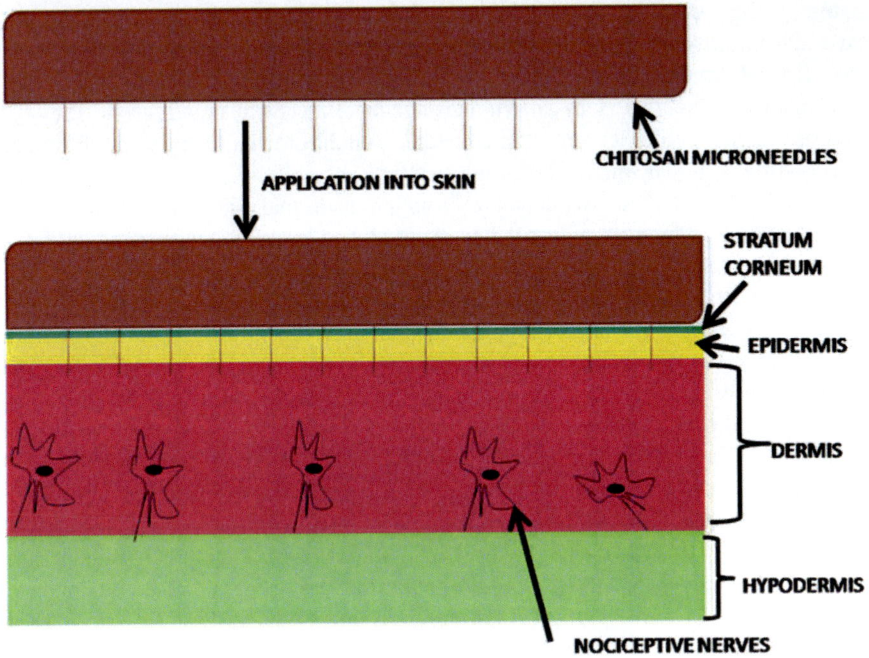

Fig. 7 Schematic representation of the application of chitosan microneedles into the skin and subsequent skin penetration

In a study microneedle based transdermal patch of thiolated chitosan, loaded with immunosuppressant drug tacrolimus is formulated and evaluated. Tacrolimus is a potent immunosuppressant drug which is in use for the treatment of rejection reactions associated with organ transplantations. The drug exhibits poor aqueous solubility and less (30%) oral bioavailability. It needs three to four times dosing and the drug is also a candidate for P-glycoprotein(P-gp) efflux pumps. A microneedle based system is developed with an aim to improve its systemic availability on transdermal application using thiolated chitosan as polymeric material. In thiolated chitosan, thiol group is attached to the primary amino group of chitosan using an amide linkage and exhibits better muco-adhesiveness, paracellular transport and gelling property when compared to chitosan. The fabrication of the system is done by moulding method and the use of thiolated chitosan provides adequate mechanical strength and flexibility to the microneedles. Ex vivo permeation studies revealed that 81% of the drug is transported across full thickness rat skin when compared with 40% achieved with the use of commercially available ointment formulation of the drug. The system could deliver the drug in a controlled manner for 48 h and is expected to improve the patient compliance and adherence when compared with conventional ointment formulation of the drug. In vivo biocompatibility studies done on rats indicate some degree of erythema within 1 h of application of the formulation but get normalized within 6 h, which indicates the safety and nontoxic nature of the system. Here the thiolated chitosan provides effective skin adhesiveness and improves paracellular transport by opening the tight junctions in between the cells. Microneedle can penetrate the SC without any pain and deliver the drug systemically. Thiolated chitosan provides adequate mechanical and tensile strength and percentage flexibility to the microneedle suitable for its intended applications [41]. Chitosan based microneedles loaded with goserelin were developed with an aim to achieve sustained drug delivery through transdermal route for the treatment of prostate cancer. Chitosan microneedles are produced by moulding method. Initially goserelin loaded chitosan hydrogel was prepared and it was then casted over PDMS mould. In vitro skin permeation studies using pig skin revealed the capability of the system to penetrate through epidermal and dermal layers. The holes created by microneedles during piecing are visualized in vivo in mice and that get healed within 7 days. The effectiveness of the system is further demonstrated by the changes in luteinizing hormone and testosterone levels in a manner suitable for use in the treatment of prostate cancer [108]. In another study chitosan based microneedle system is developed for the transdermal delivery of macromolecules by micromoulding process. Bovine serum albumin (BSA) is selected as a model drug. Initially drug loaded chitosan hydrogel is prepared and it is then casted into polydimethylsiloxane (PDMS) mould. The mechanical strength offered by chitosan provides considerable degree of insertion capability to the system. In vitro release studies revealed a sustained drug release for 8 days with 95% of drug getting released within this time period. Good transdermal penetration is exhibited when the system is subjected to skin penetration study using pork skin. Through in vivo studies conducted in rats, it was possible to visualize the puncture marks of around 200 μ size due to microneedle application and the depth of penetration was observed to be up to the superficial layer of dermis. It is also demonstrated that the partial insertion of the system is useful to deliver the drug into specific skin targets like Langerhans cells on epidermis or dendritic cells in dermis, suggesting the possibility for vaccine delivery. It is further demonstrated that the loaded drug remains stable in microneedles, which is very important, as sometimes the drug delivery system may cause conformational changes in macromolecules like proteins resulted in loss of biological activity [109]. In another study chitosan microneedle by mould technique is developed for BSA delivery. The formulation exhibited good mechanical and tensile strength aided by the use of chitosan at appropriate concentration. In vitro release studies revealed that 95% of BSA is released within 2 days when tested using rat skin. The formulation is found to be biocompatible when tested in vivo as revealed by acute dermal toxicity studies conducted on rats [110].

3.6 Chitosan Based Dendrimers

Dendrimers are polymeric drug delivery systems composed of highly branched, mono-dispersed macromolecules. These are denoted as smart novel drug carriers because of the properties unique to them. Dendrimers have well-defined stable molecular structure with sufficient internal cavities inside it. This provides high drug loading efficacy and it is possible to load both hydrophilic and lipophilic drugs into dendrimeric carriers. The common materials used for the construction of dendrimeric drug carriers include poly(amidoamine) (PAMAM), polyesters, poly propylene and poly (L-Lysine). Among these the most attractive material is PAMAM, because of the high drug loading efficacy achieved with it due to the presence of abundance of amino and carboxylic functional groups exposed externally to its dendrimeric structure. These groups can act as the sites for further functionalizations so as to achieve specific objectives like targetability, specificity, etc.

Even though the system is blessed with unique advantages, the nonspecific nature and strong interactions of its positively charged amine group with negatively charged cells causes cell lyses and accompanying toxic reactions. Through interaction with lipid by layer of cells, dendrimers can increase the cell membrane permeability with loss of integrity of cells, leakage of cellular components, cell lysis and haemolysis. The number of functionalization and modifications such as alkylation, pegylation, glycosylation has been tried with an aim to improve its biocompatibility and reduce toxicities. Some attempts have been made by using chitosan with PAMAM dendrimer to reduce its toxicity, improve its efficacy and targetability [111–115].

A novel technology to reduce the toxicity of PMAM dendrimer is demonstrated using zwitterionic chitosan (ZIC) developed by Karen C Liu et al. [116]. In the study they conjugated succinic anhydride with primary amine group of chitosan, resulting in ZIC, which exhibits negative charge at high pH and positive charge at low pH. At pH 7.4 ZIC is negatively charged and is conjugated with PAMAM dendrimer, which helps to mask the cationic charge of PAMAM and thus will not interact with normal cells. But when the system is exposed to acidic tumour environment, its charge reverses, losses the protection offered by ZIC and thus PAMAM can interact with tumour cells, causing toxicity. So the system offers a way of passive targeting to tumour cells, as at the acidic pH at tumour site, it becomes cationic in charge and interacts with tumour cells causing cell lysis [116]. In another study methotrexate loaded chitosan nanoparticles, conjugated with PAMAM dendrimer is prepared and tested for its in vitro tumour lytic property using tumour cell lines. The system exhibited improved anti-tumour efficacy when compared with free drug solution [117]. However, both of these systems have not been tried for transdermal applications.

A PAMAM dendrimer based transdermal patch of NSAID meloxicam is prepared by solvent casting method using chitosan, HPMC and PVA as polymers with an aim to overcome the GI toxicity associated with the oral use of the drug (Fig. 8).

The backing membrane is prepared using 2% polyvinyl alcohol (PVA). Drug reservoir compartment is composed of the drug meloxicam dissolved in methanol, HPMC dibutyl phthalate dispersion and PAMAM dendrimer. PMAM dendrimer is added with an aim to achieve the desired level of drug release and penetration. The rate controlling membrane is composed of chitosan. The use of chitosan in combination with HPMC and PVA resulted in a patch with smooth appearance, good

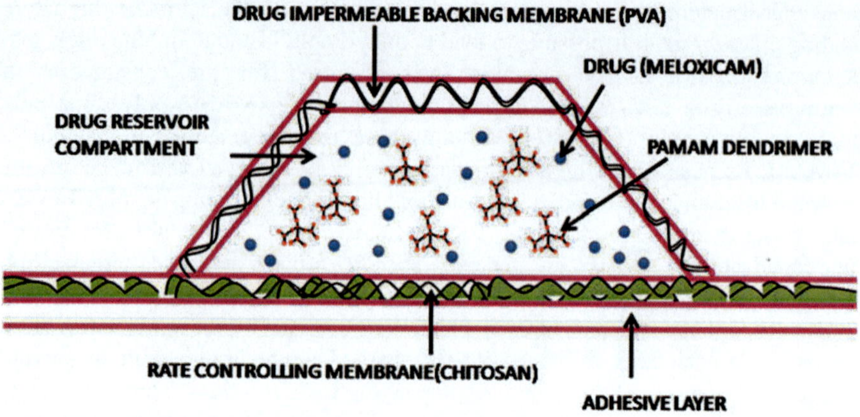

Fig. 8 The transdermal patch of meloxicam with PAMAM dendrimer in drug reservoir compartment and chitosan as rate controlling membrane

content uniformity and minimum weight variation. High folding endurance confirms that the patch will retain its integrity on application into the skin. The purpose of use of PAMAM is to provide adequate permeability and prompt release of the dug meloxicam, which is loaded at the drug reservoir compartment, i.e., the gelling property and viscosity contributed by chitosan control the release of loaded drug meloxicam, whereas PAMAM ensures that adequate amount is released due to its solubilizing effect on meloxicam to be useful for the treatment of rheumatoid arthritis and other inflammatory diseases [112].

3.7 *Chitosan Based Nanocarriers*

3.7.1 Chitosan Based Nanogel

Nanogel is a three-dimensional polymeric system consisting of ionic or non-ionic chains which are usually dispersed in a medium. It can retain large quantities of water and swell/shrink to a certain dimension. Due to the good stability, large surface area and high drug loading capacity, nanogels are increasingly applied in drug delivery research fields [102]. Extensive researches have been carried out to investigate the potential of nanogel prepared with chitosan for wound healing activity. In another study, silver sulphadiazine (SSD) was loaded in chitosan based nanogel for wound healing application (Fig. 9) [118]. The in vivo evaluation of the SSD loaded chitosan nanogel as well as 1% marketed SSD cream was carried out and compared. It is observed that the control animal group and the marketed cream treated animals showed significantly lower wound healing property when compared with SSD nanogel treated group on 5th & 10th days, respectively.

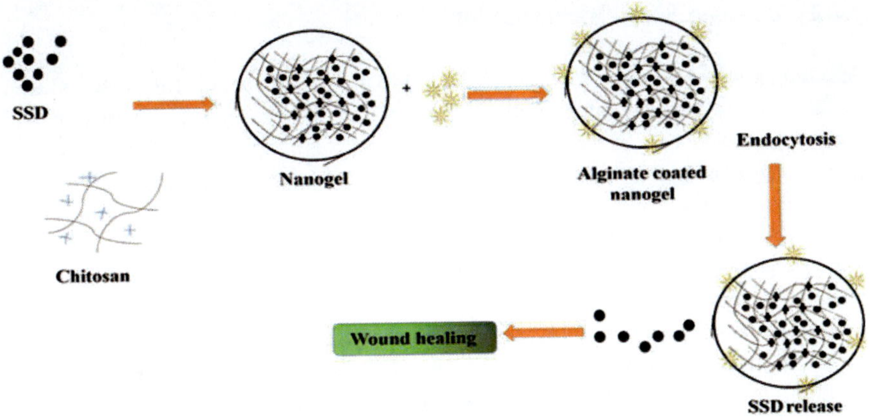

Fig. 9 Silver sulphadiazine (SSD) loaded chitosan based nanogel for wound healing

Table 5 Chitosan based nanogel systems for transdermal drug delivery

Sl. No.	Active moiety	Therapeutic activity	Benefits achieved
1	Silver sulphadiazine	Wound healing	Higher therapeutic efficacy in vivo, in comparison with marketed formulation
2	Ketoprofen	Anti-inflammatory	Maximum anti-inflammatory action compared with the conventional gel
3	Ibuprofen	Anti-inflammatory	High permeation rate
4	Rifaximin	Antibacterial	pH mediated swelling with good bio-adhesion

Ketoprofen loaded chitosan nanogel by transdermal route has been reported for pain management [119]. The in vivo studies showed that the percentage inhibition of oedema by ketoprofen loaded chitosan nanogel (74.0% after 8 h) was found to be greater when compared with the control drug suspension and ketoprofen loaded nanoemulsion. Amos et al. [120] used chitosan-ibuprofen-gellan ternary nanogel for demonstrating the strong interaction of protonated amino group of chitosan with carboxylate ion of ibuprofen to form a product having a particle size nearly 15 nm, which is 324 times less compared to the pure ibuprofen (4,580 nm). Ex vivo permeation studies using pig skin revealed that the chitosan-ibuprofen-gellan nanogel is more permeable than the control ibuprofen-gellan hydrogel and can be considered as an ideal candidate for transdermal delivery of ibuprofen [120]. In a study, chitosan (CS) nanogel loaded with rifaximin was prepared and exhibited pH-responsive swelling (at acidic pH) with enhanced bio-adhesion and good anti-oxidant activity [121]. The details of various chitosan based nanogel systems with benefits achieved in transdermal drug delivery are shown in Table 5.

3.7.2 Chitosan Based Nanoemulsion

An emulsion is a biphasic system in which one phase is dispersed (dispersed phase) in other phase (dispersion medium). The main difference between emulsion and nanoemulsion lies in the size and shape of particles dispersed in the continuous phase. Nanoemulsions contain nanosized globules of particle size ranging from 10 to 1,000 nm. They are either water in oil or oil in water dispersions in a transparent form which are stabilized by a surfactant. Recently they are used as effective nanocarriers for delivering lipophilic moieties to various sites. For attaining a long circulation time and to attain a better site specific effect, the globular surface can be modified using hydrophilic polymer such as modified chitosan. Various researchers have reported chitosan as an excellent emulsifying agent to prepare stable w/o/w multiple emulsions. Due to the structural speciality, chitosan is inactive at air/solution interface, but adsorbs at oil/water interface exhibiting high level mechanical and electrostatic stability to the droplets. The droplet size can be controlled by adjusting the chitosan/oil ratio. One of the most reported properties of chitosan is its antibacterial effect which was first proposed by Allan and Hadwiger in 1979 [122]. It is reported that chitosan based nanoemulsions of *Citrus sinensis* essential oil (cn-CSEO) having antimicrobial activity is a better alternative compared to the use of CSEO in suspension to achieve antimicrobial effect [123]. Chitosan oleate (CS-OA) is suitable to stabilize nanoemulsion loaded with α-tocopherol, which is easily dispersible in aqueous media and therefore in wound fluids [124]. Another study reported is chitosan-modified nanoemulsion containing piplartine as a new strategy for the treatment of skin cancer [125]. Sabah et.al developed chitosan-coated naringenin nanoemulsion for topical application in wound healing application [126]. In another study, o/w nanoemulsion of Mint essential oil and aqueous Parsley extract using chitosan as a natural polymer was found to be a promising candidate with good antibacterial activity against *Escherichia coli*. The cytocompatibility of the system was determined by using HEK239 human cell lines and observed that the preparation exhibited minimum toxicity in the selected cell lines. The in vitro results showed that the chitosan in the nanoemulsion controlled the size of the particles, improved the stability of the product and provided good antibacterial activity to the nanoemulsion [127]. Ae-Jin et.al developed capsaicin-loaded nanoemulsions stabilized with alginate and chitosan as a bioactive ingredient delivery system with good physical stability [128]. The benefits achieved by various chitosan based nanoemulsions for transdermal drug delivery are listed in Table 6.

3.7.3 Chitosan Based Nanolipid Carrier (NLC)

NLCs are colloidal systems consisting of solid lipids and liquid lipids. These are promising nanocarriers used to deliver hydrophilic and lipophilic moieties to the various sites of the body. Since it is a lipid carrier, it possesses good degree of penetration through biological barriers. Physical instability is the main drawback of

Table 6 List of various chitosan based nanoemulsions for transdermal drug delivery

Sl. No.	Active moiety	Method of preparation	Benefits achieved
1	*Citrus sinensis*	Catastrophic phase inversion method	Greater antimicrobial activity
2	α-Tocopherol	Low-energy emulsification technique	Highly stable product with better wound healing effect
3	Piplartine	Sonication	High degree of cytotoxicity (~2.8-fold higher) of piplartine, when loaded in the chitosan nanoemulsion compared to piplartine solution against melanoma cells
4	Naringenin	Low-energy emulsification method	Stimulated the skin regeneration and accelerated the wound healing
5	Mint oil and parsley extract	Homogenization	Significant antibacterial effect against *E. coli* with minimum cytotoxicity in HEK239 human cell lines
6	Capsaicin	Self-assembly method	Attained highly stabilized formulation

Table 7 Reported benefits of various chitosan based nanolipid carriers for transdermal route

Sl. No.	Active moiety	Method of preparation	Benefits achieved
1	Dutasteride	Hair growth	A 20-fold increase in drug concentration was attained and effective for hair growth in vitro
2	Albendazole	Anthelmintic	Improved adhesion with complete eradication of infection compared to the uncoated ones
3	Vitamin D	Supplement	Attained highly stabilized formulation

this lipid system which is reported [129]. Several researchers have investigated to correct this issue by coating the lipid surface with natural or synthetic polymers through electrostatic deposition [130, 131]. Norhayati et.al successfully synthesized stearic acid-chitosan conjugate (CSO-SA) solution and used to coat the dutasteride-loaded nanostructured lipid carriers (DST-NLCs) for topical application. Upon adding CSO-SA (5%) to DST-NLCs, the particle size and PDI are increased, and the NLC attained a positive charge. These CSO-SA coated dutasteride NLC is superior in terms of stability, safety and charge mediated drug delivery compared to the uncoated NLC and is recommendable for topical delivery to promote hair growth [132]. Rania et.al developed albendazole encapsulated chitosan coated nanostructured lipid carriers for the treatment of Trichinella spiralis [133]. In another study, vitamin D loaded chitosan coated NLC was prepared by melt-emulsification followed by chitosan coating through electrostatic deposition. The physical stability of the chitosan coated particles was evaluated by light backscattering for a period of 60 days (at 25°C). It is proved that the coating with chitosan contributes steric stabilization and resistance to deformation upon change in temperature (20–50°C) compared with the uncoated particles [134]. Various chitosan based nanolipid carriers for transdermal drug delivery are listed in Table 7.

3.7.4 Chitosan Based Nanoparticles

Chitosan nanoparticles were first synthesized by Ohya and coworkers in 1994 [135]. They developed chitosan nanoparticles by emulsification and cross linking method for the intravenous delivery of 5-fluorouracil. Many techniques like emulsion-droplet coalescence, emulsion solvent diffusion, reverse micellar, ionic gelation, polyelectrolyte complexation, spray drying and de solvation methods are reported for entrapping the pharmaceutically active molecules in chitosan nanoparticles for effective drug delivery applications. Chitosan nanoparticles offer sustained/controlled drug release and better permeation which results in enhanced drug absorption.

TPP chitosan (TCs) nanoparticles loaded with quercetin were prepared for topical use against photo-toxicity caused by UV-B radiation [136]. These nanoparticles are prepared by ionic gelation method and have a particle size of 183.63 ± 1.52 nm and can easily permeate through the epidermal layers with efficient uptake by HaCaT cells (Fig. 10). The protective effect of quercetin loaded TCs after UVB radiation is demonstrated in vivo and the immunoblotting studies revealed better inhibitory effects of NF-κB/COX-2 signalling pathways by downregulating the phosphorylation of IkB-α and the expression of COX-2 when compared with control drug Quercetin. Thus, Quercetin loaded TCs nanoparticles have a promising therapeutic potential topically to prevent UVB radiation-induced skin damage. The schematic representation of permeation of Quercetin loaded chitosan nanoparticles through the epidermal layers of skin is shown in Fig. 10.

Chitosan-curcumin nanoparticles were prepared by microwave technology and were found to be effective for the treatment of skin burn wounds [137]. In another study curcumin loaded chitosan nanoparticle was developed for the treatment of diabetic wound [138]. The prepared chitosan nanoparticles were loaded with

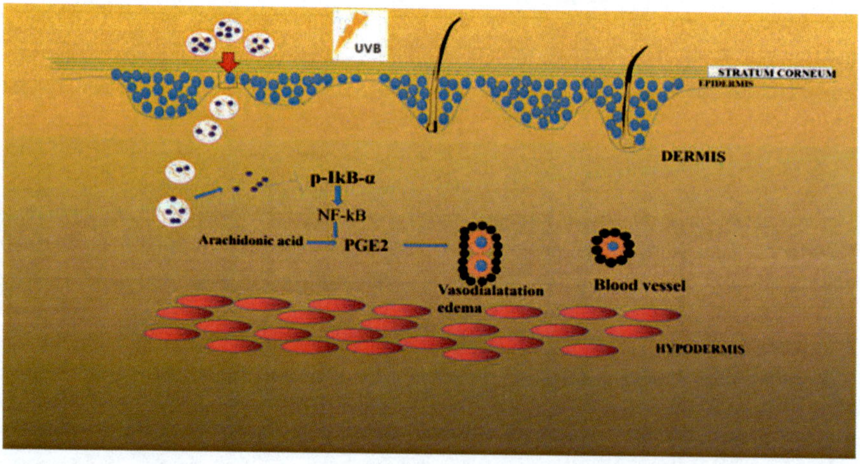

Fig. 10 Permeation of quercetin loaded chitosan nanoparticles for the treatment of UVB radiation-induced skin damage

curcumin to improve stability and solubility followed by impregnation of prepared nanoparticles into collagen scaffold (nanohybrid scaffold) for better tissue regeneration. In vivo studies proved the faster wound healing effect of the formulation compared with the control and placebo scaffold group. This study confirmed synergistic effect of curcumin (anti-oxidant, anti-inflammatory), chitosan (sustained drug release and wound healing) and collagen (scaffold with a wound healing effect) as a better treatment option for diabetic wounds. Salma et al. [139] also worked with curcumin chitosan nanoparticles with an aim to improve the transdermal penetration of curcumin. Using confocal laser scanning microscopy it was proved that these nanoparticles penetrate through appendageal route and demonstrated the diffusion of curcumin to the deeper skin layers after localization of the nanoparticles within the hair follicles [139]. Yerikala et al. designed Ramipril loaded chitosan nanoparticles dispersed in carbopol gel for transdermal application in order to improve the bioavailability of Ramipril [140]. Tao Wang et al. worked with a chitosan nanoparticle loaded calcium alginate hydrogel to regulate inflammation and neovascularization for accelerated wound healing [141]. In vitro studies proved that the formulation exhibited good antibacterial activity even at very low concentration. The in vivo studies in mice showed that the formulation accelerates the generation of ROS production thereby stimulating the production and secretion of IL-6 in vascular endothelial cells (VEC), indicating the pro-inflammatory activity of the formulation. Moreover, the formulation exhibited VEC invasion, metastasis and neovascularization in order to accelerate wound healing. Their results suggested that chitosan nanoparticle loaded calcium alginate hydrogel is a promising system for wound healing. In another work, rosmarinic acid loaded chitosan nanoparticles were incorporated into carbopol gel for wound healing application [142]. He evaluated the drug release properties and wound healing activity of rosmarinic acid loaded chitosan nanoparticles in rats. The formulation showed a sustained release profile without any signs of erythema or oedema on rat skin in vivo and the developed nanoparticle loaded gel showed an ability for fast wound healing effect when compared with the ordinary rosmarinic acid gel. In a study, propranolol-HCl loaded with chitosan nanoparticles, which is then incorporated into muco-adhesive gel was formulated and evaluated for transdermal application [143]. The potential of lecithin/chitosan nanoparticles (NPs) as colloidal nanosystem for the delivery of melatonin by transdermal route is investigated [144]. Transdermal patches of insulin loaded chitosan nanoparticles were developed for the effective delivery of insulin across the epithelial surfaces [145]. The list of various chitosan based nanoparticles developed for transdermal drug delivery is given in Table 8.

4 Summary

A review on transdermal delivery approaches based on chitosan is presented in this chapter. Chitosan, a natural polymer which is derived from chitin, has wide range of applications in pharmaceutical and biomedical fields. At present most of the

Table 8 Benefits of chitosan based nanoparticles for transdermal drug delivery

Sl. No.	Active moiety	Method of preparation	Benefits achieved
1	Quercetin	Anti-inflammatory	Significant effect in inhibiting NF-κB/COX-2 signalling pathway by downregulating the phosphorylation of IkB-α and the expression of COX-2
2	Curcumin	Wound healing	Improved solubility, good cellular internalization and better tissue permeation
3	Ramipril	Anti-hypertensive	Exhibited thixotropic behaviour with a sustained drug release property
4	Rosmarinic acid	Wound healing	Controlled drug release with high wound healing efficiency
5	Propranolol HCl	Anti-hypertensive	Attained a thixotropic behaviour with a controlled release profile and improved systemic bioavailability of propranolol
6	Insulin	Antidiabetic	Effectively delivers the insulin in a controlled rate

pharmaceutical researches are investigated on the benefit of chitosan based drug delivery systems for the treatment of various diseases. The ease of chemical modification, muco-adhesiveness, non-antigenic property, pH dependent swelling, biodegradability and drug release characteristics of chitosan make it a promising candidate for effective drug delivery. The positively charged chitosan can interact with negatively charged biological macromolecules on the skin which results in greater skin permeation. The cationic charge of chitosan is also useful for providing antimicrobial activity by interacting with microbial cells having anionic charge. Chitosan is reported to possess antioxidant property as well as wound healing activity. The anti-inflammatory and pro-inflammatory activity of chitosan depends upon the properties like size, purity, degree of acetylation and molecular weight. These natural and biological properties together with high level of safety make chitosan a promising polymer for the pharmaceutical industry for present and future applications. It is clearly demonstrated that the effective systemic delivery of pharmaceutical moieties can be achieved through the use of chitosan based transdermal devices. However further studies are needed to improve the physical and chemical properties of chitosan derivatives by adopting newer chemical modification methods thereby enhancing the potential of chitosan based drug delivery devices.

References

1. Antosova Z, Mackova M, Kral V, Macek T (2009) Therapeutic application of peptides and proteins: parenteral forever? Trends Biotechnol 27:628–635
2. Grenha A, Carrion-Recio D, Teijeiro-Osorio D, Seijo B, Remuñán-López C (2008) Nano and micro-particulate carriers for pulmonary drug delivery. In: Ravi Kumar MNV (ed) Handbook of particulate drug delivery, vol 2. American Scientific Publishers, Valencia, pp 165–192

3. DiMasi JA, Hansen RW, Grabowski HG (2003) The price of innovation: new estimates of drug development costs. J Health Econ 22:151–185
4. Langer R (1998) Drug delivery and targeting. Nature 392:5–10
5. Andreas BS, Sarah D (2012) Chitosan-based drug delivery systems. Eur J Pharm Biopharm 81:463–469
6. Eugene K, Lee YL (2003) Implantable applications of chitin and chitosan. Biomaterials 24:2339–2234
7. Ehrlich H, Steck E (2010) Three-dimensional chitin-based scaffolds from Verongida sponges (Demospongiae: Porifera). Part II: biomimetic potential and applications. Int J Biol Macromol 47:141–145
8. Mohandas A, Rangasamy J (2021) Nanocurcumin and arginine entrapped injectable chitosan hydrogel for restoration of hypoxia induced endothelial dysfunction. Int J Biol Macromol 166:471–482
9. Nivedhitha Sundaram M, Deepthi S, Mony U, Shalumon KT, Chen JP, Jayakumar R (2019) Chitosan hydrogel scaffold reinforced with twisted poly(L lactic acid) aligned microfibrous bundle to mimic tendon extracellular matrix. Int J Biol Macromol 122:37–44
10. Mohandas A, Sun W, Nimal TR, Shankarappa SA, Hwang NS, Jayakumar R (2018) Injectable chitosan-fibrin/nanocurcumin composite hydrogel for the enhancement of angiogenesis. Res Chem Intermed 44:4873–4887
11. Anjana J, Mohandas A, Seethalakshmy S, Suresh MK, Menon R, Biswas R, Jayakumar R (2018) Bi-layered nanocomposite bandages for controlling microbial infections and overproduction of matrix metalloproteinase activity. Int J Biol Macromol 110:124–132
12. Deepthi S, Nivedhitha Sundaram M, Deepti Kadavan J, Jayakumar R (2016) Layered chitosan-collagen hydrogel/aligned PLLA nanofiber construct for flexor tendon regeneration. Carbohydr Polym 153:492–500
13. Vignesh S, Sivashanmugam A, Annapoorna M, Janarthanan R, Subramania I, Shantikumar VN, Jayakumar R (2018) Injectable deferoxamine nanoparticles loaded chitosan-hyaluronic acid coacervate hydrogel for therapeutic angiogenesis. Colloids Surf B Biointerfaces 161:129–138
14. Anisha BS, Sankar D, Mohandas A, Chennazhi KP, Nair SV, Jayakumar R (2013) Chitosan-hyaluronan/nano chondroitin sulfate ternary composite sponges for medical use. Carbohydr Polym 92:1470–1476
15. Deepthi S, Venkatesan J, Kim SK, Bumgardner JD, Jayakumar R (2016) An overview of chitin or chitosan/nano ceramic composite scaffolds for bone tissue engineering. Int J Biol Macromol 93:1338–1353
16. Rajitha P, Gopinath D, Biswas R, Sabitha M, Jayakumar R (2016) Chitosan nanoparticles in drug therapy of infectious and inflammatory diseases. Expert Opin Drug Deliv 13:1177–1194
17. Ravi Kumar MN (2000) A review of chitin and chitosan applications. React Funct Polym 4:1–27
18. Zikakis JP (1984) Chitin, chitosan and related enzymes, negatively charged lipids, fats and bile acids. Academic Press, Orlando, p 97
19. Mass WA, Mass A, Tighe B (1998) A review of biodegradable polymers: uses, current developments in the synthesis and characterization of biodegradable polyesters. Blends of biodegradable polymers and recent advances in biodegradation. Polym Int 47:89
20. Mohamed NA, Sabaa MW, El-Ghandour AH, Abdel-Aziz MM, Abdel-Gawad OF (2013) Quarternized N-substituted carboxymethyl chitosan derivatives as antimicrobial agents. Int J Biol Macromol 60:156–164
21. Wang W, Meng Q, Li Q, Liu J, Zhou M, Jin Z, Zhao K (2020) Chitosan derivatives and their application in biomedicine. Int J Mol Sci 21:1–26
22. Li H, Zhang Z, Bao X, Xu G, Yao P (2018) Fatty acid and quaternary ammonium modified chitosan nanoparticles for insulin delivery. Colloids Surf B Biointerfaces 170:136–143

23. Milosavljević NB, Milašinović NZ, Popović IG, Filipović JM, Kalagasidis Krušić MT (2011) Preparation and characterization of pH-sensitive hydrogels based on chitosan, itaconic acid and methacrylic acid. Polym Int 60:443–452
24. Shastri DH (2017) Thiolated chitosan: a boon to ocular delivery of therapeutics. MOJBB 3:34–37
25. Tian Q, Zhang CN, Wang XH, Wang W, Huang W, Cha RT, Wang CH, Yuan Z, Liu M, Wan HY, Tang H (2010) Glycyrrhetinic acid-modified chitosan/poly(ethylene glycol) nanoparticles for liver-targeted delivery. Biomaterials 31:4748–4756
26. Zhao P, Wang H, Yu M, Liao Z, Wang X, Zhang F, Ji W, Wu B, Han J, Zhang H, Wang H, Chang J, Niu R (2012) Paclitaxel loaded folic acid targeted nanoparticles of mixed lipid-shell and polymer-core: in vitro and in vivo evaluation. Eur J Pharm Biopharm 81:248–256
27. Kumar Singh Yadav H, Shivakumar HG (2012) In vitro and in vivo evaluation of pH-sensitive hydrogels of carboxymethyl chitosan for intestinal delivery of theophylline. ISRN Pharm 2012:1–9
28. Sugimoto M, Morimoto M, Sashiwa H, Saimoto H, Shigemasa Y (1998) Preparation and characterization of water-soluble chitin and chitosan derivatives. Carbohydr Polym 36:49–59
29. Qin Y, Xing R, Liu S, Li K, Meng X, Li R, Cui J, Li B, Li P (2012) Novel thiosemicarbazone chitosan derivatives: preparation, characterization, and antifungal activity. Carbohydr Polym 87:2664–2670
30. Piegat A, Goszczyńska A, Idzik T, Niemczyk A (2019) The importance of reaction conditions on the chemical structure of N, O-acylated chitosan derivatives. Molecules 24:1–16
31. Deepthi S, Jeevitha K, Nivedhitha Sundaram M, Chennazhi KP, Jayakumar R (2015) Chitosan-hyaluronic acid hydrogel coated poly(caprolactone) multiscale bilayer scaffold for ligament regeneration. Chem Eng Technol 260:478–485
32. Mohandas A, Anisha BS, Chennazhi KP, Jayakumar R (2015) Chitosan-hyaluronic acid/VEGF loaded fibrin nanoparticles composite sponges for enhancing angiogenesis in wounds. Colloids Surf B Biointerfaces 127:105–113
33. Sundaram MN, Sowmya S, Deepthi S, Bumgardener JD, Jayakumar R (2016) Bilayered construct for simultaneous regeneration of alveolar bone and periodontal ligament. J Biomed Mater Res B Appl Biomater 104:761–770
34. Sivashanmugam A, Charoenlarp P, Deepthi S, Rajendran A, Nair SV, Iseki S, Jayakumar R (2017) Injectable shear-thinning CaSO4/FGF-18-incorporated chitin-PLGA hydrogel enhances bone regeneration in mice cranial bone defect model. ACS Appl Mater Interfaces 9:42639–42652
35. Deepthi S, Abdul Gafoor AA, Sivashanmugam A, Nair SV, Jayakumar R (2016) Nanostrontium ranelate incorporated injectable hydrogel enhanced matrix production supporting chondrogenesis in vitro. J Mater Chem B 4:4092–4103
36. Debasish S, Sarmila S, Priyanka M, Sasmal S, Nayak PL (2009) Chitosan: a new versatile bio-polymer for various applications. Des Monomers Polym 12:377–404
37. Badri W, Eddabra R, Fessi H, Elaissari A (2014) Biodegradable polymer based nanoparticles: dermal and transdermal drug delivery. J Colloid Sci Biotechnol 3:141–149
38. Ahmed TA, Aljaeid BM (2016) Preparation, characterization, and potential application of chitosan, chitosan derivatives, and chitosan metal nanoparticles in pharmaceutical drug delivery. Drug Des Dev Ther 10:483–507
39. Maji P, Gandhi A, Jana S, Maji N (2013) Preparation and characterization of maleic anhydride cross-linked chitosan-polyvinyl alcohol hydrogel matrix transdermal patch. J PharmaSciTech 2:62–67
40. Saidin NM, Anuar NK, Meor Mohd Affandi MMR (2018) Roles of polysaccharides in transdermal drug delivery system and future prospects. J Appl Pharm Sci 8:141–157
41. Ahmad Z, Khan MI, Siddique MI, Sarwar HS, Shahnaz G, Hussain SZ, Bukhari NI, Hussain I, Sohail MF (2020) Fabrication and characterization of thiolated chitosan microneedle patch for transdermal delivery of tacrolimus. AAPS Pharm Sci Tech 21:1–12

42. Allena RT, Yadav HKS, Sandina S, Sarat Chandra Prasad M (2012) Preparation and evaluation of transdermal patches of metformin hydrochloride using natural polymer for sustained release. Int J Pharm Pharm Sci 4:297–302
43. Sadhasivam L, Dey N, Francis AP, Devasena T (2015) Transdermal patches of chitosan nanoparticles for insulin delivery. Int J Pharm Pharm Sci 7:84–88
44. Laxman JR, Pallavi A, Vyankatrao YA, Varsha GG, Vyankatrao PM, Nawaj SS (2020) The pharmaceutical application of sulfoxy amine chitosan in design, development and evaluation of transdermal drug delivery of gliclazide. J Pharm Res Int 32:145–161
45. Yadav AV, Urade MN (2019) Formulation and evaluation of chitosan based transdermal patches of lornoxicam for prolonged drug release and to study the effect of permeation enhancer. Indian J Pharm Educ Res 53:88–96
46. Tran TTD, Tran PHL (2019) Controlled release film forming systems in drug delivery: the potential for efficient drug delivery. Pharmaceutics 11:1–16
47. Jahit IS, Nazmi NN, Isa MI, Sarbon NM (2016) Preparation and physical properties of gelatin/CMC/chitosan composite films as affected by drying temperature. Int Food Res J 23:1068–1074
48. Crouvisier-Urion K, Bodart PR, Winckler P, Raya J, Gougeon RD, Cayot P, Domenek S, Debeaufort F, Karbowiak T (2016) Biobased composite films from chitosan and lignin: antioxidant activity related to structure and moisture. ACS Sustain Chem Eng 4:6371–6381
49. Nithya S, Nimal TR, Baranwal G, Suresh MK, Anju CP, Anil Kumar V, Gopi Mohan C, Jayakumar R, Biswas R (2018) Preparation, characterization and efficacy of lysostaphin-chitosan gel against *Staphylococcus aureus*. Int J Biol Macromol 110:157–166
50. Elbi S, Nimal TR, Rajan VK, Baranwal G, Biswas R, Jayakumar R, Sathianarayanan S (2017) Fucoidan coated ciprofloxacin loaded chitosan nanoparticles for the treatment of intracellular and biofilm infections of Salmonella. Colloids Surf B Biointerfaces 160:40–47
51. Thattaruparambil Raveendran N, Mohandas A, Ramachandran Menon R, Somasekharan Menon A, Biswas R, Jayakumar R (2019) Ciprofloxacin-and fluconazole-containing fibrin-nanoparticle-incorporated chitosan bandages for the treatment of polymicrobial wound infections. ACS Appl Bio Mater 2:243–254
52. Mohandas A, Deepthi S, Biswas R, Jayakumar R (2018) Chitosan based metallic nanocomposite scaffolds as antimicrobial wound dressings. Bioact Mater 3:267–277
53. Nimal TR, Baranwal G, Bavya MC, Biswas R, Jayakumar R (2016) Anti-staphylococcal activity of injectable nano tigecycline/chitosan-PRP composite hydrogel using *Drosophila melanogaster* model for infectious wounds. ACS Appl Mater Interfaces 8:22074–22083
54. Martínez-Camacho AP, Cortez-Rocha MO, Ezquerra-Brauer JM, Graciano-Verdugo AZ, Rodriguez-Félix F, Castillo-Ortega MM, Yépiz-Gómez MS, Plascencia-Jatomea M (2010) Chitosan composite films: thermal, structural, mechanical and antifungal properties. Carbohydr Polym 82:305–315
55. Chen L, Tang CY, Ning NY, Wang CY, Fu Q, Zhang Q (2009) Preparation and properties of chitosan/lignin composite films. Chin J Polym Sci 27:739–746
56. Yang J, Kwon GJ, Hwang K, Kim DY (2018) Cellulose-chitosan antibacterial composite films prepared from LiBr solution. Polymers 10:1–7
57. Tanabe T, Okitsu N, Tachibana A, Yamauchi K (2002) Preparation and characterization of keratin-chitosan composite film. Biomaterials 23:817–825
58. Xu YX, Kim KM, Hanna MA, Nag D (2005) Chitosan-starch composite film: preparation and characterization. Ind Crop Prod 21:185–192
59. Maria VD, Bernal C, Francois NJ (2016) Development of biodegradable films based on chitosan/glycerol blends suitable for biomedical applications. J Tissue Sci Eng 7:1–9
60. Srithar S, Rao SS (2019) Capsicum-extract blended chitosan composite films and studying their antibacterial properties. Bull Mater Sci 42:1–5
61. Yuvaraj S, Rajeswari N (2018) Chitosan/modified banana epidermis starch composite films for food packaging applications. J Appl Packag Res 10:14–27

62. Rahmi (2018) Preparation of chitosan composite film using activated carbon from oil-palm empty fruit bunch for Cd^{2+} removal from waters. IOP Conf Ser Mater Sci Eng 434:1–8
63. Thakur G, Singh A, Singh I (2016) Formulation and evaluation of transdermal composite films of chitosan-montmorillonite for the delivery of curcumin. Int J Pharm Investig 6:23–31
64. Paul S, Jayan A, Sasikumar CS (2015) Physical, chemical and biological studies of gelatin/chitosan based transdermal films with embedded silver nanoparticles. Asian Pacific J Trop Dis 5:975–986
65. Oh DW, Kang JH, Lee HJ, Han SD, Kang MH, Kwon YH, Jun JH, Kim DW, Rhee YS, Kim JY, Park ES, Park CW (2017) Formulation and in vitro/in vivo evaluation of chitosan-based film forming gel containing ketoprofen. Drug Deliv 24:1056–1066
66. Behera DK, Mishra K (2019) Formulation and evaluation of chitosan-polypyrrole nanocomposites for controlled release of anticancer drug doxorubicin. Int J App Pharm 11:247–253
67. Wahid A, Sridhar BK, Shivakumar S (2008) Preparation and evaluation of transdermal drug delivery system of etoricoxib using modified chitosan. Indian J Pharm Sci 70:455–460
68. Hima BT, Vidyavathi M, Kavitha K, Sastry T, Suresh KR (2010) Preparation and evaluation of ciprofloxacin loaded chitosan-gelatin composite films for wound healing activity. Int J Drug Deliv 2:173–182
69. Cui Z, Zheng Z, Lin L, Si J, Wang Q, Peng X, Chen W (2018) Electrospinning and crosslinking of polyvinyl alcohol/chitosan composite nanofiber for transdermal drug delivery. Adv Polym Technol 37:1917–1928
70. Bavarsad N, Kouchak M, Mohamadipour P, Sadeghi-Nejad B (2016) Preparation and physicochemical characterization of topical chitosan-based film containing griseofulvin-loaded liposomes. J Adv Pharm Technol Res 7:91–98
71. Devi PM, Menda JP, Reddy T, Deepika R, Sastry TP (2014) Preparation and characterization of wound healing composites of chitosan, *Aloe vera* and *Calendula officinalis* – a comparative study. Am J Phytomed Clin Ther 2:61–76
72. Khan G, Yadav SK, Patel RR, Nath G, Bansal M, Mishra B (2016) Development and evaluation of biodegradable chitosan films of metronidazole and levofloxacin for the management of periodontitis. AAPS PharmSciTech 17:1312–1325
73. Kouchak M, Ameri A, Naseri B, Kargar Boldaji S (2014) Chitosan and polyvinyl alcohol composite films containing nitrofurazone: preparation and evaluation. Iran J Basic Med Sci 17:14–20
74. Saleem MA, Naheed MD, Murali YD, Jaydeep P, Dhaval M (2012) Preparation and evaluation of mupirocin loaded polymer composite films. J Drug Deliv Ther 2:167–172
75. Can AS, Erdal MS, Güngör S, Özsoy Y (2013) Optimization and characterization of chitosan films for transdermal delivery of ondansetron. Molecules 18:5455–5471
76. Hemant KSY, Shivakumar HG (2010) Development of chitosan acetate films for transdermal delivery of propranolol hydrochloride. Trop J Pharm Res 9:197–203
77. Jeon S, Yoo CY, Park SN (2015) Improved stability and skin permeability of sodium hyaluronate-chitosan multilayered liposomes by Layer-by-Layer electrostatic deposition for quercetin delivery. Colloids Surf B Biointerfaces 129:7–14
78. Dong W, Ye J, Wang W, Yang Y, Wang H, Sun T, Gao L, Liu Y (2020) Self-assembled lecithin/chitosan nanoparticles based on phospholipid complex: a feasible strategy to improve entrapment efficiency and transdermal delivery of poorly lipophilic drug. Int J Nanomed 15:5629–5643
79. Park SN, Jo NR, Jeon SH (2014) Chitosan-coated liposomes for enhanced skin permeation of resveratrol. J Ind Eng Chem 20:1481–1485
80. Arora R, Jain CP (2016) Advances in niosome as a drug carrier: a review. Asian J Pharm 1:29–39
81. Lakshmi PK, Bhaskaran S (2011) Phase II study of topical niosomal urea gel – an adjuvant in the treatment of psoriasis. Int J Pharm Sci Rev Res 7:1–7

82. Auda SH, Fathalla D, Fetih G, El-Badry M, Shakeel F (2016) Niosomes as transdermal drug delivery system for celecoxib: in vitro and in vivo studies. Polym Bull 73:1229–1245
83. Osanloo M, Assadpour S, Mehravaran A, Abastabar M, Akhtari J (2018) Niosome-loaded antifungal drugs as an effective nanocarrier system: a mini review. Curr Med Mycol 4:31–36
84. Lakshmi PK, Devi GS, Bhaskaran S, Sacchidanand S (2007) Niosomal methotrexate gel in the treatment of localized psoriasis: phase I and phase II studies. Indian J Dermatol Venereol Leprol 73:157–161
85. Abdelbary AA, Abou Ghaly MHH (2015) Design and optimization of topical methotrexate loaded niosomes for enhanced management of psoriasis: application of Box–Behnken design, in-vitro evaluation and in-vivo skin deposition study. Int J Pharm 485:235–243
86. Sohrabi S, Haeri A, Mahboubi A, Mortazavi A, Dadashzadeh S (2016) Chitosan gel-embedded moxifloxacin niosomes: an efficient antimicrobial hybrid system for burn infection. Int J Biol Macromol 85:625–633
87. Tiwari RK, Chauhan NS, Yogesh HS (2010) Ethosomes: a potential carries for transdermal drug. Int J Drug Dev Res 2:448–452
88. Kumar L, Verma S, Singh K, Prasad DN, Jain AK (2016) Ethanol based vesicular carriers in transdermal drug delivery: nanoethosomes and transethosomes in focus. NanoWorld J 2:41–51
89. Nasr S, Rady M, Gomaa I, Syrovet T, Simmet T, Fayad W, Kader MA (2019) Ethosomes and lipid-coated chitosan nanocarriers for skin delivery of a chlorophyll derivative: a potential treatment of squamous cell carcinoma by photodynamic therapy. Int J Pharm 568:1–11
90. Niu XQ, Zhang DP, Bian Q, Feng XF, Li H, Rao YF, Shen YM, Geng FN, Yuan AR, Ying XY, Gao JQ (2019) Mechanism investigation of ethosomes transdermal permeation. Int J Pharm X 1:1–13
91. Rabia S, Khaleeq N, Batool S, Dar MJ, Kim DW, Din FU, Khan GM (2020) Rifampicin-loaded nanotransferosomal gel for treatment of cutaneous leishmaniasis: passive targeting via topical route. Nanomedicine 15:183–203
92. Opatha SAT, Titapiwatanakun V, Chutoprapat R (2020) Transfersomes: a promising nanoencapsulation technique for transdermal drug delivery. Pharmaceutics 12:1–23
93. Fernández-García R, Lalatsa A, Statts L, Bolás-Fernández F, Ballesteros MP, Serrano DR (2020) Transferosomes as nanocarriers for drugs across the skin: quality by design from lab to industrial scale. Int J Pharm 573:1–43
94. Rai S, Pandey V, Rai G (2017) Transfersomes as versatile and flexible nano-vesicular carriers in skin cancer therapy: the state of the art. Nano Rev Exp 8:1–17
95. Mengoni T, Adrian M, Pereira S, Santos-Carballal B, Kaiser M, Goycoolea FM (2017) A chitosan-based liposome formulation enhances the in vitrowound healing efficacy of substance P neuropeptide. Pharmaceutics 9:1–17
96. Al-najjar BY, Hussain SA (2017) Chitosan microspheres for the delivery of chemotherapeutic agents: paclitaxel as a model. Asian J Pharm Clin Res 10:1–5
97. Shivashankar M, Mandal BK (2013) Design and evaluation of chitosan-based novel pHSensitive drug carrier for sustained release of cefixime. Trop J Pharm Res 12:155–161
98. Zhao F, Zhang H, Wu Y, Wang D, Zhang Y (2020) Preparation and performance evaluation of polymeric microspheres used for profile control of low-permeability reservoirs. J Chem 2020:1–11
99. Mandal A, Majumdar A, Malviya N (2018) Formulation and evaluation of chitosan microspheres for the topical delivery of itraconazole. J Pharm Res 12:890–894
100. Mitra A, Dey B (2011) Chitosan microspheres in novel drug delivery systems. Indian J Pharm Sci 73:355–366
101. Huang G, Liu Y, Chen L (2017) Chitosan and its derivatives as vehicles for drug delivery. Drug Deliv 24:108–113
102. Liu T, Luo G, Xing M (2020) Biomedical applications of polymeric microneedles for transdermal therapeutic delivery and diagnosis: current status and future perspectives. Adv Ther 3:1–15

103. Ahmed Saeed AL, Japairai K, Mahmood S, Hamed Almurisi S, Reddy Venugopal J, Rebhi Hilles A, Azmana M, Raman S (2020) Current trends in polymer microneedle for transdermal drug delivery. Int J Pharm 587:1–14
104. Demir YK, Akan Z, Kerimoglu O (2013) Characterization of polymeric microneedle arrays for transdermal drug delivery. PLoS One 8:1–9
105. Chandrasekharan A, Hwang YJ, Seong KY, Park S, Kim S, Yang SY (2019) Acid-treated water-soluble chitosan suitable for microneedle-assisted intracutaneous drug delivery. Pharmaceutics 11:1–14
106. Sadeqi A, Nejad HR, Kiaee G, Sonkusale S (2018) Cost-effective fabrication of chitosan microneedles for transdermal drug delivery. Annu Int Conf IEEE Eng Med Biol Soc 2018:5737–5740
107. Chen MY, Chen YY, Tsai HT, Tzai TS, Chen MC, Tsai YS (2017) Transdermal delivery of luteinizing hormone-releasing hormone with chitosan microneedles: a promising tool for androgen deprivation therapy. Anticancer Res 37:6791–6797
108. Chen MC, Ling MH, Lai KY, Pramudityo E (2012) Chitosan microneedle patches for sustained transdermal delivery of macromolecules. Biomacromolecules 13:4022–4031
109. Badhe RV, Adkine D, Godse A (2020) Development of PLA and BSA layered coated chitosan microneedles using novel bees wax mould. Turk J Pharm Sci 17:27–35
110. Mei CC, Ling M-H, Lai K-Y, Pramudityo E (2012) Chitosan microneedle patches for sustained transdermal delivery of macromolecules. Biomacromolecules 13:4022–4031
111. Xu Q, Wang C-H, Wayne Pack D (2010) Polymeric carriers for gene delivery: chitosan and poly(amidoamine) dendrimers. Curr Pharm Des 16:2350–2368
112. Duan XD, Ji CJ, Nie L (2015) Formulation and development of dendrimer-based transdermal patches of meloxicam for the management of arthritis. Trop J Pharm Res 14:583–590
113. Tripathy S, Das MK (2013) Dendrimers and their applications as novel drug delivery carriers. J Appl Pharm Sci 3:142–149
114. Wang J, He H, Cooper RC, Gui Q, Yang H (2019) Drug-conjugated dendrimer hydrogel enables sustained drug release via a self-cleaving mechanism. Mol Pharm 16:1874–1880
115. Liu KC, Yeo Y (2013) Zwitterionic chitosan-polyamidoamine dendrimer complex nanoparticles as a ph-sensitive drug carrier. Mol Pharm 10:1695–1704
116. Karen CL, Yoon Y (2013) Zwitterionic chitosan–polyamidoamine dendrimer complex nanoparticles as a pH-sensitive drug carrier. Mol Pharm 10:1695–1704
117. Leng ZH, Zhuang QF, Li YC, He Z, Chen Z, Huang SP, Jia HY, Zhou JW, Liu Y, Du LB (2013) Polyamidoamine dendrimer conjugated chitosan nanoparticles for the delivery of methotrexate. Carbohydr Polym 98:1173–1178
118. El-Fekya GS, El-Bannac ST, El-Bahyc GS, Abdelrazeke EM, Mustafa K (2017) Alginate coated chitosan nanogel for the controlled topical delivery of silver sulfadiazine. Carbohydr Polym 177:194–202
119. Inayat BP, Rashmi D, Wahid A (2019) Formulation and evaluation of ketoprofen loaded chitosan nanogel for pain management: ex-vivo and In-vivo study. Ars Pharm 60:101–108
120. Amos A, Sureya I, Adeola KM (2015) Ex vivo skin permeation and retention studies on chitosan-ibuprofen-gellan ternary nanogel prepared by in situ ionic gelation technique – a tool for controlled transdermal delivery of ibuprofen. Int J Pharm 490:1–2
121. Samah AA, Ali D, Kaixiang Z, Dongmei C, Chao L, Kuiyu M, Muhammad KM, Saeed A, Lingli H, Shuyu X (2020) Designing, structural determination and biological effects of rifaximin loaded chitosan-carboxymethyl chitosan nanogel. Carbohydr Polym 248:116782
122. Yilmaz Atay H (2020) Antibacterial activity of chitosan-based systems. Funct Chitosan:457–489
123. Roberta B, Elisa P, Daniel B, de Rayssa Julliane C (2010) Chitosan nanoemulsions of cold-pressed orange essential oil to preserve fruit juices. Int J Food Microbiol 331:108786
124. Bonferoni MC, Riva F, Invernizzi A, Dellera E, Sandri G, Rossi S, Marrubini G, Bruni G, Vigani B, Caramella C, Ferrari F (2018) Alpha tocopherol loaded chitosan oleate

nanoemulsions for wound healing. Evaluation on cell lines and ex vivo human biopsies, and stabilization in spray dried Trojan microparticles. Eur J Pharm Biopharm 123:31–41
125. Giacone DV, Dartora VFMC, de Matos JKR, Passos JS, Miranda DAG, de Oliveira EA, Lopes LB (2020) Effect of nanoemulsion modification with chitosan and sodium alginate on the topical delivery and efficacy of the cytotoxic agent piplartine in 2D and 3D skin cancer models. Int J Biol Macromol 165:1055–1065
126. Sabah HA, Bapi G, Anoop BN, Hira C, Manisha P, Shah Jigar N, Katharigatta NV (2020) Development and optimization of naringenin-loaded chitosan-coated nanoemulsion for topical therapy in wound healing. Pharmaceutics 12:893
127. Najmeh FL, Negar MK (2019) Preparation and characterization of O/W nanoemulsion with Mint essential oil and Parsley aqueous extract and the presence effect of chitosan. Nanomed Res J 4:48–55
128. Ae-Jin C, Chul-Jin K, Yong-Jin C, Jae-Kwan H, Chong-Tai K (2011) Characterization of capsaicin-loaded nanoemulsions stabilized with alginate and chitosan by self-assembly. Food Bioproc Tech 4:1119–1126
129. Neda N, Hadi V, Parvin Zakeri M (2015) Solid lipid nanoparticles and nanostructured lipid carriers: structure, preparation and application. Adv Pharm Bull 5:305–313
130. Randy Chi Fai C, Tzi Bun N, Jack Ho W, Wai Yee C (2015) Chitosan: an update on potential biomedical and pharmaceutical applications. Mar Drugs 13:5156–5186
131. Chandy T, Sharma CP (1990) Chitosan-as a biomaterial. Biomater Artif Cells Artif Organs 18:1–24
132. Norhayati NM, Khalid S, Satyanarayana SMG (2017) Preparation and characterization of dutasteride-loaded nanostructured lipid carriers coated with stearic acid-chitosan oligomer for topical delivery. Eur J Pharm Biopharm 117:372–384
133. Eid RK, Ashour DS, Essa EA, El Maghraby GM, Arafa MF (2020) Chitosan coated nanostructured lipid carriers for enhanced in vivo efficacy of albendazole against Trichinella spiralis. Carbohydr Polym 232:115826
134. Renata SR, Rabelo Isabela FO, Vanessa MD, Ana SP, Miriam DH (2018) Chitosan coated nanostructured lipid carriers (NLCs) for loading vitamin D: a physical stability study. Int J Biol Macromol 119:902–912
135. Ana G (2012) Chitosan nanoparticles: a survey of preparation methods. J Drug Target 20:291–300
136. Wenhao N, Li D, Houjie C, Fahim UK, Lu Y, Xinbing S, Xiaojun S (2018) Topical use of quercetin-loaded chitosan nanoparticles against ultraviolet B radiation. Front Pharmacol 9:826
137. Hafiz MB, Mohd Cairul MA, Shiow Fern N, .Haliza K, Sheefat US, Nauman RK (2020) Formulation and evaluation of microwave-modified chitosan-curcumin nanoparticles – a promising nanomaterials platform for skin tissue regeneration applications following burn wounds. Polymers 12:2608.
138. Veera Venkata Satyanarayana RK, Gowthamarajan K, Siddhartha VT, Sai Sandeep M, Radhakrishna K, Ashish Devidas W, Shashank M, Kalidhindi Rama SR, Rajkumar M (2016) Curcumin loaded chitosan nanoparticles impregnated into collagen-alginate scaffolds for diabetic wound healing. Int J Biol Macromol 93:1519–1529
139. Salma MA, Rania MH, Omaima AS (2018) Tracking the transdermal penetration pathways of optimized curcumin-loaded chitosan nanoparticles via confocal laser scanning microscopy. Int J Biol Macromol 108:753–764
140. Yerikala R, Vadhireddy S (2017) Transdermal patch of Ramipril loaded chitosan nanoparticles dispersed in carbopol gel. J Drug Deliv Therapeut 7:56–65
141. Tao W, Yan Z, Yaping S, Yijie S, Fang L, Chang S, Liang Z (2018) Chitosan nanoparticles loaded hydrogels promote skin wound healing through the modulation of reactive oxygen species, artificial cells. Nanomed Biotechnol 46:S138–S149

142. Taha UW, Syed NR, Nisar AK (2019) Rosmarinic acid loaded chitosan nanoparticles for wound healing in rats. Int J Pharm Sci:1138–1147
143. Raida AK, Jingyuan W, Angel En-Miao C, Amy Moon-Jung K, Stephanie Sze ML, Joohee Y (2016) Transdermal delivery of propranolol hydrochloride through chitosan nanoparticles dispersed in mucoadhesive gel. Carbohydr Polym 153:176–186
144. Hafner A, Lovric J, Pepic I, Filipovic-Grcic J (2011) Lecithin/chitosan nanoparticles for transdermal delivery of melatonin. J Microencapsul 28:807–815
145. Lingheswar S, Arul Prakash F (2015) Transdermal patches of chitosan nanoparticles for insulin delivery. Int J Pharm Pharm Sci 7:84–88

Adv Polym Sci (2021) 288: 447–468
https://doi.org/10.1007/12_2021_95

Published online: 25 June 2021

Delivery of Biomolecules Using Chitosan Wound Dressings

Georg M. Guebitz, Alessandro Pellis, and Gibson S. Nyanhongo

Contents

Abstract The role of wound dressings has significantly transformed from by providing a simple protective function to dressings that actively participate in every stage of the wound healing process. Chitosan is one of the most widely exploited materials for the synthesis of many functional wound dressings that actively participate at every wound healing stage from facilitating blood clotting to remodelling. In addition to its natural wound healing properties, chitosan is also a versatile polymer that is compatible with many drugs. This has witnessed the increase in the application of chitosan wound dressings for the delivery of a wide variety of structurally different biomolecules that promote wound healing. This

G. M. Guebitz and G. S. Nyanhongo (✉)
Institute of Environmental Biotechnology, University of Natural Resources and Life Sciences, Tulln, Austria

Austrian Centre for Industrial Biotechnology (ACIB), Tulln an der Donau, Austria

Faculty of Science, Department of Biotechnology, University of Johannesburg, Johannesburg, South Africa
e-mail: g.nyanhongo@boku.ac.at

A. Pellis
Institute of Environmental Biotechnology, University of Natural Resources and Life Sciences, Tulln, Austria

chapter reviews the different biomolecules that have been incorporated into different forms of designed chitosan wound dressing materials.

Keywords Antimicrobial · Hemostatic · Inflammation · Wound dressings

1 Bioactive Wound Dressings

The skin is the largest body organ covering the whole organism and acting as a protective barrier against harmful chemicals, materials, pathogens and providing homeostasis, immune surveillance, and sensory surveillance [1]. Damage to the skin results in a wound. In order to restore the skin, a series of processes divided into four time-dependent phases, namely: coagulation and hemostasis, inflammation, proliferation, and remodelling occur. However, the smooth progression of these series of events and successful wound repair depend on the size and depth of the wound, successful protection of the wound from microbial infection, sufficient supply of oxygenation, venous development, local tension and pressure, and many other systemic factors such as age, nutrition, gender, stress, underlying disease conditions, ischemia, and immune conditions of the patient [2]. To manage and speed up the smooth progression of the wound healing process, medication accompanied by the use of wound dressing materials has traditionally been used primarily to protect the wound from further damage or from microbial infection using dry cellulose fibers/cotton materials. However, the pioneering work of Winter in 1962 [3] demonstrated that a moist wound environment accelerates the wound healing process when compared to dry gauze wound dressings (e.g., cotton dressing) [4]. Winter's discovery has transformed the role of wound dressing materials in less than 40 years compared to the developments achieved in the previous 2000 years [5, 6], from simple passive protective materials to bioactive/interactive materials that actively participate in the wound healing process [7]. Advances in moist wound dressings have witnessed their active participation in all the four time-dependent wound healing phases (coagulation and hemostasis, inflammation, proliferation, and remodelling), even in problematic chronic wounds (diabetic, pressure ulcers, skin tears and surgical wounds, eruption wounds, e.g. chicken pox and shingles, etc.), as well as helping in reducing pain and discomfort [8].

Advances in moist wound dressing materials have also witnessed the evolution from synthetic based materials to biobased materials derived from naturally occurring biopolymers such as chitosan, collagen, gelatin, cellulose, alginate, hyaluronic acid, elastin, etc. [4]. Of all these biobased derived wound dressing materials, chitosan is emerging as one of the most versatile wound dressing polysaccharides due its unique $-NH_2$ groups which gives it a positively charged amino groups that electrostatically interact with the anions on the surface of red blood cells, leading to their intensive aggregation, thereby actively participating in the blood clotting process and quickly stopping bleeding. Furthermore, the backbone of chitosan

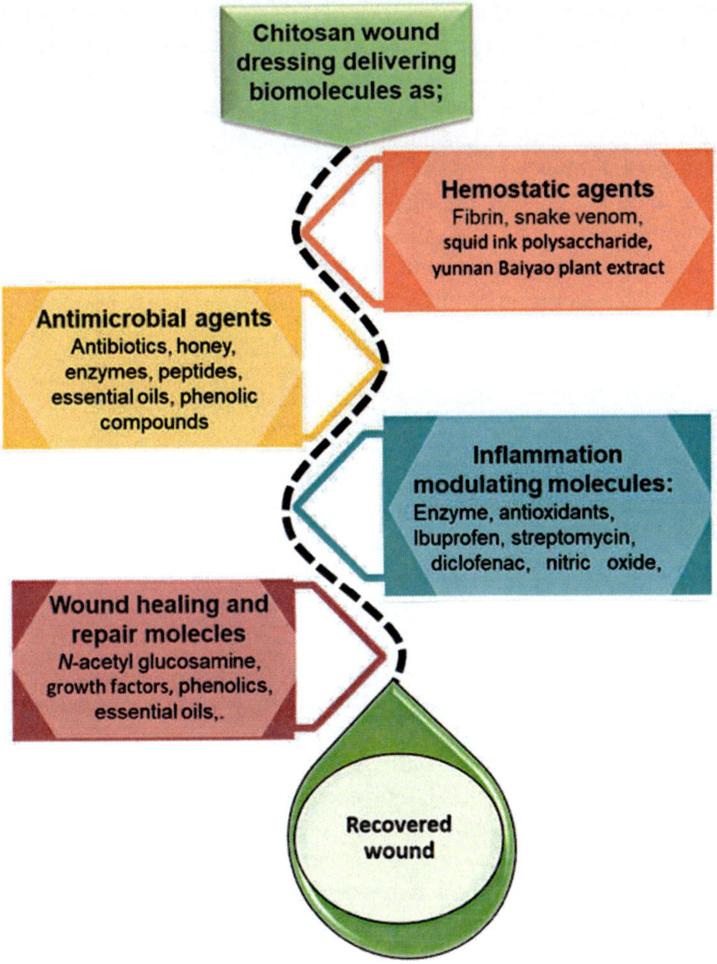

Fig. 1 Shows the remarkable role of chitosan wound dressing in delivering different classes of biomolecules that help protect the wound and help every step of the wound healing repair process

consisting of β-(1–4) linked N-acetyl-β-D-glucosamine units naturally enables it to actively participate in cell to cell and cell to matrix interactions, promote cell proliferation and migration, and mediate the release of cytokines and growth factor signaling molecules, thereby modulating many biological activities [9]. In addition, the antimicrobial properties, hemostatic effects, and most important its ability to be molded into a versatile drug delivery system have made chitosan unique among other naturally occurring polysaccharide based wound dressings. Chitosan-based wound dressings have witnessed designs ranging from nanoparticle, filaments, powders, granules, hydrogels, bandages, patches, membranes, sponges, and different chitosan-based composite wound dressings with other polymers, e.g. cotton, gelatin, or synthetic polymers [10], that actively deliver wound healing promoting biomolecules as summarized in Fig. 1. Due to their structural similarity with the skin, chitosan-based wound dressings are also increasingly being incorporated into asymmetric wound dressings [10]. Asymmetric wound dressings are designed in such a way that they resemble the skin where the outer layer protects the body from external attack and the inner part biochemical and physiological processes that actively promote the wound healing process [10]. Other chapters of this book discuss the roles played by chitosan in the biomedical field while this chapter specifically provides an overview of biomolecules incorporated into chitosan wound dressings in order to enhance its natural wound healing properties.

2 Chitosan Wound Dressings (CWD) as Hemostatic Delivery Platforms

Successful wound healing starts with stopping bleeding. Hemorrhage remains a leading cause of early death after trauma, accounting for over 30% deaths in civilian clinics and 50% battlefield casualties [11–13]. This is because the body's natural clotting capacity is not able to cope with the excessive hemorrhage in time for providing an effective hemostasis. As already extensively described in previous chapters, various hemostatic materials exploiting chitosan's natural properties have been developed [14] and their remarkable performance has seen them replace the traditionally used Combat Gauze [13]. In fact, chitosan's remarkable performances have since been recognized as the most versatile suitable hemostat for both civilian and military trauma patients [15]. For example, electrospun poly(caprolactone) membranes coated with various densities of chitosan oligomers induced faster hemostasis and affected the re-epithelialization and wound healing in mice [16]. However, despite chitosan's remarkable natural hemostatic attributes, when used alone as hemostatic agent it might get overwhelmed in cases of massive hemorrhage. In order to enhance the hemostatic capabilities of CWD, hemostatic enhancing molecules have been incorporated into chitosan-based wound dressings. For example, products enhancing blood clotting such as fibrin, a protein produced from fibrinogen and an important component in the coagulation process has been used to enhance blood clotting when incorporated in chitosan composite bandages or gels [17, 18]. Chan et al. produced an engineered hemostatic polymer (PolySTAT) that circulates innocuously in the blood, identifies sites of vascular injury, and promotes clot formation to stop bleeding by cross-linking the fibrin matrix within clots, mimicking the function of the transglutaminase factor XIII [19]. Nonwoven chitosan gauze impregnated with PolySTAT enhanced blood coagulation than commercially available Celox® Rapid [17]. Batroxobin, a thrombin-like enzyme produced by *Bothrops atrox* and *Bothrops moojeni* with proven activity participating in blood coagulation was incorporated into non-woven chitosan dressings where it facilitates erythrocyte aggregation, fibrin clot formation, and blood coagulation [18, 20]. In another study, a venom component from the snake *Bothrops atrox moojeni* was discovered to act as a thrombin-like enzyme in the coagulation cascade and was effective in enhancing the blood clotting process [18]. Similarly, squid ink polysaccharide which has been used for centuries in Chinese traditional medicine due to its including antitumor, antiradiation, ability to promote blood clotting, and immunomodulatory properties was shown to activate the blood coagulation factor FXII leading to the shortening of the blood coagulation time both in vitro and in vivo [21, 22]. Chitosan/sodium alginate impregnated with yunnan Baiyao plant extract had better hemostatic effects than chitosan/sodium alginate wound dressings alone [23]. *Yunnan Baiyao* is an herb used in oriental countries as surgical sealant and hemostat [23]. These studies show that incorporating biomolecules into CWD greatly enhances its hemostatic capabilities.

3 CWD as Antimicrobial Activity Delivery Systems

Although Chitosan wound dressings provide an ideal moist environment for the wound healing process and to some extent have antimicrobial properties, the moist environment makes it attractive for continuous attack and ultimately infection of the wound by several microorganisms. To overcome these challenges, extensive studies have developed different strategies to prevent microbial colonization of the CWD, such as: (1) taking advantage of the inherent antimicrobial properties of chitosan, (2) reactivity of chitosan -NH_2 groups to graft versatile antimicrobial molecules onto chitosan backbone, and (3) loading the chitosan wound dressings with inorganic and organic antimicrobial agents. The natural antimicrobial properties of chitosan are caused by its positive charge that enables CWD to interact with the negatively charged bacteria membrane and/or chelate trace metals [24] and oligo-elements that are essential for bacterial growth. Several bioactive CWD are now commercially available including Axiostat®, HidroKi®, Tegasorb®, Chitopack®, and KytoCel® [10]. The amino acid, arginine has been incorporated into CWD to boost its antimicrobial activity through increasing the positive charge groups that enhance interaction with the bacterial cell wall [25]. This is because the guanidinium group of arginine is known to form intense hydrogen-bond and ionic interactions with the negatively charged cell of bacteria which enhances permeability [26]. Similarly, the reactive -NH_2 groups of chitosan have been exploited for grafting antimicrobial molecules onto chitosan backbone, e.g. many sulfonamide derivatives (sulfametoxydiazine, sulfadiazine, sulfadimethoxine, sulfamethoxazole, sulfamerazine, sulfisoxazole) with chloroacetyl chloride to boost its antimicrobial properties [10] (Fig. 2).

Another commonly adopted strategy to increase the antimicrobial activities of CWD involves incorporation of antimicrobial biomolecules especially antibiotics, enzymes, phenolics, essential oils, and antimicrobial peptides. Various antibiotics like tetracycline, penicillins, cephalosporins, aminoglycosides [27–29] have been incorporated into CWD to boost its inherent antimicrobial properties. The synthesis of complex chitosan-streptomycin gold nanoparticles and chitosan-vancomycin

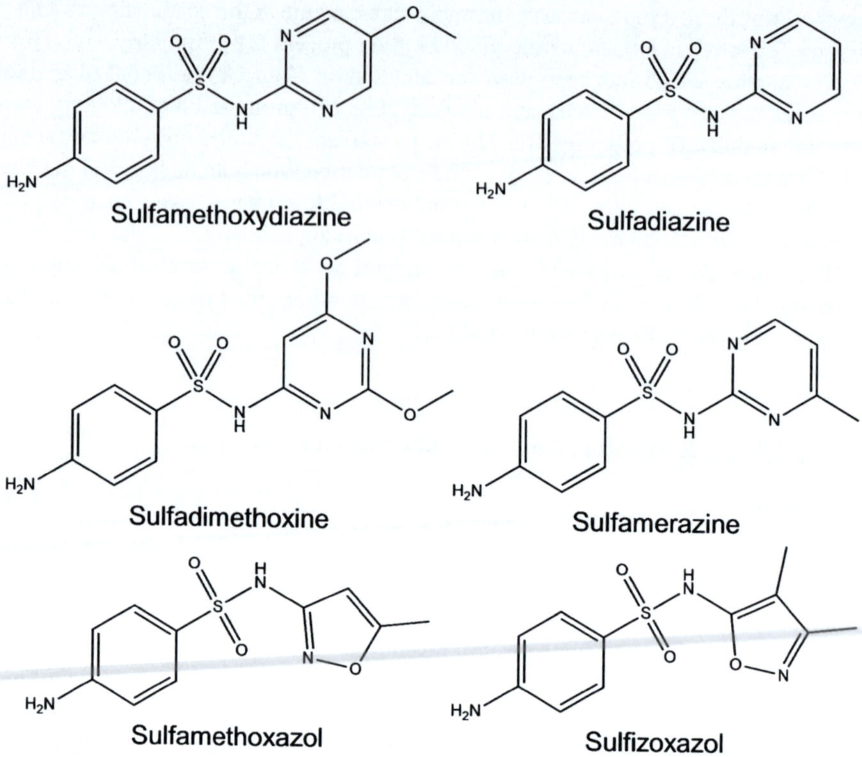

Fig. 2 Some of the sulfonamide derivatives that were grafted onto chitosan to enhance antimicrobial activity

enhanced their anti-biofilm [30] by facilitating the penetration of biomolecules into the biofilm matrix which increased their bactericidal properties even against multidrug-resistant *Staphylococcus aureus*. Hydrogels based on gellan gum and chitosan sustained release of cargoes (antibacterial drugs, tetracycline hydrochloride, and silver sulfadiazine) [31]. As summarized in Liu et al. 2018 [1], several researchers have shown that CWD and their composites are able to effectively deliver minocycline, gentamycin sulfate, chitosan–piperacillin–tazobactam, amikacin, gentamicin/ciprofloxacin, ciprofloxacin, norfloxacin, sulfadiazine, etc. that prevent infection and stimulate faster wound healing. Liu et al. 2018 [1] also reported that several researchers have shown that CWD and their composites are able to effectively deliver incorporated tetracycline hydrochloride [32], gentamycin sulfate [33], piperacillin–tazobactam [34], amikacin [35], gentamicin/ciprofloxacin [36], ciprofloxacin [37], norfloxacin [38] and provide protection from microbial invasion of the wound. Similarly, incorporation of minocycline, a broad-spectrum antibiotic effective on gram-positive and gram-negative bacteria belonging to the tetracycline family that has a broader range than other antibiotics in this family was able to control infection of burn wounds [39, 40].

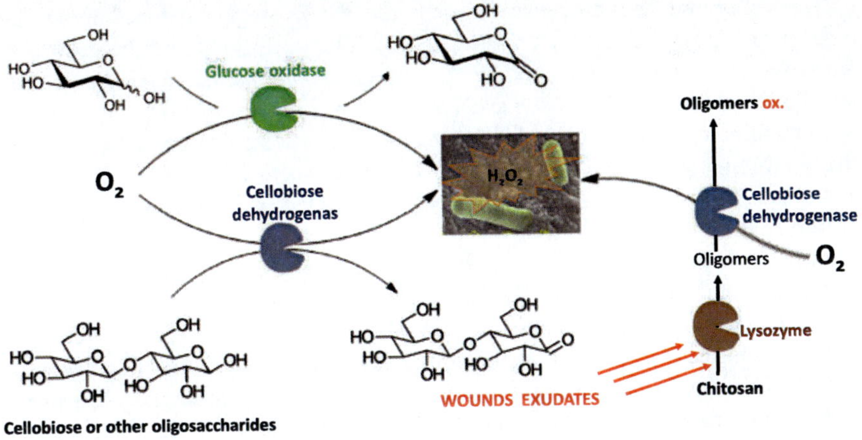

Fig. 3 CWD containing (**a**) glucose oxidase as an antimicrobial system that uses glucose to produce hydrogen peroxide while (**b**) lysozyme hydrolyzes chitosan into oligomers used by cellobiose dehydrogenase to produce hydrogen peroxide

Recently, enzymes and peptides with antimicrobial activity were incorporated into CWD. Antimicrobial peptides, for example, are new promising agents gaining scientific interests as possible solution to replace traditional antibiotics in fighting multi-drug-resistant microorganisms. The incorporation of amphiphilic peptide temporin B showed a strong and fast killing ability against Gram-positive and multidrug-resistant nosocomial bacterial species [41]. Similarly, incorporation of daptomycin into chitosan was effective as topical ocular administration in the treatment of bacterial endophthalmitis [42]. Daptomycin is a natural lipoprotein active against gram-positive bacteria. Antimicrobial enzymes have been successfully incorporated and used as antimicrobial agents in chitosan gels. Among these enzymes incorporated into CWD materials are oxidative enzymes whose ability of generating hydrogen peroxide (H_2O_2), a natural antimicrobial agent also produced by the human body has made it easy to produce in situ antimicrobial systems for application in wound dressing (Fig. 3). Glucose oxidase is the most widely used representative of oxidative enzymes producing H_2O_2 that was successfully used as an antiseptic during wound treatment [43]. Bösiger et al. produced electrospun chitosan mats impregnated with glucose oxidase for in-situ production of H_2O_2 against *E. coli* and of *S. aureus* [44]. The same colleagues successfully developed an in-situ antimicrobial generating system using a combination of cellobiose dehydrogenase and lysozyme enzymes [45, 46]. Cellobiose dehydrogenase uses cellobiose and cello-oligosaccharides including *N*-acetylglucosamine or chitooligomers as electron donors to reduce molecular oxygen to H_2O_2 Fig. 3 (a well-known antimicrobial agent) [47]. The wound dressings were designed in such a way that when applied on the wound and activated by the exudates, the N-acetylated chito-oligosaccharides were hydrolyzed by lysozyme and the resulting oligomers were converted into H_2O_2 by cellobiose dehydrogenase [46–51].

The cellobiose dehydrogenase loaded into the hydrogels reduced molecular oxygen to H_2O_2. The in situ produced H_2O_2 completely inhibited the growth of *Staphylococcus aureus*, *Bacillus subtilis*, *Pseudomonas putida*, *Escherichia coli*, and biofilm producing *Cellulosimicrobium cellulans* [52]. Nanoparticles synthesized by immobilizing lysozyme in chitosan nanoparticles were demonstrated to effectively inhibit *Pseudomonas aeruginosa*, *Klebsiella pneumoniae*, *Escherichia coli*, and *S. aureus* [53, 54]. Further, lysozyme incorporated into chitosan nanoparticles preserved the antibacterial activity of the enzyme over 3 weeks in vitro while active against *Staphylococcus epidermidis* up to 5 days of incubation [41].

Nitric oxide (NO) is a key molecule that plays important roles in human skin such as the control of homeostasis, defense, control of dermal blood flow and participates in the wound healing process. Although NO is a promising therapeutic agent, its short half-life has been challenging for its exploitation. To address this challenge, a NO- in situ producing and releasing chitosan film containing the NO precursor, *S*-nitrosoglutathione as a donor was developed as antibacterial and wound-healing promoting material in chitosan [55]. For example, Kim et al. in 2015 [55] reported a chitosan film containing *S*-nitrosoglutathione that displayed strong antibacterial activity against *Pseudomonas aeruginosa* and *Staphylococcus aureus* and also accelerated wound healing and the epithelialization process compared to the chitosan film only [55]. Similarly, incorporation of the NO donor precursor glutathione encapsulated into ultra-small chitosan nanoparticles produced NO that could be delivered transdermally [56]. In situ NO generating CWD were tenfold more effective against *Pseudomonas aeruginosa* in biofilms under both aerobic and anaerobic conditions [57]. Controlling biofilms is one of the most challenging activities in the management of wounds. The study by Choi et al. [58] demonstrated great potential application of NO generating systems in CWD for controlling biofilms. NO-releasing chitosan films were effective against methicillin-resistant *Staphylococcus aureus* biofilm-infected wounds in diabetic mice enhancing its antibacterial activity by >3 logs reduction in bacterial viability and also exhibited a three-fold higher anti-biofilm activity than the control and CS film [58]. The in vitro release study showed sustained release of NO over 3 days in simulated wound fluid [58]. CWD-NO releasing films were able to disperse biofilms and promote wound healing as evidenced by wound size reduction, re-epithelialization, and collagen deposition than in untreated CS films [58]. Recent studies are even designing more complex chitosan-nitric oxide generating systems such as copper–chitosan particles [59] and alginate/chitosan that enhance the delivery of NO [60] for wound healing applications.

Honey has been used as a remedy to cure wounds and infections for over 2000 years [61] and it is increasingly being incorporated in wound dressings. Honey mainly contains about 40% fructose, 10% maltose, and 30% glucose, enzymes, minerals, oligosaccharides, carbohydrates, and many phytochemicals [62]. Honey provides a moist wound environment, modulates inflammation, prevents swelling and reduces pain, controls odor, provokes sloughing of necrotic tissue, enhances granulation and epithelialization, and promotes wound healing with minimal scarring and also possesses antimicrobial activities attributed to its high osmolarity, pH, H_2O_2 production, and presence of other phytochemical components [62] that enhance the wound healing process. Shamloo et al. [62] produced chitosan/gelatin/PVA hydrogels integrating honey into the hydrogel matrix that helped maintain a well-structured layer of epidermis containing mature collagen and accelerated the rate of wound healing [62]. Manuka honey is highly recognized as an antimicrobial agent due to the presence of methylglyoxal that can accumulate to 1.541 mg/g [63]. CWD impregnated with manuka honey were effective against *Staphylococcus aureus* and *Escherichia coli* [63].

Plant-derived phytochemicals have traditionally been applied for treating wounds throughout humankind. Curcumin (diferuloylmethane), a yellow crystalline compound, a traditional Asian spice, possesses antineoplastic, antimicrobial, anti-inflammatory, antioxidant, and wound healing properties that are increasingly being exploited in biomedicine. Curcumin incorporated in chitosan-PVA exhibited polymers inhibited bacterial pathogens *Pasteurella multocida, Staphylococcus aureus, Escherichia coli, and Bacillus subtilis* ideal for application in burn wounds [64], while chitosan-curcumin dressings were effective in inhibiting drug resistant methicillin-resistant Staphylococcus *aureus* and *Pseudomonas aeruginosa* and an improved wound repair in a murine model [65]. Ferulic acid encapsulated chitosan nanoparticles were effective against *Candida albicans* biofilm formation [61]. Wound dressings composed of chitosan and polycaprolactone loaded with ferulic acid and resveratrol accelerated wound healing in vivo [66].

Another increasingly explored class of antimicrobial agents are plant essential oils summarized in Fig. 4. Plant-derived essential oils have been shown to be an alternative class of antimicrobial compounds that can help fight increasing microbial resistance [67]. For example, CWD loaded with quercetin and thymol improved anti-inflammatory and anesthetic properties of the films [68]. Incorporation of cinnamaldehyde in chitosan composites was effective eradicating *E. coli* and various *Pseudomonas* [69–71].

Similarly, the importance of thymol and its isomer carvacrol essential oils produced by Thymus vulgaris has been extensively reported [72] Their importance in aiding wound healing has been extensively studied and reviewed by Costa et al. [73]. Studies have demonstrated the antibacterial and antifungal, anti-inflammatory, antioxidant, skin fibroblasts stimulator, and analgesic activities [72]. Sepideh Koosehgol et al. used thymol as an antimicrobial agent in chitosan/polyethylene glycol film with excellent antimicrobial properties against both gram-positive and gram-negative bacteria [74]. Pires et al. applied thymol and β-cartone as anesthetic, anti-inflammatory, and antioxidant compounds in chitosan-alginate polyelectrolyte complexes capable to reduce bleeding by thrombus formation and had high stability against physiological fluids and/or bathing. The authors suggested that these films could potentially be used as wound dressing in low exuding wounds [75]. Carvacrol in chitosan films reduced wound areas and tissue edema, induced earlier granulation tissue formation, increased cell proliferation, increased epithelialization rates, and improved collagenization [76]. Thymol loaded chitosan nanoparticles demonstrated superior antimicrobial activity [77]. Chitosan-alginate films loaded with thymol and beta-carotene to confer anesthetic, anti-inflammatory, and antioxidant properties

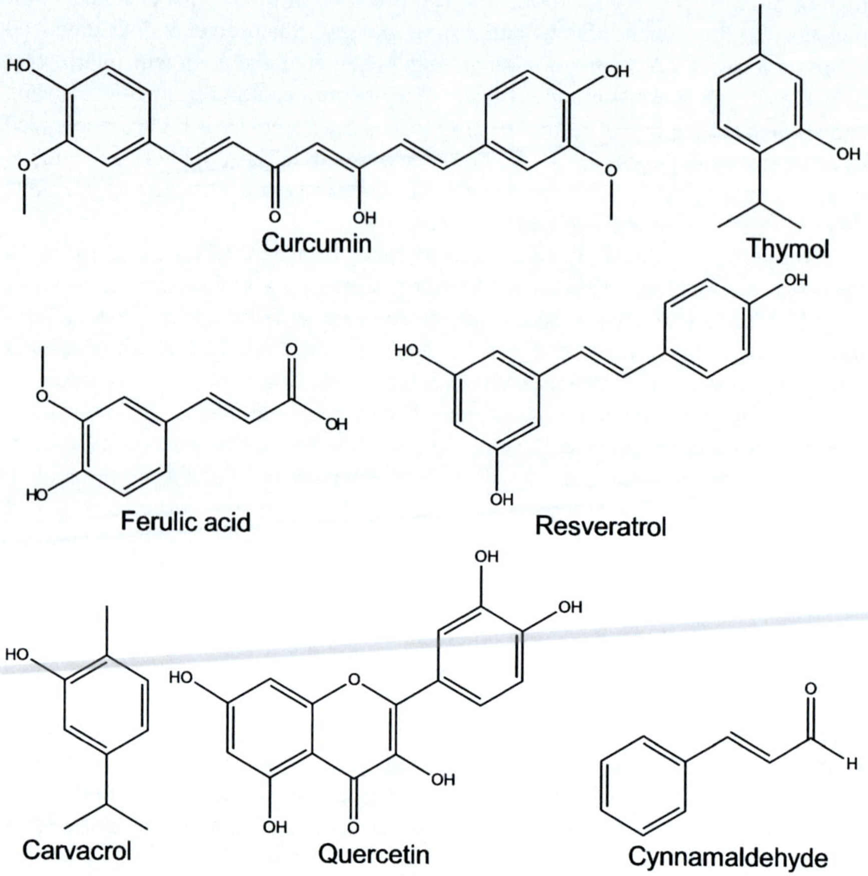

Fig. 4 Plant-derived phytochemicals and essential oil components traditionally used by mankind for wounds treatment

[75]. Another medically interesting antimicrobial agent is asiaticoside extracted from *Centella asiatica*. Asiaticoside is a triterpene glycoside with antioxidant and anti-inflammatory properties. Its incorporation in chitosan composite nanofibers enhanced wound healing in vivo. It was shown to upregulate growth factors and promote cell differentiation, among many other positive wound healing effects [78]. Chitosan impregnated with salicylic acid was effective in inhibiting *Staphylococcus aureus* bacteria as well as avoiding the formation of biofilm at the surface of the wound dressing [79]. Essential oils extracted from the leaves of *Melaleuca alternifolia* that contain a mixture of terpene hydrocarbons and tertiary alcohols incorporated into liposomes and chitosan-based electrospun nanofiber mats were effective in destroying *Staphylococcus aureus, Escherichia coli,* and *Candida albicans cells* [80].

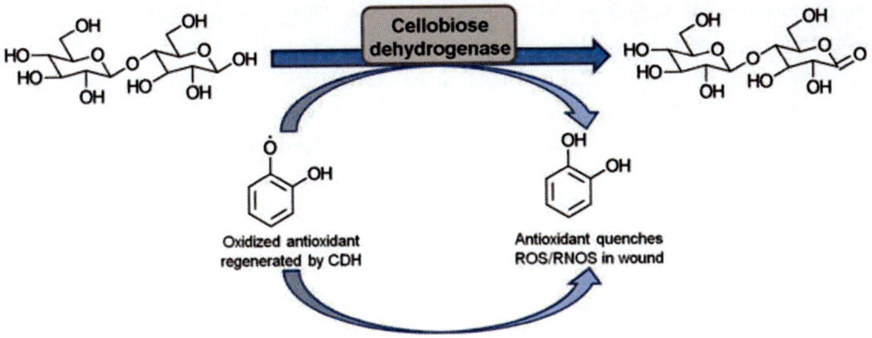

Fig. 5 CWD impregnated with cellobiose dehydrogenase (CDH) uses cellobiose to continuously reduce oxidized phenolic antioxidants to their parent compound after quenching reactive oxygen species and nitrogen centered reactive species

4 CWD Delivery of Inflammation Modulating Agents

The inflammatory phase is characterized by activation of cytokines, chemokines, growth factors that activate the immune system on the wound site and also activate the cell repair and proliferation, migration pathways. *N*-acetylglucosamine, the basic unit of chitosan is a well-recognized inflammatory regulatory biomolecule synthesized in the human body from glucose [1]. Based on this knowledge, chito-oligosaccharides with molecular weight of 5 kDa have been tested and shown to have the ability to regulate inflammation [81]. CWDs impregnated with *N*-acetylglucosamine or exposed to in vivo *and/or* lysozyme that hydrolyzed CWD and produced *N*-acetylglucosamine were shown to modulate various inflammation mediators such as interleukins and prostaglandins [82], activities of macrophages and neutrophils while promoting tissue granulation [83] and inflammatory functions of polymorphonuclear leukocytes [84]. The CWD regulatory effects help create an appropriate inflammatory microenvironment conducive for wound healing. Chitosan–PVA hydrogel containing bee honey or its venom was more effective than the standard diclofenac drug in modulating inflammation and promoting wound healing process [85]. Incorporation of ibuprofen, betamethasone sodium phosphate, streptomycin, and diclofenac in chitosan-based hydrogels had anti-inflammatory effects [1]. In addition to the above biomolecules added to CWDs, chitosan-based hydrogels impregnated with antioxidants, amino acids, vitamins, and nutrients reduced inflammation and promote wound healing [86]. Morgado and collaborators synthesized a poly(vinyl alcohol)-chitosan asymmetrical membranes with highly controlled morphology and incorporated *(S)*-ibuprofen (IBP, a non-steroidal anti-inflammatory agent) into β-cyclodextrin (β-CD) carriers. The produced membranes containing IBP-β-CD prevented scar formation, prevented excessive inflammatory response (lower number of mast cells on the tissue), and stimulated the wound regeneration increasing deposition of collagen type III facilitating full epithelialization and increasing thickness of the epithelial layer [87, 88].

Another novel multifunctional system to control inflammation was developed exploiting the versatile reactions catalyzed by cellobiose dehydrogenase. This enzyme reduces oxidized simple phenolic compounds to their original compounds by donating hydrogen atoms (Fig. 5). However, in the absence of these oxidized simple phenolics, the enzyme reduces molecular oxygen to hydrogen peroxide (Fig. 5). Its ability to reduce oxidized simple phenolics was exploited for the construction of a novel antioxidant regenerating system and its ability [47] to reduce molecular oxygen to hydrogen peroxide was exploited for developing an antimicrobial systems [47]. The antioxidant regenerating system quenched and prevented the accumulation of the most important oxygen centered reactive species produced during inflammation (NO, and super oxide and hydroxyl radicals) [47]. The antioxidant regenerating system consisting of cellobiose dehydrogenase and cellobiose regenerated catechol which then continuously quenched nitric oxide (NO), superoxide (O_2^-), and hydroxyl radicals (OH•) when incorporated into hydrogels [52]. The studies demonstrated the possibility of developing an in-built continuous antioxidant regenerating system and antimicrobial system that can be incorporated in wound dressing. An antioxidant and antibacterial polyelectrolyte wound dressing based on chitosan/hyaluronan/phosphatidylcholine dihydroquercetin showed strong free radical scavenging potency and anti-inflammatory activities of the produced films [89].

5 CWD Mediating the Wound Healing and Wound Repair Process

The wound healing phase is characterized by cell proliferation, neoangiogenesis, formation of granulation tissue, formation of extracellular matrix, and re-epithelialization. The natural presence of N-acetyl glucosamine in human dermal tissue has made it easy to develop wound dressings as there are no expected negative reactions [90]. In fact, lyzozyme mediated gradual hydrolysis of chitosan into N-acetyl-β-D-glucosamine has been shown to stimulate growth factors and the general wound healing process including increasing hyaluronic acid synthesis and preventing scar formation [82, 91].

Apart from the natural attributes of chitosan, its application in wound dressings has also proved to be efficient in delivering drugs, growth factors, stem cells, peptides, etc. that promote the wound healing process [91]. For example, sponge-like chitosan glutamate and sericin wound dressing developed to protect human fibroblasts against oxidative damage during the treatment of chronic skin ulcers improved fibroblast proliferation [81]. Loading curcumin, a well-known anti-inflammatory and antioxidant molecule in chitosan nanoparticles that were further incorporated into collagen-alginate scaffolds was an effective strategy in enhancing wound healing of diabetic wounds in adult male Wistar rats [92]. Apigenin impregnated chitosan composite hydrogel improved the wound healing in diabetic and normal wound tissues [93]. Incorporation of nerolidol, an alicyclic sesquiterpene, found in essential oils from several plants, which possesses pharmacological properties such as anti-neoplastic, antioxidant, and antimicrobial, accelerated re-epithelialization and reorganization of collagen within seventh day of treatment [94]. The authors also recommended nerolidol as an effective antimicrobial agent.

Many studies have also shown the importance of chitosan-based wound dressing in delivering growth factors. Growth factors relevant to wound healing process such as epidermal growth factor (EGF), fibroblast growth factor (FGF), transforming growth factor beta (TGF-β), platelet-derived growth factor (PDGF), and vascular endothelial growth factor (VEGF) are synthesized and secreted by fibroblasts, inflammatory cells, endothelial cells, epithelial cells, and platelets. Growth factors play an important regulatory role that includes inducing cell migration, proliferation, differentiation and promoting the synthesis of extracellular matrix (ECM) [95]. Chitosan hydrogels were effective in delivering basic fibroblast growth factor (bFGF) leading to accelerated wound healing and closure [96]. Chitosan composite hydrogels loaded with rhEGF promoted the healing of diabetic wound in rats [97] while continuous release of rhEGF for more than 3 weeks after subcutaneous implantation in rats [98]. Similarly, chitosan–polyacrylamide hydrogel loaded with EGF supported the proliferation of fibroblast cells for a longer period than that of free EGF [34]. Furthermore, sodium carboxymethyl chitosan grafted with recombinant human epidermal growth factor conjugate protects it from proteases [99]. Wound-adhesive and thermo-responsive pluronic-chitosan hydrogels containing EGF enhanced epidermal differentiation during the wound healing process through increasing local concentration of rhEGF at wound sites with maintaining keratinocytic differentiation [100]. Many studies have also explored the use of chitosan wound dressings for the delivery of FGF, a growth factor stimulating angiogenesis and demonstrated their effectiveness accelerating chronic wound healing process [101]. Adipose-derived stem cells (ADSCs) secrete several angiogenic growth factors. The incorporation of ADSCs or synovial mesenchymal stem cells into chitosan-based composite hydrogels promoted wound angiogenesis in vivo [102, 103] and was critical in promoting many wound healing phases such as cell proliferation, angiogenesis, re-epithelialization, and collagen deposition even in chronic wounds [104]. Incorporation of bFGF into chitosan membranes accelerated wound healing in genetically modified diabetic mice by enhancing proliferation of fibroblasts and increased number of capillaries were observed in both groups although development of granulation tissue was more copious [105]. Several studies have also demonstrated the ability of chitosan wound dressing to induce and accelerate tissue granulation and re-epithelialization. Chitosan–collagen hydrogel was also effective in delivering various amino acids that enhanced re-epithelialization and granulation thereby significantly accelerating the healing of diabetic wounds [106]. The incorporation of honey and curcumin in chitosan composite hydrogels increased fluid absorption capacity of the hydrogels that created a dry wound bed [107]. Lin et al. in 2015 successfully prepared a gelatin/

chitosan/epigallocatechin gallate nanoparticle incorporated in a poly(γ-glutamic acid)/gelatin hydrogel that enhanced wound regeneration and facilitated re-epithelialization [108]. Curcumin nanoformulation loaded methoxy PEG-graft-CS film increased rate of wound reduction and accelerated re-epithelialization [109]. By incorporating papain (protease enzyme obtained from *Carica papaya*) into chitin dressings prevented the formation of necrotic tissue and also [110]. Thyroxine impregnated chitosan-based dressings stimulated angiogenesis and supported fast wound healing in rats [111] by supporting epithelialization along with robust wound closure, high level of angiogenesis and fast wound healing than chitosan films alone [111]. Indeed, CWD are versatile and efficient delivery systems of many structurally and functional different biomolecules that participate in the wound healing process. Their exploitation is witnessing an exponential increase in novel CWD that are revolutionizing the wound dressing and management.

6 Conclusion

Chitin and chitosan possess inherent biological characteristics that have made them exceptionally important in the management of wounds both as active interactive molecules and as a versatile pro-wound healing delivery systems. CWD loaded with various biomolecules that participate in at every step of the wound healing process have been successfully incorporated and shown to increase the efficiency of the wound healing process. This has witnessed the number of CWD products increasing dramatically over the past decades with many already commercially available for different specialized. The massive ongoing current studies point to the possibility of more new products potent CWD delivering more effective biomolecules in the need future.

References

1. Liu H, Wang C, Li C, Qin Y, Wang Z, Yang F et al (2018) A functional chitosan-based hydrogel as a wound dressing and drug delivery system in the treatment of wound healing. RSC Adv 8:7533–7549. https://doi.org/10.1039/C7RA13510F
2. McDaniel JC, Browning KK (2014) Smoking, chronic wound healing, and implications for evidence-based practice. J Wound Ostomy Cont Nurs 41
3. Winter GD (1962) Formation of the Scab and the rate of epithelization of superficial wounds in the skin of the young domestic pig. Nature 193:193293a0. https://doi.org/10.1038/193293a0
4. Dhivya S, Padma VV, Santhini E (2015) Wound dressings – a review. Biomedicine 5:22. https://doi.org/10.7603/s40681-015-0022-9
5. Ovington LG (2007) Advances in wound dressings. Clin Dermatol 25:33–38. https://doi.org/10.1016/j.clindermatol.2006.09.003
6. Buwalda SJ, Boere KWM, Dijkstra PJ, Feijen J, Vermonden T, Hennink WE (2014) Hydrogels in a historical perspective: from simple networks to smart materials. J Control Release 190:254–273. https://doi.org/10.1016/j.jconrel.2014.03.052

7. Guebitz GM, Nyanhongo GS (2018) Enzymes as green catalysts and interactive biomolecules in wound dressing hydrogels. Trends Biotechnol 36:1040–1053. https://doi.org/10.1016/J.TIBTECH.2018.05.006
8. Vowden KVP (2011) Wound dressings: principles and practice. Surg 29:491–495. https://doi.org/10.1016/J.MPSUR.2011.06.007
9. Wang Y-H, Liu C-C, Cherng J-H, Fan G-Y, Wang Y-W, Chang S-J et al (2019) Evaluation of chitosan-based dressings in a swine model of artery-injury-related shock. Sci Rep 9:14608. https://doi.org/10.1038/s41598-019-51208-7
10. Miguel SP, Moreira AF, Correia IJ (2019) Chitosan based-asymmetric membranes for wound healing: a review. Int J Biol Macromol 127:460–475. https://doi.org/10.1016/j.ijbiomac.2019.01.072
11. Oyeniyi BT, Fox EE, Scerbo M, Tomasek JS, Wade CE, Holcomb JB (2017) Trends in 1029 trauma deaths at a level 1 trauma center: impact of a bleeding control bundle of care. Injury 48:5–12. https://doi.org/10.1016/j.injury.2016.10.037
12. Gruen RL, Brohi K, Schreiber M, Balogh ZJ, Pitt V, Narayan M et al (2012) Haemorrhage control in severely injured patients. Lancet 380:1099–1108. https://doi.org/10.1016/S0140-6736(12)61224-0
13. Khan MA, Mujahid M (2019) A review on recent advances in chitosan based composite for hemostatic dressings. Int J Biol Macromol 124:138–147. https://doi.org/10.1016/j.ijbiomac.2018.11.045
14. Li D, Chen J, Wang X, Zhang M, Li C, Zhou J (2020) Recent advances on synthetic and polysaccharide adhesives for biological hemostatic applications. Front Bioeng Biotechnol 8:926. https://doi.org/10.3389/fbioe.2020.00926
15. Wedmore I, McManus JG, Pusateri AE, Holcomb JB (2006) A special report on the chitosan-based hemostatic dressing: experience in current combat operations. J Trauma Acute Care Surg 60
16. Ho TT-P, Doan VK, Tran NM-P, Nguyen LK-K, Le AN-M, Ho MH et al (2020) Fabrication of chitosan oligomer-coated electrospun polycaprolactone membrane for wound dressing application. Mater Sci Eng C:111724. https://doi.org/10.1016/j.msec.2020.111724
17. Chan LW, Kim CH, Wang X, Pun SH, White NJ, Kim TH (2015) PolySTAT-modified chitosan gauzes for improved hemostasis in external hemorrhage. Acta Biomater 31:178–185. https://doi.org/10.1016/j.actbio.2015.11.017
18. Seon GM, Lee MH, Kwon BJ, Kim MS, Koo MA, Kim D et al (2017) Functional improvement of hemostatic dressing by addition of recombinant batroxobin. Acta Biomater 48:175–185. https://doi.org/10.1016/j.actbio.2016.10.024
19. Chan LW, Wang X, Wei H, Pozzo LD, White NJ, Pun SH (2015) A synthetic fibrin cross-linking polymer for modulating clot properties and inducing hemostasis. Sci Transl Med 7:277ra29. https://doi.org/10.1126/scitranslmed.3010383
20. Seon GM, Lee MH, Kwon BJ, Kim MS, Koo MA, Seomun Y et al (2018) Recombinant batroxobin-coated nonwoven chitosan as hemostatic dressing for initial hemorrhage control. Int J Biol Macromol 113:757–763. https://doi.org/10.1016/j.ijbiomac.2018.03.017
21. Li F, Luo P, Liu H (2018) A potential adjuvant agent of chemotherapy: sepia ink polysaccharides. Mar Drugs 16:106. https://doi.org/10.3390/md16040106
22. Huang N, Lin J, Li S, Deng Y, Kong S, Hong P et al (2018) Preparation and evaluation of squid ink polysaccharide-chitosan as a wound-healing sponge. Mater Sci Eng C 82:354–362. https://doi.org/10.1016/j.msec.2017.08.068
23. Hu Z, Zhang D-Y, Lu S-T, Li P-W, Li S-D (2018) Chitosan-based composite materials for prospective hemostatic applications. Mar Drugs 16:273. https://doi.org/10.3390/md16080273
24. Matica MA, Aachmann FL, Tøndervik A, Sletta H, Ostafe V (2019) Chitosan as a wound dressing starting material: antimicrobial properties and mode of action. Int J Mol Sci 20:5889. https://doi.org/10.3390/ijms20235889

25. Antunes BP, Moreira AF, Gaspar VM, Correia IJ (2015) Chitosan/arginine-chitosan polymer blends for assembly of nanofibrous membranes for wound regeneration. Carbohydr Polym 130:104–112. https://doi.org/10.1016/j.carbpol.2015.04.072
26. Sepahi M, Jalal R, Mashreghi M (2017) Antibacterial activity of poly-l-arginine under different conditions. Iran J Microbiol 9:103–111
27. Nguyen TV, Nguyen TTH, Wang S-L, Vo TPK, Nguyen AD (2017) Preparation of chitosan nanoparticles by TPP ionic gelation combined with spray drying, and the antibacterial activity of chitosan nanoparticles and a chitosan nanoparticle–amoxicillin complex. Res Chem Intermed 43:3527–3537. https://doi.org/10.1007/s11164-016-2428-8
28. Jamil B, Habib H, Abbasi S, Nasir H, Rahman A, Rehman A et al (2015) Cefazolin loaded chitosan nanoparticles to cure multi drug resistant gram-negative pathogens. Carbohydr Polym 136:682–691. https://doi.org/10.1016/j.carbpol.2015.09.078
29. Esmaeili A, Ghobadianpour S (2016) Vancomycin loaded superparamagnetic MnFe2O4 nanoparticles coated with PEGylated chitosan to enhance antibacterial activity. Int J Pharm 501:326–330. https://doi.org/10.1016/j.ijpharm.2016.02.013
30. Mu H, Liu Q, Niu H, Sun Y, Duan J (2016) Gold nanoparticles make chitosan–streptomycin conjugates effective towards gram-negative bacterial biofilm. RSC Adv 6:8714–8721. https://doi.org/10.1039/C5RA22803D
31. Zhang X, Pan Y, Li S, Xing L, Du S, Yuan G et al (2020) Doubly crosslinked biodegradable hydrogels based on gellan gum and chitosan for drug delivery and wound dressing. Int J Biol Macromol 164:2204–2214. https://doi.org/10.1016/j.ijbiomac.2020.08.093
32. Anjum S, Arora A, Alam MS, Gupta B (2016) Development of antimicrobial and scar preventive chitosan hydrogel wound dressings. Int J Pharm 508:92–101. https://doi.org/10.1016/j.ijpharm.2016.05.013
33. Zhang D, Zhou W, Wei B, Wang X, Tang R, Nie J et al (2015) Carboxyl-modified poly(vinyl alcohol)-crosslinked chitosan hydrogel films for potential wound dressing. Carbohydr Polym 125:189–199. https://doi.org/10.1016/j.carbpol.2015.02.034
34. Pulat M, Kahraman AS, Tan N, Gümüşderelioğlu M (2013) Sequential antibiotic and growth factor releasing chitosan-PAAm semi-IPN hydrogel as a novel wound dressing. J Biomater Sci Polym Ed 24:807–819. https://doi.org/10.1080/09205063.2012.718613
35. El-Naggar MY, Gohar YM, Sorour MA, Waheeb MG (2016) Hydrogel dressing with a nano-formula against Methicillin-resistant *Staphylococcus aureus* and *Pseudomonas aeruginosa* diabetic foot bacteria. J Microbiol Biotechnol 26:408–420. https://doi.org/10.4014/jmb.1506.06048
36. Ngadaonye JI, Geever LM, McEvoy KE, Killion J, Brady DB, Higginbotham CL (2014) Evaluation of novel antibiotic-eluting Thermoresponsive chitosan-PDEAAm based wound dressings. Int J Polym Mater Polym Biomater 63:873–883. https://doi.org/10.1080/00914037.2014.886224
37. Sinha M, Banik RM, Haldar C, Maiti P (2013) Development of ciprofloxacin hydrochloride loaded poly(ethylene glycol)/chitosan scaffold as wound dressing. J Porous Mater 20:799–807. https://doi.org/10.1007/s10934-012-9655-1
38. Mahmoud AA, Salama AH (2016) Norfloxacin-loaded collagen/chitosan scaffolds for skin reconstruction: preparation, evaluation and in-vivo wound healing assessment. Eur J Pharm Sci 83:155–165. https://doi.org/10.1016/j.ejps.2015.12.026
39. Haghi-Aminjan H, Asghari MH, Goharbari MH, Abdollahi M (2017) A systematic review on potential mechanisms of minocycline in kidney diseases. Pharmacol Rep 69:602–609. https://doi.org/10.1016/j.pharep.2017.02.001
40. Sung JH, Hwang MR, Kim JO, Lee JH, Kim II Y, Kim JH et al (2010) Gel characterisation and in vivo evaluation of minocycline-loaded wound dressing with enhanced wound healing using polyvinyl alcohol and chitosan. Int J Pharm 392:232–240. https://doi.org/10.1016/j.ijpharm.2010.03.024
41. Piras AM, Maisetta G, Sandreschi S, Gazzarri M, Bartoli C, Grassi L et al (2015) Chitosan nanoparticles loaded with the antimicrobial peptide temporin B exert a long-term antibacterial

activity in vitro against clinical isolates of Staphylococcus epidermidis. Front Microbiol 6:372. https://doi.org/10.3389/fmicb.2015.00372
42. Silva NC, Silva S, Sarmento B, Pintado M (2015) Chitosan nanoparticles for daptomycin delivery in ocular treatment of bacterial endophthalmitis. Drug Deliv 22:885–893. https://doi.org/10.3109/10717544.2013.858195
43. Kanta J (2011) The role of hydrogen peroxide and other reactive oxygen species in wound healing. Acta Med Cordoba 54:97–101. https://doi.org/10.14712/18059694.2016.28
44. Bösiger P, Tegl G, Richard IMT, Le Gat L, Huber L, Stagl V et al (2018) Enzyme functionalized electrospun chitosan mats for antimicrobial treatment. Carbohydr Polym 181:551–559. https://doi.org/10.1016/j.carbpol.2017.12.002
45. Tegl G, Öhlknecht C, Vielnascher R, Rollett A, Hofinger-Horvath A, Kosma P et al (2016) Cellobiohydrolases produce different oligosaccharides from chitosan. Biomacromolecules 17:2284–2292. https://doi.org/10.1021/acs.biomac.6b00547
46. Tegl G, Thallinger B, Beer B, Sygmund C, Ludwig R, Rollett A et al (2016) Antimicrobial cellobiose dehydrogenase-chitosan particles. ACS Appl Mater Interfaces 8. https://doi.org/10.1021/acsami.5b10801
47. Nyanhongo GS, Sygmund C, Ludwig R, Prasetyo EN, Guebitz GM (2013) An antioxidant regenerating system for continuous quenching of free radicals in chronic wounds. Eur J Pharm Biopharm 83. https://doi.org/10.1016/j.ejpb.2012.10.013
48. Ohlknecht C, Tegl G, Beer B, Sygmund C, Ludwig R, Guebitz GM et al (2017) Cellobiose dehydrogenase and chitosan-based lysozyme responsive materials for antimicrobial wound treatment. Biotechnol Bioeng 114:416–422. https://doi.org/10.1002/bit.26070
49. Huber D, Tegl G, Baumann M, Sommer E, Gorji EGEGEG, Borth N et al (2017) Chitosan hydrogel formation using laccase activated phenolics as cross-linkers. Carbohydr Polym 157:814–822. https://doi.org/10.1016/j.carbpol.2016.10.012
50. Huber D, Tegl G, Mensah A, Beer B, Baumann M, Borth N et al (2017) A dual-enzyme hydrogen peroxide generation machinery in hydrogels supports antimicrobial wound treatment. ACS Appl Mater Interfaces 9:15307–15316. https://doi.org/10.1021/acsami.7b03296
51. Tegl G, Stagl V, Mensah A, Huber D, Somitsch W, Grosse-Kracht S et al (2018) The chemo enzymatic functionalization of chitosan zeolite particles provides antioxidant and antimicrobial properties. Eng Life Sci 18:334–340. https://doi.org/10.1002/elsc.201700120
52. Nyanhongo GSGS, Sygmund C, Ludwig R, Prasetyo ENEN, Guebitz GMGMGM (2013) Synthesis of multifunctional bioresponsive polymers for the management of chronic wounds. J Biomed Mater Res Part B Appl Biomater 101(B):882–891. https://doi.org/10.1002/jbm.b.32893
53. Zhang H, Feng M, Chen S, Shi W, Wang X (2020) Incorporation of lysozyme into cellulose nanocrystals stabilized β-chitosan nanoparticles with enhanced antibacterial activity. Carbohydr Polym 236:115974. https://doi.org/10.1016/j.carbpol.2020.115974
54. Wang Y, Li S, Jin M, Han Q, Liu S, Chen X et al (2020) Enhancing the thermo-stability and anti-bacterium activity of lysozyme by immobilization on chitosan nanoparticles. Int J Mol Sci 21:1635. https://doi.org/10.3390/ijms21051635
55. Kim JO, Noh JK, Thapa RK, Hasan N, Choi M, Kim JH et al (2015) Nitric oxide-releasing chitosan film for enhanced antibacterial and in vivo wound-healing efficacy. Int J Biol Macromol 79:217–225. https://doi.org/10.1016/j.ijbiomac.2015.04.073
56. Pelegrino MT, Weller RB, Chen X, Bernardes JS, Seabra AB (2017) Chitosan nanoparticles for nitric oxide delivery in human skin. Medchemcomm 8:713–719. https://doi.org/10.1039/C6MD00502K
57. Reighard KP, Schoenfisch MH (2015) Antibacterial action of nitric oxide-releasing chitosan oligosaccharides against *Pseudomonas aeruginosa* under aerobic and anaerobic conditions. Antimicrob Agents Chemother 59:6506–6513. https://doi.org/10.1128/AAC.01208-15
58. Choi M, Hasan N, Cao J, Lee J, Hlaing SP, Yoo JW (2020) Chitosan-based nitric oxide-releasing dressing for anti-biofilm and in vivo healing activities in MRSA biofilm-infected wounds. Int J Biol Macromol 142:680–692. https://doi.org/10.1016/j.ijbiomac.2019.10.009

59. Fontana K, Ventimiglia L, Mutus B (2018) Nitric oxide generating copper–chitosan particles for wound healing applications. J Chem Technol Biotechnol 93:2093–2101. https://doi.org/10.1002/jctb.5630
60. Marcato DP, Adami FL, de Melo Barbosa R, Melo SP, Ferreira RI, de Paula L et al (2013) Development of a sustained-release system for nitric oxide delivery using alginate/chitosan nanoparticles. Curr Nanosci 9:1–7. https://doi.org/10.2174/1573413711309010003
61. Gethin G, Cowman S (2009) Retracted: Manuka honey vs. hydrogel – a prospective, open label, multicentre, randomised controlled trial to compare desloughing efficacy and healing outcomes in venous ulcers. J Clin Nurs 18:466–474. https://doi.org/10.1111/j.1365-2702.2008.02558.x
62. Shamloo A, Aghababaie Z, Afjoul H, Jami M, Bidgoli MR, Vossoughi M et al (2020) Fabrication and evaluation of chitosan/gelatin/PVA hydrogel incorporating honey for wound healing applications: an in vitro, in vivo study. Int J Pharm 592:120068. https://doi.org/10.1016/j.ijpharm.2020.120068
63. Sasikala L, Rathinamoorthy R, Dhurai B (2018) Optimization of process conditions for chitosan-manuka honey film as wound contact layer for wound dressings. Wound Med 23:11–21. https://doi.org/10.1016/j.wndm.2018.09.007
64. Abbas M, Hussain T, Arshad M, Ansari AR, Irshad A, Nisar J et al (2019) Wound healing potential of curcumin cross-linked chitosan/polyvinyl alcohol. Int J Biol Macromol 140:871–876. https://doi.org/10.1016/j.ijbiomac.2019.08.153
65. Krausz AE, Adler BL, Cabral V, Navati M, Doerner J, Charafeddine RA et al (2015) Curcumin-encapsulated nanoparticles as innovative antimicrobial and wound healing agent. Nanomed Nanotechnol Biol Med 11:195–206. https://doi.org/10.1016/j.nano.2014.09.004
66. Poornima B, Korrapati PS (2017) Fabrication of chitosan-polycaprolactone composite nanofibrous scaffold for simultaneous delivery of ferulic acid and resveratrol. Carbohydr Polym 157:1741–1749. https://doi.org/10.1016/j.carbpol.2016.11.056
67. Solórzano-Santos F, Miranda-Novales MG (2012) Essential oils from aromatic herbs as antimicrobial agents. Curr Opin Biotechnol 23:136–141. https://doi.org/10.1016/j.copbio.2011.08.005
68. Dias AMA, Braga MEM, Seabra IJ, Ferreira P, Gil MH, De Sousa HC (2011) Development of natural-based wound dressings impregnated with bioactive compounds and using supercritical carbon dioxide. Int J Pharm 408:9–19. https://doi.org/10.1016/j.ijpharm.2011.01.063
69. Ranjith R, Balraj S, Ganesh J, John Milton MC (2019) Therapeutic agents loaded chitosan-based nanofibrous mats as potential wound dressings: a review. Mater Today Chem 12:386–395. https://doi.org/10.1016/j.mtchem.2019.03.008
70. Rieger KA, Schiffman JD (2014) Electrospinning an essential oil: Cinnamaldehyde enhances the antimicrobial efficacy of chitosan/poly(ethylene oxide) nanofibers. Carbohydr Polym 113:561–568. https://doi.org/10.1016/j.carbpol.2014.06.075
71. Rieger KA, Birch NP, Schiffman JD (2016) Electrospinning chitosan/poly(ethylene oxide) solutions with essential oils: correlating solution rheology to nanofiber formation. Carbohydr Polym 139:131–138. https://doi.org/10.1016/j.carbpol.2015.11.073
72. Moeini A, Pedram P, Makvandi P, Malinconico M, Gomez d'Ayala G (2020) Wound healing and antimicrobial effect of active secondary metabolites in chitosan-based wound dressings: a review. Carbohydr Polym 233:115839. https://doi.org/10.1016/j.carbpol.2020.115839
73. Costa MF, Durço AO, Rabelo TK, de Barreto RSS, Guimarães AG (2019) Effects of Carvacrol, Thymol and essential oils containing such monoterpenes on wound healing: a systematic review. J Pharm Pharmacol 71:141–155. https://doi.org/10.1111/jphp.13054
74. Koosehgol S, Ebrahimian-Hosseinabadi M, Alizadeh M, Zamanian A (2017) Preparation and characterization of in situ chitosan/polyethylene glycol fumarate/thymol hydrogel as an effective wound dressing. Mater Sci Eng C 79:66–75. https://doi.org/10.1016/j.msec.2017.05.001
75. Pires ALR, de Azevedo ML, Dias AMA, de Sousa HC, Moraes ÂM, Braga MEM (2018) Towards wound dressings with improved properties: effects of poly(dimethylsiloxane) on

chitosan-alginate films loaded with thymol and beta-carotene. Mater Sci Eng C 93:595–605. https://doi.org/10.1016/j.msec.2018.08.005
76. Barreto R, Albuquerque-Junior RL, Pereira-Filho RN, Lima P, Nunes R, Barreto A et al (2015) Effect of chitosan film containing carvacrol, a phenolic monoterpene, on wound healing in rats. FASEB J 29:773.4. https://doi.org/10.1096/fasebj.29.1_supplement.773.4
77. Sotelo-Boyás M, Correa-Pacheco Z, Bautista-Baños S, Gómez Y (2017) Release study and inhibitory activity of thyme essential oil-loaded chitosan nanoparticles and nanocapsules against foodborne bacteria. Int J Biol Macromol 103:409–414. https://doi.org/10.1016/j.ijbiomac.2017.05.063
78. Qiu J, Yu L, Zhang X, Wu Q, Wang D, Wang X et al (2015) Asiaticoside attenuates lipopolysaccharide-induced acute lung injury via down-regulation of NF-κB signaling pathway. Int Immunopharmacol 26:181–187. https://doi.org/10.1016/j.intimp.2015.03.022
79. Figueira DR, Miguel SP, de Sá KD, Correia IJ (2016) Production and characterization of polycaprolactone- hyaluronic acid/chitosan- zein electrospun bilayer nanofibrous membrane for tissue regeneration. Int J Biol Macromol 93:1100–1110. https://doi.org/10.1016/j.ijbiomac.2016.09.080
80. Ge Y, Tang J, Fu H, Fu Y, Wu Y (2019) Characteristics, controlled-release and antimicrobial properties of tea tree oil liposomes-incorporated chitosan-based electrospun nanofiber mats. Fibers Polym 20:698–708. https://doi.org/10.1007/s12221-019-1092-1
81. Ahmed S, Ikram S (2016) Chitosan based scaffolds and their applications in wound healing. Achiev Life Sci 10:27–37. https://doi.org/10.1016/j.als.2016.04.001
82. Jayakumar R, Prabaharan M, Sudheesh Kumar PT, Nair SV, Tamura H (2011) Biomaterials based on chitin and chitosan in wound dressing applications. Biotechnol Adv 29:322–337. https://doi.org/10.1016/j.biotechadv.2011.01.005
83. Takei T, Nakahara H, Ijima H, Kawakami K (2012) Synthesis of a chitosan derivative soluble at neutral pH and gellable by freeze-thawing, and its application in wound care. Acta Biomater 8:686–693. https://doi.org/10.1016/j.actbio.2011.10.005
84. Ji Q, Deng J, Yu X, Xu Q, Wu H, Pan J (2013) Modulation of pro-inflammatory mediators in LPS-stimulated human periodontal ligament cells by chitosan and quaternized chitosan. Carbohydr Polym 92:824–829. https://doi.org/10.1016/j.carbpol.2012.09.043
85. Amin MA, Abdel-Raheem IT (2014) Accelerated wound healing and anti-inflammatory effects of physically cross linked polyvinyl alcohol–chitosan hydrogel containing honey bee venom in diabetic rats. Arch Pharm Res 37:1016–1031. https://doi.org/10.1007/s12272-013-0308-y
86. Abioye AO, Issah S, Kola-Mustapha AT (2015) Ex vivo skin permeation and retention studies on chitosan-ibuprofen-gellan ternary nanogel prepared by in situ ionic gelation technique - a tool for controlled transdermal delivery of ibuprofen. Int J Pharm 490:112–130. https://doi.org/10.1016/j.ijpharm.2015.05.030
87. Morgado PI, Miguel SP, Correia IJ, Aguiar-Ricardo A (2017) Ibuprofen loaded PVA/chitosan membranes: a highly efficient strategy towards an improved skin wound healing. Carbohydr Polym 159:136–145. https://doi.org/10.1016/j.carbpol.2016.12.029
88. Morgado PI, Lisboa PF, Ribeiro MP, Miguel SP, Simões PC, Correia IJ et al (2014) Poly(vinyl alcohol)/chitosan asymmetrical membranes: highly controlled morphology toward the ideal wound dressing. J Memb Sci 469:262–271. https://doi.org/10.1016/j.memsci.2014.06.035
89. Hassan MA, Tamer TM, Valachová K, Omer AM, El-Shafeey M, Mohy Eldin MS et al (2020) Antioxidant and antibacterial polyelectrolyte wound dressing based on chitosan/hyaluronan/phosphatidylcholine dihydroquercetin. Int J Biol Macromol 166:18–31. https://doi.org/10.1016/j.ijbiomac.2020.11.119
90. Archana D, Singh BK, Dutta J, Dutta PK (2013) In vivo evaluation of chitosan-PVP-titanium dioxide nanocomposite as wound dressing material. Carbohydr Polym 95:530–539. https://doi.org/10.1016/j.carbpol.2013.03.034
91. Singh R, Shitiz K, Singh A (2017) Chitin and chitosan: biopolymers for wound management. Int Wound J 14:1276–1289. https://doi.org/10.1111/iwj.12797

92. Karri VVSR, Kuppusamy G, Talluri SV, Mannemala SS, Kollipara R, Wadhwani AD et al (2016) Curcumin loaded chitosan nanoparticles impregnated into collagen-alginate scaffolds for diabetic wound healing. Int J Biol Macromol 93:1519–1529. https://doi.org/10.1016/j.ijbiomac.2016.05.038
93. Shukla R, Kashaw SK, Jain AP, Lodhi S (2016) Fabrication of Apigenin loaded gellan gum–chitosan hydrogels (GGCH-HGs) for effective diabetic wound healing. Int J Biol Macromol 91:1110–1119. https://doi.org/10.1016/j.ijbiomac.2016.06.075
94. Ferreira MOG, Leite LLR, de Lima IS, Barreto HM, Nunes LCC, Ribeiro AB et al (2016) Chitosan hydrogel in combination with Nerolidol for healing wounds. Carbohydr Polym 152:409–418. https://doi.org/10.1016/j.carbpol.2016.07.037
95. Guo S, DiPietro LA (2010) Factors affecting wound healing. J Dent Res 89:219–229. https://doi.org/10.1177/0022034509359125
96. Park CJ, Clark SG, Lichtensteiger CA, Jamison RD, Johnson AJW (2009) Accelerated wound closure of pressure ulcers in aged mice by chitosan scaffolds with and without bFGF. Acta Biomater 5:1926–1936. https://doi.org/10.1016/j.actbio.2009.03.002
97. Sukumar N, Ramachandran T, Kalaiarasi H, Sengottuvelu S (2015) Characterization and in vivo evaluation of silk hydrogel with enhancement of dextrin, rhEGF, and alginate beads for diabetic Wistar albino wounded rats. J Text Inst 106:133–140. https://doi.org/10.1080/00405000.2014.906100
98. Amsden B (2015) Novel biodegradable polymers for local growth factor delivery. Eur J Pharm Biopharm 97:318–328. https://doi.org/10.1016/j.ejpb.2015.06.008
99. Hajimiri M, Shahverdi S, Esfandiari MA, Larijani B, Atyabi F, Rajabiani A et al (2016) Preparation of hydrogel embedded polymer-growth factor conjugated nanoparticles as a diabetic wound dressing. Drug Dev Ind Pharm 42:707–719. https://doi.org/10.3109/03639045.2015.1075030
100. Choi JS, Yoo HS (2010) Pluronic/chitosan hydrogels containing epidermal growth factor with wound-adhesive and photo-crosslinkable properties. J Biomed Mater Res Part A 95A:564–573. https://doi.org/10.1002/jbm.a.32848
101. Wang W, Lin S, Xiao Y, Huang Y, Tan Y, Cai L et al (2008) Acceleration of diabetic wound healing with chitosan-crosslinked collagen sponge containing recombinant human acidic fibroblast growth factor in healing-impaired STZ diabetic rats. Life Sci 82:190–204. https://doi.org/10.1016/j.lfs.2007.11.009
102. Cheng NC, Lin WJ, Ling TY, Young TH (2017) Sustained release of adipose-derived stem cells by thermosensitive chitosan/gelatin hydrogel for therapeutic angiogenesis. Acta Biomater 51:258–267. https://doi.org/10.1016/j.actbio.2017.01.060
103. Liu H, Ding J, Wang C, Wang J, Wang Y, Yang M et al (2015) Intra-articular transplantation of allogeneic BMMSCs rehabilitates cartilage injury of antigen-induced arthritis. Tissue Eng Part A 21:2733–2743. https://doi.org/10.1089/ten.tea.2014.0666
104. Li M, Ke Q-F, Tao S-C, Guo S-C, Rui B-Y, Guo Y-P (2016) Fabrication of hydroxyapatite/chitosan composite hydrogels loaded with exosomes derived from miR-126-3p overexpressed synovial mesenchymal stem cells for diabetic chronic wound healing. J Mater Chem B 4:6830–6841. https://doi.org/10.1039/C6TB01560C
105. Mizuno K, Yamamura K, Yano K, Osada T, Saeki S, Takimoto N et al (2003) Effect of chitosan film containing basic fibroblast growth factor on wound healing in genetically diabetic mice. J Biomed Mater Res Part A 64A:177–181. https://doi.org/10.1002/jbm.a.10396
106. Xiao Y, Reis LA, Feric N, Knee EJ, Gu J, Cao S et al (2016) Diabetic wound regeneration using peptide-modified hydrogels to target re-epithelialization. Proc Natl Acad Sci 113: E5792–E5801. https://doi.org/10.1073/pnas.1612277113
107. Momin M, Kurhade S, Khanekar P, Mhatre S (2016) Novel biodegradable hydrogel sponge containing curcumin and honey for wound healing. J Wound Care 25:364–372. https://doi.org/10.12968/jowc.2016.25.6.364

108. Lin Y-H, Lin J-H, Li T-S, Wang S-H, Yao C-H, Chung W-Y et al (2016) Dressing with epigallocatechin gallate nanoparticles for wound regeneration. Wound Repair Regen 24:287–301. https://doi.org/10.1111/wrr.12372
109. Li X, Nan K, Li L, Zhang Z, Chen H (2012) In vivo evaluation of curcumin nanoformulation loaded methoxy poly(ethylene glycol)-graft-chitosan composite film for wound healing application. Carbohydr Polym 88:84–90. https://doi.org/10.1016/j.carbpol.2011.11.068
110. Singh D, Singh R (2012) Papain incorporated chitin dressings for wound debridement sterilized by gamma radiation. Radiat Phys Chem 81:1781–1785. https://doi.org/10.1016/j.radphyschem.2012.06.010
111. Shahzadi L, Bashir M, Tehseen S, Zehra M, Mehmood A, Chaudhry AA et al (2020) Thyroxine impregnated chitosan-based dressings stimulate angiogenesis and support fast wounds healing in rats: potential clinical candidates. Int J Biol Macromol 160:296–306. https://doi.org/10.1016/j.ijbiomac.2020.05.127